# FINITE ELEMENT ANALYSIS

# FINITE ELEMENT ANALYSIS

*Theory and Application with ANSYS*

**Saeed Moaveni**
*Minnesota State University, Mankato*

PRENTICE HALL, Upper Saddle River, New Jersey 07458

**Library of Congress Cataloging-in-Publication Data**

Moaveni, Saeed.
Finite element analysis. Theory and application with ANSYS
p. cm.
Includes bibliographical references and index.
ISBN 0-13-785098-0
1. Finite element method—Data processing. 2. ANSYS (Computer system) I. Title
TA347.F5M62 1999
620'.001'51535—dc21 98-31163
CIP

Executive editor: *BILL STENQUIST*
Production editor: *CAROLE SURACI*
Editor-in-chief: *MARCIA HORTON*
Managing editor: *EILEEN CLARK*
Marketing manager: *DANNY HOYT*
Assistant vice president of production and manufacturing: *DAVID W. RICCARDI*
Cover designer: *BRUCE KENSELAAR*
Copy editor: *ABIGAIL BAKER*
Manufacturing buyer: *PAT BROWN*
Editorial assistant: *MEG WEIST*



Printed in the United States of America

10 9 8 7 6 5 4

ISBN 0-13-785098-0

Prentice-Hall International (UK) Limited, *London*
Prentice-Hall of Australia Pty. Limited, *Sydney*
Prentice-Hall Canada, Inc., *Toronto*
Prentice-Hall Hispanoamericana, S.A., *Mexico*
Prentice-Hall of India Private Limited, *New Delhi*
Prentice-Hall of Japan, Inc., *Tokyo*
Simon & Schuster Asia Pte. Ltd., *Singapore*
Editora Prentice-Hall do Brasil, Ltda., *Rio de Janeiro*

*To my mother and father*

# Contents

# Preface

There are many good textbooks already in existence that cover the theory of finite element methods for advanced students. However, none of these books incorporate ANSYS as an integral part of their materials to introduce finite element modeling to undergraduate students and the newcomers. In recent years, the use of finite element analysis as a design tool has grown rapidly. Easy to use comprehensive packages such as ANSYS, a general-purpose finite element computer program, have become common tools in the hands of design engineers. Unfortunately, many engineers who lack the proper training or understanding of the underlying concepts have been using these tools. This introductory book is written to assist engineering students and practicing engineers new to the field of finite element modeling to gain a clear understanding of the basic concepts. The text offers insight into the theoretical aspects of finite element analysis and also covers some practical aspects of modeling. Great care has been exercised to avoid overwhelming students with theory. Yet enough theoretical background is offered to allow individuals to use ANSYS intelligently and effectively. ANSYS is an integral part of this text. In each chapter, the relevant basic theory is discussed first and demonstrated using simple problems with hand-calculations. These problems are followed by examples which are solved using ANSYS. Exercises in the text are also presented in this manner. Some exercises require manual calculations while others, more complex in nature, require the use of ANSYS. The simpler hand-calculation problems will enhance students' understanding of the concepts by encouraging them to go through the necessary steps in a finite element analysis. Design problems are also included at the end of Chapters 2, 4, 7, 8, and 10.

Various sources of error that can contribute to wrong results are discussed. A good engineer must always find ways to check the results. While experimental testing of models may be the best way, such testing may be expensive or time consuming. Therefore, whenever possible, throughout this text emphasis is placed on doing a "sanity check" to verify one's Finite Element Analysis (FEA). A section at the end of each appropriate chapter is devoted to possible approaches for verifying ANSYS results.

Another unique feature of this book is that the last chapter is devoted to the introduction of design optimization and parametric programming with ANSYS.

The book is organized into 11 chapters. Chapter 1 reviews basic ideas in finite element analysis. Common formulations, such as direct, potential energy, and weighted residual methods are discussed. Chapter 2 deals with the analysis of trusses because trusses offer economical solutions to many engineering structural problems. An overview of ANSYS program is given in Chapter 2 so that students can begin to use ANSYS right away. Chapter 3 lays the foundation for analysis of one-dimensional problems by introducing one-dimensional linear, quadratic, and cubic elements. Global, local and natural coordinate systems are also discussed in detail in Chapter 3. An introduction to isoparametric formulation and numerical integration by Gauss-Legendre formulae are also presented in Chapter 3. Chapter 4 considers Galerkin formulation of one-dimensional heat transfer and fluid problems. Minimum total potential energy of solid mechanics problems are also discussed in Chapter 4. Two-dimensional linear and higher order elements are introduced in Chapter 5. Gauss-Legendre formulae for two-dimensional integrals are also presented in Chapter 5. In Chapter 6, the essential capabilities and the organization of the ANSYS program are covered. The basic steps in creating and analyzing a model with ANSYS is discussed in detail. Chapter 7 includes the analysis of two-dimensional heat transfer problems. Chapter 8 provides analysis of torsion of noncircular shafts, beams, frames, and plane stress problems. In Chapter 9, two dimensional ideal fluid mechanics problems are analyzed. Direct formulation of the piping network problems and underground seepage flow are also discussed. Chapter 10 provides a discussion of three-dimensional elements and formulations. This chapter also presents basic ideas regarding top-down and bottom-up solid modeling methods. Design optimization ideas and parametric programming are discussed in Chapter 11. Each chapter begins by stating the objectives and concludes by summarizing what the reader should have gained from studying that chapter.

The examples which are solved using ANSYS show in great detail how to use ANSYS to model and analyze a variety of engineering problems. Chapter 6 is also written in such manner that it can be taught right away if the instructor sees the need to start with ANSYS at the beginning of the course.

A brief review of appropriate fundamental principles in solid mechanics, heat transfer, and fluid mechanics is also provided throughout the book. Additionally, when appropriate, the students are warned about becoming too quick to generate finite element models for problems for which there exist simple analytical solutions. Mechanical and thermophysical properties of some common materials used in engineering are given in Appendices A and B.

Finally, I am planning to maintain a website at http://www.prenhall.com/Moaveni for the following purposes: (1) to share any changes in the upcoming versions of

ANSYS; (2) to share additional information on upcoming text revisions including transient analysis of mechanical and thermal problems and other useful topics that you would like to see covered in the next addition. I am also planning to expand the optimization chapter. Examples with error estimation calculations are also planned; (3) to provide additional homework problems and design problems; and (4) to post, at the website, any corrections that are brought to my attention. The website will be accessible to all students.

***Saeed Moaveni***

# Acknowledgments

I would like to express my sincere gratitude to Mr. Raymond Browell of ANSYS, Inc. for providing the photographs for the cover of this book. Descriptions for the cover photographs are given on page 7. I would also like to thank ANSYS, Inc. for giving me permission to adapt material from various ANSYS documents, related to capabilities and the organization of ANSYS. The essential capabilities and organization of ANSYS are covered in Chapters 2, 6, 10, and 11.

As I have mentioned in the Preface, there are many good published books in finite element analysis. When writing this book, several of these books were consulted. They are cited at the end of each appropriate chapter. The reader can benefit from referring to these books and articles .

I would also like to thank Dr. Nancy Mackenzie of Minnesota State University who made valuable editing suggestions during the first draft of the manuscript. I am also grateful to Mr. Bill Stenquist and Ms. Carole Suraci of Prentice Hall for their assistance in the preparation of this book.

CHAPTER 1

# Introduction

The finite element method is a numerical procedure that can be used to obtain solutions to a large class of engineering problems involving stress analysis, heat transfer, electromagnetism, and fluid flow. This book was written to help you gain a clear understanding of the fundamental concepts of finite element modeling. Having a clear understanding of the basic concepts will enable you to use a general-purpose finite element software, such as ANSYS, effectively. ANSYS is an integral part of this text. In each chapter, the relevant basic theory behind each respective concept is discussed first. This discussion is followed by examples that are solved using ANSYS. Throughout this text, emphasis is placed on methods by which you may verify your findings from finite element analysis (FEA). In addition, at the end of particular chapters, a section is devoted to the approaches you should consider to verify results generated by using ANSYS.

Some of the exercises provided in this text require manual calculations. The purpose of these exercises is to enhance your understanding of the concepts by encouraging you to go through the necessary steps of finite element analysis. This book is also written in such a way that it can serve as a reference text for readers who may already be design engineers who are beginning to get involved in finite element modeling and need to know the underlying concepts of FEA.

The objective of this chapter is to introduce you to basic concepts in finite element formulation, including direct formulation, the minimum potential energy theorem, and the weighted residual methods. The main topics of Chapter 1 include the following:

**1.1** Engineering Problems

**1.2** Numerical Methods

**1.3** A Brief History of the Finite Element Method and ANSYS

**1.4** Basic Steps in the Finite Element Method

**1.5** Direct Formulation

**1.6** Minimum Total Potential Energy Formulation

**1.7** Weighted Residual Formulations

**1.8** Verification of Results

**1.9** Understanding the Problem

## 1.1 ENGINEERING PROBLEMS

In general, engineering problems are mathematical models of physical situations. Mathematical models are differential equations with a set of corresponding boundary and initial conditions. The differential equations are derived by applying the fundamental laws and principles of nature to a system or a control volume. These governing equations represent balance of mass, force, or energy. When possible, the exact solution of these equations renders detailed behavior of a system under a given set of conditions. The analytical solutions are composed of two parts: (1) a homogenous part and (2) a particular part. In any given engineering problem, there are two sets of parameters that influence the way in which a system behaves. First, there are those parameters that provide information regarding the *natural behavior* of a given system. These parameters include properties such as modulus of elasticity, thermal conductivity, and viscosity. Table 1.1 summarizes the physical properties that define the natural characteristics of various problems.

On the other hand, there are parameters that produce *disturbances* in a system. These types of parameters are summarized in Table 1.2. Examples of these parameters include external forces, moments, temperature difference across a medium, and pressure difference in fluid flow.

The system characteristics as shown in Table 1.1 dictate the natural behavior of a system, and they always appear in the *homogenous part of the solution* of a governing differential equation. In contrast, the parameters that cause the disturbances appear in the *particular solution*. It is important to understand the role of these parameters in finite element modeling in terms of their respective appearances in stiffness or conductance matrices and load or forcing matrices. The system characteristics will always show up in the stiffness matrix, conductance matrix, or resistance matrix, whereas the disturbance parameters will always appear in the load matrix.

## 1.2 NUMERICAL METHODS

There are many practical engineering problems for which we cannot obtain exact solutions. This inability to obtain an exact solution may be attributed to either the complex nature of governing differential equations or the difficulties that arise from dealing with the boundary and initial conditions. To deal with such problems, we resort to numerical approximations. In contrast to analytical solutions, which show the exact behavior of a system at any point within the system, numerical solutions approximate exact solutions only at discrete points, called nodes. The first step of any numerical procedure is discretization. This process divides the medium of interest into a number of small subregions and nodes. There are two common classes of numerical methods: (1) *finite difference methods* and (2) *finite element methods*. With finite difference methods, the differential equation is written for each node, and the derivatives are replaced by *difference equations*. This approach results in a set of simultaneous linear equations. Although finite difference methods are easy to understand and employ in simple problems, they become difficult to apply to problems with complex geometries or complex boundary conditions. This situation is also true for problems with nonisotropic material properties.

TABLE 1.1 Physical properties characterizing various engineering systems

| Problem Type | Examples of Parameters That Characterize a System |
|---|---|
| **Solid Mechanics Examples** | |
| a truss | modulus of elasticity, $E$ |
| an elastic plate | modulus of elasticity, $E$ |
| a beam | modulus of elasticity, $E$; second moment of area, $I$ |
| a shaft | modulus of rigidity, $G$; polar moment of inertia of the area, $J$ |
| **Heat Transfer Examples** | |
| 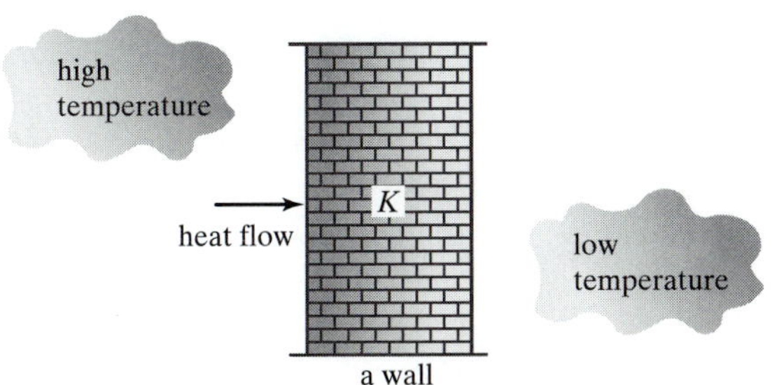  a wall | thermal conductivity, $K$ |

TABLE 1.1 *(cont.)* Physical properties characterizing various engineering systems

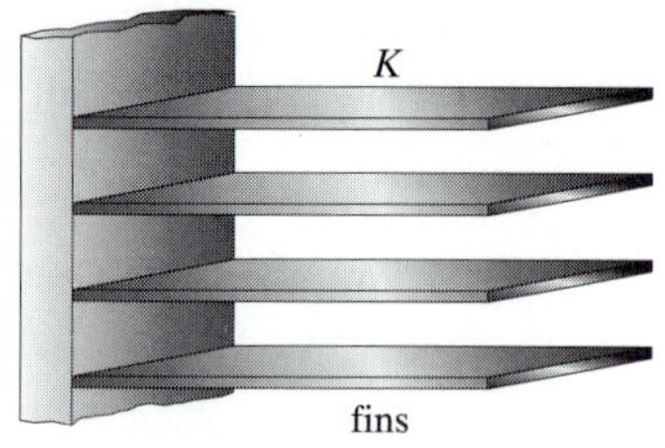


thermal conductivity, $K$

**Fluid Flow Examples**

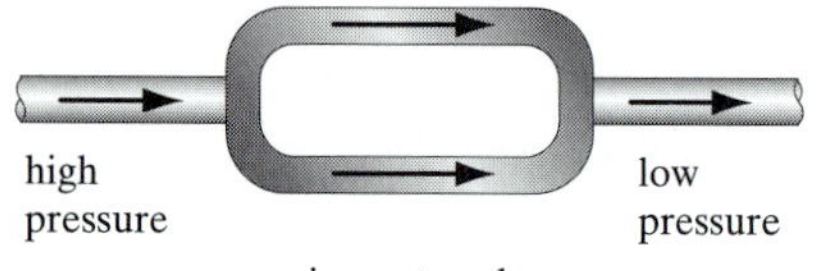


pipe networks

viscosity, $\mu$; relative roughness, $e$

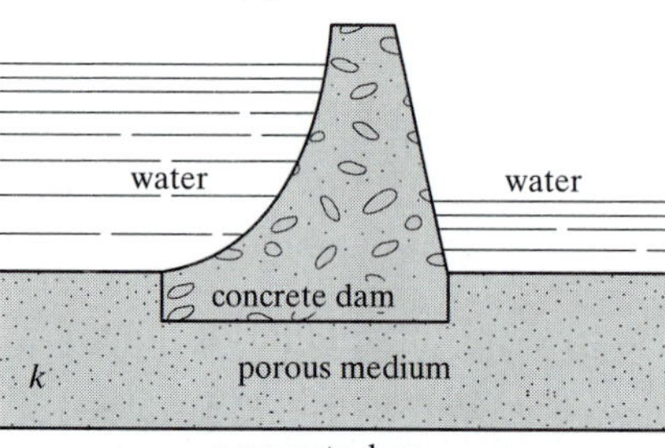


a concrete dam

soil permeability, $k$

**Electrical and Magnetism Problems**

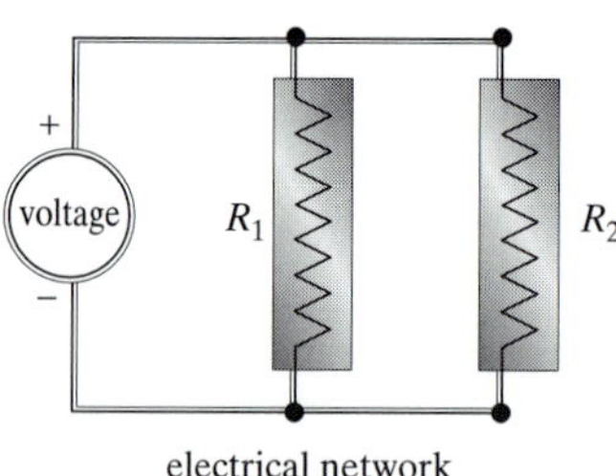


electrical network

resistance, $R$

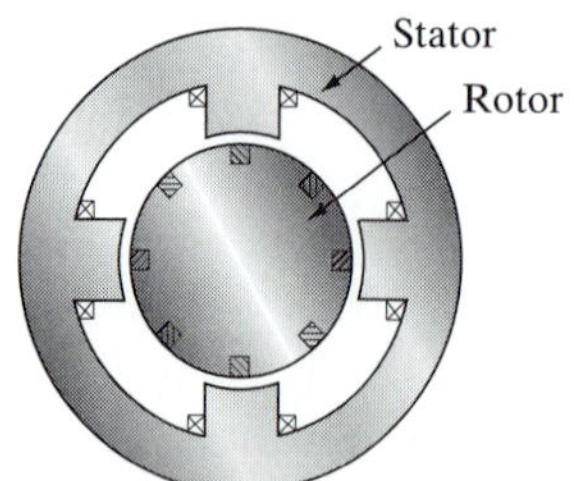


magnetic field of an electric motor

permeability, $\mu$

TABLE 1.2 Parameters causing disturbances in various engineering systems

| Problem Type | Examples of Parameters That Produce Disturbances in a System |
|---|---|
| Solid Mechanics | external forces and moments; support excitation |
| Heat Transfer | temperature difference; heat input |
| Fluid Flow and Pipe Networks | pressure difference; rate of flow |
| Electrical Network | voltage difference |

In contrast, the finite element method uses *integral formulations* rather than difference equations to create a system of algebraic equations. Moreover, an approximate continuous function is assumed to represent the solution for each element. The complete solution is then generated by connecting or assembling the individual solutions, allowing for continuity at the interelemental boundaries.

## 1.3 A BRIEF HISTORY* OF THE FINITE ELEMENT METHOD AND ANSYS

The finite element method is a numerical procedure that can be applied to obtain solutions to a variety of problems in engineering. Steady, transient, linear, or nonlinear problems in stress analysis, heat transfer, fluid flow, and electromagnetism problems may be analyzed with finite element methods. The origin of the modern finite element method may be traced back to the early 1900s, when some investigators approximated and modeled elastic continua using discrete equivalent elastic bars. However, *Courant* (*1943*) has been credited with being the first person to develop the finite element method. In a paper published in the early 1940s, Courant used piecewise polynomial interpolation over triangular subregions to investigate torsion problems.

The next significant step in the utilization of finite element methods was taken by Boeing in the 1950s when Boeing, followed by others, used triangular stress elements to model airplane wings. Yet, it was not until 1960 that Clough made the term "finite element" popular. During the 1960s, investigators began to apply the finite element method to other areas of engineering, such as heat transfer and seepage flow problems. *Zienkiewicz and Cheung (1967)* wrote the first book entirely devoted to the finite element method in 1967. In 1971, ANSYS was released for the first time.

ANSYS is a comprehensive general-purpose finite element computer program that contains over 100,000 lines of code. ANSYS is capable of performing static, dynamic, heat transfer, fluid flow, and electromagnetism analyses. ANSYS has been a leading FEA program for well over 20 years. The current version of ANSYS has a completely new look, with multiple windows incorporating Graphical User Interface (GUI), pull-down menus, dialog boxes, and a tool bar. Today, you will find ANSYS in use in many engineering fields, including aerospace, automotive, electronics, and nuclear. In order to use ANSYS or any other "canned" FEA computer program intelligently, it is imper-

*See Cook et al. (1989) for more detail.

ative that one first fully understands the underlying basic concepts and limitations of the finite element methods.

ANSYS is a very powerful and impressive engineering tool that may be used to solve a variety of problems. However, a user without a basic understanding of the finite element methods will find himself or herself in the same predicament as a computer technician with access to many impressive instruments and tools, but who cannot fix a computer because he or she does not understand the inner workings of a computer!

## 1.4 BASIC STEPS IN THE FINITE ELEMENT METHOD

The basic steps involved in any finite element analysis consist of the following:

**Preprocessing Phase**

1. Create and discretize the solution domain into finite elements; that is, subdivide the problem into nodes and elements.
2. Assume a shape function to represent the physical behavior of an element; that is, an approximate continuous function is assumed to represent the solution of an element.
3. Develop equations for an element.
4. Assemble the elements to present the entire problem. Construct the global stiffness matrix.
5. Apply boundary conditions, initial conditions, and loading.

**Solution Phase**

6. Solve a set of linear or nonlinear algebraic equations simultaneously to obtain nodal results, such as displacement values at different nodes or temperature values at different nodes in a heat transfer problem.

**Postprocessing Phase**

7. Obtain other important information. At this point, you may be interested in values of principal stresses, heat fluxes, etc.

In general, there are several approaches to formulating finite element problems: (1) *Direct Formulation*, (2) *The Minimum Total Potential Energy Formulation*, and (3) *Weighted Residual Formulations*. Again, it is important to note that the basic steps involved in any finite element analysis, regardless of how we generate the finite element model, will be the same as those listed above.

## 1.5 DIRECT FORMULATION

The following problem illustrates the steps and the procedure involved in direct formulation.

**TABLE 1.3** Examples of the capabilities of ANSYS*

A V6 engine used in front-wheel-drive automobiles was analyzed using ANSYS heat transfer capabilities. The analyses were conducted by Analysis & Design Appl. Co. Ltd. (ADAPCO) on behalf of a major U.S. automobile manufacturer to improve product performance. Contours of thermal stress in the engine block are shown in the figure above.

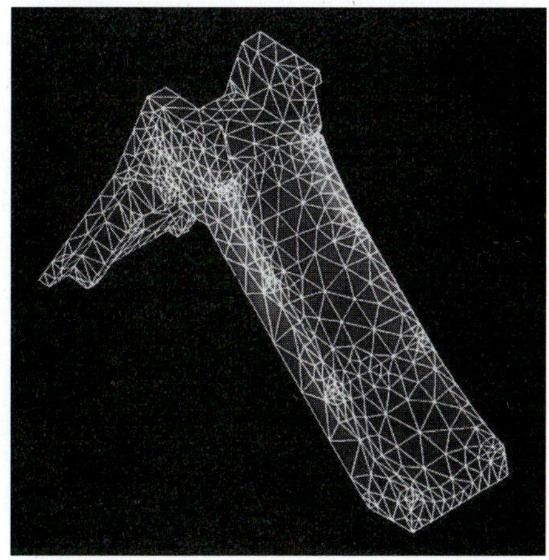

Large deflection capabilities of ANSYS were utilized by engineers at Today's Kids, a toy manufacturer, to confirm failure locations on the company's play slide, shown in the figure above, when the slide is subjected to overload. This nonlinear analysis capability is required to detect these stresses because of the product's structural behavior.

Electromagnetic capabilities of ANSYS, which include the use of both vector and scalar potentials interfaced through a specialized element, as well as a three-dimensional graphics representation of far-field decay through infinite boundary elements, are depicted in this analysis of a bath plate, shown in the figure above.

Isocontours are used to depict the intensity of the H-field. Structural Analysis Engineering Corporation used ANSYS to determine the natural frequency of a rotor in a disk-brake assembly. In this analysis, 50 modes of vibration, which are considered to contribute to brake sequel, were found to exist in the light-truck brake rotor.

*Photographs courtesy of ANSYS, Inc., Canonsburg, PA.

## EXAMPLE 1.1

Consider a bar with a variable cross section supporting a load $P$, as shown in Figure 1.1. The bar is fixed at one end and carries the load $P$ at the other end. Let us designate the width of the bar at the top by $w_1$, at the bottom by $w_2$, its thickness by $t$, and its length by $L$. The bar's modulus of elasticity will be denoted by $E$. We are interested in determining how much the bar will deflect at various points along its length when it is subjected to the load $P$. We will neglect the weight of the bar in the following analysis, assuming that the applied load is considerably larger than the weight of the bar:

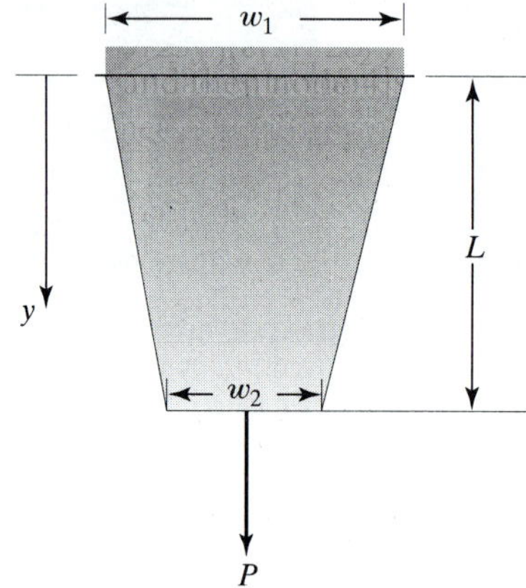


**FIGURE 1.1** A bar under axial loading.

### Preprocessing Phase

1. *Discretize the solution domain into finite elements.*
   We begin by subdividing the problem into nodes and elements. In order to highlight the basic steps in a finite element analysis, we will keep this problem simple and, thus, represent it by a model that has five nodes and four elements, as shown in Figure 1.2. However, note that we can increase the accuracy of our results by gen-

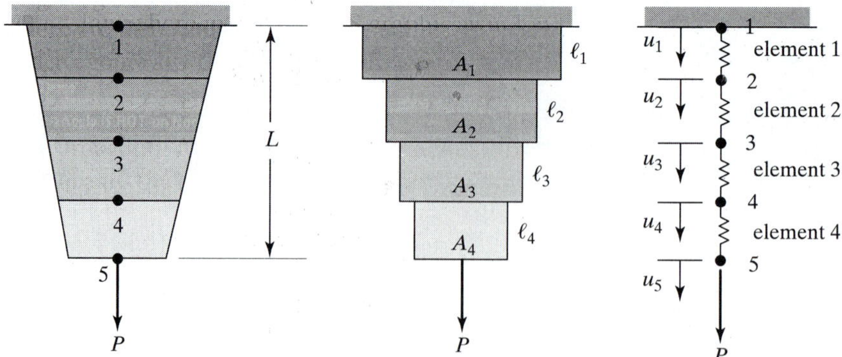


**FIGURE 1.2** Subdividing the bar into elements and nodes.

erating a model with additional nodes and elements. This task is left as an exercise for you to complete. (See Problem 1 at the end of this chapter.) The given bar is modeled using four individual segments, with each segment having a uniform cross section. The cross-sectional area of each element is represented by an average area of the cross sections at the nodes that make up (define) the element. This model is shown in Figure 1.2.

2. *Assume a solution that approximates the behavior of an element.*
In order to study the behavior of a typical element, let's consider the deflection of a solid member with a uniform cross section $A$ that has a length $\ell$ when subjected to a force $F$, as shown in Figure 1.3.

The average stress $\sigma$ in the member is given by

$$\sigma = \frac{F}{A} \tag{1.1}$$

The average normal strain $\varepsilon$ of the member is defined as the change in length $\Delta\ell$ per unit original length $\ell$ of the member:

$$\varepsilon = \frac{\Delta\ell}{\ell} \tag{1.2}$$

Over the elastic region, the stress and strain are related by Hooke's Law, according to the equation

$$\sigma = E\varepsilon \tag{1.3}$$

where $E$ is the modulus of elasticity of the material. Combining Eqs. (1.1), (1.2), and (1.3) and simplifying, we have

$$F = \left(\frac{AE}{\ell}\right)\Delta\ell \tag{1.4}$$

Note that Eq. (1.4) is similar to the equation for a linear spring, $F = kx$. Therefore, a centrally loaded member of uniform cross section may be modeled as a spring with an equivalent stiffness of

$$k_{eq} = \frac{AE}{\ell} \tag{1.5}$$

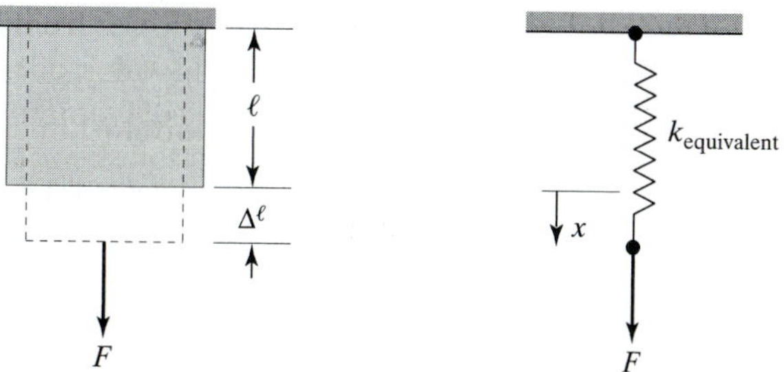


**FIGURE 1.3** A solid member of uniform cross section subjected to a force $F$.

Turning our attention to Example Problem 1.1, we note once again that the bar's cross section varies in the $y$-direction. As a first approximation, we model the bar as a series of centrally loaded members with different cross sections, as shown in Figure 1.2. Thus, the bar is represented by a model consisting of four elastic springs (elements) in series, and the elastic behavior of an element is modeled by an equivalent linear spring according to the equation

$$f = k_{eq}(u_{i+1} - u_i) = \frac{A_{avg}E}{\ell}(u_{i+1} - u_i) = \frac{(A_{i+1} + A_i)E}{2\ell}(u_{i+1} - u_1) \qquad (1.6)$$

where the equivalent element stiffness is given by

$$k_{eq} = \frac{(A_{i+1} + A_i)E}{2\ell} \qquad (1.7)$$

$A_i$ and $A_{i+1}$ are the cross-sectional areas of the member at nodes $i$ and $i + 1$, respectively, and $\ell$ is the length of the element. Employing the above model, let us consider the forces acting on each node. The free body diagram of nodes, which shows the forces acting on nodes 1 through 5 of this model, is depicted in Figure 1.4.

Static equilibrium requires that the sum of the forces acting on each node be zero. This requirement creates the following five equations:

$$\begin{aligned}
\text{node 1:}\quad & R_1 - k_1(u_2 - u_1) = 0 \\
\text{node 2:}\quad & k_1(u_2 - u_1) - k_2(u_3 - u_2) = 0 \\
\text{node 3:}\quad & k_2(u_3 - u_2) - k_3(u_4 - u_3) = 0 \\
\text{node 4:}\quad & k_3(u_4 - u_3) - k_4(u_5 - u_4) = 0 \\
\text{node 5:}\quad & k_4(u_5 - u_4) - P = 0
\end{aligned} \qquad (1.8)$$

Rearranging the equilibrium equations given by Eq. (1.8) by separating the reaction force $R_1$ and the applied external force $P$ from the internal forces, we have:

$$\begin{array}{rrrrrrrrl}
k_1u_1 & -k_1u_2 & & & & & & & = -R_1 \\
-k_1u_1 & +k_1u_2 & +k_2u_2 & -k_2u_3 & & & & & = 0 \\
 & & -k_2u_2 & +k_2u_3 & +k_3u_3 & -k_3u_4 & & & = 0 \\
 & & & & -k_3u_3 & +k_3u_4 & +k_4u_4 & -k_4u_5 & = 0 \\
 & & & & & & -k_4u_4 & +k_4u_5 & = P
\end{array} \qquad (1.9)$$

Presenting the equilibrium equations of Eq. (1.9) in a matrix form, we have:

$$\begin{bmatrix} k_1 & -k_1 & 0 & 0 & 0 \\ -k_1 & k_1 + k_2 & -k_2 & 0 & 0 \\ 0 & -k_2 & k_2 + k_3 & -k_3 & 0 \\ 0 & 0 & -k_3 & k_3 + k_4 & -k_4 \\ 0 & 0 & 0 & -k_4 & k_4 \end{bmatrix} \begin{Bmatrix} u_1 \\ u_2 \\ u_3 \\ u_4 \\ u_5 \end{Bmatrix} = \begin{Bmatrix} -R_1 \\ 0 \\ 0 \\ 0 \\ P \end{Bmatrix} \qquad (1.10)$$

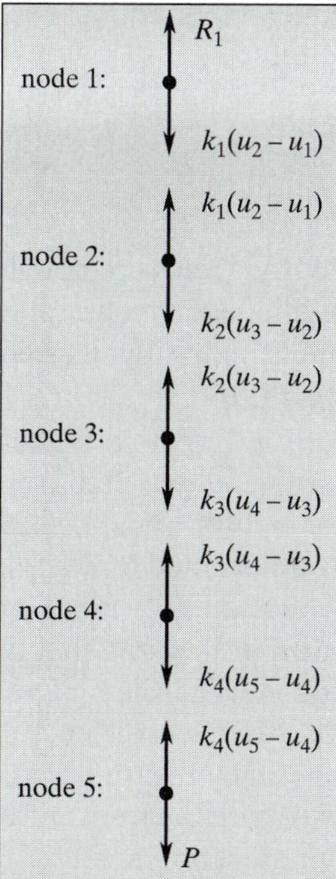


**FIGURE 1.4** Free body diagram of the nodes in Example 1.1.

It is also important to distinguish between the reaction forces and the applied loads in the load matrix. Therefore, the matrix relation of Eq. (1.10) can be written as:

$$\begin{Bmatrix} -R_1 \\ 0 \\ 0 \\ 0 \\ 0 \end{Bmatrix} = \begin{bmatrix} k_1 & -k_1 & 0 & 0 & 0 \\ -k_1 & k_1 + k_2 & -k_2 & 0 & 0 \\ 0 & -k_2 & k_2 + k_3 & -k_3 & 0 \\ 0 & 0 & -k_3 & k_3 + k_4 & -k_4 \\ 0 & 0 & 0 & -k_4 & k_4 \end{bmatrix} \begin{Bmatrix} u_1 \\ u_2 \\ u_3 \\ u_4 \\ u_5 \end{Bmatrix} - \begin{Bmatrix} 0 \\ 0 \\ 0 \\ 0 \\ P \end{Bmatrix} \tag{1.11}$$

We can readily show that under additional nodal loads and other fixed boundary conditions, the relationship given by Eq. (1.11) can be put into the general form

$$\{\mathbf{R}\} = [\mathbf{K}]\{\mathbf{u}\} - \{\mathbf{F}\} \tag{1.12}$$

which stands for

$$\{\textbf{reaction matrix}\} = [\textbf{stiffness matrix}]\{\textbf{displacement matrix}\} - \{\textbf{load matrix}\}$$

Turning our attention to Example 1.1 again, we find that because the bar is fixed at the top, the displacement of node 1 is zero. Thus, the first row of the system of equations given by Eq. (1.10) should read $u_1 = 0$. Thus, application of the boundary condition leads to the following matrix equation:

$$\begin{bmatrix} 1 & 0 & 0 & 0 & 0 \\ -k_1 & k_1 + k_2 & -k_2 & 0 & 0 \\ 0 & -k_2 & k_2 + k_3 & -k_3 & 0 \\ 0 & 0 & -k_3 & k_3 + k_4 & -k_4 \\ 0 & 0 & 0 & -k_4 & k_4 \end{bmatrix} \begin{Bmatrix} u_1 \\ u_2 \\ u_3 \\ u_4 \\ u_5 \end{Bmatrix} = \begin{Bmatrix} 0 \\ 0 \\ 0 \\ 0 \\ P \end{Bmatrix} \tag{1.13}$$

The solution of the above matrix yields the nodal displacement values. In the next section, we will develop the general elemental stiffness matrix and discuss the construction of the global stiffness matrix by inspection.

**3.** *Develop equations for an element.*
Because each of the elements in Example 1.1 has two nodes, and with each node we have associated a displacement, we need to create two equations for each element. These equations must involve nodal displacements and the element's stiffness. Consider the internally transmitted forces $f_i$ and $f_{i+1}$ and the end displacements $u_i$ and $u_{i+1}$ of an element, which are shown in Figure 1.5.

Static equilibrium conditions require that the sum of $f_i$ and $f_{i+1}$ be zero. Note that the sum of $f_i$ and $f_{i+1}$ is zero regardless of which representation of Figure 1.5 is selected. However, for the sake of consistency in the forthcoming derivation, we will use the representation given by Figure 1.5(b), so that $f_i$ and $f_{i+1}$ are given in the positive $y$-direction. Thus, we write the transmitted forces at nodes $i$ and $i + 1$ according to the following equations:

$$\begin{aligned} f_i &= k_{eq}(u_i - u_{i+1}) \\ f_{i+1} &= k_{eq}(u_{i+1} - u_i) \end{aligned} \tag{1.14}$$

Equation (1.14) can be expressed in a matrix form by

$$\begin{Bmatrix} f_i \\ f_{i+1} \end{Bmatrix} = \begin{bmatrix} k_{eq} & -k_{eq} \\ -k_{eq} & k_{eq} \end{bmatrix} \begin{Bmatrix} u_i \\ u_{i+1} \end{Bmatrix} \tag{1.15}$$

**4.** *Assemble the elements to present the entire problem.*
Applying the elemental description given by Eq. (1.15) to all elements and assembling them (putting them together) will lead to the formation of the global stiffness matrix. The stiffness matrix for element (1) is given by

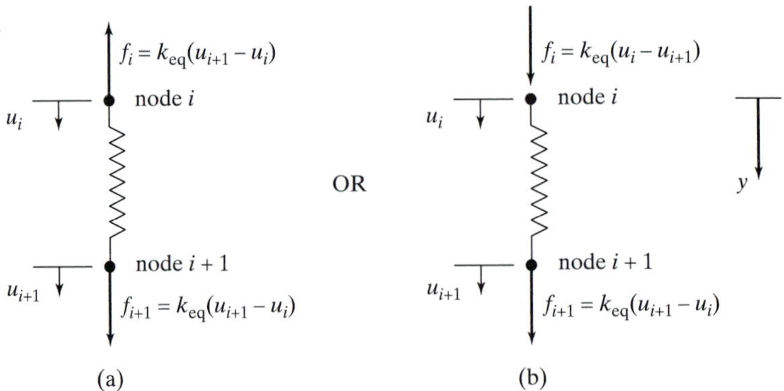


**FIGURE 1.5** Internally transmitted forces through an arbitrary element.

$$[\mathbf{K}]^{(1)} = \begin{bmatrix} k_1 & -k_1 \\ -k_1 & k_1 \end{bmatrix}$$

and its position in the global stiffness matrix is given by

$$[\mathbf{K}]^{(1G)} = \begin{bmatrix} k_1 & -k_1 & 0 & 0 & 0 \\ -k_1 & k_1 & 0 & 0 & 0 \\ 0 & 0 & 0 & 0 & 0 \\ 0 & 0 & 0 & 0 & 0 \\ 0 & 0 & 0 & 0 & 0 \end{bmatrix} \begin{matrix} u_1 \\ u_2 \\ u_3 \\ u_4 \\ u_5 \end{matrix}$$

The nodal displacement matrix is shown alongside the position of element 1 in the global stiffness matrix to aid us to observe the contribution of a node to its neighboring elements. Similarly, for elements (2), (3), and (4), we have

$$[\mathbf{K}]^{(2)} = \begin{bmatrix} k_2 & -k_2 \\ -k_2 & k_2 \end{bmatrix}$$

and its position in the global matrix

$$[\mathbf{K}]^{(2G)} = \begin{bmatrix} 0 & 0 & 0 & 0 & 0 \\ 0 & k_2 & -k_2 & 0 & 0 \\ 0 & -k_2 & k_2 & 0 & 0 \\ 0 & 0 & 0 & 0 & 0 \\ 0 & 0 & 0 & 0 & 0 \end{bmatrix} \begin{matrix} u_1 \\ u_2 \\ u_3 \\ u_4 \\ u_5 \end{matrix}$$

$$[\mathbf{K}]^{(3)} = \begin{bmatrix} k_3 & -k_3 \\ -k_3 & k_3 \end{bmatrix}$$

and its position in the global matrix

$$[\mathbf{K}]^{(3G)} = \begin{bmatrix} 0 & 0 & 0 & 0 & 0 \\ 0 & 0 & 0 & 0 & 0 \\ 0 & 0 & k_3 & -k_3 & 0 \\ 0 & 0 & -k_3 & k_3 & 0 \\ 0 & 0 & 0 & 0 & 0 \end{bmatrix} \begin{matrix} u_1 \\ u_2 \\ u_3 \\ u_4 \\ u_5 \end{matrix}$$

and

$$[\mathbf{K}]^{(4)} = \begin{bmatrix} k_4 & -k_4 \\ -k_4 & k_4 \end{bmatrix}$$

and its position in the global matrix

$$[\mathbf{K}]^{(4G)} = \begin{bmatrix} 0 & 0 & 0 & 0 & 0 \\ 0 & 0 & 0 & 0 & 0 \\ 0 & 0 & 0 & 0 & 0 \\ 0 & 0 & 0 & k_4 & -k_4 \\ 0 & 0 & 0 & -k_4 & k_4 \end{bmatrix} \begin{matrix} u_1 \\ u_2 \\ u_3 \\ u_4 \\ u_5 \end{matrix}$$

The final global stiffness matrix is obtained by assembling, or adding, together each element's position in the global stiffness matrix:

$$[\mathbf{K}]^{(G)} = [\mathbf{K}]^{(1G)} + [\mathbf{K}]^{(2G)} + [\mathbf{K}]^{(3G)} + [\mathbf{K}]^{(4G)}$$

$$[\mathbf{K}]^{(G)} = \begin{bmatrix} k_1 & -k_1 & 0 & 0 & 0 \\ -k_1 & k_1 + k_2 & -k_2 & 0 & 0 \\ 0 & -k_2 & k_2 + k_3 & -k_3 & 0 \\ 0 & 0 & -k_3 & k_3 + k_4 & -k_4 \\ 0 & 0 & 0 & -k_4 & k_4 \end{bmatrix} \tag{1.16}$$

Note that the global stiffness matrix obtained using elemental description, as given by Eq. (1.16), is identical to the global stiffness matrix we obtained earlier from the analysis of the free body diagrams of the nodes, as given by the left hand side of Eq. (1.10).

5. *Apply boundary conditions and loads.*
The bar is fixed at the top, which leads to the boundary condition $u_1 = 0$. The external load $P$ is applied at node 5. Applying these conditions results in the following set of linear equations.

$$\begin{bmatrix} 1 & 0 & 0 & 0 & 0 \\ -k_1 & k_1 + k_2 & -k_2 & 0 & 0 \\ 0 & -k_2 & k_2 + k_3 & -k_3 & 0 \\ 0 & 0 & -k_3 & k_3 + k_4 & -k_4 \\ 0 & 0 & 0 & -k_4 & k_4 \end{bmatrix} \begin{Bmatrix} u_1 \\ u_2 \\ u_3 \\ u_4 \\ u_5 \end{Bmatrix} = \begin{Bmatrix} 0 \\ 0 \\ 0 \\ 0 \\ P \end{Bmatrix} \tag{1.17}$$

Again, note that the first row of the matrix in Eq. (1.17) must contain a 1 followed by four 0's to read $u_1 = 0$, the given boundary condition. Also note that in solid mechanics problems, the finite element formulation will always lead to the following general form:

**[stiffness matrix]{displacement matrix} = {load matrix}**

## Solution Phase

6. *Solve a system of algebraic equations simultaneously.*
In order to obtain numerical values of the nodal displacements, let us assume that $E = 10.4 \times 10^6$ lb/in$^2$ (aluminum), $w_1 = 2$ in, $w_2 = 1$ in, $t = 0.125$ in, $L = 10$ in, and $P = 1000$ lb. You may consult Table 1.4 while working toward the solution.

TABLE 1.4 Properties of the elements in Example 1.1

| Element | Nodes | | Average cross-sectional area (in$^2$) | Length (in) | Modulus of elasticity (lb/in$^2$) | Element's stiffness coefficient (lb/in) |
|---|---|---|---|---|---|---|
| 1 | 1 | 2 | 0.234375 | 2.5 | $10.4 \times 10^6$ | $975 \times 10^3$ |
| 2 | 2 | 3 | 0.203125 | 2.5 | $10.4 \times 10^6$ | $845 \times 10^3$ |
| 3 | 3 | 4 | 0.171875 | 2.5 | $10.4 \times 10^6$ | $715 \times 10^3$ |
| 4 | 4 | 5 | 0.140625 | 2.5 | $10.4 \times 10^6$ | $585 \times 10^3$ |

The variation of the cross-sectional area of the bar in the $y$-direction can be expressed by:

$$A(y) = \left(w_1 + \left(\frac{w_2 - w_1}{L}\right)y\right)t = \left(2 + \frac{(1-2)}{10}y\right)(0.125) = 0.25 - 0.0125y \quad (1.18)$$

Using Eq. (1.18), we can compute the cross-sectional areas at each node:

$A_1 = 0.25 \text{ in}^2$ $\quad A_2 = 0.25 - 0.0125(2.5) = 0.21875 \text{ in}^2$

$A_3 = 0.25 - 0.0125(5.0) = 0.1875 \text{ in}^2$ $\quad A_4 = 0.25 - 0.0125(7.5) = 0.15625 \text{ in}^2$

$A_5 = 0.125 \text{ in}^2$

Next, the equivalent stiffness coefficient for each element is computed from the equations

$$k_{eq} = \frac{(A_{i+1} + A_i)E}{2\ell}$$

$$k_1 = \frac{(0.21875 + 0.25)(10.4 \times 10^6)}{2(2.5)} = 975 \times 10^3 \frac{\text{lb}}{\text{in}}$$

$$k_2 = \frac{(0.1875 + 0.21875)(10.4 \times 10^6)}{2(2.5)} = 845 \times 10^3 \frac{\text{lb}}{\text{in}}$$

$$k_3 = \frac{(0.15625 + 0.1875)(10.4 \times 10^6)}{2(2.5)} = 715 \times 10^3 \frac{\text{lb}}{\text{in}}$$

$$k_4 = \frac{(0.125 + 0.15625)(10.4 \times 10^6)}{2(2.5)} = 585 \times 10^3 \frac{\text{lb}}{\text{in}}$$

and the elemental matrices are

$$[\mathbf{K}]^{(1)} = \begin{bmatrix} k_1 & -k_1 \\ -k_1 & k_1 \end{bmatrix} = 10^3 \begin{bmatrix} 975 & -975 \\ -975 & 975 \end{bmatrix}$$

$$[\mathbf{K}]^{(2)} = \begin{bmatrix} k_2 & -k_2 \\ -k_2 & k_2 \end{bmatrix} = 10^3 \begin{bmatrix} 845 & -845 \\ -845 & 845 \end{bmatrix}$$

$$[\mathbf{K}]^{(3)} = \begin{bmatrix} k_3 & -k_3 \\ -k_3 & k_3 \end{bmatrix} = 10^3 \begin{bmatrix} 715 & -715 \\ -715 & 715 \end{bmatrix}$$

$$[\mathbf{K}]^{(4)} = \begin{bmatrix} k_4 & -k_4 \\ -k_4 & k_4 \end{bmatrix} = 10^3 \begin{bmatrix} 585 & -585 \\ -585 & 585 \end{bmatrix}$$

Assembling the elemental matrices leads to the generation of the global stiffness matrix:

$$[\mathbf{K}]^{(G)} = 10^3 \begin{bmatrix} 975 & -975 & 0 & 0 & 0 \\ -975 & 975 + 845 & -845 & 0 & 0 \\ 0 & -845 & 845 + 715 & -715 & 0 \\ 0 & 0 & -715 & 715 + 585 & -585 \\ 0 & 0 & 0 & -585 & 585 \end{bmatrix}$$

Applying the boundary condition $u_1 = 0$ and the load $P = 1000$ lb, we get

$$10^3\begin{bmatrix} 1 & 0 & 0 & 0 & 0 \\ -975 & 1820 & -845 & 0 & 0 \\ 0 & -845 & 1560 & -715 & 0 \\ 0 & 0 & -715 & 1300 & -585 \\ 0 & 0 & 0 & -585 & 585 \end{bmatrix} \begin{Bmatrix} u_1 \\ u_2 \\ u_3 \\ u_4 \\ u_5 \end{Bmatrix} = \begin{Bmatrix} 0 \\ 0 \\ 0 \\ 0 \\ 10^3 \end{Bmatrix}$$

Because in the second row, the $-975$ coefficient gets multiplied by $u_1 = 0$, we need only to solve the following $4 \times 4$ matrix:

$$10^3\begin{bmatrix} 1820 & -845 & 0 & 0 \\ -845 & 1560 & -715 & 0 \\ 0 & -715 & 1300 & -585 \\ 0 & 0 & -585 & 585 \end{bmatrix} \begin{Bmatrix} u_2 \\ u_3 \\ u_4 \\ u_5 \end{Bmatrix} = \begin{Bmatrix} 0 \\ 0 \\ 0 \\ 10^3 \end{Bmatrix}$$

The displacement solution is: $u_1 = 0$, $u_2 = 0.001026$ in, $u_3 = 0.002210$ in, $u_4 = 0.003608$ in, and $u_5 = 0.005317$ in.

### Postprocessing Phase

**7.** *Obtain other information.*

For Example 1.1, we may be interested in obtaining other information, such as the average normal stresses in each element. These values can be determined from the equation

$$\sigma = \frac{f}{A_{\text{avg}}} = \frac{k_{\text{eq}}(u_{i+1} - u_i)}{A_{\text{avg}}} = \frac{\dfrac{A_{\text{avg}}E}{\ell}(u_{i+1} - u_i)}{A_{\text{avg}}} = E\left(\frac{u_{i+1} - u_i}{\ell}\right) \tag{1.19}$$

Since the displacements of different nodes are known, Eq. (1.19) could have been obtained directly from the relationship between the stresses and strains,

$$\sigma = E\varepsilon = E\left(\frac{u_{i+1} - u_i}{\ell}\right) \tag{1.20}$$

Employing Eq. (1.20) in Example 1.1, we compute the average normal stress for each element as

$$\sigma^{(1)} = E\left(\frac{u_2 - u_1}{\ell}\right) = \frac{(10.4 \times 10^6)(0.001026 - 0)}{2.5} = 4268\,\frac{\text{lb}}{\text{in}^2}$$

$$\sigma^{(2)} = E\left(\frac{u_3 - u_2}{\ell}\right) = \frac{(10.4 \times 10^6)(0.002210 - 0.001026)}{2.5} = 4925\,\frac{\text{lb}}{\text{in}^2}$$

$$\sigma^{(3)} = E\left(\frac{u_4 - u_3}{\ell}\right) = \frac{(10.4 \times 10^6)(0.003608 - 0.002210)}{2.5} = 5816\,\frac{\text{lb}}{\text{in}^2}$$

$$\sigma^{(4)} = E\left(\frac{u_5 - u_4}{\ell}\right) = \frac{(10.4 \times 10^6)(0.005317 - 0.003608)}{2.5} = 7109\,\frac{\text{lb}}{\text{in}^2}$$

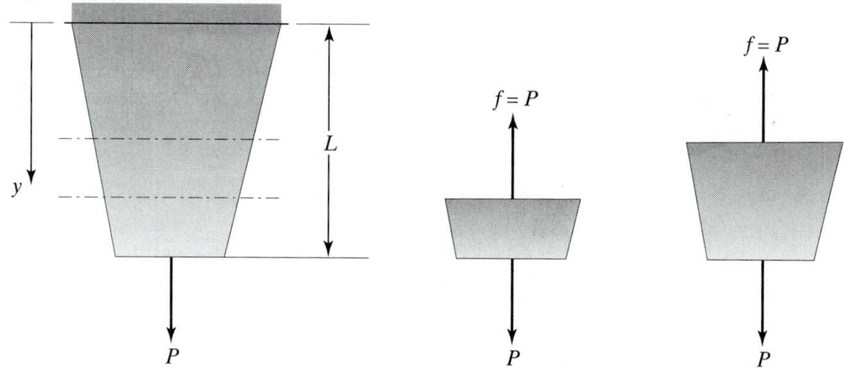


**FIGURE 1.6** The internal forces in Example 1.1.

In Figure 1.6, we note that for the given problem, regardless of where we cut a section through the bar, the internal force at the section is equal to 1000 lb. So,

$$\sigma^{(1)} = \frac{f}{A_{avg}} = \frac{1000}{0.234375} = 4267 \frac{\text{lb}}{\text{in}^2}$$

$$\sigma^{(2)} = \frac{f}{A_{avg}} = \frac{1000}{0.203125} = 4923 \frac{\text{lb}}{\text{in}^2}$$

$$\sigma^{(3)} = \frac{f}{A_{avg}} = \frac{1000}{0.171875} = 5818 \frac{\text{lb}}{\text{in}^2}$$

$$\sigma^{(4)} = \frac{f}{A_{avg}} = \frac{1000}{0.140625} = 7111 \frac{\text{lb}}{\text{in}^2}$$

Ignoring the errors we get from rounding off our answers, we find that these results are identical to the element stresses computed from the displacement information. This comparison tells us that our displacement calculations are good for this problem.

**Reaction Forces** For Example 1.1, the reaction force may be computed in a number of ways. First, referring to Figure 1.4, we note that the statics equilibrium at node 1 requires

$$R_1 = k_1(u_2 - u_1) = 975 \times 10^3(0.001026 - 0) = 1000 \text{ lb}$$

The statics equilibrium for the entire bar also requires that

$$R_1 = P = 1000 \text{ lb}$$

As you may recall, we can also compute the reaction forces from the general reaction equation

$$\{\mathbf{R}\} = [\mathbf{K}]\{\mathbf{u}\} - \{\mathbf{F}\}$$

or

$$\{\textbf{reaction matrix}\} = [\textbf{stiffness matrix}]\{\textbf{displacement matrix}\} - \{\textbf{load matrix}\}$$

Because Example 1.1 is a simple problem, we do not actually need to go through the matrix operations in the aforementioned general equation to compute the reaction forces. However, as a demonstration, the procedure is shown here. From the general equation, we get

$$\begin{Bmatrix} R_1 \\ R_2 \\ R_3 \\ R_4 \\ R_5 \end{Bmatrix} = 10^3 \begin{bmatrix} 975 & -975 & 0 & 0 & 0 \\ -975 & 1820 & -845 & 0 & 0 \\ 0 & -845 & 1560 & -715 & 0 \\ 0 & 0 & -715 & 1300 & -585 \\ 0 & 0 & 0 & -585 & 585 \end{bmatrix} \begin{Bmatrix} 0 \\ 0.001026 \\ 0.002210 \\ 0.003608 \\ 0.005317 \end{Bmatrix} - \begin{Bmatrix} 0 \\ 0 \\ 0 \\ 0 \\ 10^3 \end{Bmatrix}$$

where $R_1, R_2, R_3, R_4$, and $R_5$ represent the reactions forces at nodes 1 through 5, respectively. Performing the matrix operation, we have

$$\begin{Bmatrix} R_1 \\ R_2 \\ R_3 \\ R_4 \\ R_5 \end{Bmatrix} = \begin{Bmatrix} -1000 \\ 0 \\ 0 \\ 0 \\ 0 \end{Bmatrix}$$

The negative value of $R_1$ simply means that the direction of the reaction force is up (because we assumed that the positive $y$-direction points down). Of course, as expected, the outcome is the same as in our earlier calculations because the rows of the above matrix represent the static equilibrium conditions at each node. Next, we will consider finite element formulation of a heat transfer problem.

## EXAMPLE 1.2

A typical exterior frame wall (made up of 2 × 4 studs) of a house contains the materials shown in the table below. Let us assume an inside room temperature of 70°F and an outside air temperature of 20°F, with an exposed area of 150 ft$^2$. We are interested in determining the temperature distribution through the wall.

| Items | Resistance hr•ft$^2$•F/Btu | $U$-factor Btu/hr•ft$^2$•F |
|---|---|---|
| 1. Outside film resistance (winter, 15-mph wind) | 0.17 | 5.88 |
| 2. Siding, wood (1/2 × 8 lapped) | 0.81 | 1.23 |
| 3. Sheathing (1/2 in regular) | 1.32 | 0.76 |
| 4. Insulation batt (3 − 31/2 in) | 11.0 | 0.091 |
| 5. Gypsum wall board (1/2 in) | 0.45 | 2.22 |
| 6. Inside film resistance (winter) | 0.68 | 1.47 |

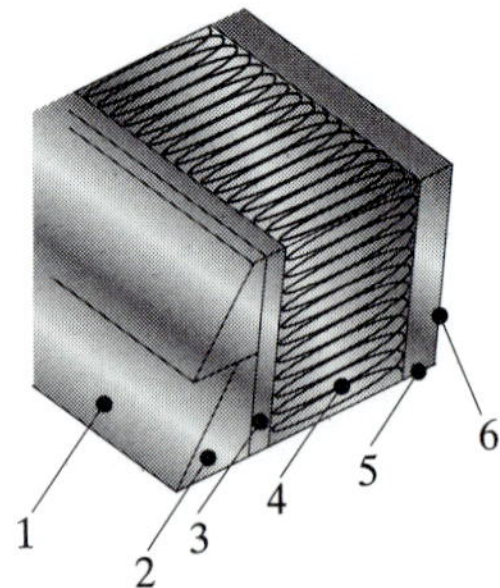

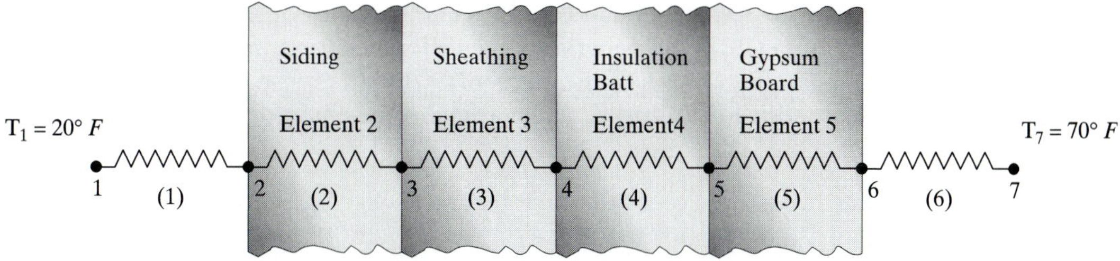


**FIGURE 1.7** Finite element model of Example 1.2.

## Preprocessing Phase

1. *Discretize the solution domain into finite elements.*
   We will represent this problem by a model that has seven nodes and six elements, as shown in Figure 1.7.
2. *Assume a solution that approximates the behavior of an element.*
   For Example 1.2, there are two modes of heat transfer (conduction and convection) that we must first understand before we can proceed with formulating the conductance matrix and the thermal load matrix. The steady state thermal behavior of the elements (2), (3), (4), and (5) may be modeled using Fourier's Law. When there exists a temperature gradient in a medium, conduction heat transfer occurs, as shown in Figure 1.8. The energy is transported from the high-temperature region to the low-temperature region by molecular activities. The heat transfer rate is given by Fourier's Law:

$$q_X = -kA\frac{\partial T}{\partial X} \tag{1.21}$$

$q_X$ is the $X$-component of the heat transfer rate, $k$ is the thermal conductivity of the medium, $A$ is the area, and $\frac{\partial T}{\partial X}$ is the temperature gradient. The minus sign in

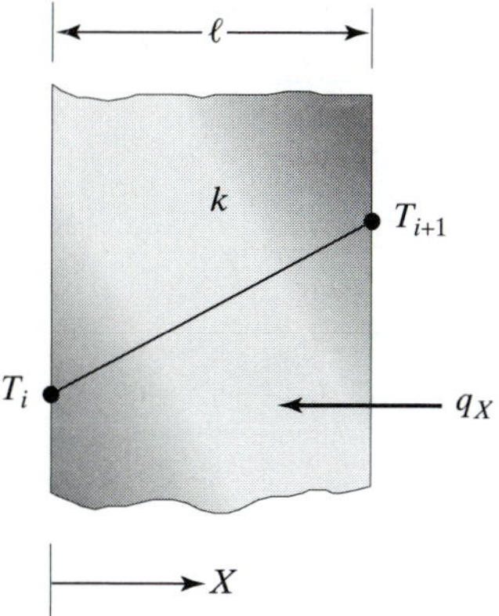


**FIGURE 1.8** Heat transfer in a medium by conduction.

Eq. (1.21) is due to the fact that heat flows in the direction of decreasing temperature. Equation (1.21) can be written in a difference form in terms of the spacing between the nodes (length of the element) $\ell$ and the respective temperatures of the nodes $i$ and $i + 1$, $T_i$ and $T_{i+1}$, according to the equation

$$q = \frac{kA(T_{i+1} - T_i)}{\ell} \tag{1.22}$$

In the field of heat transfer, it is also common to write Eq. (1.22) in terms of the thermal transmittance coefficient $U$, or, as it is often called, the $U$-factor ($U = \frac{k}{\ell}$). The $U$-factor represents thermal transmission through a unit area and has the units of Btu/hr·ft$^2$·F. It is the reciprocal of thermal resistance. So,

$$q = UA(T_{i+1} - T_i) \tag{1.23}$$

The steady state thermal behavior of elements (1) and (6) may be modeled using Newton's Law of Cooling. Convection heat transfer occurs when a fluid in motion comes into contact with a surface whose temperature differs from the moving fluid. The overall heat transfer rate between the fluid and the surface is governed by Newton's Law of Cooling, according to the equation

$$q = hA(T_s - T_f) \tag{1.24}$$

where $h$ is the heat transfer coefficient, $T_s$ is the surface temperature, and $T_f$ represents the temperature of the moving fluid. Newton's Law of Cooling can also be written in terms of the $U$-factor, such that

$$q = UA(T_s - T_f) \tag{1.25}$$

where $U = h$, and it represents the reciprocal of thermal resistance due to convection boundary conditions. Under steady state conduction, the application of energy balance to a surface requires that the energy transferred to this surface via conduction must be equal to the energy transfer by convection. This principle,

$$-kA\frac{\partial T}{\partial X} = hA[T_s - T_f] \tag{1.26}$$

is depicted in Figure 1.9.

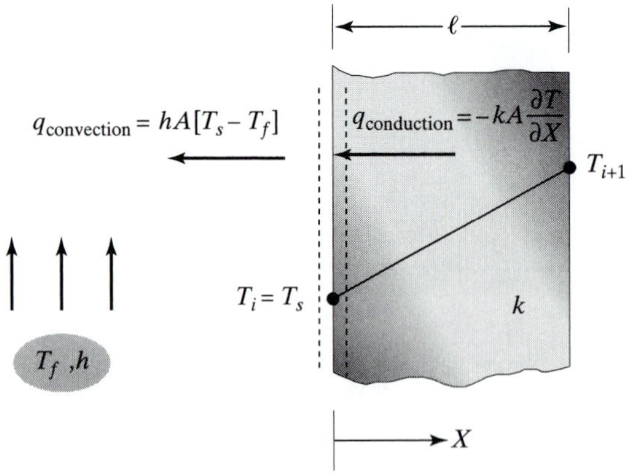


**FIGURE 1.9** Energy balance at a surface with a convective heat transfer.

Now that we understand the two modes of heat transfer involved in this problem, we can apply the energy balance to the various surfaces of the wall, starting with the wall's exterior surface located at node 2. The heat loss through the wall due to conduction must equal the heat loss to the surrounding cold air by convection. That is,

$$U_2 A(T_3 - T_2) = U_1 A(T_2 - T_1)$$

The application of energy balance to surfaces located at nodes 3, 4, and 5 yields the equations

$$U_3 A(T_4 - T_3) = U_2 A(T_3 - T_2)$$

$$U_4 A(T_5 - T_4) = U_3 A(T_4 - T_3)$$

$$U_5 A(T_6 - T_5) = U_4 A(T_5 - T_4)$$

For the interior surface of the wall, located at node 6, the heat loss by convection of warm air is equal to the heat transfer by conduction through the gypsum board, according to the equation

$$U_6 A(T_7 - T_6) = U_5 A(T_6 - T_5)$$

Separating the known temperatures from the unknown temperatures, we have:

$$\begin{array}{cccccccc}
+(U_1 + U_2)AT_2 & -U_2 AT_3 & & & & = & U_1 AT_1 \\
-U_2 AT_2 & +(U_2 + U_3)AT_3 & -U_3 AT_4 & & & = & 0 \\
 & -U_3 AT_3 & +(U_3 + U_4)AT_4 & -U_4 AT_5 & & = & 0 \\
 & & -U_4 AT_4 & +(U_4 + U_5)AT_5 & -U_5 AT_6 & = & 0 \\
 & & & -U_5 AT_5 & +(U_5 + U_6)AT_6 & = & U_6 AT_7
\end{array}$$

The above relationships can be represented in matrix form as

$$A\begin{bmatrix} U_1 + U_2 & -U_2 & 0 & 0 & 0 \\ -U_2 & U_2 + U_3 & -U_3 & 0 & 0 \\ 0 & -U_3 & U_3 + U_4 & -U_4 & 0 \\ 0 & 0 & -U_4 & U_4 + U_5 & -U_5 \\ 0 & 0 & 0 & -U_5 & U_5 + U_6 \end{bmatrix}\begin{Bmatrix} T_2 \\ T_3 \\ T_4 \\ T_5 \\ T_6 \end{Bmatrix} = \begin{Bmatrix} U_1 AT_1 \\ 0 \\ 0 \\ 0 \\ U_6 AT_7 \end{Bmatrix} \quad (1.27)$$

Note that the relationship given by Eq. (1.27) was developed by applying the conservation of energy to the surfaces located at nodes 2, 3, 4, 5, and 6. Next, we will consider the elemental formulation of this problem, which will lead to the same results.

**3.** *Develop equations for an element.*
In general, for conduction problems, the heat transfer rates, $q_i$ and $q_{i+1}$, and the nodal temperatures, $T_i$ and $T_{i+1}$, for an element are related according to the equations

$$q_i = \frac{kA}{\ell}(T_i - T_{i+1})$$

$$q_{i+1} = \frac{kA}{\ell}(T_{i+1} - T_i) \quad (1.28)$$

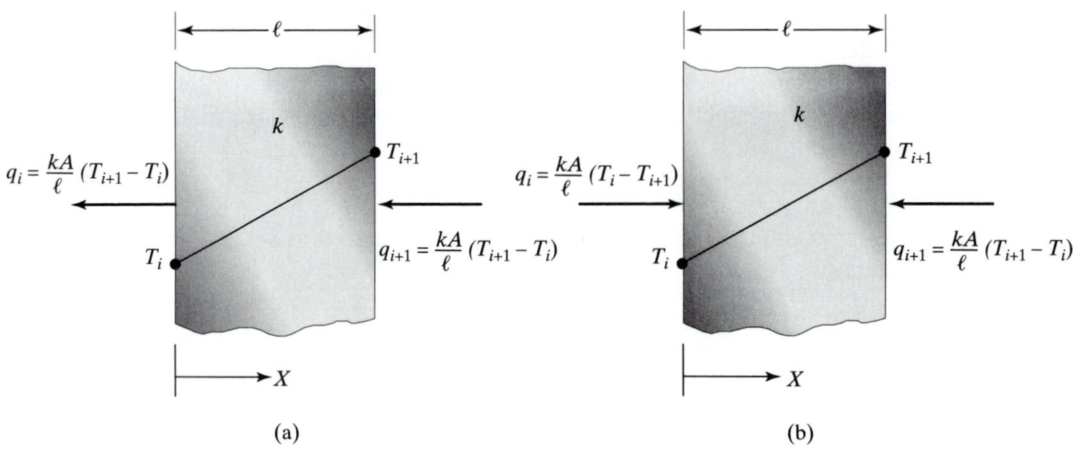


**FIGURE 1.10** Heat flow through nodes $i$ and $i + 1$.

The heat flow through nodes $i$ and $i + 1$ is depicted in Figure 1.10.

Because each of the elements in Example 1.2 has two nodes, and we have associated a temperature with each node, we want to create two equations for each element. These equations must involve nodal temperatures and the element's thermal conductivity or $U$-factor, based on Fourier's Law. Under steady state conditions, the application of the conservation of energy requires that the sum of $q_i$ and $q_{i+1}$ into an element be zero; that is, the energy flowing into node $i + 1$ must be equal to the energy flowing out of node $i$. Note that the sum of $q_i$ and $q_{i+1}$ is zero regardless of which representation of Figure 1.10 is selected. However, for the sake of consistency in the forthcoming derivation, we will use the representation given by Figure 1.10(b). Elemental description given by Eq. (1.28) can be expressed in matrix form by

$$\begin{Bmatrix} q_i \\ q_{i+1} \end{Bmatrix} = \frac{kA}{\ell}\begin{bmatrix} 1 & -1 \\ -1 & 1 \end{bmatrix}\begin{Bmatrix} T_i \\ T_{i+1} \end{Bmatrix} \tag{1.29}$$

The thermal conductance matrix for an element is

$$[\mathbf{K}]^{(e)} = \frac{kA}{\ell}\begin{bmatrix} 1 & -1 \\ -1 & 1 \end{bmatrix} \tag{1.30}$$

The conductance matrix can also be written in terms of the $U$-factor $\left(U = \frac{k}{\ell}\right)$:

$$[\mathbf{K}]^{(e)} = UA\begin{bmatrix} 1 & -1 \\ -1 & 1 \end{bmatrix} \tag{1.31}$$

Similarly, under steady state conditions, the application of the conservation of energy to the nodes of a convective element gives

$$q_i = hA(T_i - T_{i+1})$$
$$q_{i+1} = hA(T_{i+1} - T_i) \tag{1.32}$$

Equation (1.32) expressed in a matrix form is

$$\begin{Bmatrix} q_i \\ q_{i+1} \end{Bmatrix} = hA\begin{bmatrix} 1 & -1 \\ -1 & 1 \end{bmatrix}\begin{Bmatrix} T_i \\ T_{i+1} \end{Bmatrix}$$

The thermal conductance matrix for a convective element then becomes

$$[\mathbf{K}]^{(e)} = hA\begin{bmatrix} 1 & -1 \\ -1 & 1 \end{bmatrix} \tag{1.33}$$

Equation (1.33) can also be written in terms of the $U$-factor ($U = h$):

$$[\mathbf{K}]^{(e)} = UA\begin{bmatrix} 1 & -1 \\ -1 & 1 \end{bmatrix} \tag{1.34}$$

**4.** *Assemble the elements to present the entire problem.*
Applying the elemental description given by Eqs. (1.31) and (1.34) to all of the elements in Example 1.2 and assembling leads to the formation of the global stiffness matrix. So,

$$[\mathbf{K}]^{(1)} = A\begin{bmatrix} U_1 & -U_1 \\ -U_1 & U_1 \end{bmatrix}$$

and its position in the global matrix is

$$[\mathbf{K}]^{(1G)} = A\begin{bmatrix} U_1 & -U_1 & 0 & 0 & 0 & 0 & 0 \\ -U_1 & U_1 & 0 & 0 & 0 & 0 & 0 \\ 0 & 0 & 0 & 0 & 0 & 0 & 0 \\ 0 & 0 & 0 & 0 & 0 & 0 & 0 \\ 0 & 0 & 0 & 0 & 0 & 0 & 0 \\ 0 & 0 & 0 & 0 & 0 & 0 & 0 \\ 0 & 0 & 0 & 0 & 0 & 0 & 0 \end{bmatrix}\begin{matrix} T_1 \\ T_2 \\ T_3 \\ T_4 \\ T_5 \\ T_6 \\ T_7 \end{matrix}$$

The nodal temperature matrix is shown along with the global thermal conductance matrix to help you observe the contribution of a node to its neighboring elements:

$$[\mathbf{K}]^{(2)} = A\begin{bmatrix} U_2 & -U_2 \\ -U_2 & U_2 \end{bmatrix} \text{ and } [\mathbf{K}]^{(2G)} = A\begin{bmatrix} 0 & 0 & 0 & 0 & 0 & 0 & 0 \\ 0 & U_2 & -U_2 & 0 & 0 & 0 & 0 \\ 0 & -U_2 & U_2 & 0 & 0 & 0 & 0 \\ 0 & 0 & 0 & 0 & 0 & 0 & 0 \\ 0 & 0 & 0 & 0 & 0 & 0 & 0 \\ 0 & 0 & 0 & 0 & 0 & 0 & 0 \\ 0 & 0 & 0 & 0 & 0 & 0 & 0 \end{bmatrix}\begin{matrix} T_1 \\ T_2 \\ T_3 \\ T_4 \\ T_5 \\ T_6 \\ T_7 \end{matrix}$$

$$[\mathbf{K}]^{(3)} = A\begin{bmatrix} U_3 & -U_3 \\ -U_3 & U_3 \end{bmatrix} \text{ and } [\mathbf{K}]^{(3G)} = A\begin{bmatrix} 0 & 0 & 0 & 0 & 0 & 0 & 0 \\ 0 & 0 & 0 & 0 & 0 & 0 & 0 \\ 0 & 0 & U_3 & -U_3 & 0 & 0 & 0 \\ 0 & 0 & -U_3 & U_3 & 0 & 0 & 0 \\ 0 & 0 & 0 & 0 & 0 & 0 & 0 \\ 0 & 0 & 0 & 0 & 0 & 0 & 0 \\ 0 & 0 & 0 & 0 & 0 & 0 & 0 \end{bmatrix}\begin{matrix} T_1 \\ T_2 \\ T_3 \\ T_4 \\ T_5 \\ T_6 \\ T_7 \end{matrix}$$

$$[\mathbf{K}]^{(4)} = A\begin{bmatrix} U_4 & -U_4 \\ -U_4 & U_4 \end{bmatrix} \text{ and } [\mathbf{K}]^{(4G)} = A\begin{bmatrix} 0 & 0 & 0 & 0 & 0 & 0 & 0 \\ 0 & 0 & 0 & 0 & 0 & 0 & 0 \\ 0 & 0 & 0 & 0 & 0 & 0 & 0 \\ 0 & 0 & 0 & U_4 & -U_4 & 0 & 0 \\ 0 & 0 & 0 & -U_4 & U_4 & 0 & 0 \\ 0 & 0 & 0 & 0 & 0 & 0 & 0 \\ 0 & 0 & 0 & 0 & 0 & 0 & 0 \end{bmatrix}\begin{matrix} T_1 \\ T_2 \\ T_3 \\ T_4 \\ T_5 \\ T_6 \\ T_7 \end{matrix}$$

$$[\mathbf{K}]^{(5)} = A\begin{bmatrix} U_5 & -U_5 \\ -U_5 & U_5 \end{bmatrix} \text{ and } [\mathbf{K}]^{(5G)} = A\begin{bmatrix} 0 & 0 & 0 & 0 & 0 & 0 & 0 \\ 0 & 0 & 0 & 0 & 0 & 0 & 0 \\ 0 & 0 & 0 & 0 & 0 & 0 & 0 \\ 0 & 0 & 0 & 0 & 0 & 0 & 0 \\ 0 & 0 & 0 & 0 & U_5 & -U_5 & 0 \\ 0 & 0 & 0 & 0 & -U_5 & U_5 & 0 \\ 0 & 0 & 0 & 0 & 0 & 0 & 0 \end{bmatrix}\begin{matrix} T_1 \\ T_2 \\ T_3 \\ T_4 \\ T_5 \\ T_6 \\ T_7 \end{matrix}$$

$$[\mathbf{K}]^{(6)} = A\begin{bmatrix} U_6 & -U_6 \\ -U_6 & U_6 \end{bmatrix} \text{ and } [\mathbf{K}]^{(6G)} = A\begin{bmatrix} 0 & 0 & 0 & 0 & 0 & 0 & 0 \\ 0 & 0 & 0 & 0 & 0 & 0 & 0 \\ 0 & 0 & 0 & 0 & 0 & 0 & 0 \\ 0 & 0 & 0 & 0 & 0 & 0 & 0 \\ 0 & 0 & 0 & 0 & 0 & 0 & 0 \\ 0 & 0 & 0 & 0 & 0 & U_6 & -U_6 \\ 0 & 0 & 0 & 0 & 0 & -U_6 & U_6 \end{bmatrix}\begin{matrix} T_1 \\ T_2 \\ T_3 \\ T_4 \\ T_5 \\ T_6 \\ T_7 \end{matrix}$$

The global conductance matrix is

$$[\mathbf{K}]^{(G)} = [\mathbf{K}]^{(1G)} + [\mathbf{K}]^{(2G)} + [\mathbf{K}]^{(3G)} + [\mathbf{K}]^{(4G)} + [\mathbf{K}]^{(5G)} + [\mathbf{K}]^{(6G)}$$

$$[\mathbf{K}]^{(G)} = A\begin{bmatrix} U_1 & -U_1 & 0 & 0 & 0 & 0 & 0 \\ -U_1 & U_1 + U_2 & -U_2 & 0 & 0 & 0 & 0 \\ 0 & -U_2 & U_2 + U_3 & -U_3 & 0 & 0 & 0 \\ 0 & 0 & -U_3 & U_3 + U_4 & -U_4 & 0 & 0 \\ 0 & 0 & 0 & -U_4 & U_4 + U_5 & -U_5 & 0 \\ 0 & 0 & 0 & 0 & -U_5 & U_5 + U_6 & -U_6 \\ 0 & 0 & 0 & 0 & 0 & -U_6 & U_6 \end{bmatrix} \quad (1.35)$$

5. *Apply boundary conditions and thermal loads.*
For the given problem, the exterior of the wall is exposed to a known air temperature $T_1$, and the room temperature, $T_7$, is also known. Thus, we want the first row to read $T_1 = 20°F$ and the last row to read $T_7 = 70°F$. So, we have

$$A\begin{bmatrix} 1/A & 0 & 0 & 0 & 0 & 0 & 0 \\ -U_1 & U_1 + U_2 & -U_2 & 0 & 0 & 0 & 0 \\ 0 & -U_2 & U_2 + U_3 & -U_3 & 0 & 0 & 0 \\ 0 & 0 & -U_3 & U_3 + U_4 & -U_4 & 0 & 0 \\ 0 & 0 & 0 & -U_4 & U_4 + U_5 & -U_5 & 0 \\ 0 & 0 & 0 & 0 & -U_5 & U_5 + U_6 & -U_6 \\ 0 & 0 & 0 & 0 & 0 & 0 & 1/A \end{bmatrix}\begin{Bmatrix} T_1 \\ T_2 \\ T_3 \\ T_4 \\ T_5 \\ T_6 \\ T_7 \end{Bmatrix} = \begin{Bmatrix} 20°F \\ 0 \\ 0 \\ 0 \\ 0 \\ 0 \\ 70°F \end{Bmatrix} \quad (1.36)$$

Note that the finite element formulation of heat transfer problems will always lead to an equation of the form

$$[\mathbf{K}]\{\mathbf{T}\} = \{\mathbf{q}\}$$

$$[\textbf{conductance matrix}]\{\textbf{temperature matrix}\} = \{\textbf{heat flow matrix}\}$$

Also note that for Example 1.2, the heat transfer rate through each element was caused by temperature differences across the nodes of a given element. Thus, the external nodal heat flow values are zero in the heat flow matrix. An example of a situation in which external nodal heat values are not zero is a heating strip attached to a solid surface (e.g., the base of a pressing iron); for such a situation, the external nodal heat value is equal to the amount of heat being generated by the heating strip over the surface. Turning our attention to the matrices given by Eq. (1.36), and incorporating the known boundary conditions into rows 2 and 6 of the conductance matrix, we can reduce Eq. (1.36) to

$$A\begin{bmatrix} U_1 + U_2 & -U_2 & 0 & 0 & 0 \\ -U_2 & U_2 + U_3 & -U_3 & 0 & 0 \\ 0 & -U_3 & U_3 + U_4 & -U_4 & 0 \\ 0 & 0 & -U_4 & U_4 + U_5 & -U_5 \\ 0 & 0 & 0 & -U_5 & U_5 + U_6 \end{bmatrix}\begin{Bmatrix} T_2 \\ T_3 \\ T_4 \\ T_5 \\ T_6 \end{Bmatrix} = \begin{Bmatrix} U_1 A T_1 \\ 0 \\ 0 \\ 0 \\ U_6 A T_7 \end{Bmatrix}$$

Keep in mind that the above matrix was obtained by assembling the elemental description and applying the boundary conditions. Moreover, the results of this approach are identical to the relations we obtained earlier by balancing the heat flows at the nodes, as given by Eq. (1.27). This equality in the outcome is expected because the elemental formulations are based on the application of energy balance as well.

Referring to the original global matrix, substituting for the $U$-values, and employing the given boundary conditions, we have

$$150\begin{bmatrix} \frac{1}{150} & 0 & 0 & 0 & 0 & 0 & 0 \\ -5.88 & 5.88+1.23 & -1.23 & 0 & 0 & 0 & 0 \\ 0 & -1.23 & 1.23+0.76 & -0.76 & 0 & 0 & 0 \\ 0 & 0 & -0.76 & 0.76+0.091 & -0.091 & 0 & 0 \\ 0 & 0 & 0 & -0.091 & 0.091+2.22 & -2.22 & 0 \\ 0 & 0 & 0 & 0 & -2.22 & 2.22+1.47 & -1.47 \\ 0 & 0 & 0 & 0 & 0 & 0 & \frac{1}{150} \end{bmatrix} \begin{Bmatrix} T_1 \\ T_2 \\ T_3 \\ T_4 \\ T_5 \\ T_6 \\ T_7 \end{Bmatrix} = \begin{Bmatrix} 20°\text{F} \\ 0 \\ 0 \\ 0 \\ 0 \\ 0 \\ 70°\text{F} \end{Bmatrix}$$

Simplifying, we obtain

$$\begin{bmatrix} 7.11 & -1.23 & 0 & 0 & 0 \\ -1.23 & 1.99 & -0.76 & 0 & 0 \\ 0 & -0.76 & 0.851 & -0.091 & 0 \\ 0 & 0 & -0.091 & 2.311 & -2.22 \\ 0 & 0 & 0 & -2.22 & 3.69 \end{bmatrix} \begin{Bmatrix} T_2 \\ T_3 \\ T_4 \\ T_5 \\ T_6 \end{Bmatrix} = \begin{Bmatrix} (5.88)(20) \\ 0 \\ 0 \\ 0 \\ (1.47)(70) \end{Bmatrix}$$

**Solution Phase**

**6.** *Solve a system of algebraic equations simultaneously.*
Solving the previous matrix yields the temperature distribution along the wall:

$$\begin{Bmatrix} T_1 \\ T_2 \\ T_3 \\ T_4 \\ T_5 \\ T_6 \\ T_7 \end{Bmatrix} = \begin{Bmatrix} 20.00 \\ 20.59 \\ 23.41 \\ 27.97 \\ 66.08 \\ 67.64 \\ 70.00 \end{Bmatrix} °\text{C}$$

For problems similar to the type discussed here, the knowledge of temperature distribution within the wall is important in determining where condensation may occur in the wall and, thus, where should one place a vapor barrier to avoid moisture condensation. To demonstrate this concept, let us assume that moisture can diffuse through the gypsum board and that the inside air has a relative humidity of 40%. With the help of a psychometric chart, using a dry bulb temperature of 70°F and the value $\phi = 40\%$, we identify the condensation temperature to be 44°F. Therefore, the water vapor in the air at any surface whose temperature is 44°F or below will condense. In the absence of a vapor barrier, the water vapor in the air will condense somewhere between surface 5 and 4 for the assumed conditions in this problem.

**Postprocessing Phase**

**7.** *Obtain other information.*
For this example, we may be interested in obtaining other information, such as heat loss through the wall. Such information is important in computing the heat load for

a building. Because we have assumed steady state conditions, the heat loss through the wall should be equal to the heat transfer through each element. This value can be determined from the equation

$$q = UA(T_{i+1} - T_i) \tag{1.37}$$

The heat transfer through each element is:

$$q = UA(T_{i+1} - T_i) = (1.47)(150)(70 - 67.64) = (2.22)(150)(67.64 - 66.08) = \cdots$$

$$= (5.88)(150)(20.59 - 20) = 520 \frac{\text{Btu}}{\text{hr}}$$

We also could have calculated the heat loss through the wall using the overall $U$-factor in the following manner:

$$q = U_{\text{overall}} A(T_{\text{inside}} - T_{\text{outside}}) = \frac{1}{\Sigma \text{ Resistance}} A(T_{\text{inside}} - T_{\text{outside}})$$

$$= (0.0693)(150)(70 - 20) = 520 \frac{\text{Btu}}{\text{hr}}$$

This problem is just another example of how we can generate finite element models using the direct method.

## A Torsional Problem: Direct Formulation

### EXAMPLE 1.3

Consider the torsion of circular shafts, shown in Figure 1.11. Recall from your previous study of the mechanics of materials that the angle of twist $\theta$ for a shaft with a constant cross-sectional area with a polar moment of inertia $J$ and length $\ell$, made of homogenous material with a shear modulus of elasticity $G$, subject to a torque $T$ is given by

$$\theta = \frac{T\ell}{JG}$$

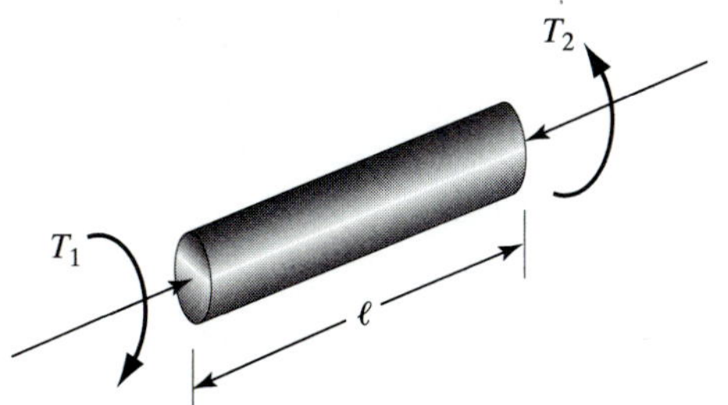


**FIGURE 1.11** A torsion of circular shaft.

Using direct formulation, equilibrium conditions, and

$$\theta = \frac{T\ell}{JG}$$

we can show that for an element comprising of two nodes, the stiffness matrix, the angle of twists, and the torques are related according to the equation

$$\frac{JG}{\ell}\begin{bmatrix} 1 & -1 \\ -1 & 1 \end{bmatrix}\begin{Bmatrix} \theta_1 \\ \theta_2 \end{Bmatrix} = \begin{Bmatrix} T_1 \\ T_2 \end{Bmatrix} \tag{1.38}$$

We will discuss torsional problems in much more detail in Chapter 8. For now, let us consider a shaft that is made of two parts, as shown in Figure 1.12. Part $AB$ is made of material with a shear modulus of elasticity of $G_{AB} = 3.9 \times 10^6$ lb/in$^2$ and has a diameter of 1.5 in. Segment $BC$ is made of slightly different material with a shear modulus of elasticity of $G_{BC} = 4.0 \times 10^6$ lb/in$^2$ and with a diameter of 1 in. The shaft is fixed at both ends. A torque of 200 lb·ft is applied at $D$. Using three elements, let us determine the angle of twist at $D$ and $B$, and the torsional reactions at the boundaries.

We will represent this problem by a model that has four nodes at $A$, $B$, $C$, and $D$, respectively, and three elements $(AD, DB, BC)$.

The polar moment of inertia for each element is given by:

$$J_1 = J_2 = \frac{1}{2}\pi r^4 = \frac{1}{2}\pi\left(\frac{1.5}{2}\text{ in}\right)^4 = 0.497\text{ in}^4$$

$$J_3 = \frac{1}{2}\pi r^4 = \frac{1}{2}\pi\left(\frac{1.0}{2}\text{ in}\right)^4 = 0.0982\text{ in}^4$$

The stiffness matrix for each element is computed from Eq. (1.38) as

$$[\mathbf{K}]^{(e)} = \frac{JG}{\ell}\begin{bmatrix} 1 & -1 \\ -1 & 1 \end{bmatrix}$$

So, for element (1), the stiffness matrix is

$$[\mathbf{K}]^{(1)} = \frac{(3.9 \times 10^6\text{ lb/in}^2)(0.497\text{ in}^4)}{(12 \times 2.5)\text{ in}}\begin{bmatrix} 1 & -1 \\ -1 & 1 \end{bmatrix} = \begin{bmatrix} 64610 & -64610 \\ -64610 & 64610 \end{bmatrix}\text{lb}\cdot\text{in}$$

and its position in the global stiffness matrix is

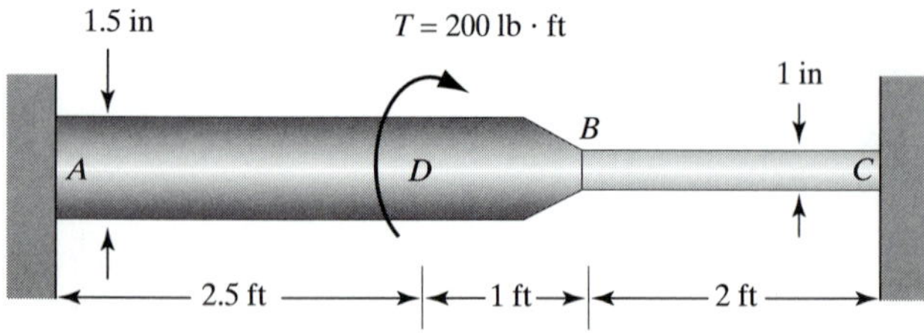


**FIGURE 1.12** A schematic of the shaft in Example 1.3.

$$[\mathbf{K}]^{(1G)} = \begin{bmatrix} 64610 & -64610 & 0 & 0 \\ -64610 & 64610 & 0 & 0 \\ 0 & 0 & 0 & 0 \\ 0 & 0 & 0 & 0 \end{bmatrix} \begin{matrix} \theta_1 \\ \theta_2 \\ \theta_3 \\ \theta_4 \end{matrix}$$

Similarly, for elements (2) and (3), their respective stiffness matrices and positions in the global stiffness matrix are as follows:

$$[\mathbf{K}]^{(2)} = \frac{(3.9 \times 10^6 \text{ lb/in}^2)(0.497 \text{ in}^4)}{(12 \times 1.0) \text{ in}} \begin{bmatrix} 1 & -1 \\ -1 & 1 \end{bmatrix} = \begin{bmatrix} 161525 & -161525 \\ -161525 & 161525 \end{bmatrix} \text{lb} \cdot \text{in}$$

$$[\mathbf{K}]^{(2G)} = \begin{bmatrix} 0 & 0 & 0 & 0 \\ 0 & 161525 & -161525 & 0 \\ 0 & -161525 & 161525 & 0 \\ 0 & 0 & 0 & 0 \end{bmatrix} \begin{matrix} \theta_1 \\ \theta_2 \\ \theta_3 \\ \theta_4 \end{matrix}$$

$$[\mathbf{K}]^{(3)} = \frac{(4.0 \times 10^6 \text{ lb/in}^2)(0.0982 \text{ in}^4)}{(12 \times 2.0) \text{ in}} \begin{bmatrix} 1 & -1 \\ -1 & 1 \end{bmatrix} = \begin{bmatrix} 16367 & -16367 \\ -16367 & 16367 \end{bmatrix} \text{lb} \cdot \text{in}$$

$$[\mathbf{K}]^{(3G)} = \begin{bmatrix} 0 & 0 & 0 & 0 \\ 0 & 0 & 0 & 0 \\ 0 & 0 & 16367 & -16367 \\ 0 & 0 & -16367 & 16367 \end{bmatrix} \begin{matrix} \theta_1 \\ \theta_2 \\ \theta_3 \\ \theta_4 \end{matrix}$$

The final global matrix is obtained simply by assembling, or adding, elemental descriptions:

$$[\mathbf{K}]^{(G)} = [\mathbf{K}]^{(1G)} + [\mathbf{K}]^{(2G)} + [\mathbf{K}]^{(3G)}$$

$$[\mathbf{K}]^{(G)} = \begin{bmatrix} 64610 & -64610 & 0 & 0 \\ -64610 & 64610 + 161525 & -161525 & 0 \\ 0 & -161525 & 161525 + 16367 & -16367 \\ 0 & 0 & -16367 & 16367 \end{bmatrix}$$

Applying the fixed boundary conditions at points $A$ and $C$ and applying the external torque, we have:

$$\begin{bmatrix} 1 & 0 & 0 & 0 \\ -64610 & 226135 & -161525 & 0 \\ 0 & -161525 & 177892 & -16367 \\ 0 & 0 & 0 & 1 \end{bmatrix} \begin{Bmatrix} \theta_1 \\ \theta_2 \\ \theta_3 \\ \theta_4 \end{Bmatrix} = \begin{Bmatrix} 0 \\ -(200 \times 12) \text{ lb} \cdot \text{in} \\ 0 \\ 0 \end{Bmatrix}$$

Solving the above set of equations, we obtain

$$\begin{Bmatrix} \theta_1 \\ \theta_2 \\ \theta_3 \\ \theta_4 \end{Bmatrix} = \begin{Bmatrix} 0 \\ -0.03020 \text{ rad} \\ -0.02742 \text{ rad} \\ 0 \end{Bmatrix}$$

The reaction moments at boundaries $A$ and $C$ can be determined as follows:

$$\{\mathbf{R}\} = [\mathbf{K}]\{\boldsymbol{\theta}\} - \{\mathbf{T}\}$$

$$\begin{Bmatrix} R_A \\ R_D \\ R_B \\ R_C \end{Bmatrix} = \begin{bmatrix} 64610 & -64610 & 0 & 0 \\ -64610 & 226135 & -161525 & 0 \\ 0 & -161525 & 177892 & -16367 \\ 0 & 0 & -16367 & 16367 \end{bmatrix} \begin{Bmatrix} 0 \\ -0.03020 \text{ rad} \\ -0.02742 \text{ rad} \\ 0 \end{Bmatrix} - \begin{Bmatrix} 0 \\ -(200 \times 12) \text{ lb}\cdot\text{in} \\ 0 \\ 0 \end{Bmatrix}$$

$$\begin{Bmatrix} R_A \\ R_D \\ R_B \\ R_C \end{Bmatrix} = \begin{Bmatrix} 1951 \text{ lb}\cdot\text{in} \\ 0 \\ 0 \\ 449 \text{ lb}\cdot\text{in} \end{Bmatrix}$$

Note that the sum of $R_A$ and $R_C$ is equal to the applied torque of 2400 lb·in. Also note that the change in the diameter of the shafts will give rise to stress concentrations that are not accounted for by the model we used here.

---

### EXAMPLE 1.4

A steel plate is subjected to an axial load, as shown in Figure 1.13. Approximate the deflections and average stresses along the plate. The plate is 1/16 in thick and has a modulus of elasticity $E = 29 \times 10^6$ lb/in$^2$.

We may model this problem using four nodes and four elements, as shown in Figure 1.13. Next, we compute the equivalent stiffness coefficient for each element:

$$k_1 = \frac{A_1 E}{\ell_1} = \frac{(5)(0.0625)(29 \times 10^6)}{1} = 9{,}062{,}500 \text{ lb/in}$$

$$k_2 = k_3 = \frac{A_2 E}{\ell_2} = \frac{(2)(0.0625)(29 \times 10^6)}{4} = 906{,}250 \text{ lb/in}$$

$$k_4 = \frac{A_4 E}{\ell_4} = \frac{(5)(0.0625)(29 \times 10^6)}{2} = 4{,}531{,}250 \text{ lb/in}$$

The stiffness matrix for element (1) is

$$[\mathbf{K}]^{(1)} = \begin{bmatrix} k_1 & -k_1 \\ -k_1 & k_1 \end{bmatrix}$$

and its position in the global stiffness matrix is

$$[\mathbf{K}]^{(1G)} = \begin{bmatrix} k_1 & -k_1 & 0 & 0 \\ -k_1 & k_1 & 0 & 0 \\ 0 & 0 & 0 & 0 \\ 0 & 0 & 0 & 0 \end{bmatrix} \begin{matrix} u_1 \\ u_2 \\ u_3 \\ u_4 \end{matrix}$$

Similarly, the respective stiffness matrices and positions in the global stiffness matrix for elements (2), (3), and (4) are

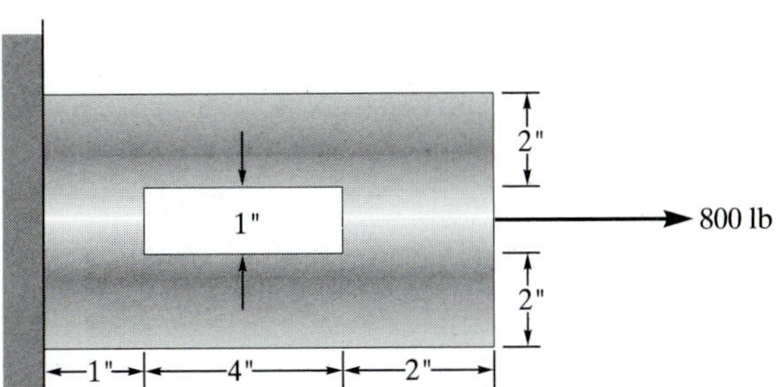


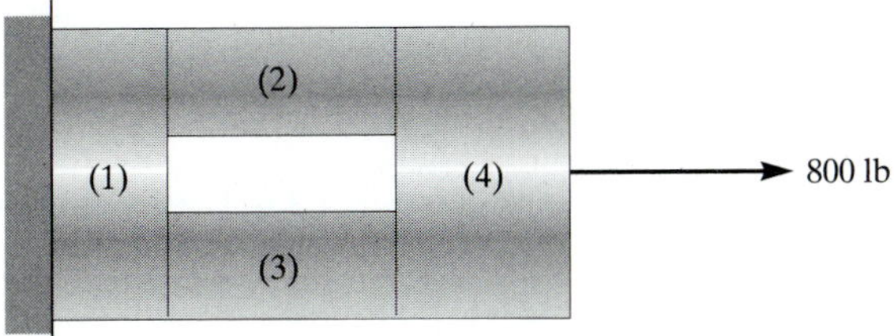


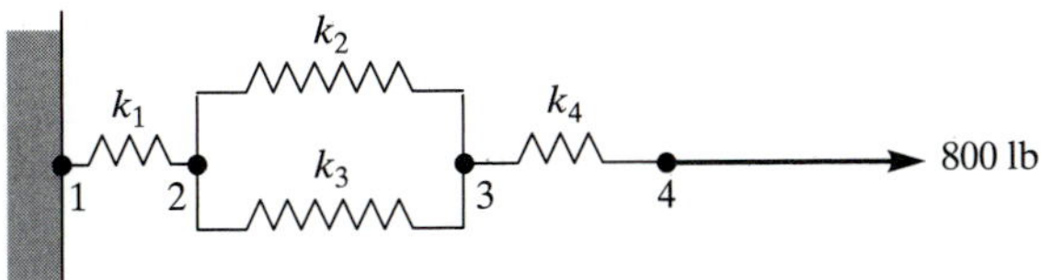


**FIGURE 1.13** A schematic of the steel plate in Example 1.4.

$$[\mathbf{K}]^{(2)} = \begin{bmatrix} k_2 & -k_2 \\ -k_2 & k_2 \end{bmatrix} \qquad [\mathbf{K}]^{(2G)} = \begin{bmatrix} 0 & 0 & 0 & 0 \\ 0 & k_2 & -k_2 & 0 \\ 0 & -k_2 & k_2 & 0 \\ 0 & 0 & 0 & 0 \end{bmatrix} \begin{matrix} u_1 \\ u_2 \\ u_3 \\ u_4 \end{matrix}$$

$$[\mathbf{K}]^{(3)} = \begin{bmatrix} k_3 & -k_3 \\ -k_3 & k_3 \end{bmatrix} \qquad [\mathbf{K}]^{(3G)} = \begin{bmatrix} 0 & 0 & 0 & 0 \\ 0 & k_3 & -k_3 & 0 \\ 0 & -k_3 & k_3 & 0 \\ 0 & 0 & 0 & 0 \end{bmatrix} \begin{matrix} u_1 \\ u_2 \\ u_3 \\ u_4 \end{matrix}$$

$$[\mathbf{K}]^{(4)} = \begin{bmatrix} k_4 & -k_4 \\ -k_4 & k_4 \end{bmatrix} \qquad [\mathbf{K}]^{(4G)} = \begin{bmatrix} 0 & 0 & 0 & 0 \\ 0 & 0 & 0 & 0 \\ 0 & 0 & k_4 & -k_4 \\ 0 & 0 & -k_4 & k_4 \end{bmatrix} \begin{matrix} u_1 \\ u_2 \\ u_3 \\ u_4 \end{matrix}$$

The final global matrix is obtained simply by assembling, or adding, the individual elemental matrices:

$$[\mathbf{K}]^{(G)} = [\mathbf{K}]^{(1G)} + [\mathbf{K}]^{(2G)} + [\mathbf{K}]^{(3G)} + [\mathbf{K}]^{(4G)}$$

$$[\mathbf{K}]^{(G)} = \begin{bmatrix} k_1 & -k_1 & 0 & 0 \\ -k_1 & k_1 + k_2 + k_3 & -k_2 - k_3 & 0 \\ 0 & -k_2 - k_3 & k_2 + k_3 + k_4 & -k_4 \\ 0 & 0 & -k_4 & k_4 \end{bmatrix}$$

Substituting for the elements' respective stiffness coefficients, the global stiffness matrix becomes:

$$[\mathbf{K}]^{(G)} = \begin{bmatrix} 9{,}062{,}500 & -9{,}062{,}500 & 0 & 0 \\ -9{,}062{,}500 & 10{,}875{,}000 & -1{,}812{,}500 & 0 \\ 0 & -1{,}812{,}500 & 6{,}343{,}750 & -4{,}531{,}250 \\ 0 & 0 & -4{,}531{,}250 & 4{,}531{,}250 \end{bmatrix}$$

Applying the boundary condition $u_1 = 0$ and the load to node 4, we obtain

$$\begin{bmatrix} 1 & 0 & 0 & 0 \\ -9{,}062{,}500 & 10{,}875{,}000 & -1{,}812{,}500 & 0 \\ 0 & -1{,}812{,}500 & 6{,}343{,}750 & -4{,}531{,}250 \\ 0 & 0 & -4{,}531{,}250 & 4{,}531{,}250 \end{bmatrix} \begin{Bmatrix} u_1 \\ u_2 \\ u_3 \\ u_4 \end{Bmatrix} = \begin{Bmatrix} 0 \\ 0 \\ 0 \\ 800 \end{Bmatrix}$$

Solving the system of equations yields the displacement solution as

$$\begin{Bmatrix} u_1 \\ u_2 \\ u_3 \\ u_4 \end{Bmatrix} = \begin{Bmatrix} 0 \\ 8.827 \times 10^{-5} \\ 5.296 \times 10^{-4} \\ 7.062 \times 10^{-4} \end{Bmatrix} \text{in}$$

and the stresses in each element are:

$$\sigma^{(1)} = E\left(\frac{u_2 - u_1}{\ell}\right) = \frac{(29 \times 10^6)(8.827 \times 10^{-5} - 0)}{1} = 2560 \frac{\text{lb}}{\text{in}^2}$$

$$\sigma^{(2)} = \sigma^{(3)} = E\left(\frac{u_3 - u_2}{\ell}\right) = \frac{(29 \times 10^6)(5.296 \times 10^{-4} - 8{,}827 \times 10^{-5})}{4} = 3200 \frac{\text{lb}}{\text{in}^2}$$

$$\sigma^{(4)} = E\left(\frac{u_4 - u_3}{\ell}\right) = \frac{(29 \times 10^6)(7.062 \times 10^{-4} - 5.296 \times 10^{-4})}{2} = 2560 \frac{\text{lb}}{\text{in}^2}$$

Note that the model used to analyze this problem consisted of springs in parallel as well as in series. The two springs in parallel could have been combined and represented by a single spring having a stiffness equal to $k_2 + k_3$. Also note that because of the hole, the abrupt changes in the cross section of the strip will give rise to stress concentrations with values exceeding those average values we computed here. After you study plane-stress finite element formulation (discussed in Chapter 8), you will revisit this problem (see Problem 8.13) and be asked to solve it using ANSYS. Furthermore, you will be asked to plot the components of the stress distributions in the plate and, thus, identify the location and magnitude of maximum stresses.

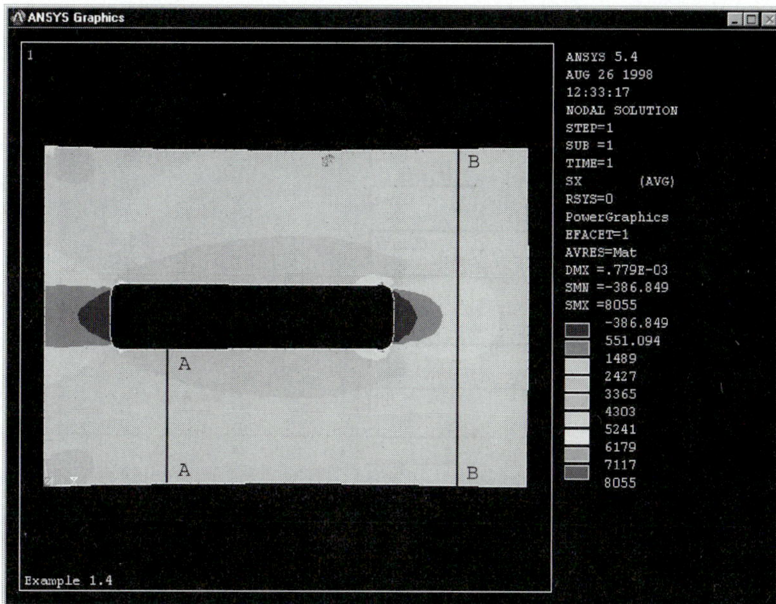


**FIGURE 1.14** The $x$-component of stress distribution for the plate in Example 1.4, as computed by ANSYS.

To give you just a taste of what is to come in Chapter 8 and also to shed more light on our discussion about the stress concentration regions, we have solved Example 1.4 using ANSYS and have determined the $x$-component of the stress distribution in the plate, as shown in Figure 1.14. In the results shown in Figure 1.14, the load was applied as a pressure over the entire right surface of the bar. Note the variation of the stresses at section $A$–$A$ from approximately 3000 psi to 3500 psi. At section $B$–$B$, the $x$-component of the stresses varies from approximately 2300 psi to 2600 psi. These values are not that far off from the average stress values obtained using the direct model. Also note that the maximum and minimum stress values given by ANSYS could change, depending upon how we apply the load to the bar, especially in the regions near the point of load application and the regions near the hole. Keeping in mind Example 1.4 and Figure 1.13, remember that in a real situation, the load would be applied over an area, not at a single point. Thus, remember that how you apply the external load to your finite element model will influence the stress distribution results, particularly in the region near where the load is applied. This principle is especially true in Example 1.4 because it deals with a short plate with a hole.

## 1.6 MINIMUM TOTAL POTENTIAL ENERGY FORMULATION

The minimum total potential energy formulation is a common approach in generating finite element models in solid mechanics. External loads applied to a body will cause the body to deform. During the deformation, the work done by the external forces is stored

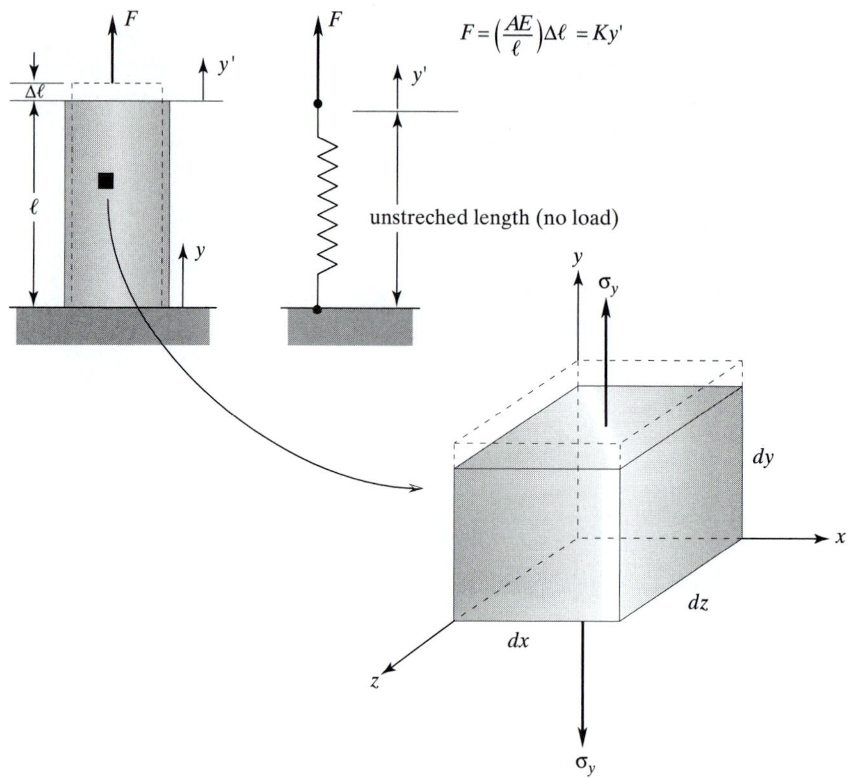


**FIGURE 1.15** The elastic behavior of a member subjected to a central load.

in the material in the form of elastic energy, called strain energy. Let us consider the strain energy in a solid member when it is subjected to a central force $F$, as shown in Figure 1.15.

Also shown in Figure 1.15 is a piece of material from the member in the form of differential volume and the normal stresses acting on the surfaces of this volume. Earlier, it was shown that the elastic behavior of the member may be modeled as a linear spring. When the member is stretched by a differential amount $dy'$, the stored energy in the material is

$$d\Lambda = \int_0^{y'} F dy' = \int_0^{y'} ky'dy' = \frac{1}{2}ky'^2 = \left(\frac{1}{2}ky'\right)y' \tag{1.39}$$

We can write Eq. (1.39) in terms of the normal stress and strain:

$$d\Lambda = \frac{1}{2}\overbrace{(ky')}^{\text{elastic force}}\ y' = \left(\frac{1}{2}\sigma_y dx\, dz\right)\overbrace{\varepsilon dy}^{y'} = \frac{1}{2}\sigma\varepsilon\, dV$$

Therefore, for a member or an element under axial loading, the strain energy $\Lambda^{(e)}$ is given by

$$\Lambda^{(e)} = \int d\Lambda = \int_V \frac{\sigma\varepsilon}{2}\,dV = \int_V \frac{E\varepsilon^2}{2}\,dV \tag{1.40}$$

where $V$ is the volume of the member. The total potential energy $\Pi$ for a body consisting of $n$ elements and $m$ nodes is the difference between the total strain energy and the work done by the external forces:

$$\Pi = \sum_{e=1}^{n} \Lambda^{(e)} - \sum_{i=1}^{m} F_i u_i \tag{1.41}$$

The minimum total potential energy principle simply states that for a stable system, the displacement at the equilibrium position occurs such that the value of the system's total potential energy is a minimum.

$$\frac{\partial \Pi}{\partial u_i} = \frac{\partial}{\partial u_i}\sum_{e=1}^{n} \Lambda^{(e)} - \frac{\partial}{\partial u_i}\sum_{i=1}^{m} F_i u_i = 0 \quad \text{for } i = 1, 2, 3, \ldots, n \tag{1.42}$$

Let us turn our attention back to Example 1.1. The strain energy for an arbitrary element $(e)$ can be determined from Eq. (1.40) as

$$\Lambda^{(e)} = \int_V \frac{E\varepsilon^2}{2}\,dV = \frac{A_{\text{avg}}E}{2\ell}\left(u_{i+1}^2 + u_i^2 - 2u_{i+1}u_i\right) \tag{1.43}$$

where $\varepsilon = (u_{i+1} - u_i)/\ell$ was substituted for the axial strain. Minimizing the strain energy with respect to $u_i$ and $u_{i+1}$ leads to

$$\frac{\partial \Lambda^{(e)}}{\partial u_i} = \frac{A_{\text{avg}}E}{\ell}(u_i - u_{i+1}) \tag{1.44}$$

$$\frac{\partial \Lambda^{(e)}}{\partial u_{i+1}} = \frac{A_{\text{avg}}E}{\ell}(u_{i+1} - u_i)$$

and, in matrix form,

$$\begin{Bmatrix} \dfrac{\partial \Lambda^{(e)}}{\partial u_i} \\ \dfrac{\partial \Lambda^{(e)}}{\partial u_{i+1}} \end{Bmatrix} = \begin{bmatrix} k_{\text{eq}} & -k_{\text{eq}} \\ -k_{\text{eq}} & k_{\text{eq}} \end{bmatrix} \begin{Bmatrix} u_i \\ u_{i+1} \end{Bmatrix} \tag{1.45}$$

where $k_{\text{eq}} = (A_{\text{avg}}E)/\ell$. Minimizing the work done by the external forces at nodes $i$ and $i + 1$ of an arbitrary element $(e)$, we get

$$\frac{\partial}{\partial u_i}(F_i u_i) = F_i \tag{1.46}$$

$$\frac{\partial}{\partial u_{i+1}}(F_{i+1}u_{i+1}) = F_{i+1}$$

For Example 1.1, the minimum total potential energy formulation leads to a global stiffness matrix that is identical to the one obtained from direct formulation:

$$[\mathbf{K}]^{(G)} = \begin{bmatrix} k_1 & -k_1 & 0 & 0 & 0 \\ -k_1 & k_1 + k_2 & -k_2 & 0 & 0 \\ 0 & -k_2 & k_2 + k_3 & -k_3 & 0 \\ 0 & 0 & -k_3 & k_3 + k_4 & -k_4 \\ 0 & 0 & 0 & -k_4 & k_4 \end{bmatrix}$$

Furthermore, application of the boundary condition and the load results in:

$$\begin{bmatrix} 1 & 0 & 0 & 0 & 0 \\ -k_1 & k_1 + k_2 & -k_2 & 0 & 0 \\ 0 & -k_2 & k_2 + k_3 & -k_3 & 0 \\ 0 & 0 & -k_3 & k_3 + k_4 & -k_4 \\ 0 & 0 & 0 & -k_4 & k_4 \end{bmatrix} \begin{Bmatrix} u_1 \\ u_2 \\ u_3 \\ u_4 \\ u_5 \end{Bmatrix} = \begin{Bmatrix} 0 \\ 0 \\ 0 \\ 0 \\ P \end{Bmatrix} \quad (1.47)$$

The displacement results will be identical to the ones obtained earlier from the direct method, as given by Eq. (1.17). The concepts of strain energy and minimum total potential energy will be used to formulate solid mechanics problems in Chapters 4, 8, and 10. Therefore, spending a little extra time now to understand the basic ideas will benefit you enormously later.

### Example 1.1: Exact Solution

In this section, we will derive the exact solution to Example 1.1 and compare the finite element formulation displacement results for this problem to the exact displacement solutions. As shown in Figure 1.16, the statics equilibrium requires the sum of the forces in the $y$-direction to be zero. This requirement leads to the relation

$$P - (\sigma_{avg})A(y) = 0 \quad (1.48)$$

Once again, using Hooke's Law ($\sigma = E\varepsilon$) and substituting for the average stress in terms of the strain, we have

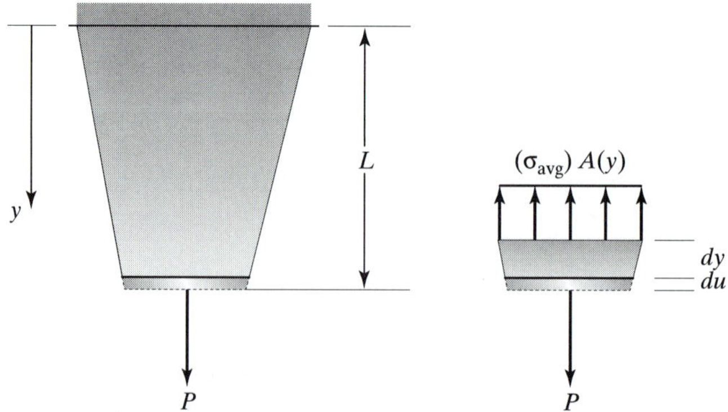


**FIGURE 1.16** The relationship between the external force $P$ and the average stresses for the bar in Example 1.1.

$$P - E\varepsilon A(y) = 0 \tag{1.49}$$

Recall that the average normal strain is the change in length *du* per unit original length of the differential segment *dy*. So,

$$\varepsilon = \frac{du}{dy}$$

If we substitute this relationship into Eq. (1.49), we now have

$$P - EA(y)\frac{du}{dy} = 0 \tag{1.50}$$

Rearranging Eq. (1.50), we get

$$du = \frac{Pdy}{EA(y)} \tag{1.51}$$

The exact solution is then obtained by integrating Eq. (1.51) over the length of the bar

$$\int_0^u du = \int_0^L \frac{Pdy}{EA(y)} \tag{1.52}$$

$$u(y) = \int_0^y \frac{Pdy}{EA(y)} = \int_0^y \frac{Pdy}{E\left(w_1 + \left(\frac{w_2 - w_1}{L}\right)y\right)t}$$

where the area is

$$A(y) = \left(w_1 + \left(\frac{w_2 - w_1}{L}\right)y\right)t$$

The deflection profile along the bar is obtained by integrating Eq. (1.52), resulting in:

$$u(y) = \frac{PL}{Et(w_2 - w_1)}\left[\ln\left(w_1 + \left(\frac{w_2 - w_1}{L}\right)y\right) - \ln w_1\right] \tag{1.53}$$

Equation (1.53) can be used to generate displacement values at various points along the bar. It is now appropriate to examine the accuracy of the direct and potential energy methods by comparing their displacement results with the exact values. Table 1.5 shows nodal displacements computed using exact, direct, and energy methods.

TABLE 1.5 Comparison of displacement results

| Location of a Point Along the Bar (in) | Results from the Exact Displacement Method (in) Eq. (1.53) | Results from the Direct Method (in) | Results from the Energy Method (in) |
|---|---|---|---|
| $y = 0$ | 0 | 0 | 0 |
| $y = 2.5$ | 0.001027 | 0.001026 | 0.001026 |
| $y = 5.0$ | 0.002213 | 0.002210 | 0.002210 |
| $y = 7.5$ | 0.003615 | 0.003608 | 0.003608 |
| $y = 10$ | 0.005333 | 0.005317 | 0.005317 |

It is clear from examination of Table 1.5 that all of the results are in good agreement with each other.

## 1.7 WEIGHTED RESIDUAL FORMULATIONS

The *weighted residual methods* are based on assuming an approximate solution for the governing differential equation. The assumed solution must satisfy the initial and boundary conditions of the given problem. Because the assumed solution is not exact, substitution of the solution into the differential equation will lead to some *residuals* or *errors*. Simply stated, each residual method requires the error to vanish over some selected intervals or at some points. To demonstrate this, concept, let's turn our attention to Example 1.1. The governing differential equation and the corresponding boundary condition for this problem are as follows:

$$A(y)E\frac{du}{dy} - P = 0 \qquad \text{subject to the boundary condition } u(0) = 0 \tag{1.54}$$

Next, we need to assume an approximate solution. Again, keep in mind that the assumed solution must satisfy the boundary condition. We choose

$$u(y) = c_1 y + c_2 y^2 + c_3 y^3 \tag{1.55}$$

where $c_1$, $c_2$, and $c_3$ are unknown coefficients. Equation (1.55) certainly satisfies the fixed boundary condition represented by $u(0) = 0$. Substitution of the assumed solution, Eq. (1.55), into the governing differential equation, Eq. (1.54), yields the error function $\mathfrak{R}$:

$$\overbrace{\left(w_1 + \left(\frac{w_2 - w_1}{L}\right)y\right)}^{A(y)} t\,E\overbrace{(c_1 + 2c_2 y + 3c_3 y^2)}^{\frac{du}{dy}} - P = \mathfrak{R} \tag{1.56}$$

Substituting for values of $w_1, w_2, L, t,$ and $E$ in Example 1.1 and simplifying, we get

$$\mathfrak{R}/E = (0.25 - 0.0125y)(c_1 + 2c_2 y + 3c_3 y^2) - 96.154 \times 10^{-6}$$

### Collocation Method

In the *collocation method* the error, or residual, function $\mathcal{R}$ is forced to be zero at as many points as there are unknown coefficients. Because the assumed solution in this example has three unknown coefficients, we will force the error function to equal zero at three points. We choose the error function to vanish at $y = L/3$, $y = 2L/3$, and $y = L$:

$$\mathcal{R}(c, y)\Big|_{y=\frac{L}{3}} = 0$$

$$\mathcal{R} = \left(0.25 - 0.0125\left(\frac{10}{3}\right)\right)\left(c_1 + 2c_2\left(\frac{10}{3}\right) + 3c_3\left(\frac{10}{3}\right)^2\right) - 96.154 \times 10^{-6} = 0$$

$$\mathcal{R}(c, y)\Big|_{y=\frac{2L}{3}} = 0$$

$$\mathcal{R} = \left(0.25 - 0.0125\left(\frac{20}{3}\right)\right)\left(c_1 + 2c_2\left(\frac{20}{3}\right) + 3c_3\left(\frac{20}{3}\right)^2\right) - 96.154 \times 10^{-6} = 0$$

$$\mathcal{R}(c, y)\Big|_{y=L} = 0$$

$$\mathcal{R} = (0.25 - 0.0125(10))(c_1 + 2c_2(10) + 3c_3(10)^2) - 96.154 \times 10^{-6} = 0$$

This procedure creates three linear equations that we can solve to obtain the unknown coefficients $c_1$, $c_2$, and $c_3$:

$$c_1 + \frac{20}{3}c_2 + \frac{100}{3}c_3 = 461.539 \times 10^{-6}$$

$$c_1 + \frac{40}{3}c_2 + \frac{400}{3}c_3 = 576.924 \times 10^{-6}$$

$$c_1 + 20c_2 + 300c_3 = 769.232 \times 10^{-6}$$

Solving the above equations yields $c_1 = 423.0776 \times 10^{-6}$, $c_2 = 21.65 \times 10^{-15}$, and $c_3 = 1.153848 \times 10^{-6}$. Substitution of the $c$-coefficients into Eq. (1.55) yields the approximate displacement profile:

$$u(y) = 423.0776 \times 10^{-6}y + 21.65 \times 10^{-15}y^2 + 1.153848 \times 10^{-6}y^3 \quad (1.57)$$

In order to get an idea of how accurate the collocation approximate results are, we will compare them to the exact results later in this chapter.

### Subdomain Method

In the *subdomain method*, the integral of the error function over some selected subintervals is forced to be zero. The number of subintervals chosen must equal the number of unknown coefficients. Thus, for our assumed solution, we will have three integrals:

$$\int_0^{\frac{L}{3}} \mathcal{R}\, dy = 0 \quad (1.58)$$

$$\int_0^{\frac{L}{3}} [(0.25 - 0.0125y)(c_1 + 2c_2 y + 3c_3 y^2) - 96.154 \times 10^{-6}] dy = 0$$

$$\int_{\frac{L}{3}}^{\frac{2L}{3}} \mathcal{R}\, dy = 0$$

$$\int_{\frac{L}{3}}^{\frac{2L}{3}} [(0.25 - 0.0125y)(c_1 + 2c_2 y + 3c_3 y^2) - 96.154 \times 10^{-6}] dy = 0$$

$$\int_{\frac{2L}{3}}^{L} \mathcal{R}\, dy = 0$$

$$\int_{\frac{2L}{3}}^{L} [(0.25 - 0.0125y)(c_1 + 2c_2 y + 3c_3 y^2) - 96.154 \times 10^{-6}] dy = 0$$

Integration of Eqs. (1.58) results in three linear equations that we can solve to obtain the unknown coefficients $c_1, c_2$, and $c_3$:

$$763.88889 \times 10^{-3} c_1 + 2.4691358c_2 + 8.1018519c_3 = 320.513333 \times 10^{-6}$$

$$0.625c_1 + 6.1728395c_2 + 47.4537041c_3 = 3.2051333 \times 10^{-4}$$

$$0.4861111c_1 + 8.0246917c_2 + 100.694444c_3 = 3.2051333 \times 10^{-4}$$

Solving the above equations yields $c_1 = 391.35088 \times 10^{-6}, c_2 = 6.075 \times 10^{-6}$, and $c_3 = 809.61092 \times 10^{-9}$. Substitution of the $c$-coefficients into Eq. (1.55) yields the approximate displacement profile:

$$u(y) = 391.35088 \times 10^{-6} y + 6.075 \times 10^{-6} y^2 + 809.61092 \times 10^{-9} y^3 \qquad (1.59)$$

We will compare the displacement results obtained from the subdomain method to the exact results later in this chapter.

### Galerkin Method

The *Galerkin method* requires the error to be orthogonal to some weighting functions $\Phi_i$, according to the integral

$$\int_a^b \Phi_i \mathcal{R}\, dy = 0 \qquad i = 1, 2, \ldots, N \qquad (1.60)$$

The weighting functions are chosen to be members of the approximate solution. Because there are three unknowns in the assumed approximate solution for Example 1.1, we need to generate three equations. Recall that the assumed solution is $u(y) = c_1 y + c_2 y^2 + c_3 y^3$; thus, the weighting functions are then selected to be $\Phi_1 = y$, $\Phi_2 = y^2$, and $\Phi_3 = y^3$. This selection leads to the following equations:

$$\int_0^L y[(0.25 - 0.0125y)(c_1 + 2c_2 y + 3c_3 y^2) - 96.154 \times 10^{-6}] dy = 0 \qquad (1.61)$$

$$\int_0^L y^2[(0.25 - 0.0125y)(c_1 + 2c_2 y + 3c_3 y^2) - 96.154 \times 10^{-6}] dy = 0$$

$$\int_0^L y^3[0.25 - 0.0125y)(c_1 + 2c_2 y + 3c_3 y^2) - 96.154 \times 10^{-6}] dy = 0$$

Integration of Eqs. (1.61) results in three linear equations that we can solve to obtain the unknown coefficients $c_1$, $c_2$, and $c_3$:

$$8.333333c_1 + 104.1666667c_2 + 1125c_3 = 0.0048077$$

$$52.083333c_1 + 750c_2 + 8750c_3 = 0.0320513333$$

$$375c_1 + 5833.3333c_2 + 71428.57143c_3 = 0.240385$$

Solving the above equations yields $c_1 = 400.642 \times 10^{-6}$, $c_2 = 4.006 \times 10^{-6}$, and $c_3 = 0.935 \times 10^{-6}$. Substitution of the $c$-coefficients into Eq. (1.55) yields the approximate displacement profile:

$$u(y) = 400.642 \times 10^{-6}\, y + 4.006 \times 10^{-6}\, y^2 + 0.935 \times 10^{-6}\, y^3 \qquad (1.62)$$

We will compare the displacement results obtained from the Galerkin method to the exact results later in this chapter.

### Least-Squares Method

The *least-squares method* requires the error to be minimized with respect to the unknown coefficients in the assumed solution, according to the relationship

$$\text{Minimize}\left(\int_a^b \mathcal{R}^2\, dy\right)$$

which leads to

$$\int_a^b \mathcal{R}\frac{\partial \mathcal{R}}{\partial c_i}\, dy = 0 \qquad i = 1, 2, \ldots, N \qquad (1.63)$$

Because there are three unknowns in the approximate solution of Example 1.1, Eq. (1.63) generates three equations. Recall that the error function is

$$\mathcal{R}/E = (0.25 - 0.0125y)(c_1 + 2c_2 y + 3c_3 y^2) - 96.154 \times 10^{-6}$$

Differentiating the error function with respect to $c_1$, $c_2$, and $c_3$ and substituting into Eq. (1.63), we have:

$$\int_0^{10} \overbrace{[(0.25 - 0.0125y)(c_1 + 2c_2 y + 3c_3 y^2) - 96.154 \times 10^{-6}]}^{\mathcal{R}}\overbrace{(0.25 - 0.0125y)}^{\frac{\partial \mathcal{R}}{\partial c_1}}\, dy = 0$$

$$\int_0^{10} \overbrace{[(0.25 - 0.0125y)(c_1 + 2c_2 y + 3c_3 y^2) - 96.154 \times 10^{-6}]}^{\mathcal{R}}\overbrace{(0.25 - 0.0125y)2y}^{\frac{\partial \mathcal{R}}{\partial c_2}}\, dy = 0$$

$$\int_0^{10} \overbrace{[(0.25 - 0.0125y)(c_1 + 2c_2 y + 3c_3 y^2) - 96.154 \times 10^{-6}]}^{\mathcal{R}}\overbrace{(0.25 - 0.0125y)3y^2}^{\frac{\partial \mathcal{R}}{\partial c_3}}\, dy = 0$$

Integration of the above equations results in three linear equations that we can solve to obtain the unknown coefficients $c_1, c_2$, and $c_3$:

$$0.364583333c_1 + 2.864583333c_2 + 25c_3 = 0.000180289$$

$$2.864583333c_1 + 33.333333c_2 + 343.75c_3 = 0.001602567$$

$$25c_1 + 343.75c_2 + 3883.928571c_3 = 0.015024063$$

Solving the set of equations simultaneously yields $c_1 = 389.773 \times 10^{-6}$, $c_2 = 6.442 \times 10^{-6}$, and $c_3 = 0.789 \times 10^{-6}$. Substitution of the $c$-coefficients into Eq. (1.55) yields the approximate displacement profile:

$$u(y) = 389.733 \times 10^{-6} y + 6.442 \times 10^{-6} y^2 + 0.789 \times 10^{-6} y^3 \qquad (1.64)$$

Next, we will compare the displacement results obtained from the least-squares method and the other weighted residual methods to the exact results.

## Comparison of Weighted Residual Solutions

Now we will examine the accuracy of weighted residual methods by comparing their displacement results with the exact values. Table 1.6 shows nodal displacements computed using the exact, collocation, subdomain, Galerkin, and least-squares methods.

TABLE 1.6 Comparison of weighted residual results

| Location of a Point Along the Bar (in) | Displacement Results from the Exact Solution Eq. (1.53) (in) | Displacement Results from the Collocation Method Eq. (1.57) (in) | Displacement Results from the Subdomain Method Eq. (1.59) (in) | Displacement Results from the Galerkin Method Eq. (1.62) (in) | Displacement Results from the Least-Squares Method Eq. (1.64) (in) |
|---|---|---|---|---|---|
| $y = 0$ | 0 | 0 | 0 | 0 | 0 |
| $y = 2.5$ | 0.001027 | 0.001076 | 0.001029 | 0.001041 | 0.001027 |
| $y = 5.0$ | 0.002213 | 0.002259 | 0.002209 | 0.002220 | 0.002208 |
| $y = 7.5$ | 0.003615 | 0.003660 | 0.003618 | 0.003624 | 0.003618 |
| $y = 10$ | 0.005333 | 0.005384 | 0.005330 | 0.005342 | 0.005331 |

It is clear from an examination of Table 1.6 that the results are in good agreement with each other. It is also important to note here that the primary purpose of Section 1.7 was to introduce you to the general concepts of weighted residual methods and the basic procedures in the simplest possible way. Because the Galerkin method is one of the most commonly used procedures in finite element formulations, more detail and an in-depth view of the Galerkin method will be offered later in Chapters 4 and 7. We will employ the Galerkin method to formulate one- and two-dimensional problems once you have become familiar with the ideas of one- and two-dimensional elements. Also note that in the above examples of the use of weighted residual methods, we assumed a solution that was to provide an approximate solution over the entire domain of the given problem. As you will see later, we will use piecewise solutions with the Galerkin method. That is to say, we will assume linear or nonlinear solutions that are valid only over each element and then combine, or assemble, the elemental solutions.

## 1.8 VERIFICATION OF RESULTS

In recent years, the use of finite element analysis as a design tool has grown rapidly. Easy-to-use comprehensive packages such as ANSYS have become a common tool in the hands of design engineers. Unfortunately, many engineers without the proper training or a solid understanding of the underlying concepts have been using finite element analysis. Engineers who use finite element analysis must understand the limitations of the finite element procedures. There are various sources of error that can contribute to incorrect results. They include:

1. *Wrong input data, such as physical properties and dimensions*
   This mistake can be corrected by simply listing and verifying physical properties and coordinates of nodes or keypoints (points defining the vertices of an object; they are covered in more detail in Chapters 6 and 10) before proceeding any further with the analysis.
2. *Selecting inappropriate types of elements*
   Understanding the underlying theory will benefit you the most in this respect. You need to fully grasp the limitations of a given type of element and understand to which type of problems it applies.
3. *Poor element shape and size after meshing*
   This area is a very important part of any finite element analysis. Inappropriate element shape and size will influence the accuracy of your results. It is important that the user understands the difference between free meshing (using mixed-area element shapes) and mapped meshing (using all quadrilateral area elements or all hexahedral volume elements) and the limitations associated with them. These concepts will be explained in more detail in Chapters 6 and 10.
4. *Applying wrong boundary conditions and loads*
   This step is usually the most difficult aspect of modeling. It involves taking an actual problem and estimating the loading and the appropriate boundary conditions for a finite element model. This step requires good judgment and some experience.

You must always find ways to check your results. While experimental testing of your model may be the best way to do so, it may be expensive or time consuming. You should always start by applying equilibrium conditions and energy balance to different portions of a model to ensure that the physical laws are not violated. For example, for static models, the sum of the forces acting on a free body diagram of your model must be zero. This concept will allow you to check for the accuracy of computed reaction forces. You may want to consider defining and mapping stresses along an arbitrary cross section and integrating this information. The resultant internal forces computed in this manner must balance against external forces. In a heat transfer problem under steady state conditions, apply conservation of energy to a control volume surrounding an arbitrary node. Are the energies flowing into and out of a node balanced? At the end of particular chapters in this text, a section is devoted to verifying the results of your models. In these sections, problems will be solved using ANSYS, and the steps for verifying results will be shown.

## 1.9 UNDERSTANDING THE PROBLEM

You can save lots of time and money if you first spend a little time with a piece of paper and a pencil to try to understand the problem you are planning to analyze. Before initiating numerical modeling on the computer and generating a finite element model, it is imperative that you develop a sense of or a feel for the problem. There are many questions that a good engineer will ask before proceeding with the modeling process, such as: Is the material under axial loading? Is the body under bending moments or twisting moments or a combination of the two? Do you need to worry about buckling? Can we approximate the behavior of the material with a two-dimensional model? Does heat transfer play a significant role in the problem? Which modes of heat transfer are influential? If you choose to employ FEA, "back-of-the-envelope" calculations will greatly enhance your understanding of the problem, in turn helping you to develop a good, reasonable finite element model, particularly in terms of your selection of element types. Some practicing engineers still use finite element analysis to solve a problem that could have been solved more easily by hand by someone with a good grasp of the fundamental concepts of the mechanics of materials and heat transfer.

## SUMMARY

At this point you should:

1. have a good understanding of the physical properties and the parameters that characterize the behavior of an engineering system. Examples of these properties and parameters are given in Tables 1.1 and 1.2.
2. realize that a good understanding of the fundamental concepts of the finite element method will benefit you by enabling you to use ANSYS more effectively.

3. know the seven basic steps involved in any finite element analysis, as discussed in Section 1.4.
4. understand the differences among direct formulation, minimum total potential energy formulation, and the weighted residual methods (particularly the Galerkin formulation).
5. know that it is wise to spend some time to gain a full understanding of a problem before initiating a finite element model of the problem. There may even exist a reasonable closed-form solution to the problem, and thus, you can save lots of time and money.
6. realize that you must always find a way to verify your FEA results.

## REFERENCES

ASHRAE Handbook, *Fundamental Volume*, American Society of Heating, Refrigerating, and Air-Conditioning Engineers, Atlanta, 1993.

Bickford, B. W., *A First Course in the Finite Element Method*, Burr Ridge, Richard D. Irwin, 1989.

Clough, R. W., "The Finite Element Method in Plane Stress Analysis, Proceedings of American Society of Civil Engineers, 2nd Conference on Electronic Computations," Vol. 23, Pittsburgh, 1960, pp. 345–378.

Cook, R. D., Malkus, D. S., and Plesha, M. E., *Concepts and Applications of Finite Element Analysis*, 3d. ed., John Wiley and Sons, New York, 1989.

Courant, R., "Variational Methods for the Solution of Problems of Equilibrium and Vibrations," Bulletin of the American Mathematical Society, Vol. 49, Providence, RI, 1943, pp. 1–23.

Hrennikoff, A., "Solution of Problems in Elasticity by the Framework Method," *J. Appl. Mech.*, Vol. 8, No. 4, New York, 1941, pp. A169–A175.

Levy, S., "Structural Analysis and Influence Coefficients for Delta Wings," *Journal of the Aeronautical Sciences*, Vol. 20, No.7, Easton, PA, 1953, pp. 449-454.

Patankar, S. V., *Numerical Heat Transfer and Fluid Flow*, McGraw-Hill, New York, 1991.

Zienkiewicz, O. C., and Cheung, Y. K. K., *The Finite Element Method in Structural and Continuum Mechanics*, McGraw-Hill, London, 1967.

Zienkiewicz, O. C., *The Finite Element Method*, 3d. ed., McGraw-Hill, London, 1979.

## PROBLEMS

1. Solve Example 1.1 using (1) two elements and (2) eight elements. Compare your results to the exact values.
2. A concrete table column-support with the profile shown in the accompanying figure is to carry a load of approximately 500 lb. Using the direct method discussed in Section 1.5, determine the deflection and average normal stresses along the column. Divide the column into five elements. ($E = 3.27 \times 10^3$ ksi)

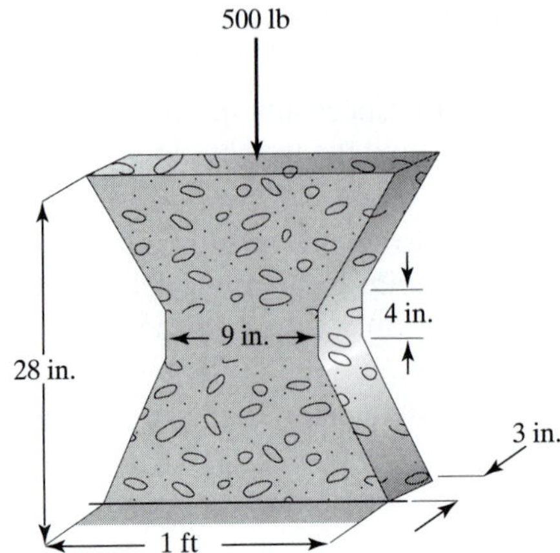


**3.** An aluminum strap with a thickness of 6 mm and the profile shown in the accompanying figure is to carry a load of 1800 N. Using the direct method discussed in Section 1.5, determine the deflection and the average normal stress along the strap. Divide the strap into three elements. This problem may be revisited again in Chapter 8, where a more in-depth analysis may be sought. ($E = 68.9$ GPa)

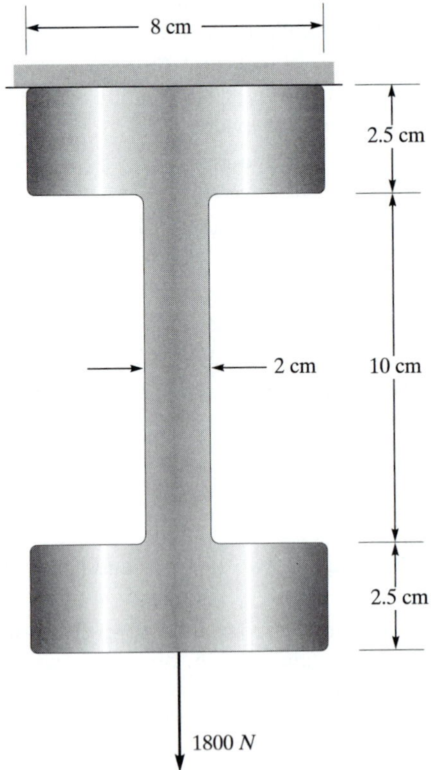

**4.** A thin steel plate with the profile shown in the accompanying figure is subjected to an axial load. Approximate the deflection and the average normal stresses along the plate using the model shown in the figure. The plate has a thickness of 0.125 in and a modulus of elasticity $E = 28 \times 10^3$ ksi. You will be asked to use ANSYS to analyze this problem again in Chapter 8.

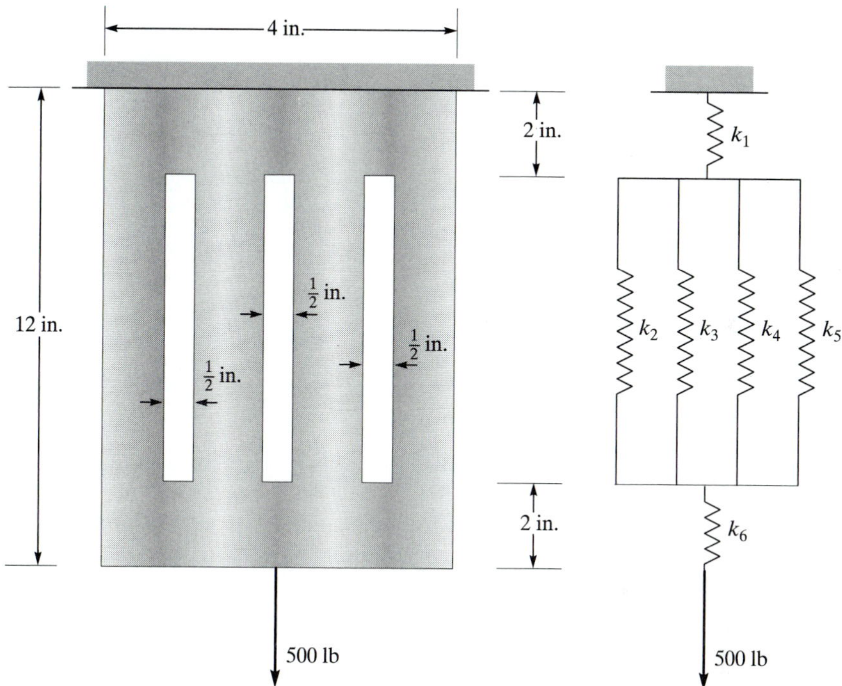


**5.** Apply the statics equilibrium conditions directly to each node of the thin steel plate (using a finite element model) in Problem 4.

**6.** For the spring system shown in the accompanying figure, determine the displacement of each node. Start by identifying the size of the global matrix. Write down elemental stiffness matrices, and show the position of each elemental matrix in the global matrix. Apply the boundary conditions and loads. Solve the set of linear equations. Also compute the reaction forces.

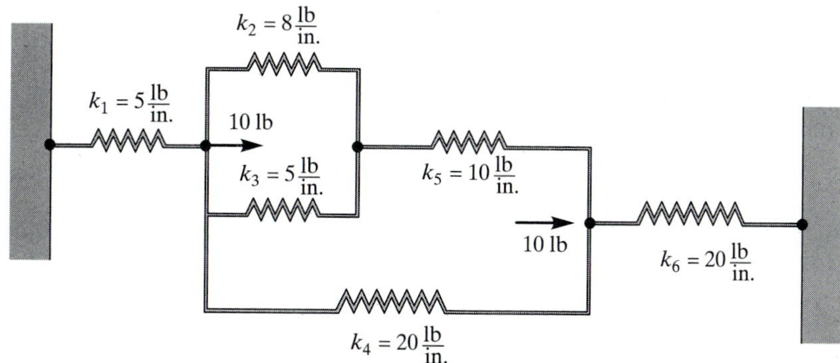

**7.** A typical exterior masonry wall of a house, shown in the accompanying figure, consists of the items in the accompanying table. Assume an inside room temperature of 68°F and an outside air temperature of 10°F, with an exposed area of 150 $ft^2$. Determine the temperature distribution through the wall. Also calculate the heat loss through the wall.

| Items | Resistance hr•ft²•F/Btu | *U*-factor Btu/hr•ft²•F |
|---|---|---|
| 1. Outside film resistance (winter, 15-mph wind) | 0.17 | 5.88 |
| 2. Face brick (4 in) | 0.44 | 2.27 |
| 3. Cement mortar (1/2 in) | 0.1 | 10.0 |
| 4. Cinder block (8 in) | 1.72 | 0.581 |
| 5. Air space (3/4 in) | 1.28 | 0.781 |
| 6. Gypsum board (1/2 in) | 0.45 | 2.22 |
| 7. Inside film resistance (winter) | 0.68 | 1.47 |

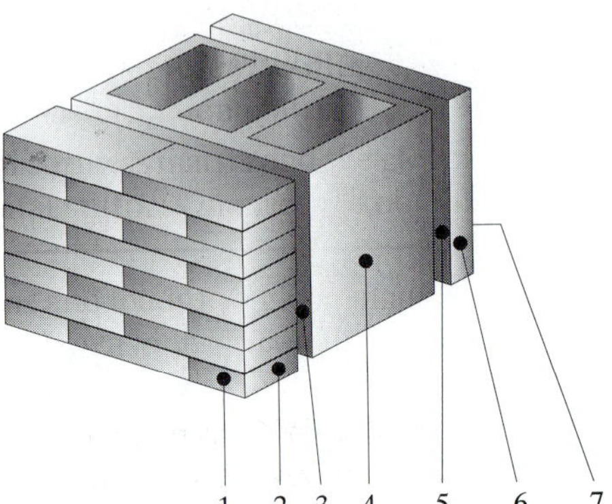


**8.** In order to increase the thermal resistance of a typical exterior frame wall, such as the one in Example 1.2, it is customary to use 2 × 6 studs instead of 2 × 4 studs to allow for placement of more insulation within the wall cavity. A typical exterior (2 × 6) frame wall of a house consists of the materials shown in the accompanying figure. Assume an inside room temperature of 68°F and an outside air temperature of 20°F, with an exposed area of 150 $ft^2$. Determine the temperature distribution through the wall.

| Items | Resistance hr•ft²•F/Btu | *U*-factor Btu/hr•ft²•F |
|---|---|---|
| 1. Outside film resistance (winter, 15-mph wind) | 0.17 | 5.88 |
| 2. Siding, wood (1/2 × 8 lapped) | 0.81 | 1.23 |
| 3. Sheathing (1/2 in regular) | 1.32 | 0.76 |
| 4. Insulation batt ($5\frac{1}{2}$ in) | 19.0 | 0.053 |
| 5. Gypsum wall board (1/2 in) | 0.45 | 2.22 |
| 6. Inside film resistance (winter) | 0.68 | 1.47 |

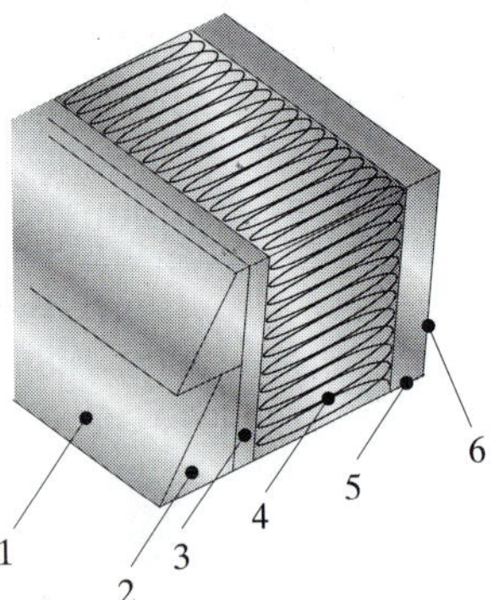


**9.** Assuming the moisture can diffuse through the gypsum board in Problem 8, where should you place a vapor barrier to avoid moisture condensation? Assume an indoor air temperature of 68°F with relative humidity of 40%.

**10.** A typical ceiling of a house consists of the items in the accompanying table. Assume an inside room temperature of 70°F and an attic air temperature of 15°F, with an exposed area of

1000 ft$^2$. Determine the temperature distribution through the wall. Also calculate heat loss through the ceiling.

| Items | Resistance hr•ft$^2$•F/Btu | $U$-factor Btu/hr•ft$^2$•F |
|---|---|---|
| 1. Inside attic film resistance | 0.68 | 1.47 |
| 2. Insulation batt (6 in) | 19 | 0.053 |
| 3. Gypsum board (1/2 in) | 0.45 | 2.22 |
| 4. Inside film resistance (winter) | 0.68 | 1.47 |

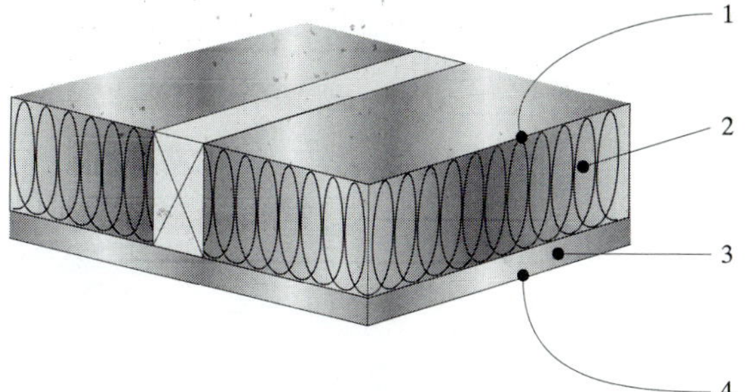


**11.** A typical $1\frac{3}{8}$-in solid wood core door exposed to winter conditions has the characteristics shown in the accompanying table. Assume an inside room temperature of 70°F and an outside air temperature of 20°F, with an exposed area of 22.5 ft$^2$. (a) Determine the inside and outside temperatures of the door's surface. (b) Determine heat loss through the door.

| Items | Resistance hr•ft$^2$•F/Btu | $U$-factor Btu/hr•ft$^2$•F |
|---|---|---|
| 1. Outside film resistance (winter, 15-mph wind) | 0.17 | 5.88 |
| 2. $1\frac{3}{8}$-in solid wood core | 0.39 | 2.56 |
| 3. Inside film resistance (winter) | 0.68 | 1.47 |

**12.** The concrete table column-support in Problem 2 is reinforced with three $\frac{1}{2}$-in steel rods, as shown in the accompanying figure. Determine the deflection and average normal stresses along the column under a load of 1000 lb. Divide the column into five elements. ($E_C$ = $3.27 \times 10^3$ ksi; $E_s = 29 \times 10^3$ ksi)

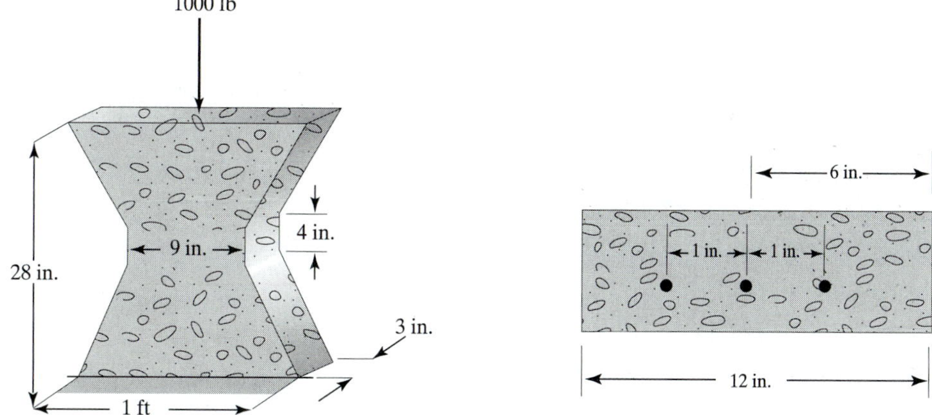


**13.** Compute the total strain energy for the concrete table column-support in Problem 12 .

**14.** A 10-in slender rod weighing 6 lb is supported by a spring with a stiffness $k = 60$ lb/in. A force $P = 35$ lb is applied to the rod at the location shown in the accompanying figure. Determine the deflection of the spring (a) by drawing a free body diagram of the rod and applying the statics equilibrium conditions, and (b) by applying the minimum total potential energy concept.

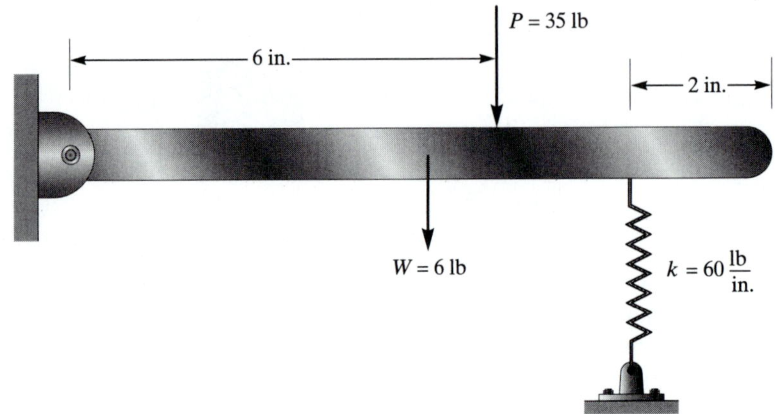


**15.** In a DC electrical circuit, Ohm's Law relates the voltage drop $V_2 - V_1$ across a resistor to a current $I$ flowing through the element and the resistance $R$ according to the equation $V_2 - V_1 = RI$.

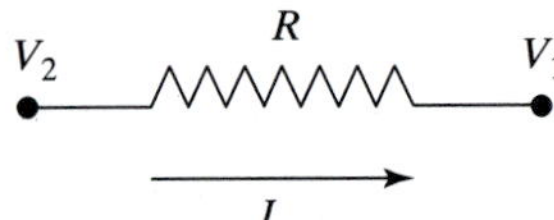


Using direct formulation, show that for a resistance element comprising two nodes, the conductance matrix, the voltage drop, and the currents are related according to the equation

$$\frac{1}{R}\begin{bmatrix} 1 & -1 \\ -1 & 1 \end{bmatrix}\begin{Bmatrix} V_1 \\ V_2 \end{Bmatrix} = \begin{Bmatrix} I_1 \\ I_2 \end{Bmatrix}$$

**16.** Use the results of Problem 15 to set up and solve for the voltage in each branch of the circuit shown in the accompanying figure.

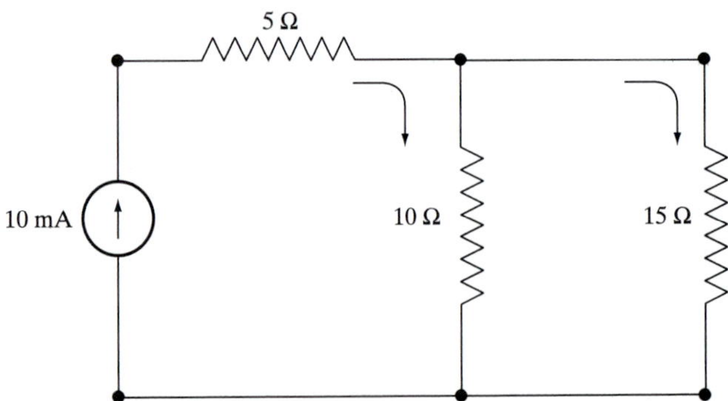

**17.** The deformation of a simply supported beam under a distributed load, shown in the accompanying figure, is governed by the relationship

$$\frac{d^2Y}{dX^2} = \frac{M(X)}{EI}$$

where $M(X)$ is the internal bending moment and is given by

$$M(X) = \frac{wX(L - X)}{2}$$

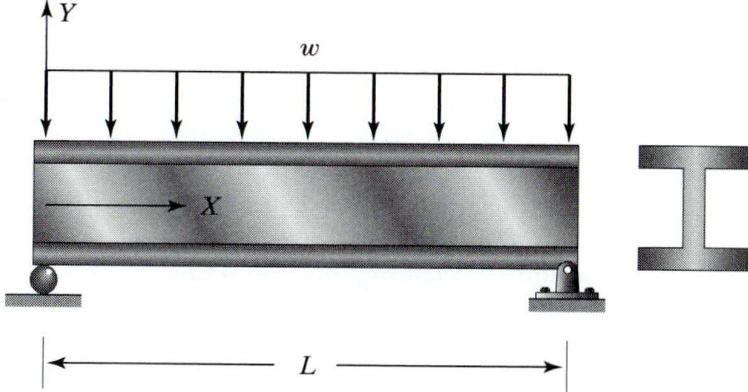


Derive the equation for the exact deflection. Assume an approximate deflection solution of the form

$$Y(X) = c_1\left[\left(\frac{X}{L}\right)^2 - \left(\frac{X}{L}\right)\right]$$

Use the following methods to evaluate $c_1$: (a) the collocation method and (b) the subdomain method. Also, using the approximate solutions, determine the maximum deflection of the beam if a W24 $\times$ 104 (wide flange shape) with a span of $L = 20$ ft supports a distributed load of $w = 5$ kips/ft.

**18.** For the example problem used throughout Section 1.7, assume an approximate solution of the form $u(y) = c_1 y + c_2 y^2 + c_3 y^3 + c_4 y^4$. Using the collocation, subdomain, Galerkin, and least-squares methods, determine the unknown coefficients $c_1, c_2, c_3$, and $c_4$. Compare your results to those obtained in Section 1.7.

**19.** The leakage flow of hydraulic fluid through the gap between a piston–cylinder arrangement may be modeled as laminar flow of fluid between infinite parallel plates, as shown in the accompanying figure. This model offers reasonable results for relatively small gaps. The differential equation governing the flow is

$$\mu \frac{d^2u}{dy^2} = \frac{dp}{dx}$$

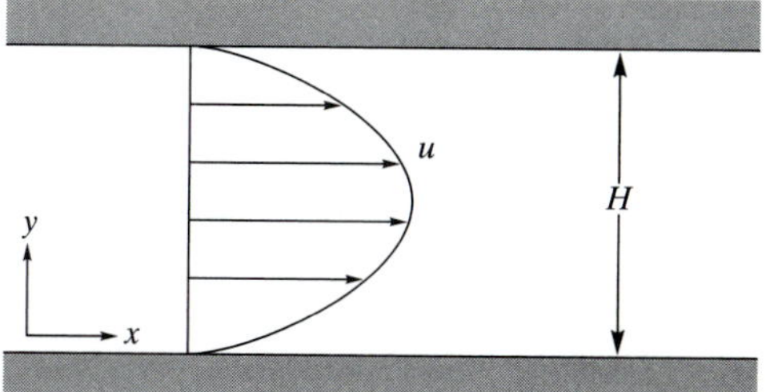

where $\mu$ is the dynamic viscosity of the hydraulic fluid, $u$ is the fluid velocity, and $\frac{dp}{dx}$ is the pressure drop and is constant. Derive the equation for the exact fluid velocities. Assume an approximate fluid velocity solution of the form $u(y) = c_1\left[\sin\left(\frac{\pi y}{H}\right)\right]$. Use the following methods to evaluate $c_1$: (a) the collocation method and (b) the subdomain method. Compare the approximate results to the exact solution.

**20.** Use the Galerkin and least-squares methods to solve Problem 19. Compare the approximate results to the exact solution.

**21.** For the cantilever beam shown in the accompanying figure, the deformation of the beam under a load $P$ is governed by the relationship

$$\frac{d^2Y}{dX^2} = \frac{M(X)}{EI}$$

where $M(X)$ is the internal bending moment and is

$$M(X) = -PX$$

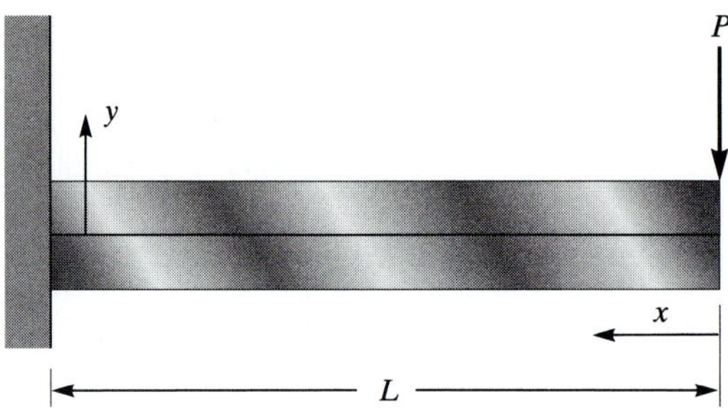


Derive the equation for the exact deflection. Assume an appropriate form of a polynomial function. Keep in mind that the assumed solution must satisfy the given boundary conditions. Use the subdomain method and the Galerkin method to solve for the unknown coefficients of the assumed solution.

**22.** A shaft is made of three parts, as shown in the accompanying figure. Parts $AB$ and $CD$ are made of the same material with a modulus of rigidity of $G = 9.8 \times 10^3$ ksi, and each has a diameter of 1.5 in. Segment $BC$ is made of a material with a modulus of rigidity of $G = 11.2 \times 10^3$ ksi and has a diameter of 1 in. The shaft is fixed at both ends. A torque of 2400 lb•in is applied at $C$. Using three elements, determine the angle of twist at $B$ and $C$ and the torsional reactions at the boundaries.

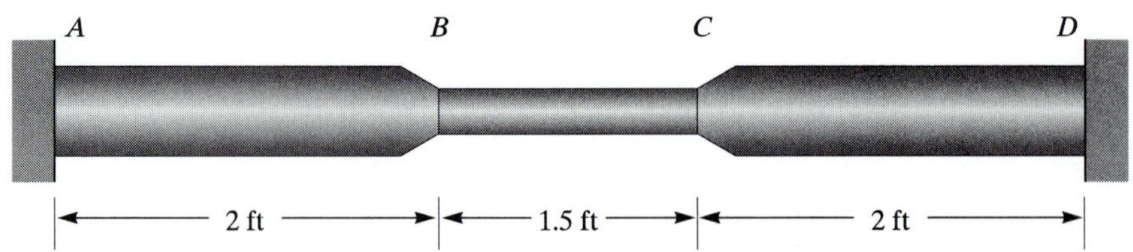

**23.** For the shaft in Problem 22, replace the torque at $C$ by two equal torques of 1500 lb•in at $B$ and $C$. Compute the angle of twist at $B$ and $C$ and the torsional reactions at the boundaries.

**24.** Consider a plate with a variable cross section supporting a load of 1500 lb, as shown in the accompanying figure. Using direct formulation, determine the deflection of the bar at locations $y = 2.5$ in, $y = 7.5$ in, and $y = 10$ in. The plate is made of a material with a modulus of elasticity $E = 10.6 \times 10^3$ ksi.

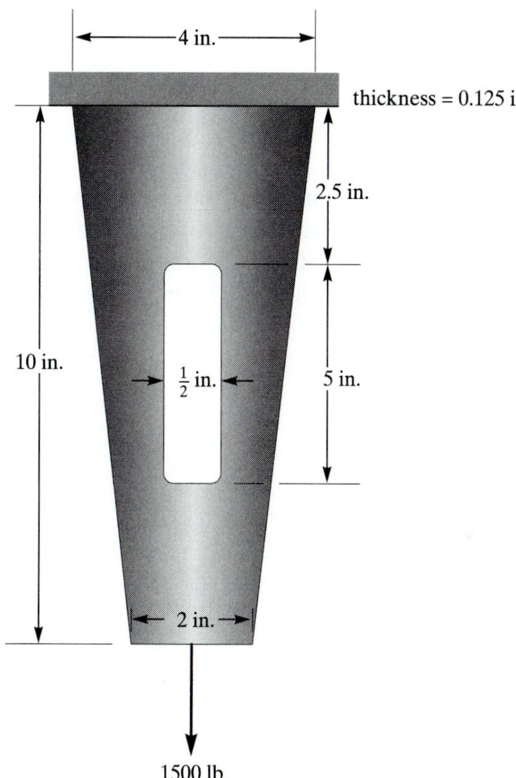

CHAPTER 2

# Trusses

The objectives of this chapter are to introduce the basic concepts in finite element formulation of trusses and to provide an overview of the ANSYS program. A major section of this chapter is devoted to the Launcher, the Graphical User Interface, and the organization of the ANSYS program. The main topics discussed in Chapter 2 include the following:

**2.1** Definition of a Truss
**2.2** Finite Element Formulation
**2.3** Space Trusses
**2.4** Overview of the ANSYS Program
**2.5** Examples Using ANSYS
**2.6** Verification of Results

## 2.1 DEFINITION OF A TRUSS

*A truss* is an engineering structure consisting of straight members connected at their ends by means of bolts, rivets, pins, or welding. The members found in trusses may consist of steel or aluminum tubes, wooden struts, metal bars, angles, and channels. Trusses offer practical solutions to many structural problems in engineering, such as power transmission towers, bridges, and roofs of buildings. A plane truss is defined as a truss whose members lie in a single plane. The forces acting on such a truss must also lie in this plane. Members of a truss are generally considered to be *two-force members*. This term means that internal forces act in equal and opposite directions along the members, as shown in Figure 2.1.

In the analysis that follows, it is assumed that the members are connected together by smooth pins and by a ball-and-socket joint in three-dimensional trusses. Moreover, it can be shown that as long as the center lines of the joining members intersect at a common point, trusses with bolted or welded joints may be treated as having smooth pins (no bending). Another important assumption deals with the way loads are applied. All loads must be applied at the joints. This assumption is true for most situations because trusses are designed in a manner such that the majority of the load is applied at the joints. Usually, the weights of members are negligible compared to those of the applied loads. However, if the weights of the members are to be considered, then half of

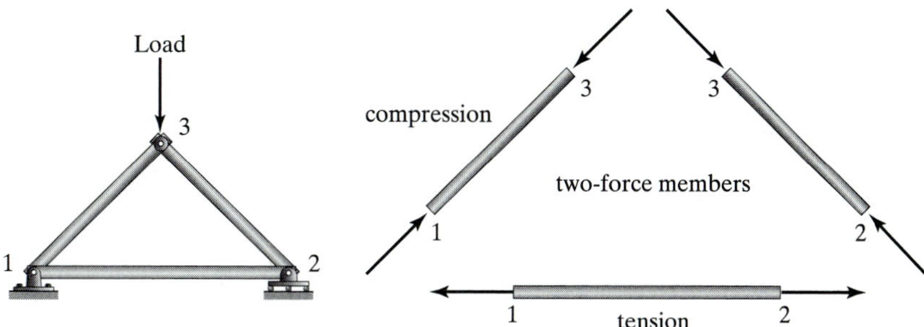


**FIGURE 2.1** A simple truss subjected to a load.

the weight of each member is applied to the connecting joints. *Statically determinate* truss problems are covered in many elementary mechanics text. This class of problems is analyzed by the methods of joints or sections. These methods do not provide information on deflection of the joints because the truss members are treated as rigid bodies. Because the truss members are assumed to be rigid bodies, *statically indeterminate* problems are impossible to analyze. The finite element method allows us to remove the rigid body restriction and solve this class of problems. Figure 2.2 depicts examples of statically determinate and statically indeterminate problems.

## 2.2 FINITE ELEMENT FORMULATION

Let us consider the deflection of a single member when it is subjected to force $F$, as shown in Figure 2.3. The forthcoming derivation of the stiffness coefficient is identical to the analysis of a centrally loaded member that was presented in Section 1.4. As a review and for the sake of continuity and convenience, the steps to derive the elements' equivalent stiffness coefficients are presented here again. Recall that the average stresses in any two-force member are given by

$$\sigma = \frac{F}{A} \tag{2.1}$$

The average strain of the member can be expressed by

$$\varepsilon = \frac{\Delta L}{L} \tag{2.2}$$

Over the elastic region, the stress and strain are related by Hooke's Law,

$$\sigma = E\varepsilon \tag{2.3}$$

Combining Eqs. (2.1), (2.2), and (2.3) and simplifying, we have

$$F = \left(\frac{AE}{L}\right)\Delta L \tag{2.4}$$

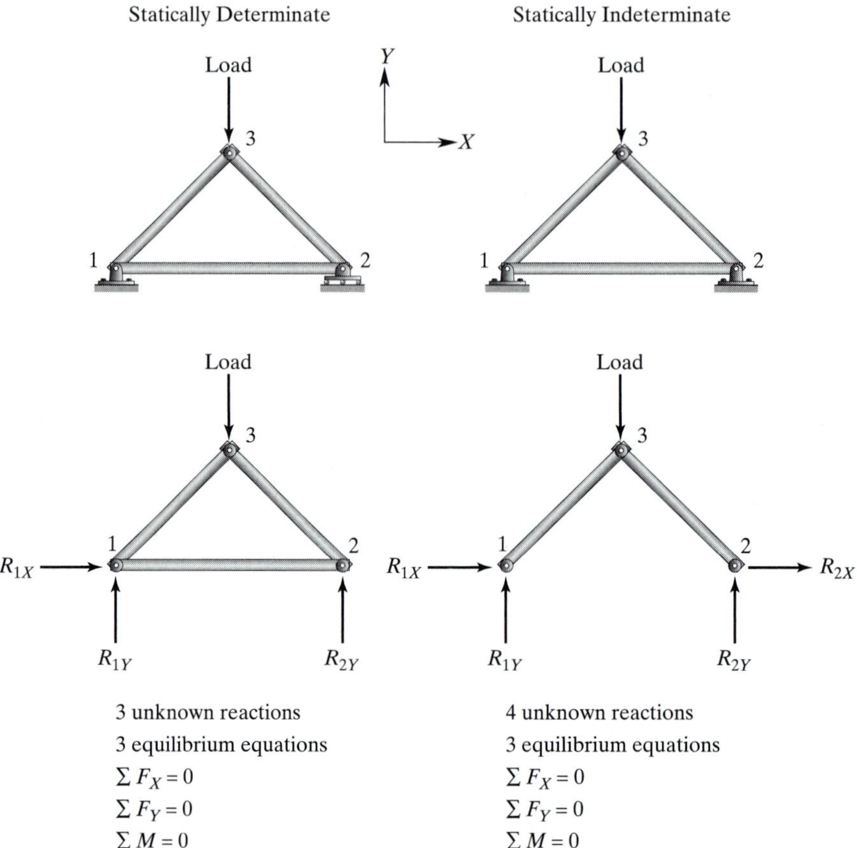


**FIGURE 2.2** Examples of statically determinate and statically indeterminate problems.

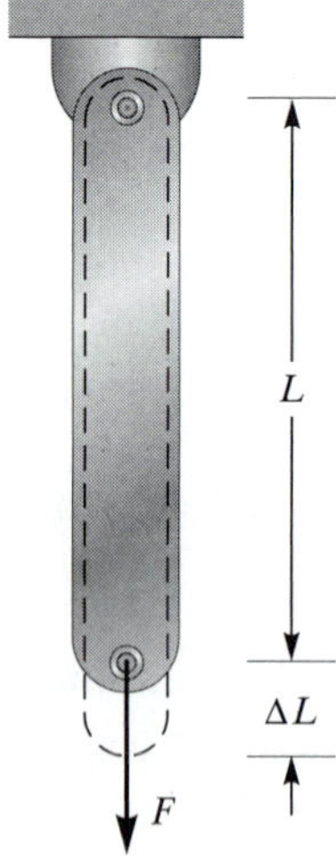


**FIGURE 2.3** A two-force member subjected to a force $F$.

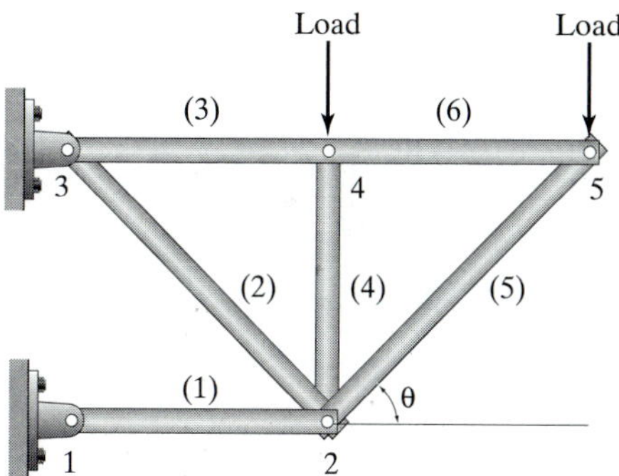


**FIGURE 2.4** A balcony truss.

Note that Eq. (2.4) is similar to the equation of a linear spring, $F = kx$. Therefore, a centrally loaded member of uniform cross section may be modeled as a spring with an equivalent stiffness of

$$k_{eq} = \frac{AE}{L} \tag{2.5}$$

A relatively small balcony truss with five nodes and six elements is shown in Figure 2.4. From this truss, consider isolating a member with an arbitrary orientation. Let us select element (5).

In general, two frames of reference will be required to describe truss problems: a *global coordinate system* and a *local frame of reference*. We choose a fixed global coordinate system, $XY$ (1) to represent the location of each joint (node) and to keep track of the orientation of each member (element), using angles such as θ; (2) to apply the constraints and the applied loads in terms of their respective global components; and (3) to represent the solution—that is, the displacement of each joint in global directions. We will also need a local, or an elemental, coordinate system to describe the two-force member behavior of individual members (elements). The relationship between the local (element) descriptions and the global descriptions is shown in Figure 2.5.

The global displacements are related to the local displacements according to the equations

$$\begin{aligned}
U_{iX} &= u_{ix}\cos\theta - u_{iy}\sin\theta \\
U_{iY} &= u_{ix}\sin\theta + u_{iy}\cos\theta \\
U_{jX} &= u_{jx}\cos\theta - u_{jy}\sin\theta \\
U_{jY} &= u_{jx}\sin\theta + u_{jy}\cos\theta
\end{aligned} \tag{2.6}$$

If we write Eqs. (2.6) in matrix form, we have

$$\{\mathbf{U}\} = [\mathbf{T}]\{\mathbf{u}\} \tag{2.7}$$

where

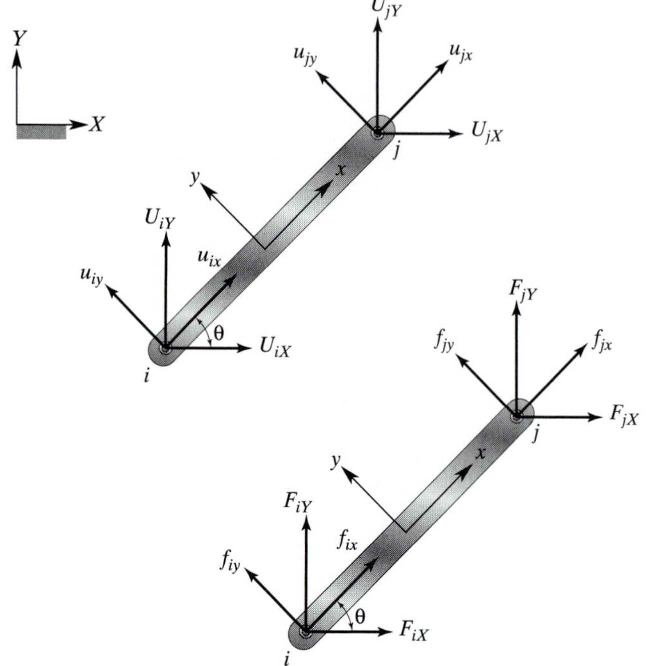


**FIGURE 2.5** Relationship between local and global coordinates.

$$\{\mathbf{U}\} = \begin{Bmatrix} U_{iX} \\ U_{iY} \\ U_{jX} \\ U_{jY} \end{Bmatrix}, [\mathbf{T}] = \begin{bmatrix} \cos\theta & -\sin\theta & 0 & 0 \\ \sin\theta & \cos\theta & 0 & 0 \\ 0 & 0 & \cos\theta & -\sin\theta \\ 0 & 0 & \sin\theta & \cos\theta \end{bmatrix}, \text{and } \{\mathbf{u}\} = \begin{Bmatrix} u_{ix} \\ u_{iy} \\ u_{jx} \\ u_{jy} \end{Bmatrix}$$

$\{\mathbf{U}\}$ and $\{\mathbf{u}\}$ represent the displacements of nodes $i$ and $j$ with respect to the global $XY$ and the local $xy$ frame of references, respectively. $[\mathbf{T}]$ is the transformation matrix that allows for the transfer of local deformations to their respective global values. In a similar way, the local and global forces may be related according to the equations

$$\begin{aligned} F_{iX} &= f_{ix}\cos\theta - f_{iy}\sin\theta \\ F_{iY} &= f_{ix}\sin\theta + f_{iy}\cos\theta \\ F_{jX} &= f_{jx}\cos\theta - f_{jy}\sin\theta \\ F_{jY} &= f_{jx}\sin\theta + f_{jy}\cos\theta \end{aligned} \tag{2.8}$$

or, in matrix form,

$$\{\mathbf{F}\} = [\mathbf{T}]\{\mathbf{f}\} \tag{2.9}$$

where

$$\{\mathbf{F}\} = \begin{Bmatrix} F_{iX} \\ F_{iY} \\ F_{jX} \\ F_{jY} \end{Bmatrix}$$

are components of forces acting at nodes $i$ and $j$ with respect to global coordinates and

$$\{\mathbf{f}\} = \begin{Bmatrix} f_{ix} \\ f_{iy} \\ f_{jx} \\ f_{jy} \end{Bmatrix}$$

represent the local components of the forces at nodes $i$ and $j$.

A general relationship between the local and the global properties was derived in the preceding steps. However, we need to keep in mind that the displacements and the forces in the local $y$-direction are zero. This fact is simply because under the two-force assumption, the members can only be stretched or shortened along their longitudinal axis (local $x$-axis). Of course, this fact also holds true for the internal forces that act only in the local $x$-direction. We do not initially set these terms equal to zero in order to maintain a general matrix description that will make the derivation of the element stiffness matrix easier. This process will become clear when we set the $y$-components of the displacements and forces equal to zero. The local internal forces and displacements are related through the stiffness matrix

$$\begin{Bmatrix} f_{ix} \\ f_{iy} \\ f_{jx} \\ f_{jy} \end{Bmatrix} = \begin{bmatrix} k & 0 & -k & 0 \\ 0 & 0 & 0 & 0 \\ -k & 0 & k & 0 \\ 0 & 0 & 0 & 0 \end{bmatrix} \begin{Bmatrix} u_{ix} \\ u_{iy} \\ u_{jx} \\ u_{jy} \end{Bmatrix} \tag{2.10}$$

where $k = k_{\text{eq}} = \dfrac{AE}{L}$, and using matrix form we can write

$$\{\mathbf{f}\} = [\mathbf{K}]\{\mathbf{u}\} \tag{2.11}$$

After substituting for $\{\mathbf{f}\}$ and $\{\mathbf{u}\}$ in terms of $\{\mathbf{F}\}$ and $\{\mathbf{U}\}$, we have

$$[\mathbf{T}]^{-1}\{\mathbf{F}\} = [\mathbf{K}][\mathbf{T}]^{-1}\{\mathbf{U}\} \tag{2.12}$$

where $[\mathbf{T}]^{-1}$ is the inverse of the transformation matrix $[\mathbf{T}]$ and is

$$[\mathbf{T}]^{-1} = \begin{bmatrix} \cos\theta & \sin\theta & 0 & 0 \\ -\sin\theta & \cos\theta & 0 & 0 \\ 0 & 0 & \cos\theta & \sin\theta \\ 0 & 0 & -\sin\theta & \cos\theta \end{bmatrix} \tag{2.13}$$

Multiplying both sides of Eq. (2.12) by $[\mathbf{T}]$ and simplifying, we obtain:

$$\{\mathbf{F}\} = [\mathbf{T}][\mathbf{K}][\mathbf{T}]^{-1}\{\mathbf{U}\} \tag{2.14}$$

Substituting for values of the $[\mathbf{T}]$, $[\mathbf{K}]$, $[\mathbf{T}]^{-1}$, and $\{\mathbf{U}\}$ matrices in Eq. (2.14) and multiplying, we are left with

$$\begin{Bmatrix} F_{iX} \\ F_{iY} \\ F_{jX} \\ F_{jY} \end{Bmatrix} = k \begin{bmatrix} \cos^2\theta & \sin\theta\cos\theta & -\cos^2\theta & -\sin\theta\cos\theta \\ \sin\theta\cos\theta & \sin^2\theta & -\sin\theta\cos\theta & -\sin^2\theta \\ -\cos^2\theta & -\sin\theta\cos\theta & \cos^2\theta & \sin\theta\cos\theta \\ -\sin\theta\cos\theta & -\sin^2\theta & \sin\theta\cos\theta & \sin^2\theta \end{bmatrix} \begin{Bmatrix} U_{iX} \\ U_{iY} \\ U_{jX} \\ U_{jY} \end{Bmatrix} \quad (2.15)$$

Equations (2.15) express the relationship between the applied forces, the element stiffness matrix $[\mathbf{K}]^{(e)}$, and the global deflection of the nodes of an arbitrary element. The stiffness matrix $[\mathbf{K}]^{(e)}$ for any member (element) of the truss is

$$[\mathbf{K}]^{(e)} = k \begin{bmatrix} \cos^2\theta & \sin\theta\cos\theta & -\cos^2\theta & -\sin\theta\cos\theta \\ \sin\theta\cos\theta & \sin^2\theta & -\sin\theta\cos\theta & -\sin^2\theta \\ -\cos^2\theta & -\sin\theta\cos\theta & \cos^2\theta & \sin\theta\cos\theta \\ -\sin\theta\cos\theta & -\sin^2\theta & \sin\theta\cos\theta & \sin^2\theta \end{bmatrix} \quad (2.16)$$

The next few steps involve assembling, or connecting, the elemental stiffness matrices, applying boundary conditions and loads, solving for displacements, and obtaining other information, such as normal stresses. These steps are best illustrated through an example problem.

## EXAMPLE 2.1

Consider the balcony truss in Figure 2.4, shown here with dimensions. We are interested in determining the deflection of each joint under the loading shown in the figure. All members are made from Douglas-fir wood with a modulus of elasticity of $E = 1.90 \times 10^6$ lb/in$^2$ and a cross-sectional area of 8 in$^2$. We are also interested in calculating average stresses in each member. First, we will solve this problem manually. Later, once we learn how to use ANSYS, we will revisit this problem and solve it using ANSYS.

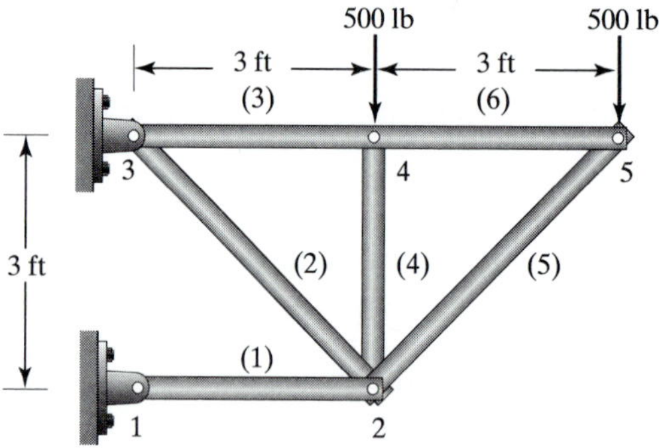


As discussed in Section 1.4, there are seven steps involved in any finite element analysis. Here, these steps are discussed again to emphasize the three phases (preprocessing, solution, and postprocessing) associated with the analysis of truss problems.

### Preprocessing Phase

1. *Discretize the problem into nodes and elements.*
   Each truss member is considered an element, and each joint connecting members is a node. Therefore, the given truss can be modeled with five nodes and six elements. Consult Table 2.1 while following the solution.

TABLE 2.1 The relationship between the elements and their corresponding nodes

| Element | Node $i$ | Node $j$ | $\theta$ See Figures 2.6–2.9 |
|---|---|---|---|
| **(1)** | 1 | 2 | 0 |
| **(2)** | 2 | 3 | 135 |
| **(3)** | 3 | 4 | 0 |
| **(4)** | 2 | 4 | 90 |
| **(5)** | 2 | 5 | 45 |
| **(6)** | 4 | 5 | 0 |

2. *Assume a solution that approximates the behavior of an element.*
As discussed in Section 2.2, we will model the elastic behavior of each element as a spring with an equivalent stiffness of $k$ as given by Eq. (2.5). Since elements (1), (3), (4), and (6) have the same length, cross-sectional area, and modulus of elasticity, the equivalent stiffness constant for these elements (members) is

$$k = \frac{AE}{L} = \frac{(8\text{ in}^2)\left(1.90 \times 10^6 \dfrac{\text{lb}}{\text{in}^2}\right)}{36\text{ in}} = 4.22 \times 10^5\text{ lb/in.}$$

The stiffness constant for elements (2) and (5) is

$$k = \frac{AE}{L} = \frac{(8\text{ in}^2)\left(1.90 \times 10^6 \dfrac{\text{lb}}{\text{in}^2}\right)}{50.9\text{ in}} = 2.98 \times 10^5\text{ lb/in.}$$

3. *Develop equations for elments.*
For elements (1), (3), and (6), the local and the global coordinate systems are aligned, which means that $\theta = 0$. This relationship is shown in Figure 2.6. Using Eq. (2.16), we find that the stiffness matrices are

$$[\mathbf{K}]^{(e)} = k\begin{bmatrix} \cos^2\theta & \sin\theta\cos\theta & -\cos^2\theta & -\sin\theta\cos\theta \\ \sin\theta\cos\theta & \sin^2\theta & -\sin\theta\cos\theta & -\sin^2\theta \\ -\cos^2\theta & -\sin\theta\cos\theta & \cos^2\theta & \sin\theta\cos\theta \\ -\sin\theta\cos\theta & -\sin^2\theta & \sin\theta\cos\theta & \sin^2\theta \end{bmatrix}$$

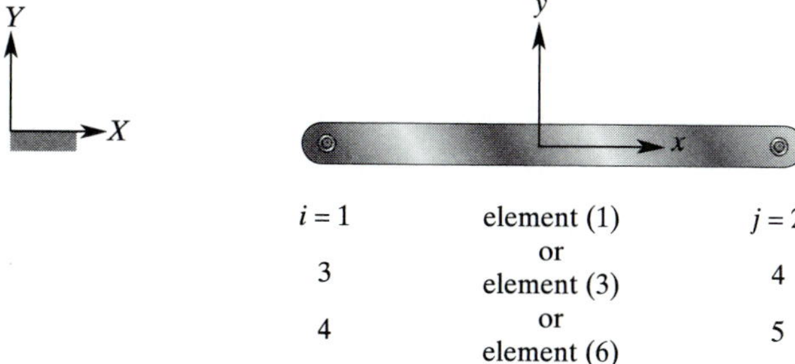


FIGURE 2.6 The orientation of the local coordinates with respect to the global coordinates for elements (1), (3), and (6).

$$[\mathbf{K}]^{(e)} = 4.22 \times 10^5 \begin{bmatrix} \cos^2(0) & \sin(0)\cos(0) & -\cos^2(0) & -\sin(0)\cos(0) \\ \sin(0)\cos(0) & \sin^2(0) & -\sin(0)\cos(0) & -\sin^2(0) \\ -\cos^2(0) & -\sin(0)\cos(0) & \cos^2(0) & \sin(0)\cos(0) \\ -\sin(0)\cos(0) & -\sin^2(0) & \sin(0)\cos(0) & \sin^2(0) \end{bmatrix}$$

$$[\mathbf{K}]^{(1)} = 4.22 \times 10^5 \begin{bmatrix} 1 & 0 & -1 & 0 \\ 0 & 0 & 0 & 0 \\ -1 & 0 & 1 & 0 \\ 0 & 0 & 0 & 0 \end{bmatrix} \begin{matrix} U_{1X} \\ U_{1Y} \\ U_{2X} \\ U_{2Y} \end{matrix}$$

and the position of element (1)'s stiffness matrix in the global matrix is

$$[\mathbf{K}]^{(1G)} = 10^5 \begin{bmatrix} 4.22 & 0 & -4.22 & 0 & 0 & 0 & 0 & 0 & 0 & 0 \\ 0 & 0 & 0 & 0 & 0 & 0 & 0 & 0 & 0 & 0 \\ -4.22 & 0 & 4.22 & 0 & 0 & 0 & 0 & 0 & 0 & 0 \\ 0 & 0 & 0 & 0 & 0 & 0 & 0 & 0 & 0 & 0 \\ 0 & 0 & 0 & 0 & 0 & 0 & 0 & 0 & 0 & 0 \\ 0 & 0 & 0 & 0 & 0 & 0 & 0 & 0 & 0 & 0 \\ 0 & 0 & 0 & 0 & 0 & 0 & 0 & 0 & 0 & 0 \\ 0 & 0 & 0 & 0 & 0 & 0 & 0 & 0 & 0 & 0 \\ 0 & 0 & 0 & 0 & 0 & 0 & 0 & 0 & 0 & 0 \\ 0 & 0 & 0 & 0 & 0 & 0 & 0 & 0 & 0 & 0 \end{bmatrix} \begin{matrix} U_{1X} \\ U_{1Y} \\ U_{2X} \\ U_{2Y} \\ U_{3X} \\ U_{3Y} \\ U_{4X} \\ U_{4Y} \\ U_{5X} \\ U_{5Y} \end{matrix}$$

Note that the nodal displacement matrix is shown alongside element (1)'s position in the global matrix to aid us in observing the location of element (1)'s stiffness matrix in the global matrix. Similarly, the stiffness matrix for element (3) is

$$[\mathbf{K}]^{(3)} = 4.22 \times 10^5 \begin{bmatrix} 1 & 0 & -1 & 0 \\ 0 & 0 & 0 & 0 \\ -1 & 0 & 1 & 0 \\ 0 & 0 & 0 & 0 \end{bmatrix} \begin{matrix} U_{3X} \\ U_{3Y} \\ U_{4X} \\ U_{4Y} \end{matrix}$$

and its position in the global matrix is

$$[\mathbf{K}]^{(3G)} = 10^5 \begin{bmatrix} 0 & 0 & 0 & 0 & 0 & 0 & 0 & 0 & 0 & 0 \\ 0 & 0 & 0 & 0 & 0 & 0 & 0 & 0 & 0 & 0 \\ 0 & 0 & 0 & 0 & 0 & 0 & 0 & 0 & 0 & 0 \\ 0 & 0 & 0 & 0 & 0 & 0 & 0 & 0 & 0 & 0 \\ 0 & 0 & 0 & 0 & 4.22 & 0 & -4.22 & 0 & 0 & 0 \\ 0 & 0 & 0 & 0 & 0 & 0 & 0 & 0 & 0 & 0 \\ 0 & 0 & 0 & 0 & -4.22 & 0 & 4.22 & 0 & 0 & 0 \\ 0 & 0 & 0 & 0 & 0 & 0 & 0 & 0 & 0 & 0 \\ 0 & 0 & 0 & 0 & 0 & 0 & 0 & 0 & 0 & 0 \\ 0 & 0 & 0 & 0 & 0 & 0 & 0 & 0 & 0 & 0 \end{bmatrix} \begin{matrix} U_{1X} \\ U_{1Y} \\ U_{2X} \\ U_{2Y} \\ U_{3X} \\ U_{3Y} \\ U_{4X} \\ U_{4Y} \\ U_{5X} \\ U_{5Y} \end{matrix}$$

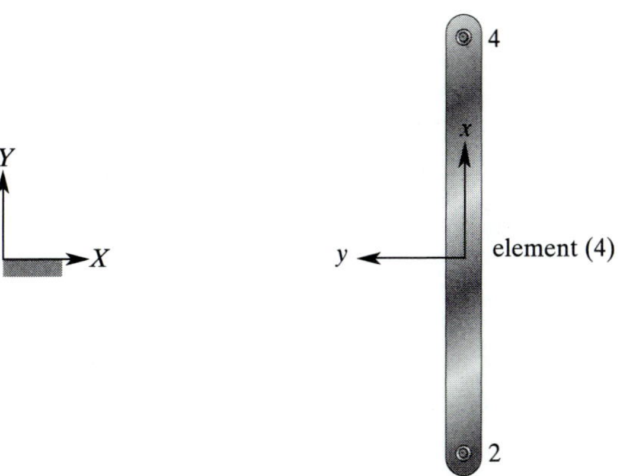


**FIGURE 2.7** The orientation of the local coordinates with respect to the global coordinates for element (4).

The stiffness matrix for element (6) is

$$[\mathbf{K}]^{(6)} = 4.22 \times 10^5 \begin{bmatrix} 1 & 0 & -1 & 0 \\ 0 & 0 & 0 & 0 \\ -1 & 0 & 1 & 0 \\ 0 & 0 & 0 & 0 \end{bmatrix} \begin{matrix} U_{4X} \\ U_{4Y} \\ U_{5X} \\ U_{5Y} \end{matrix}$$

and its position in the global matrix is

$$[\mathbf{K}]^{(6G)} = 10^5 \begin{bmatrix} 0 & 0 & 0 & 0 & 0 & 0 & 0 & 0 & 0 & 0 \\ 0 & 0 & 0 & 0 & 0 & 0 & 0 & 0 & 0 & 0 \\ 0 & 0 & 0 & 0 & 0 & 0 & 0 & 0 & 0 & 0 \\ 0 & 0 & 0 & 0 & 0 & 0 & 0 & 0 & 0 & 0 \\ 0 & 0 & 0 & 0 & 0 & 0 & 0 & 0 & 0 & 0 \\ 0 & 0 & 0 & 0 & 0 & 0 & 0 & 0 & 0 & 0 \\ 0 & 0 & 0 & 0 & 0 & 0 & 4.22 & 0 & -4.22 & 0 \\ 0 & 0 & 0 & 0 & 0 & 0 & 0 & 0 & 0 & 0 \\ 0 & 0 & 0 & 0 & 0 & 0 & -4.22 & 0 & 4.22 & 0 \\ 0 & 0 & 0 & 0 & 0 & 0 & 0 & 0 & 0 & 0 \end{bmatrix} \begin{matrix} U_{1X} \\ U_{1Y} \\ U_{2X} \\ U_{2Y} \\ U_{3X} \\ U_{3Y} \\ U_{4X} \\ U_{4Y} \\ U_{5X} \\ U_{5Y} \end{matrix}$$

For element (4), the orientation of the local coordinate system with respect to the global coordinates is shown in Figure 2.7. Thus, for element (4), $\theta = 90$, which leads to the stiffness matrix

$$[\mathbf{K}]^{(4)} = 4.22 \times 10^5 \begin{bmatrix} \cos^2(90) & \sin(90)\cos(90) & -\cos^2(90) & -\sin(90)\cos(90) \\ \sin(90)\cos(90) & \sin^2(90) & -\sin(90)\cos(90) & -\sin^2(90) \\ -\cos^2(90) & -\sin(90)\cos(90) & \cos^2(90) & \sin(90)\cos(90) \\ -\sin(90)\cos(90) & -\sin^2(90) & \sin(90)\cos(90) & \sin^2(90) \end{bmatrix}$$

$$[\mathbf{K}]^{(4)} = 4.22 \times 10^5 \begin{bmatrix} 0 & 0 & 0 & 0 \\ 0 & 1 & 0 & -1 \\ 0 & 0 & 0 & 0 \\ 0 & -1 & 0 & 1 \end{bmatrix} \begin{matrix} U_{2X} \\ U_{2Y} \\ U_{4X} \\ U_{4Y} \end{matrix}$$

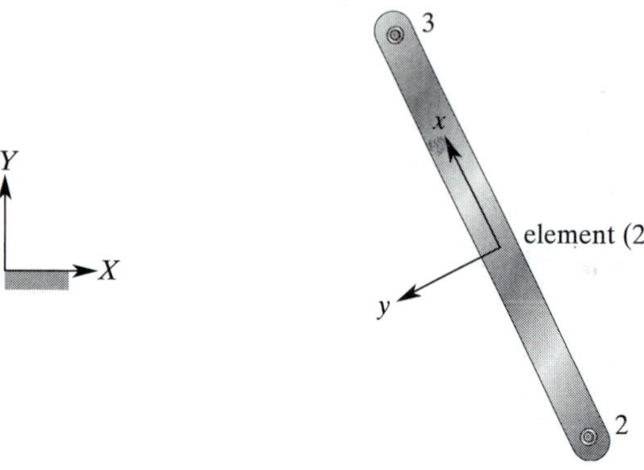


**FIGURE 2.8** The orientation of the local coordinates with respect to the global coordinates for element (2).

and its global position

$$[\mathbf{K}]^{(4G)} = 10^5 \begin{bmatrix} 0 & 0 & 0 & 0 & 0 & 0 & 0 & 0 & 0 & 0 \\ 0 & 0 & 0 & 0 & 0 & 0 & 0 & 0 & 0 & 0 \\ 0 & 0 & 0 & 0 & 0 & 0 & 0 & 0 & 0 & 0 \\ 0 & 0 & 0 & 4.22 & 0 & 0 & 0 & -4.22 & 0 & 0 \\ 0 & 0 & 0 & 0 & 0 & 0 & 0 & 0 & 0 & 0 \\ 0 & 0 & 0 & 0 & 0 & 0 & 0 & 0 & 0 & 0 \\ 0 & 0 & 0 & 0 & 0 & 0 & 0 & 0 & 0 & 0 \\ 0 & 0 & 0 & -4.22 & 0 & 0 & 0 & 4.22 & 0 & 0 \\ 0 & 0 & 0 & 0 & 0 & 0 & 0 & 0 & 0 & 0 \\ 0 & 0 & 0 & 0 & 0 & 0 & 0 & 0 & 0 & 0 \end{bmatrix} \begin{matrix} U_{1X} \\ U_{1Y} \\ U_{2X} \\ U_{2Y} \\ U_{3X} \\ U_{3Y} \\ U_{4X} \\ U_{4Y} \\ U_{5X} \\ U_{5Y} \end{matrix}$$

For element (2), the orientation of the local coordinate system with respect to the global coordinates is shown in Figure 2.8. Thus, for element (2), $\theta = 135$, yielding the stiffness matrix

$$[\mathbf{K}]^{(2)} = 2.98 \times 10^5 \begin{bmatrix} \cos^2(135) & \sin(135)\cos(135) & -\cos^2(135) & -\sin(135)\cos(135) \\ \sin(135)\cos(135) & \sin^2(135) & -\sin(135)\cos(135) & -\sin^2(135) \\ -\cos^2(135) & -\sin(135)\cos(135) & \cos^2(135) & \sin(135)\cos(135) \\ -\sin(135)\cos(135) & -\sin^2(135) & \sin(135)\cos(135) & \sin^2(135) \end{bmatrix}$$

$$[\mathbf{K}]^{(2)} = 2.98 \times 10^5 \begin{bmatrix} .5 & -.5 & -.5 & .5 \\ -.5 & .5 & .5 & -.5 \\ -.5 & .5 & .5 & -.5 \\ .5 & -.5 & -.5 & .5 \end{bmatrix} \begin{matrix} U_{2X} \\ U_{2Y} \\ U_{3X} \\ U_{3Y} \end{matrix}$$

Simplifying, we get

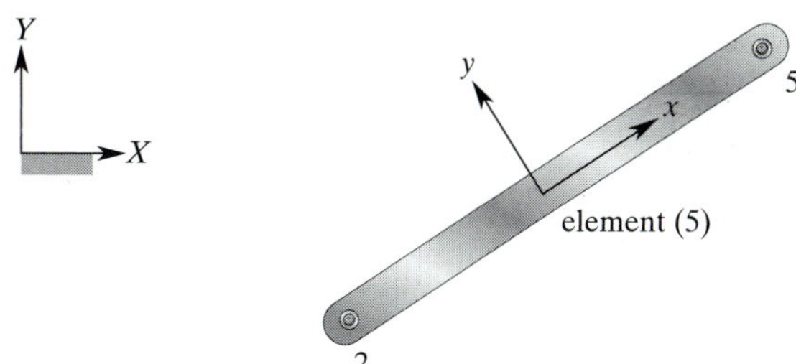


**FIGURE 2.9** The orientation of the local coordinates with respect to the global coordinates for element (5).

$$[\mathbf{K}]^{(2)} = 1.49 \times 10^5 \begin{bmatrix} 1 & -1 & -1 & 1 \\ -1 & 1 & 1 & -1 \\ -1 & 1 & 1 & -1 \\ 1 & -1 & -1 & 1 \end{bmatrix} \begin{matrix} U_{2X} \\ U_{2Y} \\ U_{3X} \\ U_{3Y} \end{matrix}$$

and its position in the global matrix is

$$[\mathbf{K}]^{(2G)} = 10^5 \begin{bmatrix} 0 & 0 & 0 & 0 & 0 & 0 & 0 & 0 & 0 & 0 \\ 0 & 0 & 0 & 0 & 0 & 0 & 0 & 0 & 0 & 0 \\ 0 & 0 & 1.49 & -1.49 & -1.49 & 1.49 & 0 & 0 & 0 & 0 \\ 0 & 0 & -1.49 & 1.49 & 1.49 & -1.49 & 0 & 0 & 0 & 0 \\ 0 & 0 & -1.49 & 1.49 & 1.49 & -1.49 & 0 & 0 & 0 & 0 \\ 0 & 0 & 1.49 & -1.49 & -1.49 & 1.49 & 0 & 0 & 0 & 0 \\ 0 & 0 & 0 & 0 & 0 & 0 & 0 & 0 & 0 & 0 \\ 0 & 0 & 0 & 0 & 0 & 0 & 0 & 0 & 0 & 0 \\ 0 & 0 & 0 & 0 & 0 & 0 & 0 & 0 & 0 & 0 \\ 0 & 0 & 0 & 0 & 0 & 0 & 0 & 0 & 0 & 0 \end{bmatrix} \begin{matrix} U_{1X} \\ U_{1Y} \\ U_{2X} \\ U_{2Y} \\ U_{3X} \\ U_{3Y} \\ U_{4X} \\ U_{4Y} \\ U_{5X} \\ U_{5Y} \end{matrix}$$

For element (5), the orientation of the local coordinate system with respect to the global coordinates is shown in Figure 2.9. Thus, for element (5), $\theta = 45$, yielding the stiffness matrix

$$[\mathbf{K}]^{(5)} = 2.98 \times 10^5 \begin{bmatrix} \cos^2(45) & \sin(45)\cos(45) & -\cos^2(45) & -\sin(45)\cos(45) \\ \sin(45)\cos(45) & \sin^2(45) & -\sin(45)\cos(45) & -\sin^2(45) \\ -\cos^2(45) & -\sin(45)\cos(45) & \cos^2(45) & \sin(45)\cos(45) \\ -\sin(45)\cos(45) & -\sin^2(45) & \sin(45)\cos(45) & \sin^2(45) \end{bmatrix}$$

$$[\mathbf{K}]^{(5)} = 2.98 \times 10^5 \begin{bmatrix} .5 & .5 & -.5 & -.5 \\ .5 & .5 & -.5 & -.5 \\ -.5 & -.5 & .5 & .5 \\ -.5 & -.5 & .5 & .5 \end{bmatrix} \begin{matrix} U_{2X} \\ U_{2Y} \\ U_{5X} \\ U_{5Y} \end{matrix}$$

and its position in the global stiffness matrix is

$$[\mathbf{K}]^{(5G)} = 10^5 \begin{bmatrix} 0 & 0 & 0 & 0 & 0 & 0 & 0 & 0 & 0 & 0 \\ 0 & 0 & 0 & 0 & 0 & 0 & 0 & 0 & 0 & 0 \\ 0 & 0 & 1.49 & 1.49 & 0 & 0 & 0 & 0 & -1.49 & -1.49 \\ 0 & 0 & 1.49 & 1.49 & 0 & 0 & 0 & 0 & -1.49 & -1.49 \\ 0 & 0 & 0 & 0 & 0 & 0 & 0 & 0 & 0 & 0 \\ 0 & 0 & 0 & 0 & 0 & 0 & 0 & 0 & 0 & 0 \\ 0 & 0 & 0 & 0 & 0 & 0 & 0 & 0 & 0 & 0 \\ 0 & 0 & 0 & 0 & 0 & 0 & 0 & 0 & 0 & 0 \\ 0 & 0 & -1.49 & -1.49 & 0 & 0 & 0 & 0 & 1.49 & 1.49 \\ 0 & 0 & -1.49 & -1.49 & 0 & 0 & 0 & 0 & 1.49 & 1.49 \end{bmatrix} \begin{matrix} U_{1X} \\ U_{1Y} \\ U_{2X} \\ U_{2Y} \\ U_{3X} \\ U_{3Y} \\ U_{4X} \\ U_{4Y} \\ U_{5X} \\ U_{5Y} \end{matrix}$$

It is worth noting again that the nodal displacements associated with each element are shown next to each element's stiffness matrix. This practice makes it easier to connect (assemble) the individual stiffness matrices into the global stiffness matrix for the truss.

4. *Assemble elements.* The global stiffness matrix is obtained by assembling, or adding together, the individual elements' matrices:

$$[\mathbf{K}]^{(G)} = [\mathbf{K}]^{(1G)} + [\mathbf{K}]^{(2G)} + [\mathbf{K}]^{(3G)} + [\mathbf{K}]^{(4G)} + [\mathbf{K}]^{(5G)} + [\mathbf{K}]^{(6G)}$$

$$[\mathbf{K}]^{(G)} = 10^5 \left[ \begin{matrix} 4.22 & 0 & -4.22 & 0 & 0 \\ 0 & 0 & 0 & 0 & 0 \\ -4.22 & 0 & 4.22 + 1.49 + 1.49 & -1.49 + 1.49 & -1.49 \\ 0 & 0 & 1.49 - 1.49 & 4.22 + 1.49 + 1.49 & 1.49 \\ 0 & 0 & -1.49 & 1.49 & 4.22 + 1.49 \\ 0 & 0 & 1.49 & -1.49 & -1.49 \\ 0 & 0 & 0 & 0 & -4.22 \\ 0 & 0 & 0 & -4.22 & 0 \\ 0 & 0 & -1.49 & -1.49 & 0 \\ 0 & 0 & -1.49 & -1.49 & 0 \end{matrix} \right.$$

$$\left. \begin{matrix} 0 & 0 & 0 & 0 & 0 \\ 0 & 0 & 0 & 0 & 0 \\ 1.49 & 0 & 0 & -1.49 & -1.49 \\ -1.49 & 0 & -4.22 & -1.49 & -1.49 \\ -1.49 & -4.22 & 0 & 0 & 0 \\ 1.49 & 0 & 0 & 0 & 0 \\ 0 & 4.22 + 4.22 & 0 & -4.22 & 0 \\ 0 & 0 & 4.22 & 0 & 0 \\ 0 & -4.22 & 0 & 4.22 + 1.49 & 1.49 \\ 0 & 0 & 0 & 1.49 & 1.49 \end{matrix} \right] \begin{matrix} U_{1X} \\ U_{1Y} \\ U_{2X} \\ U_{2Y} \\ U_{3X} \\ U_{3Y} \\ U_{4X} \\ U_{4Y} \\ U_{5X} \\ U_{5Y} \end{matrix}$$

Simplifying, we get

$$[\mathbf{K}]^{(G)} = 10^5 \begin{bmatrix} 4.22 & 0 & -4.22 & 0 & 0 & 0 & 0 & 0 & 0 & 0 \\ 0 & 0 & 0 & 0 & 0 & 0 & 0 & 0 & 0 & 0 \\ -4.22 & 0 & 7.2 & 0 & -1.49 & 1.49 & 0 & 0 & -1.49 & -1.49 \\ 0 & 0 & 0 & 7.2 & 1.49 & -1.49 & 0 & -4.22 & -1.49 & -1.49 \\ 0 & 0 & -1.49 & 1.49 & 5.71 & -1.49 & -4.22 & 0 & 0 & 0 \\ 0 & 0 & 1.49 & -1.49 & -1.49 & 1.49 & 0 & 0 & 0 & 0 \\ 0 & 0 & 0 & 0 & -4.22 & 0 & 8.44 & 0 & -4.22 & 0 \\ 0 & 0 & 0 & -4.22 & 0 & 0 & 0 & 4.22 & 0 & 0 \\ 0 & 0 & -1.49 & -1.49 & 0 & 0 & -4.22 & 0 & 5.71 & 1.49 \\ 0 & 0 & -1.49 & -1.49 & 0 & 0 & 0 & 0 & 1.49 & 1.49 \end{bmatrix}$$

5. *Apply the boundary conditions and loads.*
The following boundary conditions apply to this problem: nodes 1 and 3 are fixed; which implies that $U_{1X} = 0$, $U_{1Y} = 0$, $U_{3X} = 0$, and $U_{3Y} = 0$. Incorporating these conditions into the global stiffness matrix and applying the external loads at nodes 4 and 5 such that $F_{4Y} = -500$ lb and $F_{5Y} = -500$ lb results in a set of linear equations that must be solved simultaneously:

$$10^5 \begin{bmatrix} 1 & 0 & 0 & 0 & 0 & 0 & 0 & 0 & 0 & 0 \\ 0 & 1 & 0 & 0 & 0 & 0 & 0 & 0 & 0 & 0 \\ -4.22 & 0 & 7.2 & 0 & -1.49 & 1.49 & 0 & 0 & -1.49 & -1.49 \\ 0 & 0 & 0 & 7.2 & 1.49 & -1.49 & 0 & -4.22 & -1.49 & -1.49 \\ 0 & 0 & 0 & 0 & 1 & 0 & 0 & 0 & 0 & 0 \\ 0 & 0 & 0 & 0 & 0 & 1 & 0 & 0 & 0 & 0 \\ 0 & 0 & 0 & 0 & -4.22 & 0 & 8.44 & 0 & -4.22 & 0 \\ 0 & 0 & 0 & -4.22 & 0 & 0 & 0 & 4.22 & 0 & 0 \\ 0 & 0 & -1.49 & -1.49 & 0 & 0 & -4.22 & 0 & 5.71 & 1.49 \\ 0 & 0 & -1.49 & -1.49 & 0 & 0 & 0 & 0 & 1.49 & 1.49 \end{bmatrix} \begin{Bmatrix} U_{1X} \\ U_{1Y} \\ U_{2X} \\ U_{2Y} \\ U_{3X} \\ U_{3Y} \\ U_{4X} \\ U_{4Y} \\ U_{5X} \\ U_{5Y} \end{Bmatrix} = \begin{Bmatrix} 0 \\ 0 \\ 0 \\ 0 \\ 0 \\ 0 \\ 0 \\ -500 \\ 0 \\ -500 \end{Bmatrix}$$

Because $U_{1X} = 0, U_{1Y} = 0, U_{3X} = 0$, and $U_{3Y} = 0$, we can eliminate the first, second, fifth, and sixth rows and columns from our calculation such that we need only solve a $6 \times 6$ matrix:

$$10^5 \begin{bmatrix} 7.2 & 0 & 0 & 0 & -1.49 & -1.49 \\ 0 & 7.2 & 0 & -4.22 & -1.49 & -1.49 \\ 0 & 0 & 8.44 & 0 & -4.22 & 0 \\ 0 & -4.22 & 0 & 4.22 & 0 & 0 \\ -1.49 & -1.49 & -4.22 & 0 & 5.71 & 1.49 \\ -1.49 & -1.49 & 0 & 0 & 1.49 & 1.49 \end{bmatrix} \begin{Bmatrix} U_{2X} \\ U_{2Y} \\ U_{4X} \\ U_{4Y} \\ U_{5X} \\ U_{5Y} \end{Bmatrix} = \begin{Bmatrix} 0 \\ 0 \\ 0 \\ -500 \\ 0 \\ -500 \end{Bmatrix}$$

### Solution Phase

6. *Solve a system of algebraic equations simultaneously.*
Solving the above matrix for the unknown displacements yields $U_{2X} = -0.00355$ in, $U_{2Y} = -0.01026$ in, $U_{4X} = 0.00118$ in, $U_{4Y} = -0.0114$ in, $U_{5X} = 0.00240$ in, and $U_{5Y} = -0.0195$ in. Thus, the global displacement matrix is

$$\begin{Bmatrix} U_{1X} \\ U_{1Y} \\ U_{2X} \\ U_{2Y} \\ U_{3X} \\ U_{3Y} \\ U_{4X} \\ U_{4Y} \\ U_{5X} \\ U_{5Y} \end{Bmatrix} = \begin{Bmatrix} 0 \\ 0 \\ -0.00355 \\ -0.01026 \\ 0 \\ 0 \\ 0.00118 \\ -0.0114 \\ 0.00240 \\ -0.0195 \end{Bmatrix} \text{in.}$$

Recognize that the displacements of the nodes are given with respect to the global coordinate system.

## Postprocessing Phase

**7.** *Obtain other information.*

**Reaction Forces** As discussed in Chapter 1, the reaction forces can be computed from

$$\{\mathbf{R}\} = [\mathbf{K}]^{(G)}\{\mathbf{U}\} - \{\mathbf{F}\}$$

such that

$$\begin{Bmatrix} R_{1X} \\ R_{1Y} \\ R_{2X} \\ R_{2Y} \\ R_{3X} \\ R_{3Y} \\ R_{4X} \\ R_{4Y} \\ R_{5X} \\ R_{5Y} \end{Bmatrix} = 10^5 \begin{bmatrix} 4.22 & 0 & -4.22 & 0 & 0 & 0 & 0 & 0 & 0 & 0 \\ 0 & 0 & 0 & 0 & 0 & 0 & 0 & 0 & 0 & 0 \\ -4.22 & 0 & 7.2 & 0 & -1.49 & 1.49 & 0 & 0 & -1.49 & -1.49 \\ 0 & 0 & 0 & 7.2 & 1.49 & -1.49 & 0 & -4.22 & -1.49 & -1.49 \\ 0 & 0 & -1.49 & 1.49 & 5.71 & -1.49 & -4.22 & 0 & 0 & 0 \\ 0 & 0 & 1.49 & -1.49 & -1.49 & 1.49 & 0 & 0 & 0 & 0 \\ 0 & 0 & 0 & 0 & -4.22 & 0 & 8.44 & 0 & -4.22 & 0 \\ 0 & 0 & 0 & -4.22 & 0 & 0 & 0 & 4.22 & 0 & 0 \\ 0 & 0 & -1.49 & -1.49 & 0 & 0 & -4.22 & 0 & 5.71 & 1.49 \\ 0 & 0 & -1.49 & -1.49 & 0 & 0 & 0 & 0 & 1.49 & 1.49 \end{bmatrix}$$

$$\begin{Bmatrix} 0 \\ 0 \\ -0.00355 \\ -0.01026 \\ 0 \\ 0 \\ 0.00118 \\ -0.0114 \\ 0.00240 \\ -0.0195 \end{Bmatrix} - \begin{Bmatrix} 0 \\ 0 \\ 0 \\ 0 \\ 0 \\ 0 \\ 0 \\ -500 \\ 0 \\ -500 \end{Bmatrix}$$

Performing matrix operations, yields the reaction results

$$\begin{Bmatrix} R_{1X} \\ R_{1Y} \\ R_{2X} \\ R_{2Y} \\ R_{3X} \\ R_{3Y} \\ R_{4X} \\ R_{4Y} \\ R_{5X} \\ R_{5Y} \end{Bmatrix} = \begin{Bmatrix} 1500 \\ 0 \\ 0 \\ 0 \\ -1500 \\ 1000 \\ 0 \\ 0 \\ 0 \\ 0 \end{Bmatrix} \text{lb}$$

**Internal Forces and Normal Stresses** Now let us compute internal forces, and the average normal stresses, in each member. The member internal forces $f_{ix}$ and $f_{jx}$, which are equal and opposite in direction, are

$$\begin{aligned} f_{ix} &= k(u_{ix} - u_{jx}) \\ f_{jx} &= k(u_{jx} - u_{ix}) \end{aligned} \tag{2.17}$$

Note that the sum of $f_{ix}$ and $f_{jx}$ is zero regardless of which representation of Figure 2.10 we select. However, for the sake of consistency in the forthcoming derivation, we will use the second representation so that $f_{ix}$ and $f_{jx}$ are given in the positive local $x$-direction. In order to use Eq. (2.17) to compute the internal force in a given element, we must know the displacements of the element's end nodes, $u_{ix}$ and $u_{jx}$, with respect to the local coordinate system, $x$, $y$. Recall that the glob-

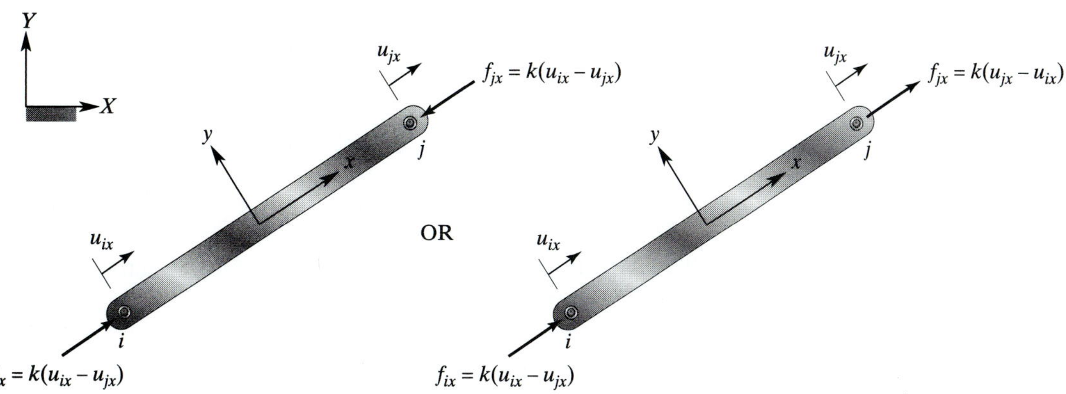


**FIGURE 2.10** Internal forces in a truss member.

al displacements are related to the local displacements through a transformation matrix, according to Eq. (2.7), repeated here for convenience,

$$\{\mathbf{U}\} = [\mathbf{T}]\{\mathbf{u}\}$$

and the local displacements in terms of the global displacements:

$$\{\mathbf{u}\} = [\mathbf{T}]^{-1}\{\mathbf{U}\}$$

$$\begin{Bmatrix} u_{ix} \\ u_{iy} \\ u_{jx} \\ u_{jy} \end{Bmatrix} = \begin{bmatrix} \cos\theta & \sin\theta & 0 & 0 \\ -\sin\theta & \cos\theta & 0 & 0 \\ 0 & 0 & \cos\theta & \sin\theta \\ 0 & 0 & -\sin\theta & \cos\theta \end{bmatrix} \begin{bmatrix} U_{iX} \\ U_{iY} \\ U_{jX} \\ U_{jY} \end{bmatrix}$$

Once the internal force in each member is computed, the normal stress in each member can be determined from the equation

$$\sigma = \frac{\text{internal force}}{area} = \frac{f}{A}$$

or alternatively, we can compute the normal stresses from

$$\sigma = \frac{f}{A} = \frac{k(u_{ix} - u_{jx})}{A} = \frac{\dfrac{AE}{L}(u_{ix} - u_{jx})}{A} = E\left(\frac{u_{ix} - u_{jx}}{L}\right) \tag{2.18}$$

As an example, let us compute the internal force and the normal stress in element (5). For element (5), $\theta = 45$, $U_{2X} = -0.00355$ in, $U_{2Y} = -0.01026$ in, $U_{5X} = 0.0024$ in, and $U_{5Y} = -0.0195$ in. First, we solve for local displacements of nodes 2 and 5 from the relation

$$\begin{Bmatrix} u_{2x} \\ u_{2y} \\ u_{5x} \\ u_{5y} \end{Bmatrix} = \begin{bmatrix} \cos 45 & \sin 45 & 0 & 0 \\ -\sin 45 & \cos 45 & 0 & 0 \\ 0 & 0 & \cos 45 & \sin 45 \\ 0 & 0 & -\sin 45 & \cos 45 \end{bmatrix} \begin{Bmatrix} -0.00355 \\ -0.01026 \\ 0.00240 \\ -0.01950 \end{Bmatrix}$$

which reveals that $u_{2x} = -0.00976$ in and $u_{5x} = -0.01209$ in. Upon substitution of these values into Eqs. (2.17) and (2.18), the internal force and the normal stress in element (5) are 696 lb and 87 lb/in$^2$, respectively. Similarly, the internal forces and stresses can be obtained for other elements.

This problem will be revisited later and solved using ANSYS. The verification of these results will also be discussed in detail later in this chapter.

## 2.3 SPACE TRUSSES

A three-dimensional truss is often called a space truss. A simple space truss has six members joined together at their ends to form a tetrahedron, as shown in Figure 2.11. We can create more complex structures by adding three new members to a simple truss. This addition should be done in a manner where one end of each new member is connected to a separate existing joint, attaching the other ends of the new members together to form a new joint. This structure is shown in Figure 2.12. As mentioned earlier, members of a truss are generally considered to be two-force members. In the analysis of space trusses, it is assumed that the members are connected together by ball-and-socket joints. It can be shown that as long as the center lines of the adjacent bolted members intersect at a common point, trusses with bolted or welded joints may also be treated under the ball-and-socket joints assumption (negligible bending moments at the joints). Another restriction deals with the assumption that all loads must be applied at the joints. This assumption is true for most situations. As stated earlier, the weights of members are usually negligible compared to the applied loads. However, if the weights of the members are to be considered, then half of the weight of each member is applied to the connecting joints.

Finite element formulation of space trusses is an extension of the analysis of plane trusses. In a space truss, the global displacement of an element is represented by six unknowns, $U_{iX}$, $U_{iY}$, $U_{iZ}$, $U_{jX}$, $U_{jY}$, and $U_{jZ}$, because each node (joint) can move in three

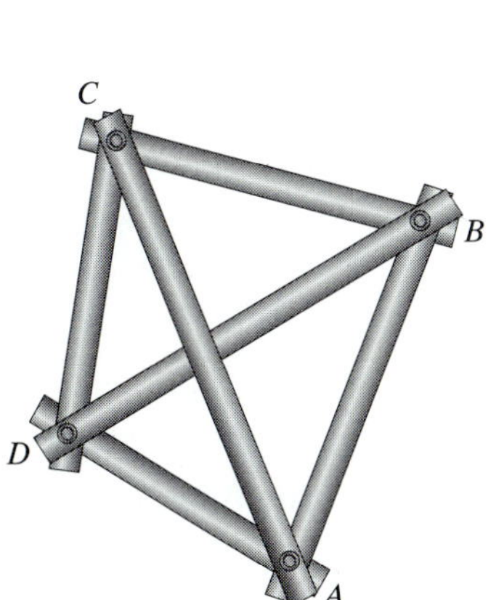


**FIGURE 2.11** A simple truss.

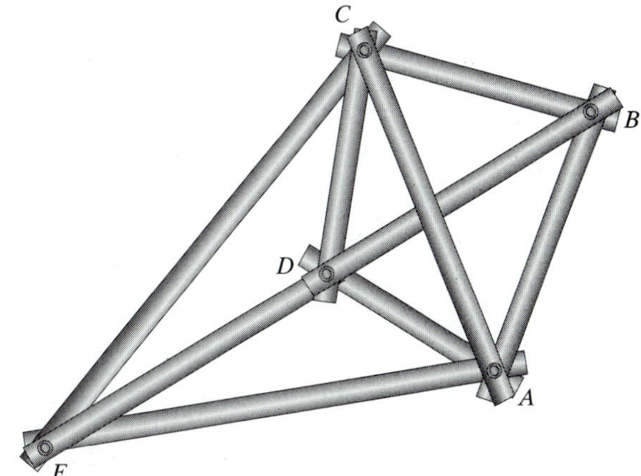


**FIGURE 2.12** Addition of new elements to a simple truss to form complex structures.

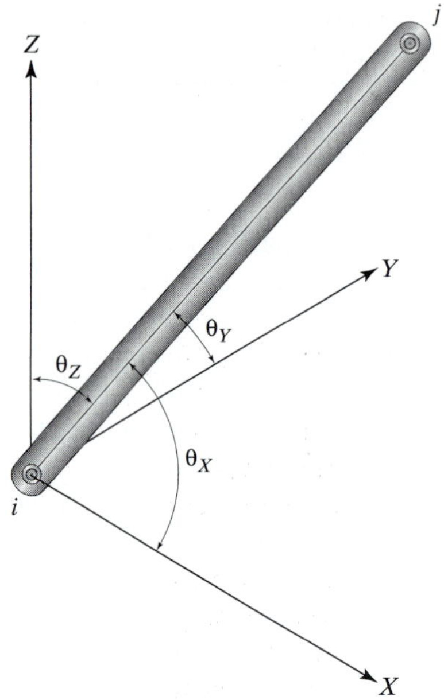

**FIGURE 2.13** The angles formed by a member with the $X$-, $Y$-, and $Z$-axis.

directions. Moreover, the angles $\theta_X$, $\theta_Y$, and $\theta_Z$ define the orientation of a member with respect to the global coordinate system, as shown in Figure 2.13.

The directional cosines can be written in terms of the difference between the coordinates of nodes $j$ and $i$ of a member and the member's length according to the relationships

$$\cos\theta_X = \frac{X_j - X_i}{L} \tag{2.19}$$

$$\cos\theta_Y = \frac{Y_j - Y_i}{L} \tag{2.20}$$

$$\cos\theta_Z = \frac{Z_j - Z_i}{L} \tag{2.21}$$

where $L$ is the length of the member and is given by

$$L = \sqrt{(X_j - X_i)^2 + (Y_j - Y_i)^2 + (Z_j - Z_i)^2} \tag{2.22}$$

The procedure for obtaining the element stiffness matrix for a space-truss member is identical to the one we followed to derive the two-dimensional truss element stiffness. We start the procedure by relating the global displacements and forces to local displacements and forces through a transformation matrix. We then make use of the two-force-member property of a member. We use a matrix relationship similar to the one given by Eq. (2.14). This relationship leads to the stiffness matrix $[\mathbf{K}]^{(e)}$ for an element. However, it is important to realize that the elemental stiffness matrix for a space-truss element is a $6 \times 6$ matrix, rather than the $4 \times 4$ matrix that we obtained for the two-dimensional truss element. For a space-truss member, the elemental stiffness matrix is

$$[\mathbf{K}]^{(e)} = k\begin{bmatrix} \cos^2\theta_X & \cos\theta_X\cos\theta_Y & \cos\theta_X\cos\theta_Z & -\cos^2\theta_X & -\cos\theta_X\cos\theta_Y & -\cos\theta_X\cos\theta_Z \\ \cos\theta_X\cos\theta_Y & \cos^2\theta_Y & \cos\theta_Y\cos\theta_Z & -\cos\theta_X\cos\theta_Y & -\cos^2\theta_Y & -\cos\theta_Y\cos\theta_Z \\ \cos\theta_X\cos\theta_Z & \cos\theta_Y\cos\theta_Z & \cos^2\theta_Z & -\cos\theta_X\cos\theta_Z & -\cos\theta_Y\cos\theta_Z & -\cos^2\theta_Z \\ -\cos^2\theta_X & -\cos\theta_X\cos\theta_Y & -\cos\theta_X\cos\theta_Z & \cos^2\theta_X & \cos\theta_X\cos\theta_Y & \cos\theta_X\cos\theta_Z \\ -\cos\theta_X\cos\theta_Y & -\cos^2\theta_Y & -\cos\theta_Y\cos\theta_Z & \cos\theta_X\cos\theta_Y & \cos^2\theta_Y & \cos\theta_Y\cos\theta_Z \\ -\cos\theta_X\cos\theta_Z & -\cos\theta_Y\cos\theta_Z & -\cos^2\theta_Z & \cos\theta_X\cos\theta_Z & \cos\theta_Y\cos\theta_Z & \cos^2\theta_Z \end{bmatrix} \quad (2.23)$$

The procedure for the assembly of individual elemental matrices for a space-truss member—applying boundary conditions, loads, and solving for displacements—is exactly identical to the one we followed for a two-dimensional truss.

## 2.4 OVERVIEW OF THE ANSYS* PROGRAM

### Entering ANSYS

This section provides a brief overview of the ANSYS program. More detailed information about how you should go about using ANSYS to model a physical Problem is provided in Chapter 6. But for now, enough information will be provided to get you started. The simplest way to enter the ANSYS program is through the ANSYS Launcher, shown in Figure 2.14. The Launcher has a menu containing push buttons that provide the choices you need to run the ANSYS program and other auxiliary programs.

When using the Launcher to enter ANSYS, follow these basic steps:

1. Activate the Launcher by issuing the command **xansys54** at the system prompt if you are running ANSYS on a Unix Platform.
2. Select the **Interactive Set Up** option from the Launcher menu by positioning the cursor of the mouse over it and clicking the left mouse button. This command brings up a dialog box containing interactive entry options.
   a. **Working directory**: This directory is the one in which the ANSYS run will be executed. If the directory displayed is not the one you want to work in, pick the "..." button to the right of the directory name and specify the desired directory.
   b. **Initial jobname**: This jobname is the one that will be used as the prefix of the file name for all files generated by the ANSYS run. Type the desired jobname in this field of the dialog box.
   c. **GUI configuration**: This command brings up a dialog box that allows you to choose the desired menu layout and font size. Do not change the default settings, but simply press **OK** on this dialog box so that the proper X resource

*Materials were adapted with permission from ANSYS documents.

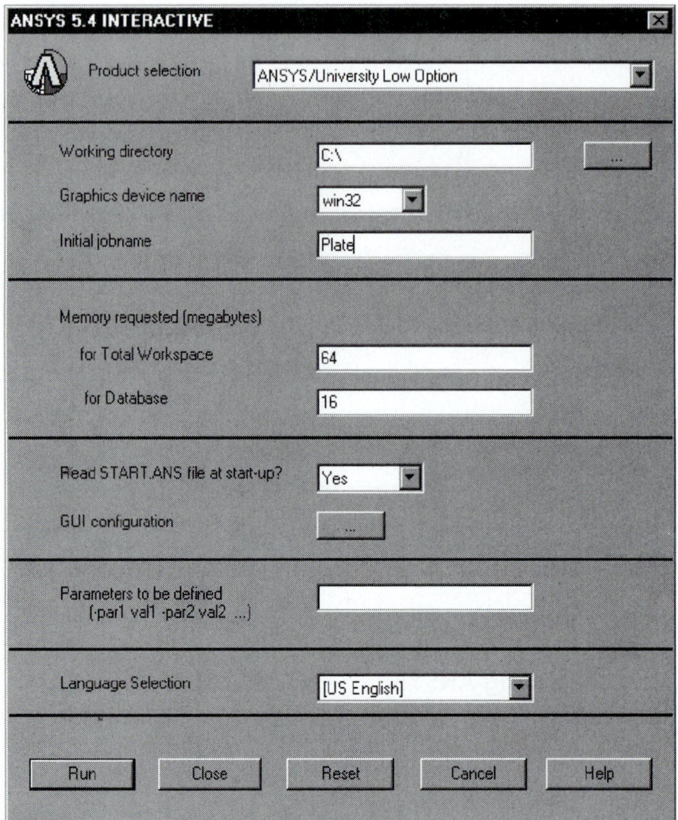


**FIGURE 2.14** The ANSYS Launcher.

file is created for the terminal you are using. This step is only required the first time you enter ANSYS. Then choose the **Interactive** Set Up dialog box.

3. Pick **ANSYS Interactive** from the Launcher menu. This command activates the ANSYS program by bringing up a window entitled ANSYS_Output. Move the mouse cursor into the window and press <Return> or <Enter> to acknowledge that you have read the disclaimer notice. The Graphical User Interface (GUI) will then be activated, and you are ready to begin.

### Program Organization

Before introducing the Graphical User Interface, we will discuss some basic concepts of the ANSYS program. The ANSYS program is organized into two levels: (1) the *Begin level* and (2) the *Processor level*. When you first enter the program, you are at the Begin level. From this level, you can enter the ANSYS processors, as shown in Figure 2.15.

You may have more or fewer processors available to you than the ones shown in Figure 2.15. The actual processors available depend on the particular ANSYS product

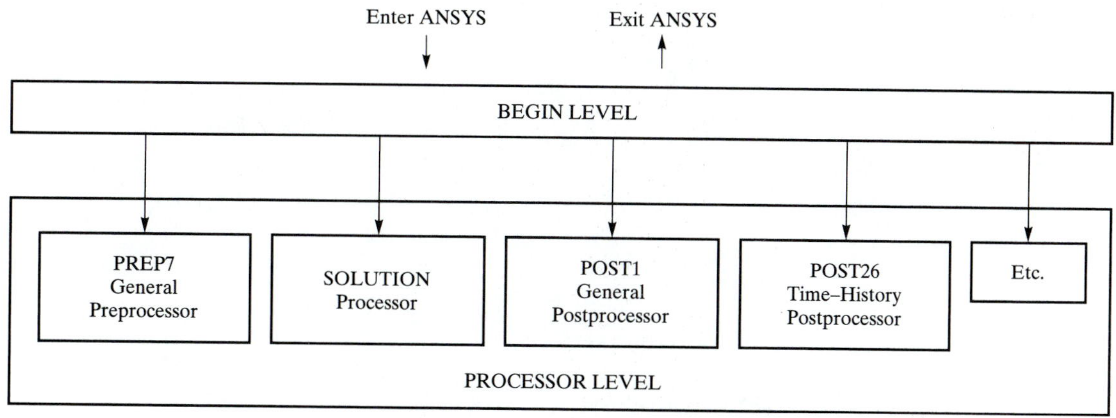


**FIGURE 2.15** The organization of ANSYS.

you have. The Begin level acts as a gateway into and out of the ANSYS program. It is also used to access certain global program controls. At the Processor level, several routines (processors) are available; each accomplishes a specific task. Most of your analysis will be done at the Processor level. A typical analysis in ANSYS involves three distinct steps:

1. *Preprocessing*: Using the **PREP7** processor, you provide data such as the geometry, materials, and element type to the program.
2. *Solution*: Using the **Solution** processor, you define the type of analysis, set boundary conditions, apply loads, and initiate finite element solutions.
3. *Postprocessing*: Using **POST1** (for static or steady state problems) or **POST26** (for transient problems), you review the results of your analysis through graphical displays and tabular listings.

You enter a processor by selecting it from the ANSYS main menu in the Graphical User Interface (GUI). You can move from one processor to another by simply choosing the processor you want from the ANSYS main menu. The next section presents a brief overview of the Graphical User Interface.

## The Graphical User Interface (GUI)

The simplest way to communicate with ANSYS is by using the ANSYS menu system, called the Graphical User Interface (GUI). The GUI provides an interface between you and the ANSYS program. The program is internally driven by ANSYS commands. However, by using the GUI, you can perform an analysis with little or no knowledge of ANSYS commands. This process works because each GUI function ultimately produces one or more ANSYS commands that are automatically executed by the program.

**Layout of the GUI** The ANSYS GUI consist of six main regions, or windows, as shown in Figure 2.16.

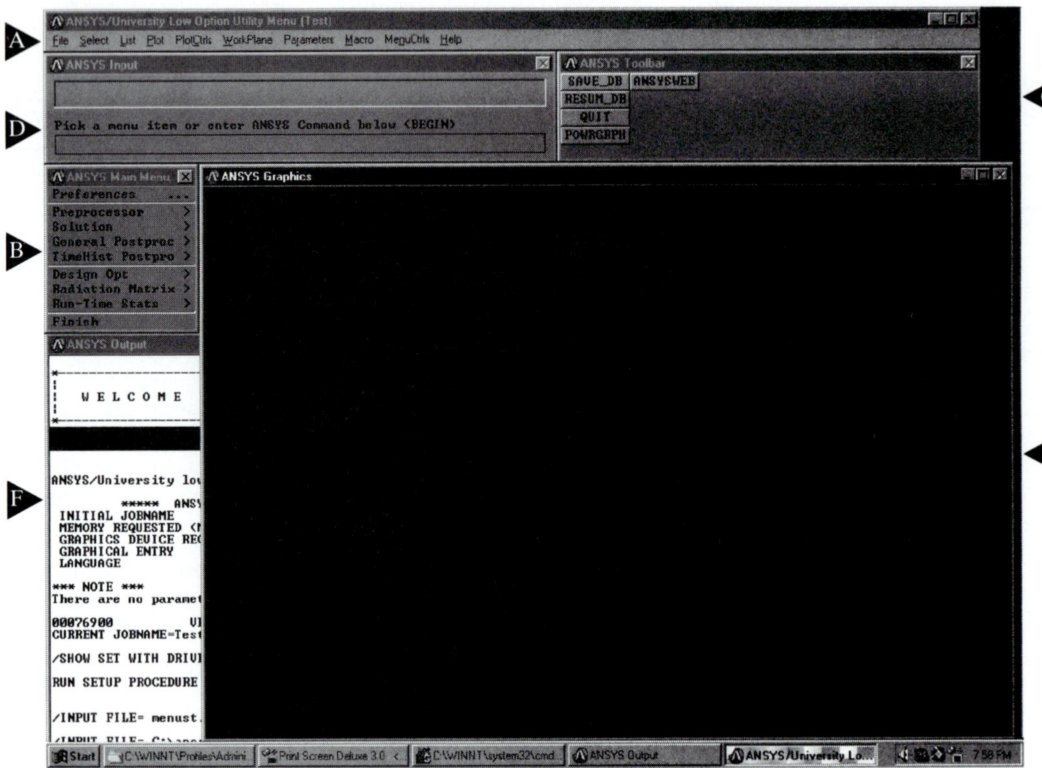


**FIGURE 2.16** The ANSYS GUI.

A **Utility Menu**: Contains utility functions that are available throughout the ANSYS session, such as file controls, selecting, and graphics controls. You will also exit the ANSYS program through this menu.

B **Main Menu**: Contains the primary ANSYS functions, organized by processors. These functions include preprocessor, solution, general postprocessor, design optimizer, etc.

C **Toolbar**: Contains push buttons that execute commonly used ANSYS commands and functions. You may add your own push buttons by defining abbreviations.

D **Input Window**: Shows program prompt messages and allows you to type in commands directly. All previously typed-in commands also appear in this window for easy reference and access.

E **Graphics Window**: A window where graphics displays are drawn.

F **Output Window**: Receives text output from the program. It is usually positioned behind the other windows and can be brought to the front when necessary.

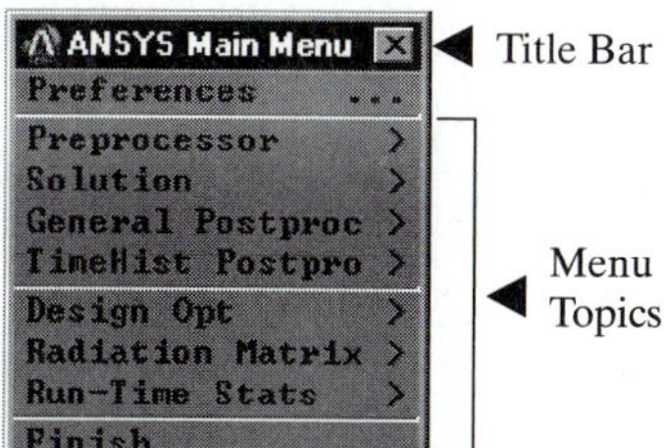


**FIGURE 2.17** The main menu.

The ANSYS main menu and the ANSYS utility menu, both of which you will use most often, are discussed next.

### The Main Menu

The main menu, shown in Figure 2.17, contains main ANSYS functions such as preprocessing, solution, and postprocessing. All functions in the main menu are "modal" with respect to each other; that is, you must complete one function before starting the next one.

Each menu topic on the main menu either brings up a submenu (indicated by a > after the topic) or performs an action. The symbol to the right of the topic indicates the action:

no symbol for immediate execution of the function

... for a dialog box

+ for a picking menu.

The left mouse button is used to select a topic from the main menu. The submenus in the main menu stay in place until you choose a different menu topic higher up in the hierarchy. If a menu topic is obscured by its submenus, you can bring it to the front by clicking anywhere in the title bar or on its border.

### The Utility Menu

The utility menu, shown in Figure 2.18, contains ANSYS utility functions such as file controls, selecting, and graphic controls. Most of these functions are "modeless;" that is, they can be executed at any time during the ANSYS session. The modeless nature of the utility menu greatly enhances the productivity and user friendliness of the GUI.

Each menu topic on the utility menu activates a pull-down menu of subtopics, which in turn will either cascade to a submenu, indicated by a > after the topic, or perform an action. The symbols used to indicate the actions are the same as for the main menu.

Clicking the left mouse button on a menu topic on the utility menu is used to "pull down" the menu topic. Pressing the left mouse button while dragging the cursor of the

**FIGURE 2.18** The utility menu.

mouse allows you to move cursor of the mouse rapidly to the desired subtopic. Releasing the mouse button while the cursor is on an "action" subtopic causes that action to be performed. Clicking the left mouse button leaves the pull-down and cascading menus in place. The menus will disappear when you click on an action subtopic or elsewhere in the GUI.

## Graphical Picking

In order to use the GUI effectively, it is important to understand graphical picking. You can use the mouse to identify model entities and coordinate locations. There are two types of graphical-picking operations: *locational* picking, where you locate the coordinates of a new point, and *retrieval* picking, where you identify existing entities. For example, creating key points by picking their locations on the working plane is a locational-picking operation, whereas picking already-existing key points to apply a load on them is a retrieval-picking operation.

Whenever you use graphical picking, the GUI brings up a picking menu. Figure 2.19 shows the picking menus for locational and retrieval picking.

The features of the picking menu that are used most frequently in upcoming examples are described in detail below.

**Picking Mode**: Allows you to pick or unpick a location or entity. You can use either these toggle buttons or the right mouse button to switch between pick and unpick modes. The mouse pointer is an up arrow for picking and a down arrow

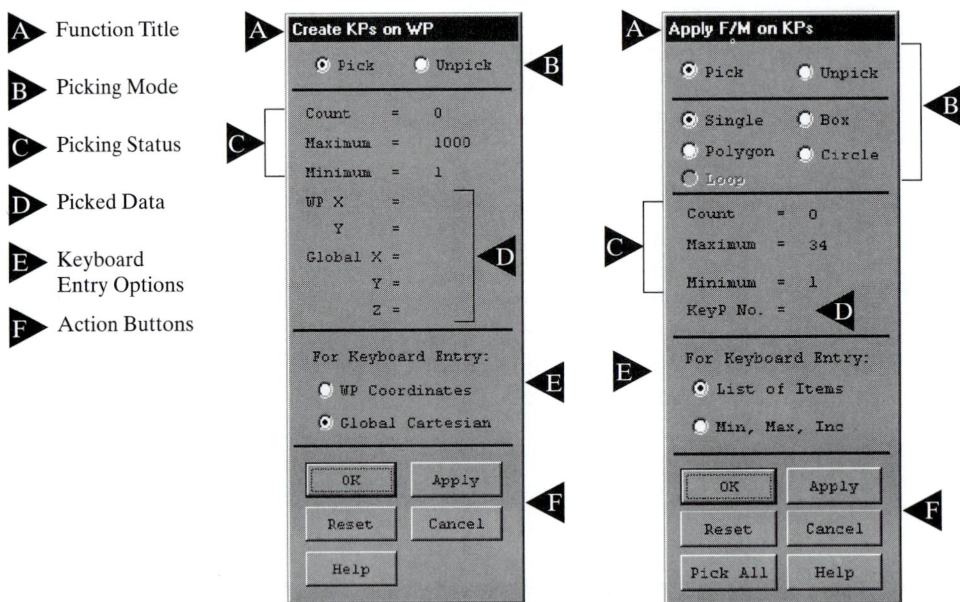


**FIGURE 2.19** Picking menu for locational and retrieval picking.

for unpicking. For retrieval picking, you also have the option to choose from single pick, box, circle, and polygon mode.

**Picked Data**: Shows information about the item being picked. For locational picking, the working plane and global Cartesian coordinates of the point are shown. For retrieval picking, this area shows the entity number. You can see this data by pressing the mouse button and dragging the cursor of the mouse into the graphics area. This procedure allows you to preview the information before releasing the mouse button and picking the item.

**Action Buttons**: This area of the menu contains buttons that take certain actions on the picked entities, as follows:

**OK**: Applies the picked items to execute the function and closes the picking menu.
**Apply**: Applies the picked items to execute the function.
**Reset**: Unpicks all picked entities.
**Cancel**: Cancels the function and closes the picking menu.
**Pick All**: Picks all entities available for retrieval picking only.
**Help**: Brings up help information for the function being performed.

**Mouse-Button Assignments for Picking** A summary of the mouse-button assignments used during a picking operation is given below:

The left button picks or unpicks the entity or location closest to the cursor of the mouse. Pressing the left mouse button and dragging the cursor of the mouse allows you to preview the items being picked or unpicked.

The middle button applies the picked items to execute the function. Its function is the same as that of the Apply button on the picking menu.

The right button toggles between pick and unpick mode. Its function is the same as that of the toggle buttons on the picking menu.

### The Help System

The ANSYS help system gives you information for virtually any component in the Graphical User Interface and any ANSYS command or concept. It can be accessed within the GUI via the help topic on the utility menu or by pressing the help button from within a dialog box. You can access a help topic by choosing from a manual's table of contents or index. Other features of the help system includes hypertext links, word search, and the ability to print out help topics. An in-depth explanation of the capabilities and the organization of the ANSYS program is offered in Chapter 6.

## 2.5 EXAMPLES USING ANSYS

In this section, ANSYS is used to solve truss problems. ANSYS offers two types of elements for the analysis of trusses: **LINK1** and **LINK8**. A two-dimensional spar, called **LINK1**, with two nodes and two degrees of freedom $(U_X, U_Y)$ at each node is commonly used to analyze plane truss problems. Input data must include node locations, cross-sectional area of the member, and modulus of elasticity. If a member is prestressed, then the initial strain should be included in the input data as well. As we learned previously in our discussion on the theory of truss element, we cannot apply surface loads to this element; thus, all loads must be applied directly at the nodes. To analyze space-truss problems, ANSYS offers a three-dimensional spar element. This element, denoted by **LINK8**, offers three degrees of freedom ($U_X$, $U_Y$, $U_Z$) at each node. Required input data is similar to **LINK1** input information. To get additional information about these elements, run the ANSYS on-line help menu.

---

### EXAMPLE 2.1 (revisited)

Consider the balcony truss from Example 2.1, as shown in the accompanying figure. We are interested in determining the deflection of each joint under the loading shown in the figure. All members are made from Douglas-fir wood with a modulus of elasticity of $E = 1.90 \times 10^6$ lb/in$^2$ and a cross-sectional area of 8 in$^2$. We can now analyze this problem using ANSYS.

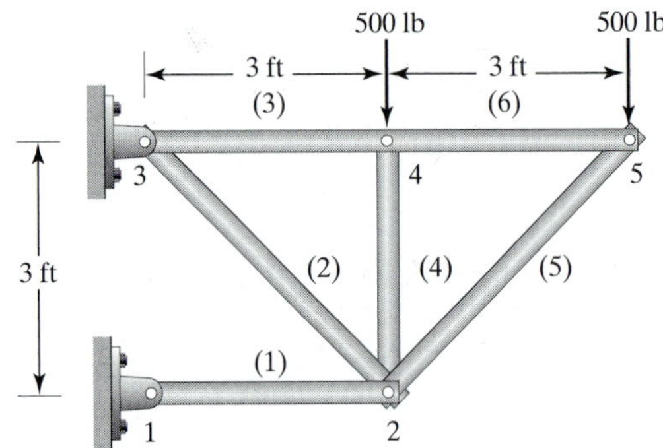


The following steps demonstrate how to create the truss geometry, choose the appropriate element type, apply boundary conditions and loads, and obtain results:

Enter the **ANSYS** program by using the Launcher.

Type **xansys54** on the command line if you are running ANSYS on a UNIX platform, or consult your system administrator for information on how to run ANSYS from your computer system's platform.

Pick **Interactive** from the Launcher menu.

Type **Truss** (or a file name of your choice) in the **Initial Jobname** entry field of the dialog box.

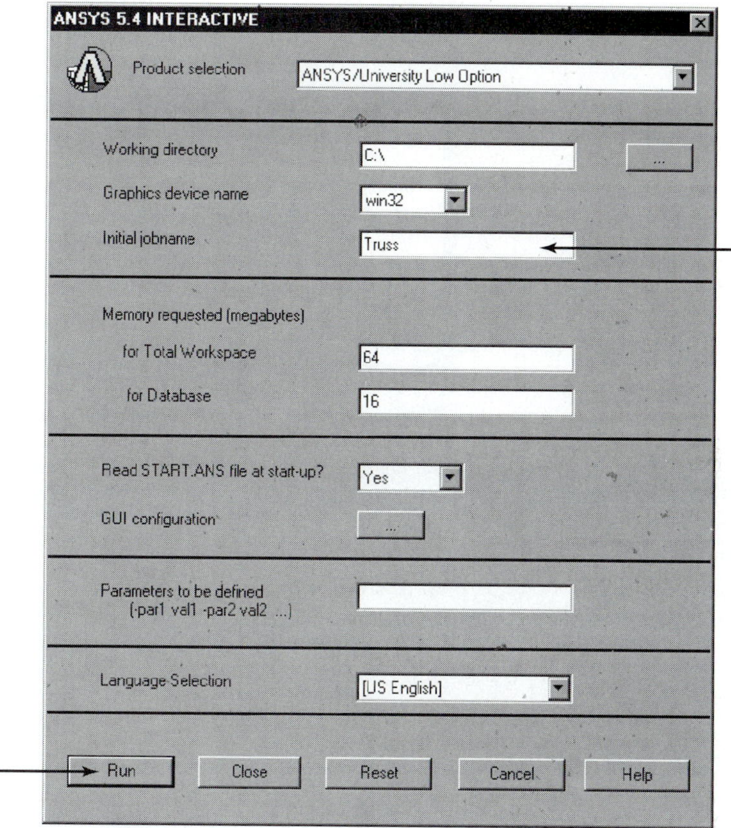


Pick **Run** to start the Graphic User Interface (GUI). A window will open with some disclaimer information. You will eventually be asked to press the <Return> key to start the graphics window and the main menu. Do so in order to proceed.

Create a title for the problem. This title will appear on ANSYS display windows to provide a simple way of identifying the displays. To create a title, issue the command

utility menu: **File** → **Change Title** ...

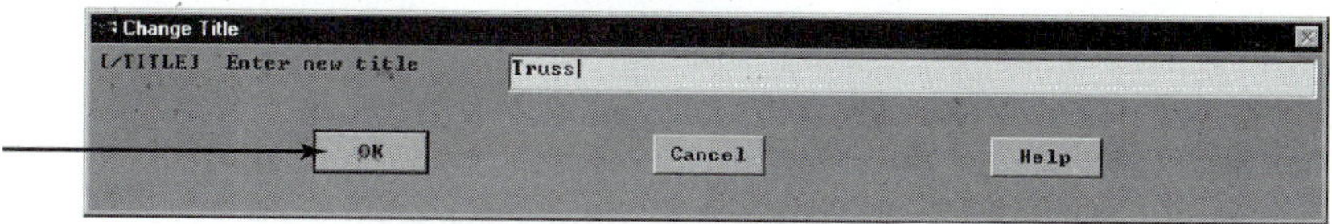


Define the element type and material properties:

main menu: **Preprocessor** → **Element type** → **Add/Edit/Delete** ...

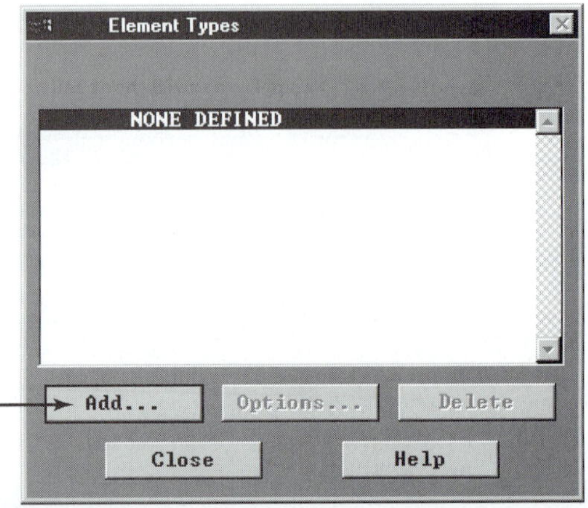


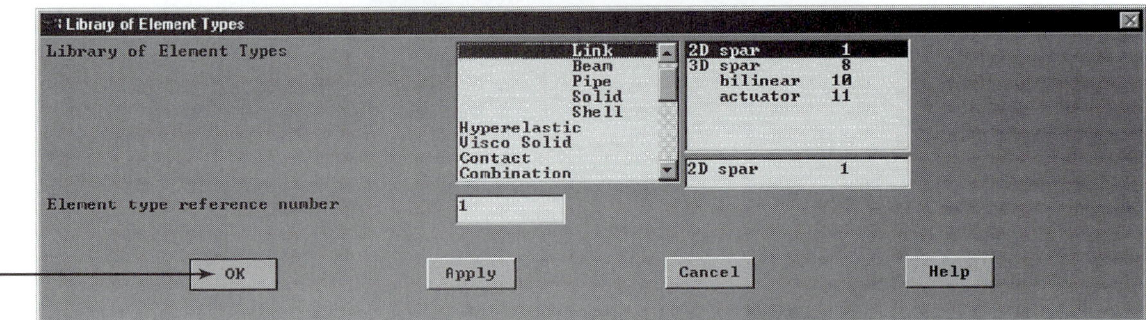


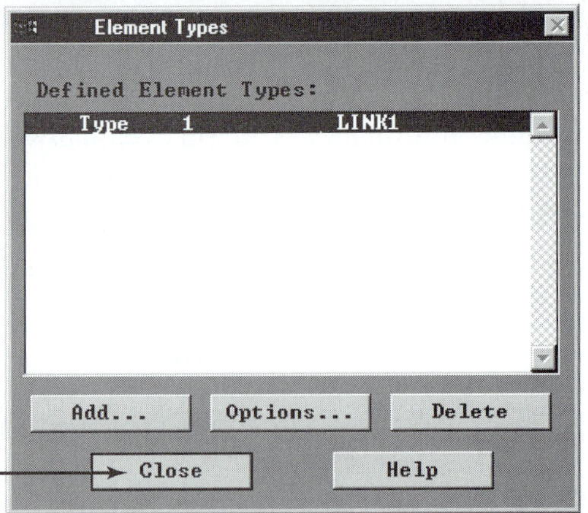


Assign the cross-sectional area of the truss members:

main menu: **Preprocessor** → **Real Constants** ...

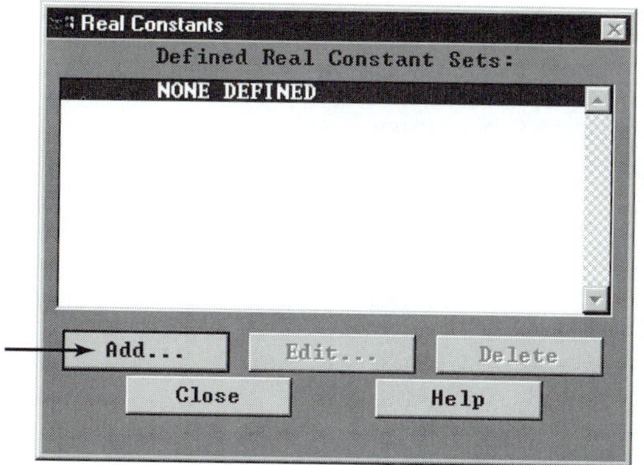


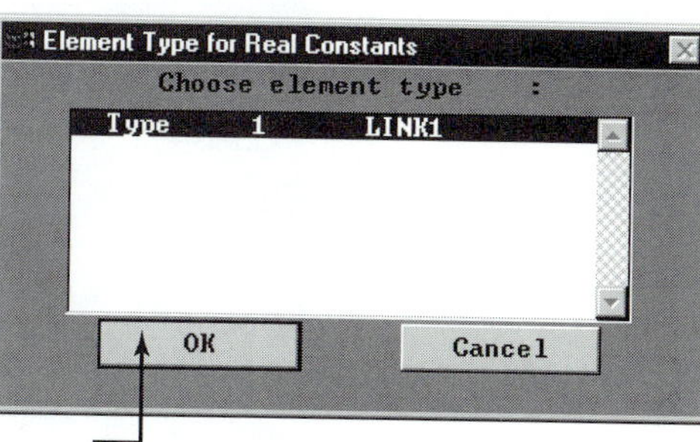


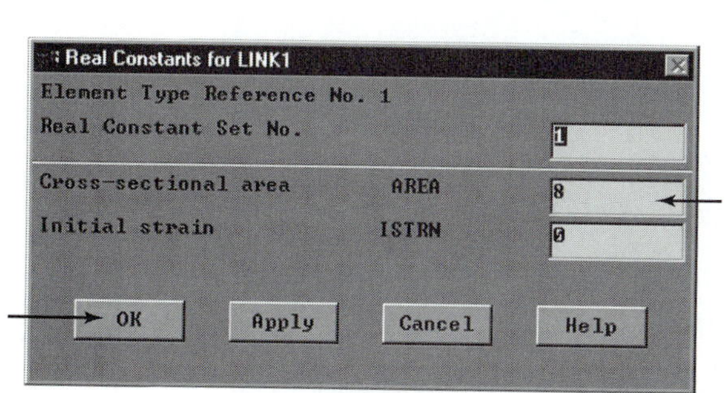


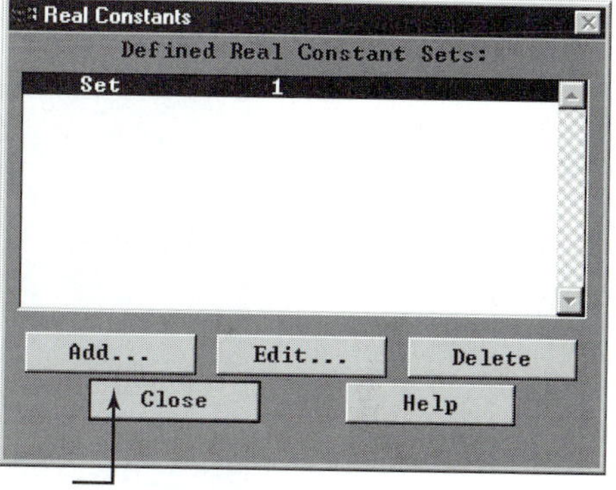


Assign the value of the modulus of elasticity:

main menu: **Preprocessor** $\rightarrow$ **Material Props** $\rightarrow$ **-Constant-Isotropic** ...

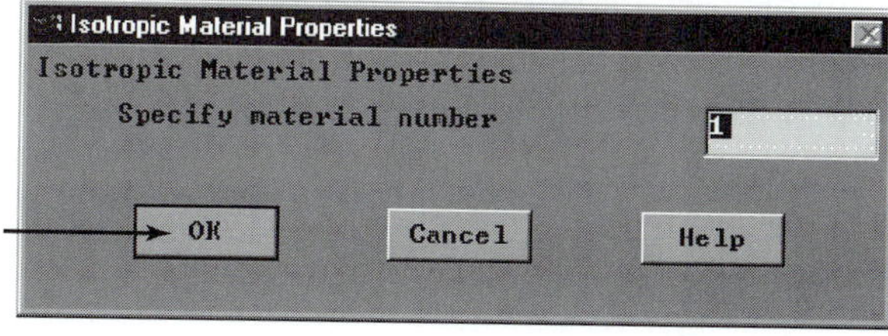

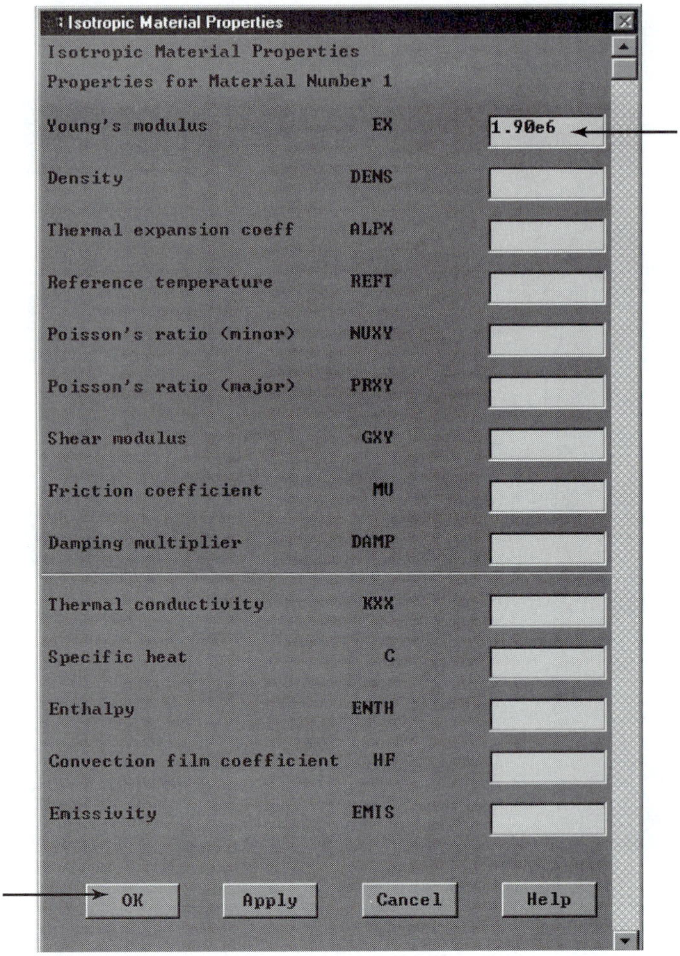

Isotropic Material Properties

Isotropic Material Properties
Properties for Material Number 1

| | | |
|---|---|---|
| Young's modulus | EX | 1.90e6 |
| Density | DENS | |
| Thermal expansion coeff | ALPX | |
| Reference temperature | REFT | |
| Poisson's ratio (minor) | NUXY | |
| Poisson's ratio (major) | PRXY | |
| Shear modulus | GXY | |
| Friction coefficient | MU | |
| Damping multiplier | DAMP | |
| Thermal conductivity | KXX | |
| Specific heat | C | |
| Enthalpy | ENTH | |
| Convection film coefficient | HF | |
| Emissivity | EMIS | |

OK Apply Cancel Help

Save the input data:

ANSYS Toolbar: **SAVE_DB**

Set up the graphics area (i.e., workplane, zoom, etc.):

utility menu: **Workplane** → **WP Settings** ...

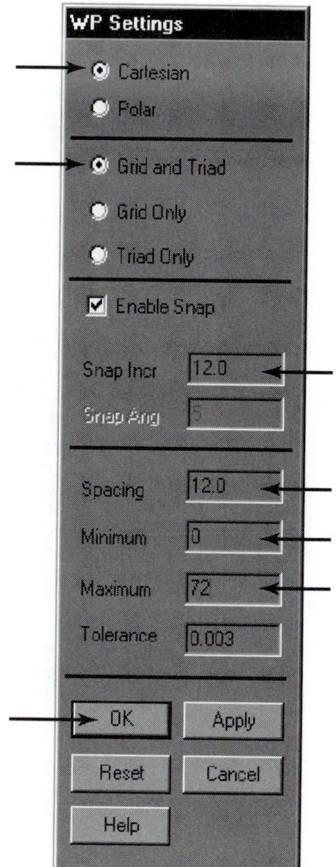


Toggle on the workplane by the following sequence:

utility menu: **Workplane** → **Display working plane**

Bring the workplane to view using the following sequence:

utility menu: **PlotCtrls** → **Pan, Zoom, Rotate** ...

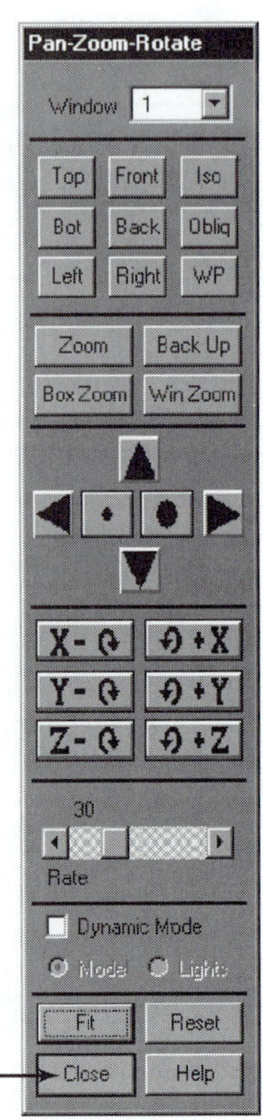


Click on the **small circle** until you bring the workplane to view. You can also use the **arrow** buttons to move the workplane in a desired direction. Then, create nodes by picking points on the workplane:

main menu: **Preprocessor** → **-Modeling-Create** → **Nodes** → **On Working Plane** +

On the workplane, pick the location of joints (nodes) and apply them:

**[WP** = 0,0]

**[WP** = 36,0]

**[WP** = 0,36]

**[WP** = 36,36]

**[WP** = 72,36]

**OK**

You may want to turn off the workplane now and turn on node numbering instead:

utility menu: **Workplane** → **Display Working Plane**

utility menu: **PlotCtrls** → **Numbering** ...

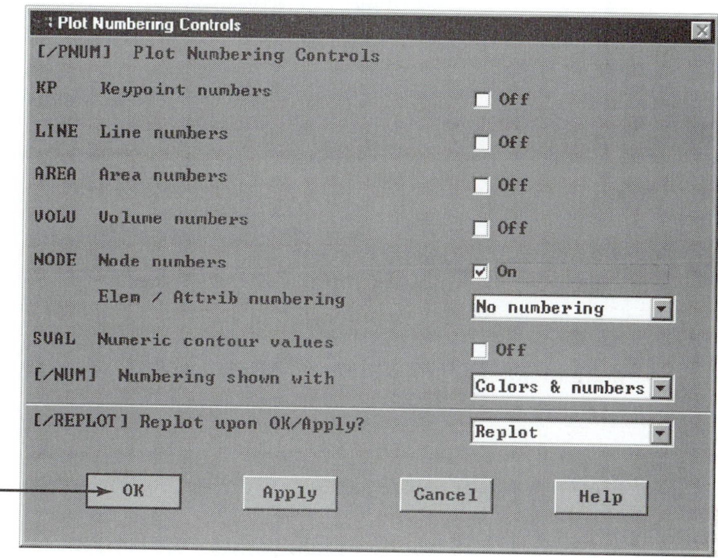


You may want to list nodes at this point in order to check your work:

utility menu: **List** → **nodes** ...

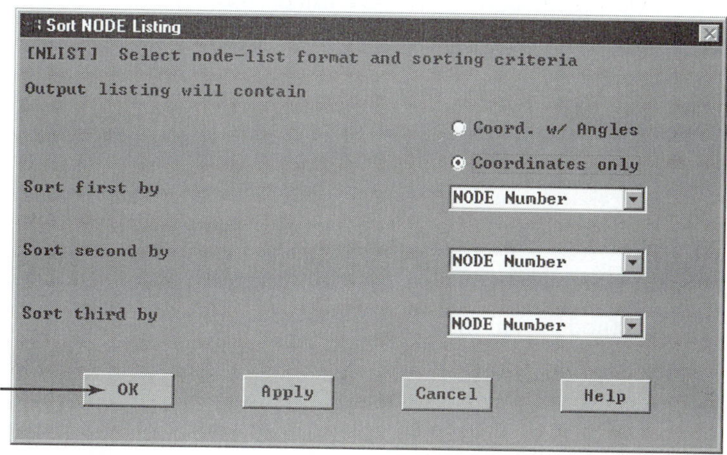

NLIST Command

File

```
LIST ALL SELECTED NODES.   DSYS=  0
SORT TABLE ON  NODE  NODE  NODE

  NODE        X                   Y                   Z
      1     .000000000000       .000000000000       .000000000000
      2     36.0000000000       .000000000000       .000000000000
      3     .000000000000       36.0000000000       .000000000000
      4     36.0000000000       36.0000000000       .000000000000
      5     72.0000000000       36.0000000000       .000000000000
```

**Close**

ANSYS Toolbar: **SAVE_DB**

Define elements by picking nodes:

main menu: **Preprocessor** → **-Modeling-Create** → **Elements** → **-Auto Numbered-Thru Nodes** +

[node 1 and then node 2]

[Use the middle button anywhere in the ANSYS graphics window to apply.]

[node 2 and then node 3]

[anywhere in the ANSYS graphics window]

[node 3 and then node 4]

[anywhere in the ANSYS graphics window]

[node 2 and then node 4]

[anywhere in the ANSYS graphics window]

[node 2 and then node 5]

[anywhere in the ANSYS graphics window]

[node 4 and then node 5]

[anywhere in the ANSYS graphics window]

**OK**

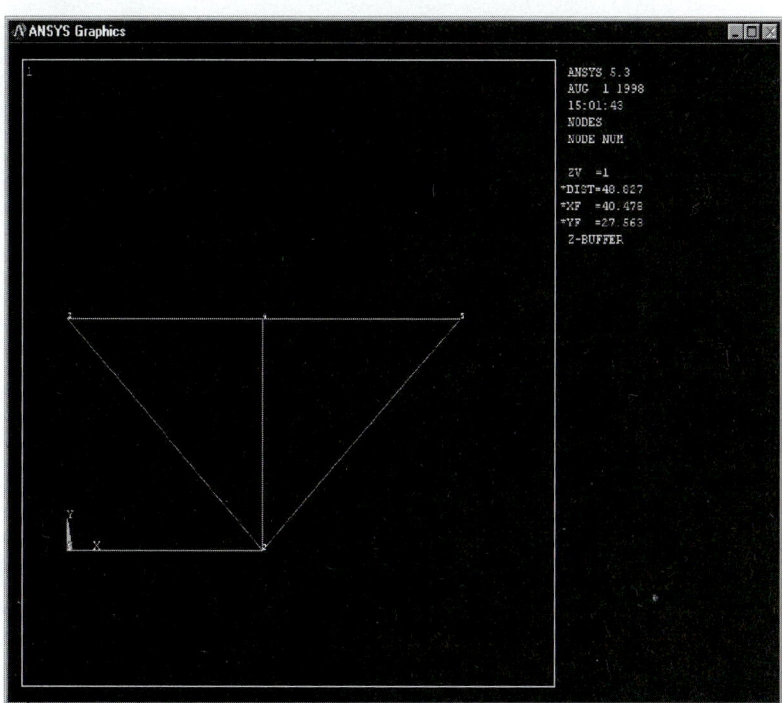


ANSYS Toolbar: **SAVE_DB**

Apply boundary conditions and loads:

main menu: **Solution** → **-Loads-Apply** → **-Structural-Displacement** → **On Nodes** +

[node 1]

[node 3]

[anywhere in the ANSYS graphics window]

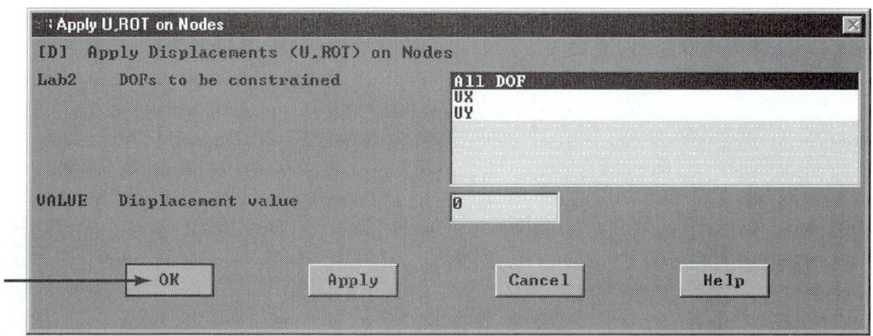


main menu: **Solutions** → **-Loads-Apply** → **-Structural-Force/Moment** → **On Nodes** +

[node 4]

[node 5]

[anywhere in the ANSYS graphics window]

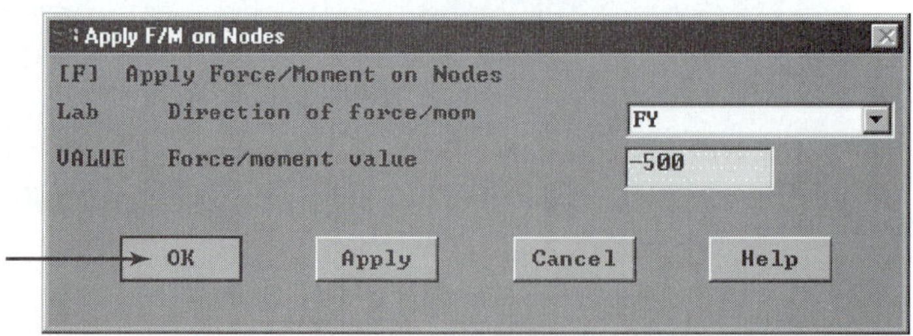


ANSYS Toolbar: **SAVE_DB**

Solve the problem:

main menu: **Solution** → **-Solve-Current LS**

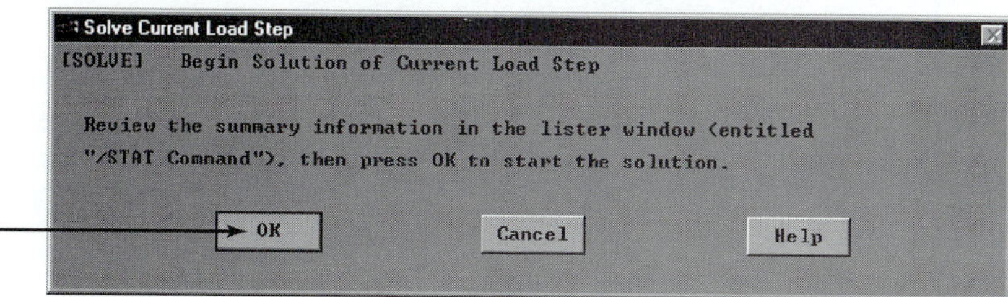


**Close** (the solution is done!) window.

**Close** (the /STAT Command) window.

For the postprocessing phase, first plot the deformed shape:

main menu: **General Postproc** → **Plot Results** → **Deformed Shape** ...

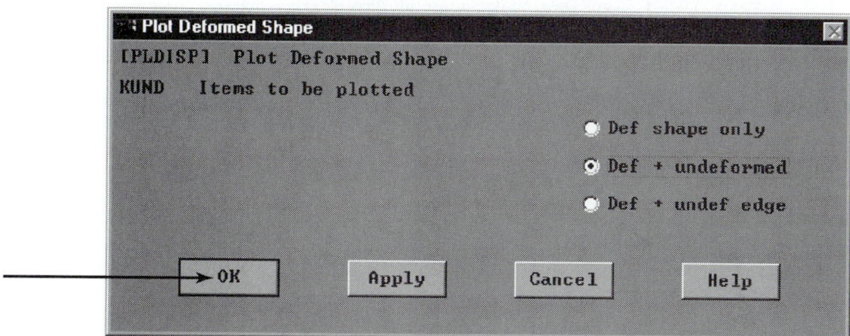


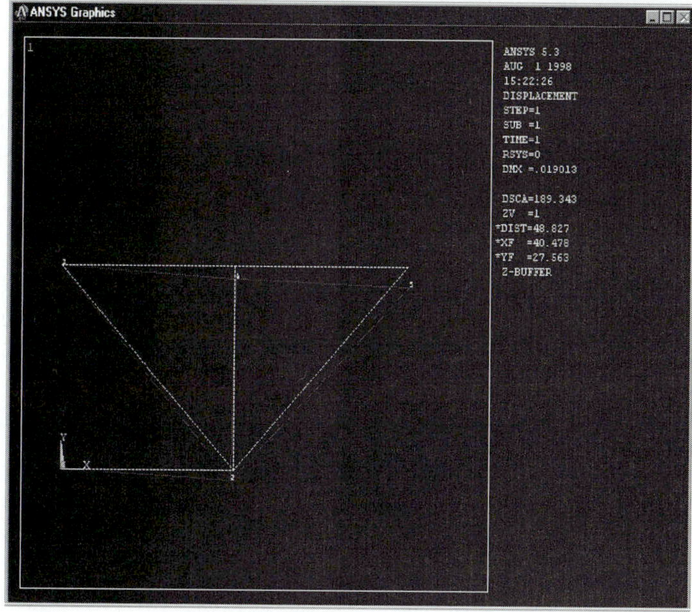

**General Postproc → List Results → Nodal Solutions ...**

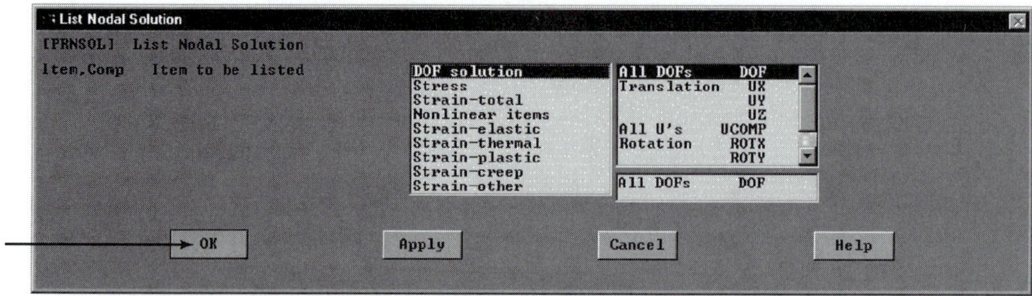


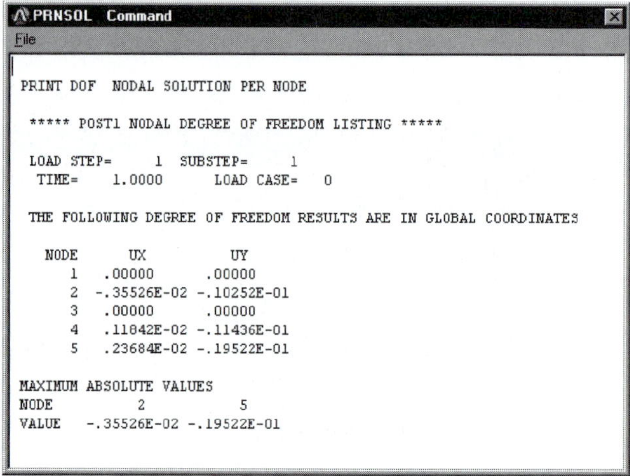

```
PRNSOL Command
File

PRINT DOF  NODAL SOLUTION PER NODE

 ***** POST1 NODAL DEGREE OF FREEDOM LISTING *****

 LOAD STEP=     1  SUBSTEP=     1
  TIME=    1.0000      LOAD CASE=   0

 THE FOLLOWING DEGREE OF FREEDOM RESULTS ARE IN GLOBAL COORDINATES

   NODE      UX          UY
      1   .00000      .00000
      2  -.35526E-02 -.10252E-01
      3   .00000      .00000
      4   .11842E-02 -.11436E-01
      5   .23684E-02 -.19522E-01

MAXIMUM ABSOLUTE VALUES
NODE          2           5
VALUE   -.35526E-02 -.19522E-01
```

To review other results, such as axial forces and axial stresses, we must copy these results into element tables. These items are obtained using *item label* and *sequence numbers*, as given in the Table 4.1–4.3 section of the ANSYS elements manual. For truss elements, the values of internal forces and stresses, which ANSYS computes from the nodal displacement results, may be looked up and assigned to user defined lables. For Example 2.1, we have assigned the internal force, as computed by ANSYS, in each member to a user defined lable "Axforce." However, note that ANSYS allows up to eight characters to define such lables. Similarly, the axial stress result for each member is assigned to the lable "Axstress." We now run the following sequence:

main menu: **General Postproc → Element Table → Define Table ...**

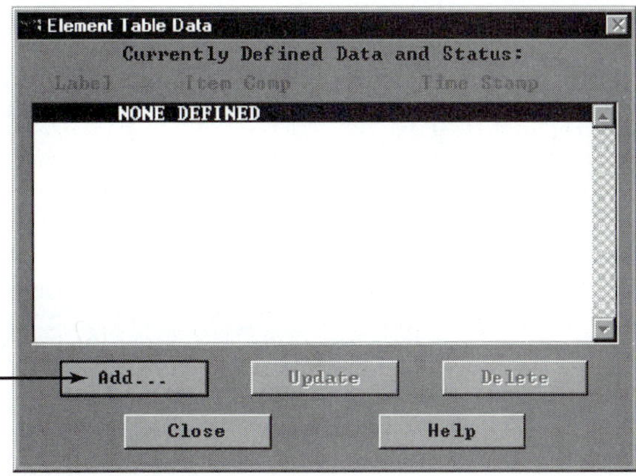
Element Table Data
Currently Defined Data and Status:
Label Item Comp Time Stamp
NONE DEFINED
Add...
Update
Delete
Close
Help

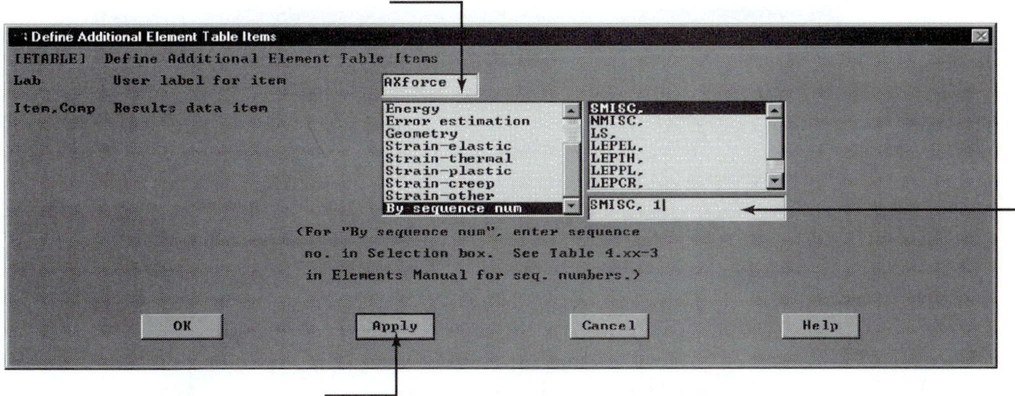
Define Additional Element Table Items
[ETABLE] Define Additional Element Table Items
Lab User label for item
AXforce
Item,Comp Results data item
Energy
Error estimation
Geometry
Strain-elastic
Strain-thermal
Strain-plastic
Strain-creep
Strain-other
By sequence num
SMISC,
NMISC,
LS,
LEPEL,
LEPTH,
LEPPL,
LEPCR,
SMISC, 1
(For "By sequence num", enter sequence
no. in Selection box. See Table 4.xx-3
in Elements Manual for seq. numbers.)
OK
Apply
Cancel
Help

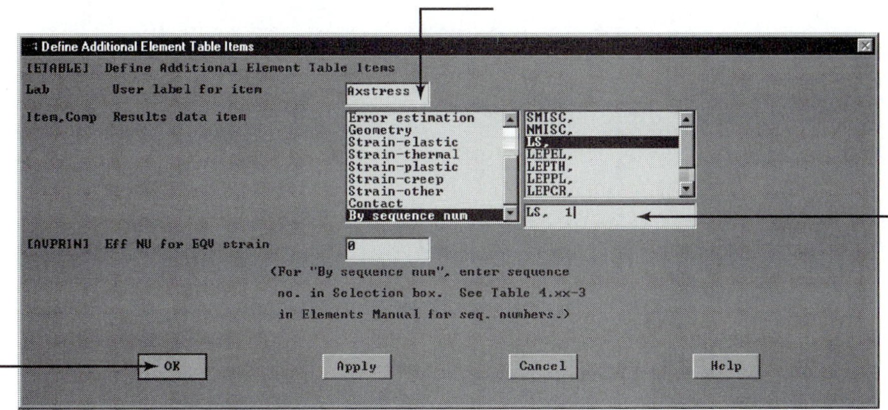
Define Additional Element Table Items
[ETABLE] Define Additional Element Table Items
Lab User label for item
Axstress
Item,Comp Results data item
Error estimation
Geometry
Strain-elastic
Strain-thermal
Strain-plastic
Strain-creep
Strain-other
Contact
By sequence num
SMISC,
NMISC,
LS,
LEPEL,
LEPTH,
LEPPL,
LEPCR,
LS, 1
[AVPRIN] Eff NU for EQV strain
0
(For "By sequence num", enter sequence
no. in Selection box. See Table 4.xx-3
in Elements Manual for seq. numbers.)
OK
Apply
Cancel
Help

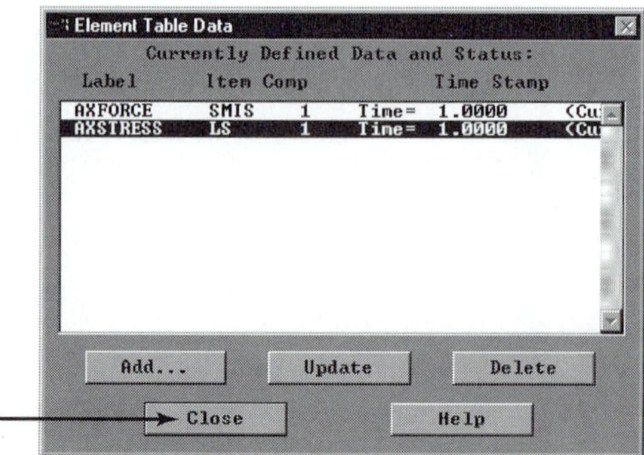


main menu: **General Postproc** → **Element Table** → **Plot Element Table** ...

or

main menu: **General Postproc** → **Element Table** → **List Element Table** ...

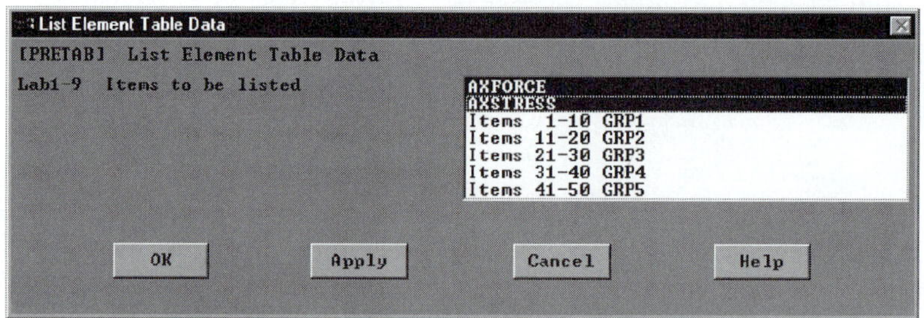


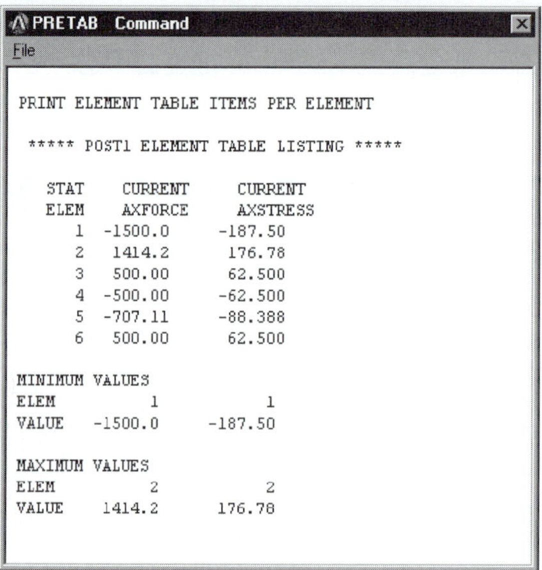


```
PRINT ELEMENT TABLE ITEMS PER ELEMENT

 ***** POST1 ELEMENT TABLE LISTING *****

   STAT     CURRENT      CURRENT
   ELEM     AXFORCE      AXSTRESS
      1   -1500.0      -187.50
      2    1414.2       176.78
      3    500.00       62.500
      4   -500.00      -62.500
      5   -707.11      -88.388
      6    500.00       62.500

MINIMUM VALUES
ELEM          1            1
VALUE   -1500.0      -187.50

MAXIMUM VALUES
ELEM          2            2
VALUE    1414.2       176.78
```

**Close**

List reaction solutions:

main menu: **General Postproc** → **List Results** → **Reaction Solu** ...

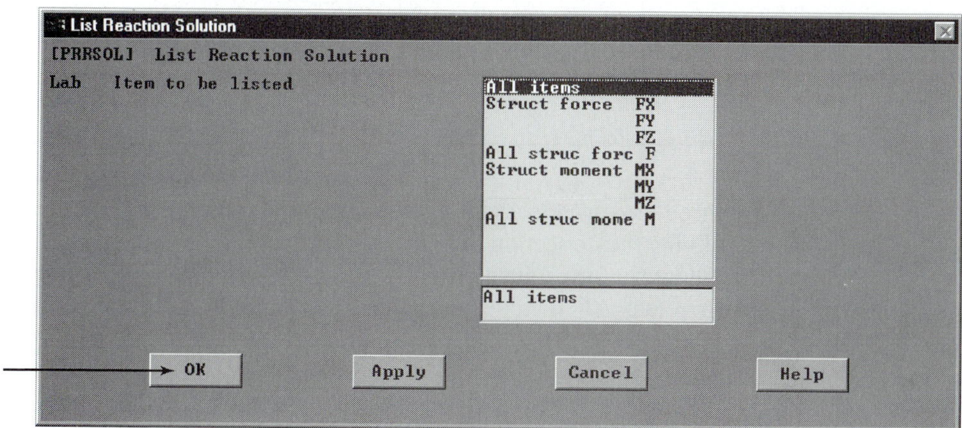


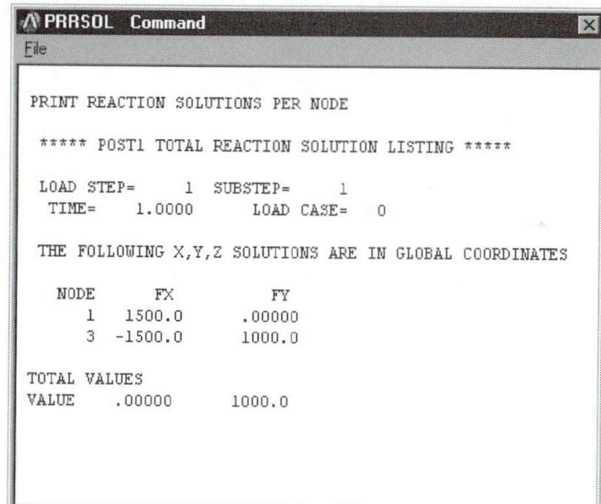
PRRSOL Command

File

```
PRINT REACTION SOLUTIONS PER NODE

 ***** POST1 TOTAL REACTION SOLUTION LISTING *****

 LOAD STEP=     1  SUBSTEP=     1
  TIME=    1.0000      LOAD CASE=   0

 THE FOLLOWING X,Y,Z SOLUTIONS ARE IN GLOBAL COORDINATES

   NODE       FX          FY
      1   1500.0      .00000
      3  -1500.0      1000.0

TOTAL VALUES
VALUE    .00000      1000.0
```

Exit ANSYS and save everything, including element tables and reaction forces:

ANSYS Toolbar: **QUIT**

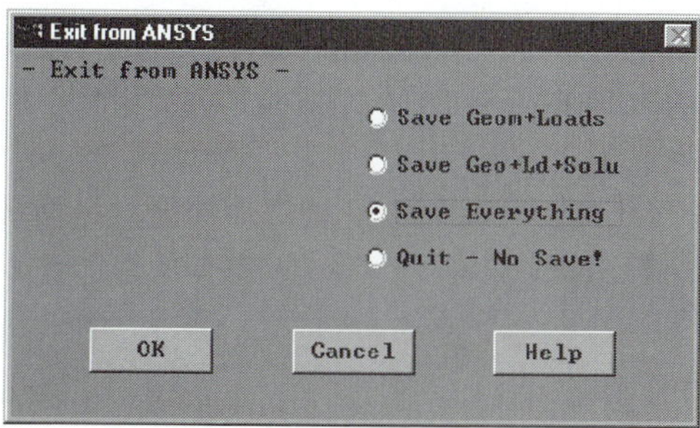

If, for any reason, you need to modify a model, first launch ANSYS and then type the **file name** of the model in the **Initial Jobname** entry field of the Interactive dialog box. Then press **Run**. From the File menu, choose **Resume Jobname.DB**. Now you have complete access to your model. You can plot nodes, elements, and so on to make certain that you have chosen the right problem.

---

## EXAMPLE 2.2

Consider the three-dimensional truss shown in the accompanying figure. We are interested in determining the deflection of joint 2 under the loading shown in the figure. The Cartesian coordinates of the joints with respect to the coordinate system shown in the figure are given in feet. All members are made from aluminum with a modulus of elasticity of $E = 10.6 \times 10^6$ lb/in$^2$ and a cross-sectional area of 1.56 in$^2$.

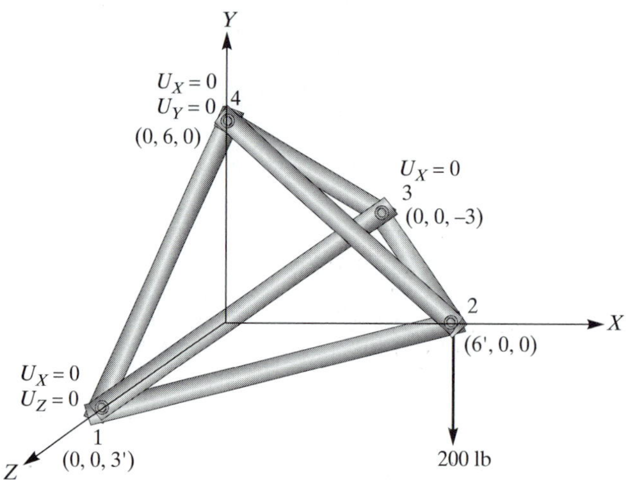


To solve this problem using ANSYS, we employ the following steps:

Enter the **ANSYS** program by using the Launcher.

Type **xansys54** on the command line if you are running ANSYS on a UNIX platform, or consult your system administrator for information on how to run ANSYS from your computer system's platform.

Pick **Interactive** from the Launcher menu.

Type **Truss3D** (or a file name of your choice) in the **Initial Jobname** entry field of the dialog box.

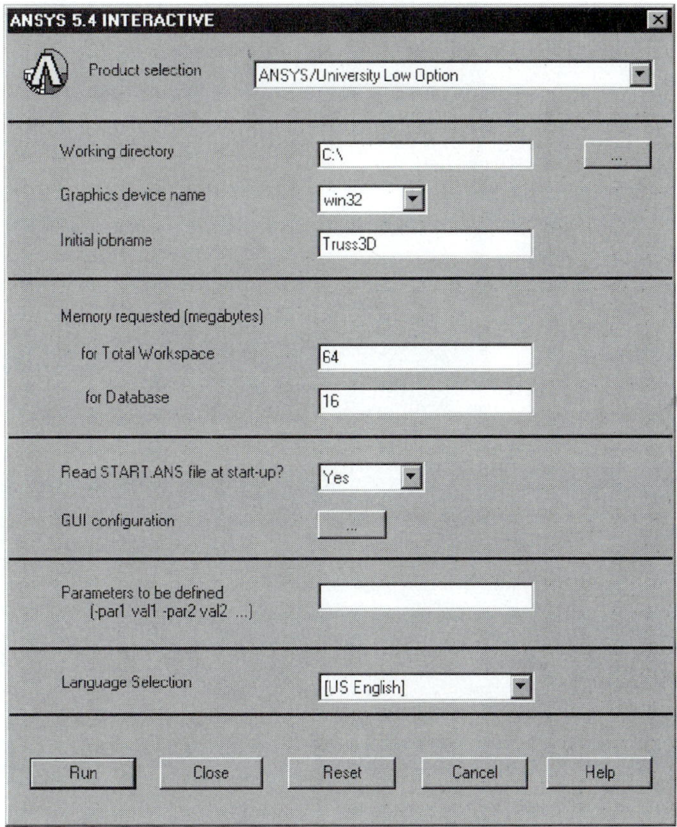


Pick **Run** to start the Graphic User Interface (GUI). A window will open with some disclaimer information. You will eventually be asked to press the <Return> key to start the graphics window and the main menu. Do so in order to proceed.

Create a title for the problem. This title will appear on ANSYS display windows to provide a simple way of identifying the displays:

utility menu: **File** → **Change Title** ...

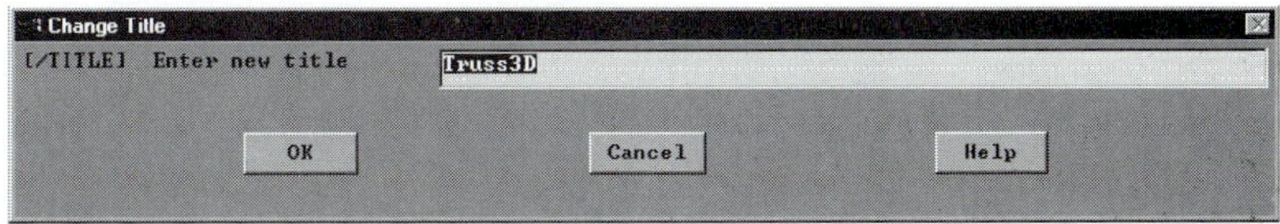


Define the element type and material properties:

main menu: **Preprocessor** → **Element Type** → **Add/Edit/Delete** ...

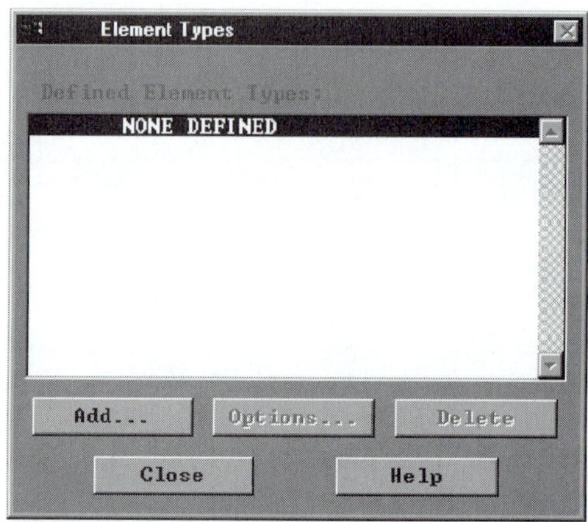


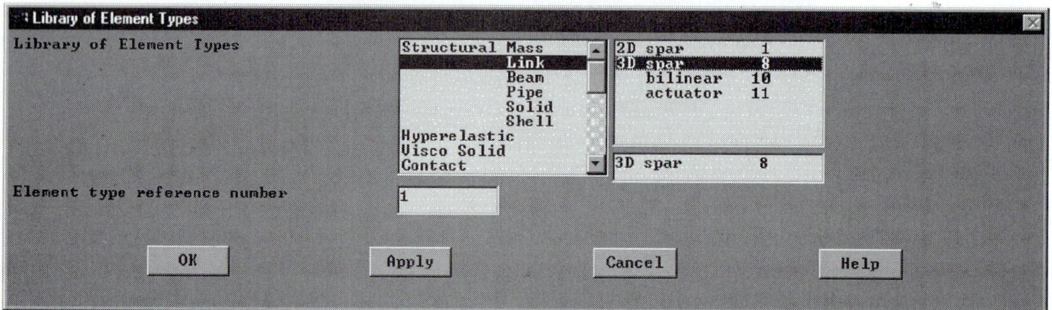


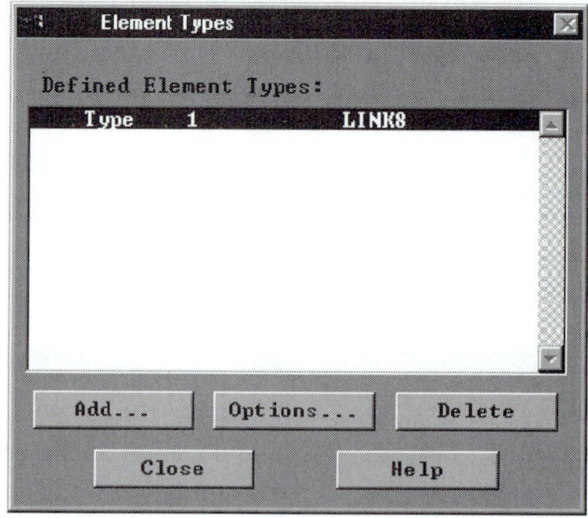


Assign the cross-sectional area of the truss members:

main menu: **Preprocessor** → **Real Constant** ...

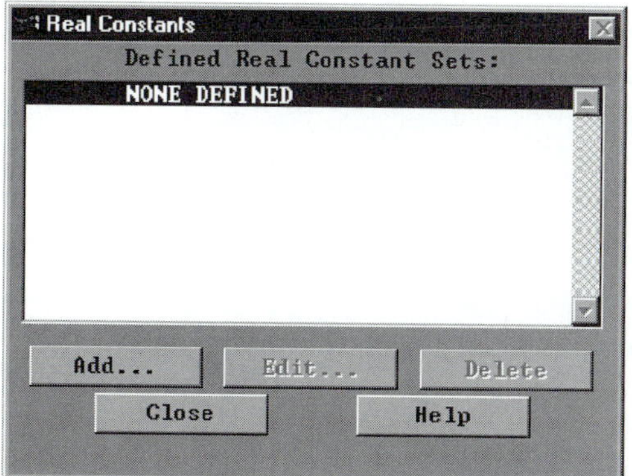


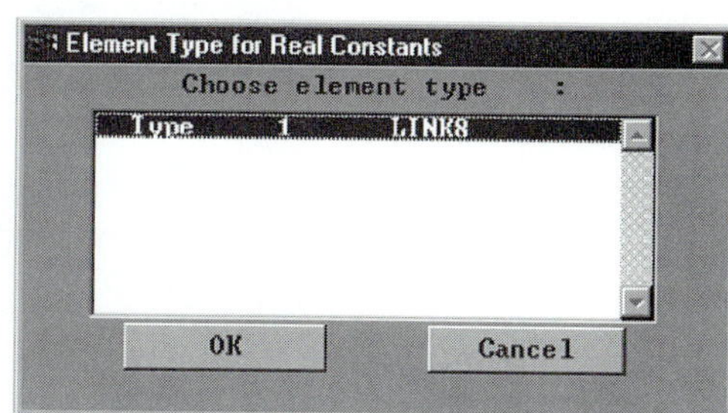


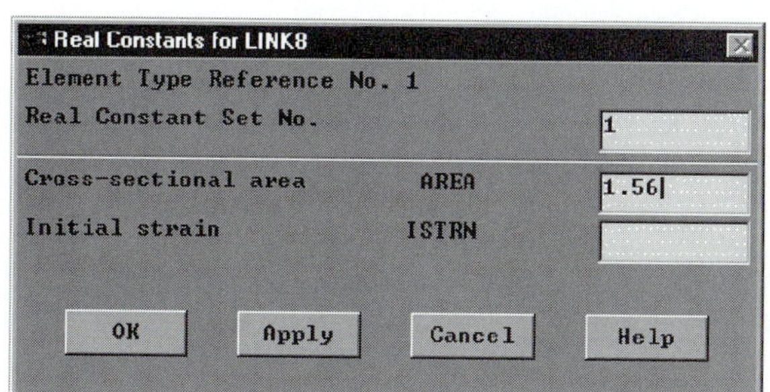


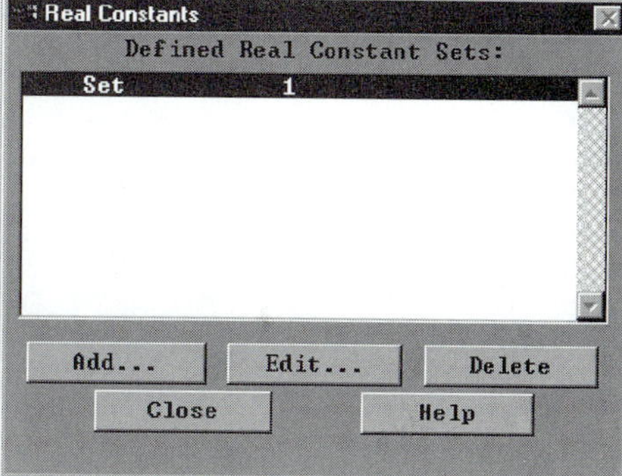


Assign the value of the modulus of elasticity:

main menu: **Preprocessor** → **Material Props** → **-Constant-Isotropic** ...

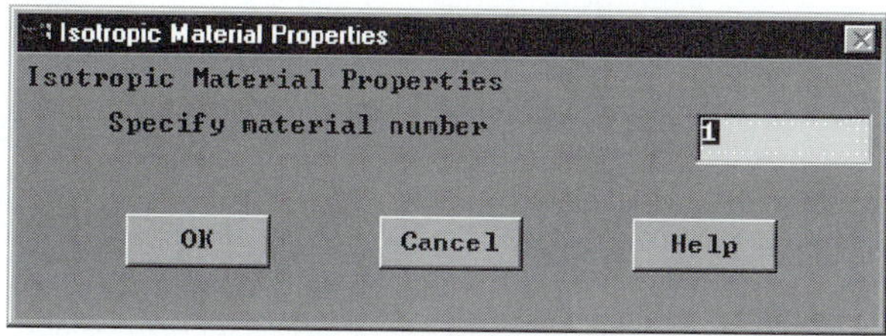

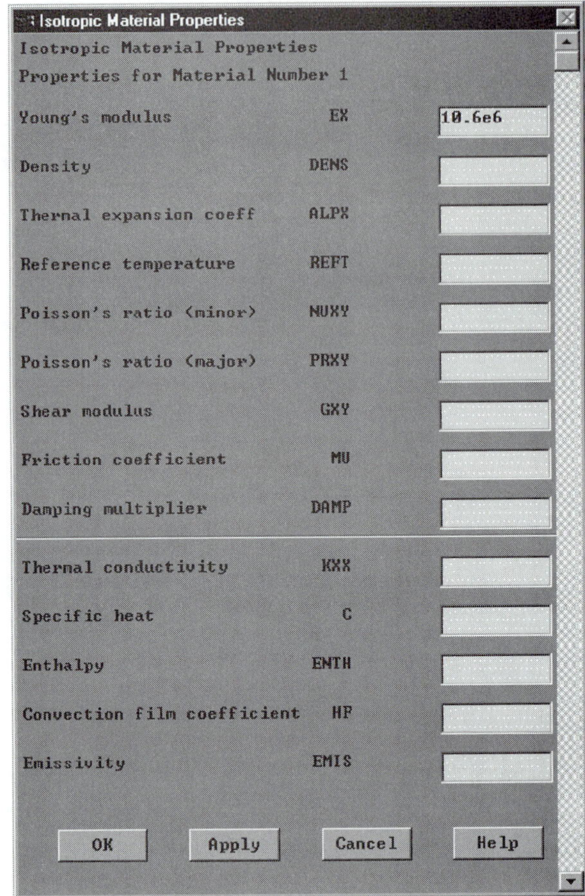


ANSYS Toolbar: **SAVE_DB**

Create nodes in active coordinate system:

main menu: **Preprocessor** → **-Modeling-Create** → **Nodes** → **In Active Cs** ...

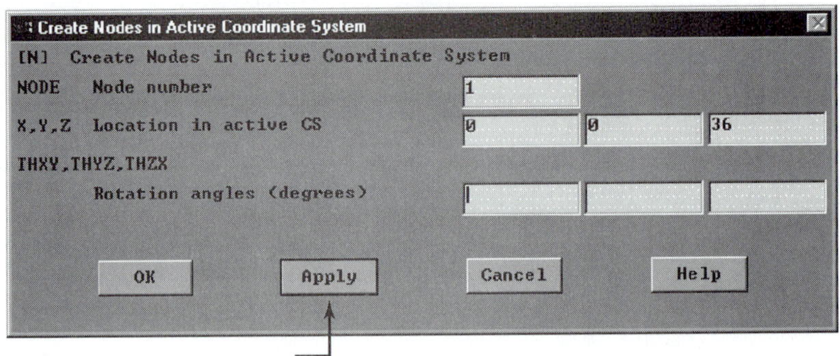

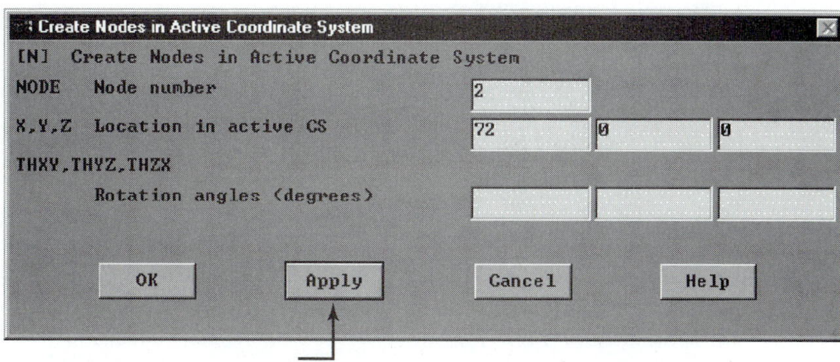


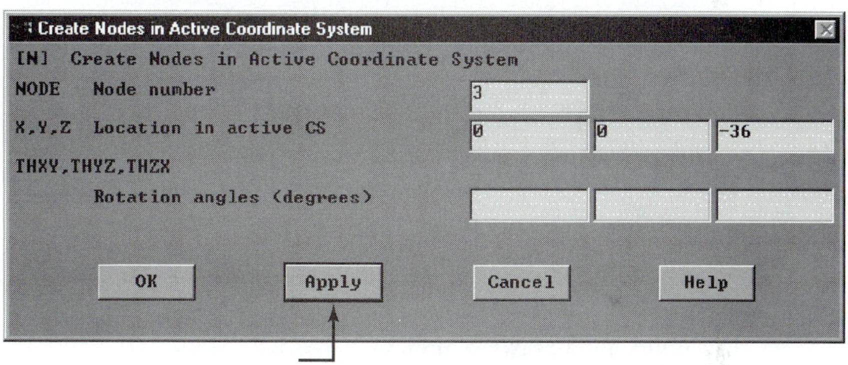


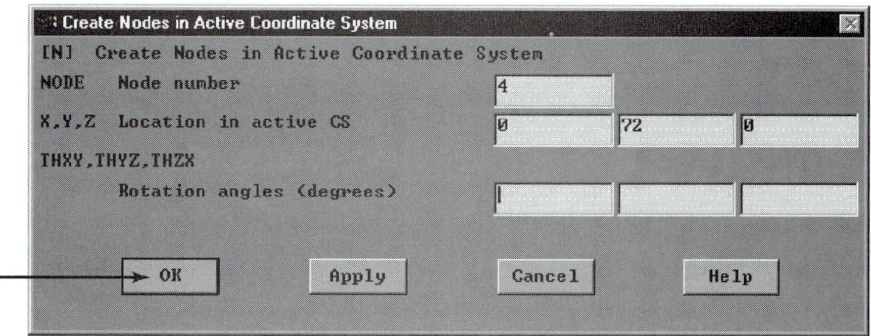


You may want to turn on node numbering:

utility menu: **PlotCtrls** → **Numbering** …

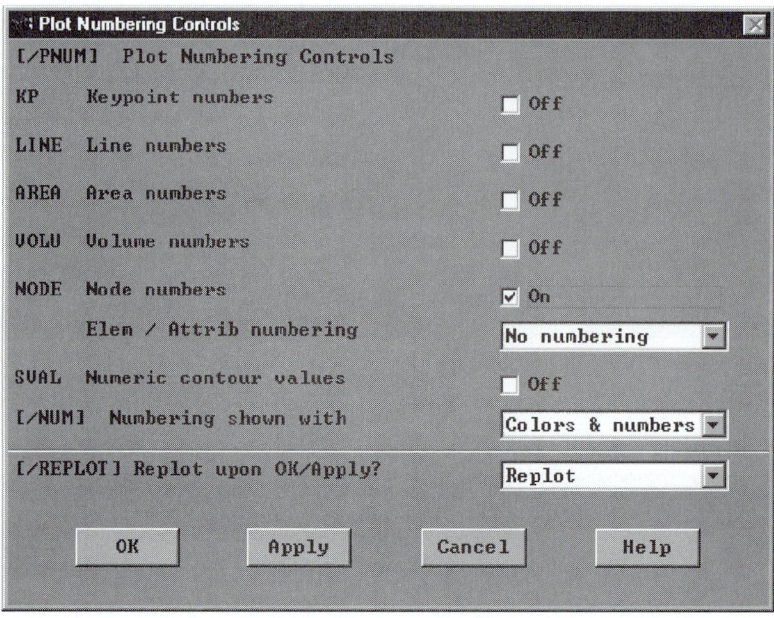


You may want to list nodes at this point in order to check your work:

utility menu: **List** → **nodes** ...

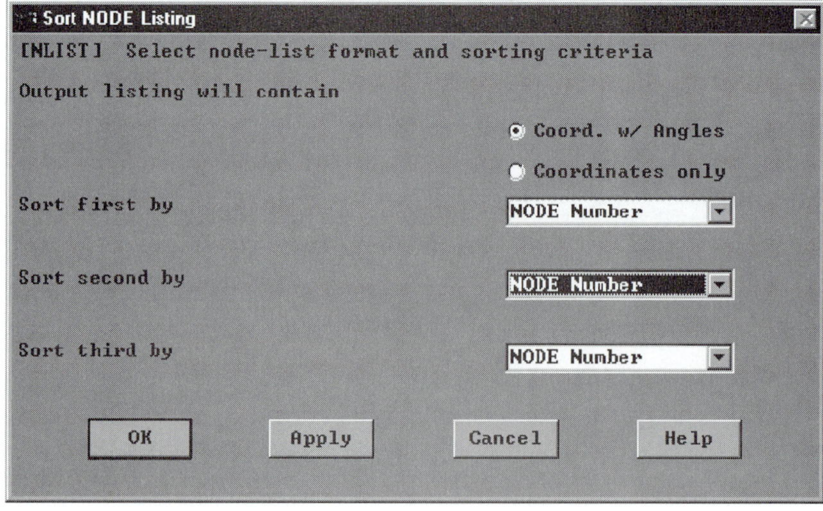

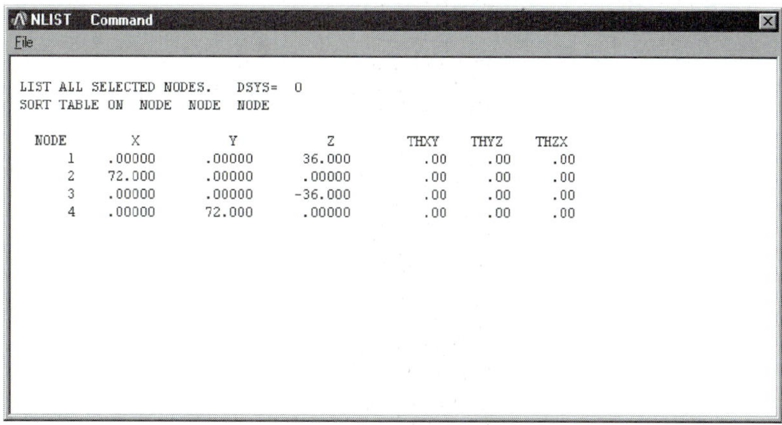
NLIST Command

File

LIST ALL SELECTED NODES. DSYS= 0
SORT TABLE ON NODE NODE NODE

| NODE | X | Y | Z | THXY | THYZ | THZX |
|---|---|---|---|---|---|---|
| 1 | .00000 | .00000 | 36.000 | .00 | .00 | .00 |
| 2 | 72.000 | .00000 | .00000 | .00 | .00 | .00 |
| 3 | .00000 | .00000 | -36.000 | .00 | .00 | .00 |
| 4 | .00000 | 72.000 | .00000 | .00 | .00 | .00 |

**Close**

ANSYS Toolbar: **SAVE_DB**

Define elements by picking nodes. But first set the view angle:

utility menu: **PlotCtrls** → **Pan, Zoom, Rotate** ...

Select the oblique or isometric viewing.

main menu: **Preprocessor** → **-Modeling-Create** → **Elements** → **-Auto Numbered-Thru Nodes** +

[node 1 and then node 2]

[Use the middle button anywhere in the ANSYS graphics window to apply.]

[node 1 and then node 3]

[anywhere in the ANSYS graphics window]

[node 1 and then node 4]

[anywhere in the ANSYS graphics window]

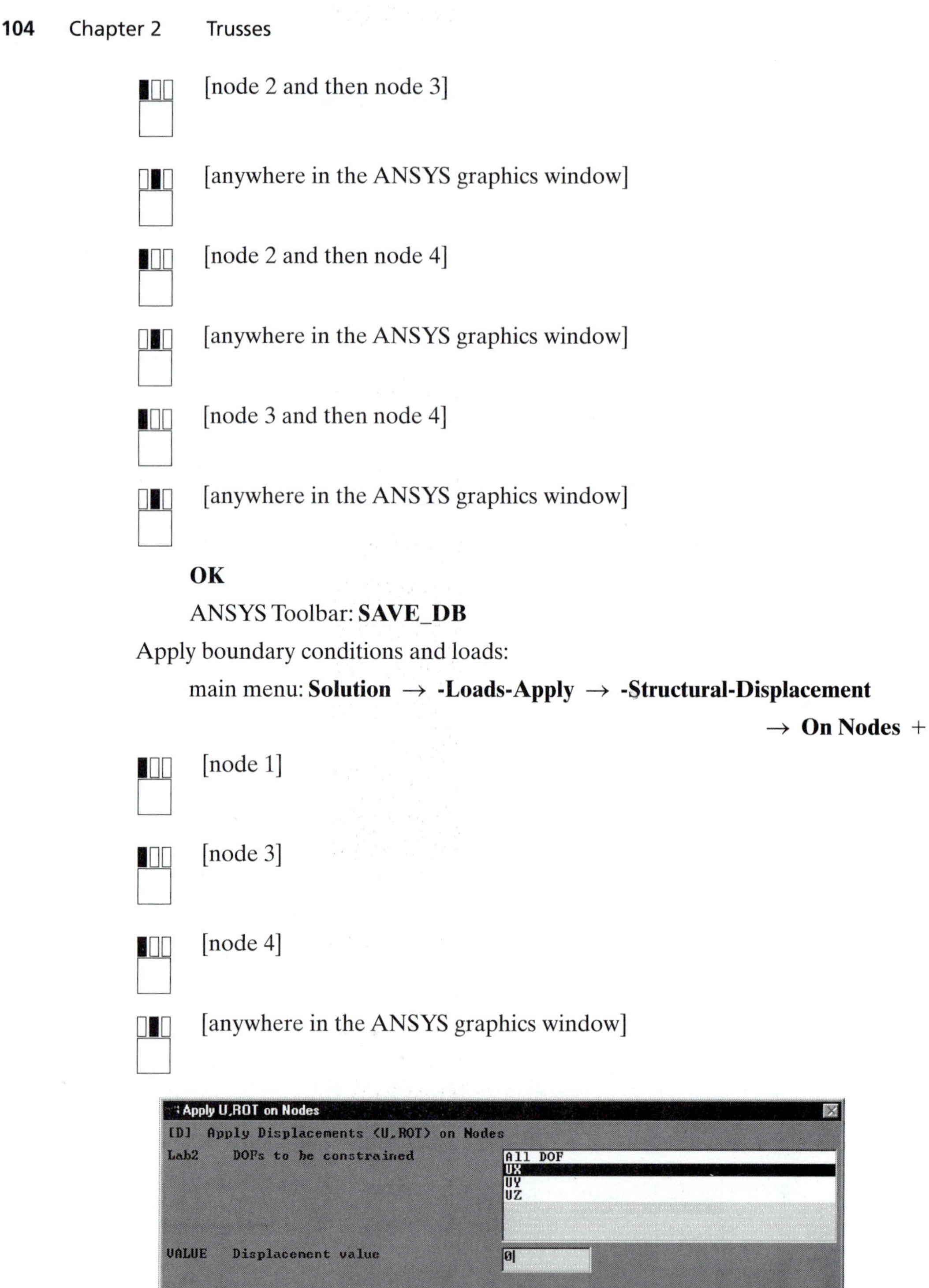

[node 2 and then node 3]

[anywhere in the ANSYS graphics window]

[node 2 and then node 4]

[anywhere in the ANSYS graphics window]

[node 3 and then node 4]

[anywhere in the ANSYS graphics window]

**OK**

ANSYS Toolbar: **SAVE_DB**

Apply boundary conditions and loads:

main menu: **Solution** → **-Loads-Apply** → **-Structural-Displacement** → **On Nodes** +

[node 1]

[node 3]

[node 4]

[anywhere in the ANSYS graphics window]

main menu: **Solution** → **-Loads-Apply** → **-Structural-Displacement** → **On Nodes** +

[node 1]

[anywhere in the ANSYS graphics window]

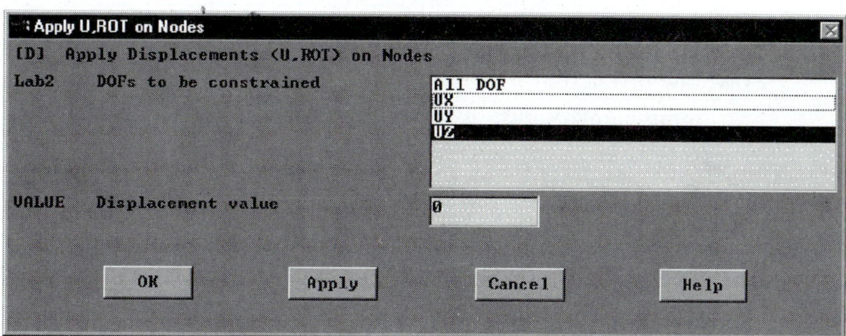


main menu: **Solution** → **-Loads-Apply** → **-Structural-Displacement** → **On Nodes** +

[node 4]

[anywhere in the ANSYS graphics window]

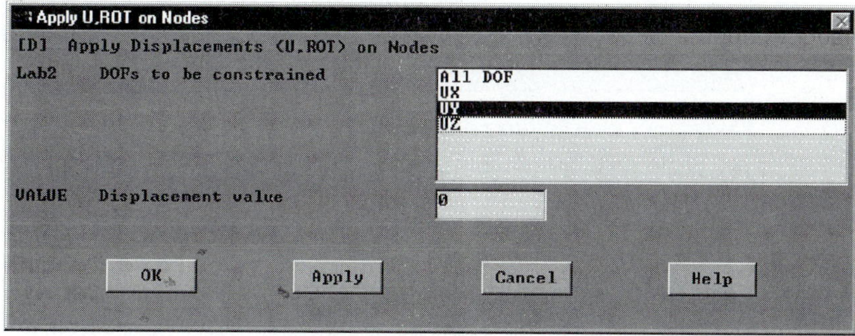


main menu: **Solution** → **-Loads-Apply** → **-Structural-Force/Moment** → **On Nodes** +

[node 2]

[anywhere in the ANSYS graphics window]

Apply F/M on Nodes
[F] Apply Force/Moment on Nodes
Lab Direction of force/mom FY
VALUE Force/moment value -200
OK Apply Cancel Help

ANSYS Toolbar: **SAVE_DB**

Solve the problem:

main menu: **Solution** → **-Solve-Current LS**

**OK**

**Close** (the solution is done!) window.

**Close** (the /STAT Command) window.

Now we run the postprocessing phase by listing nodal solutions (displacements):

main menu: **General Postproc** → **List Results** → **Nodal Solu** ...

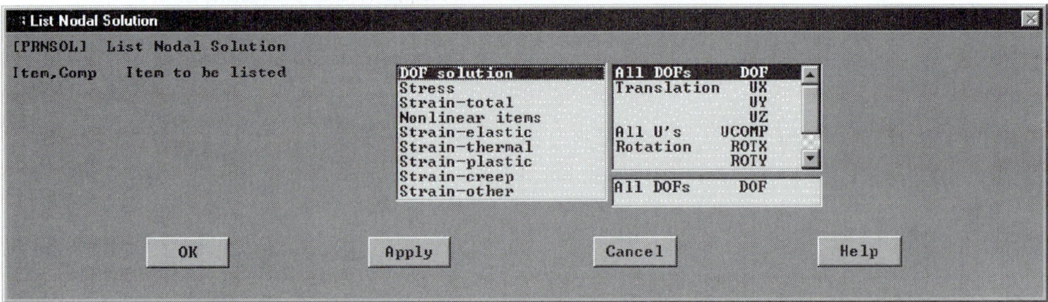


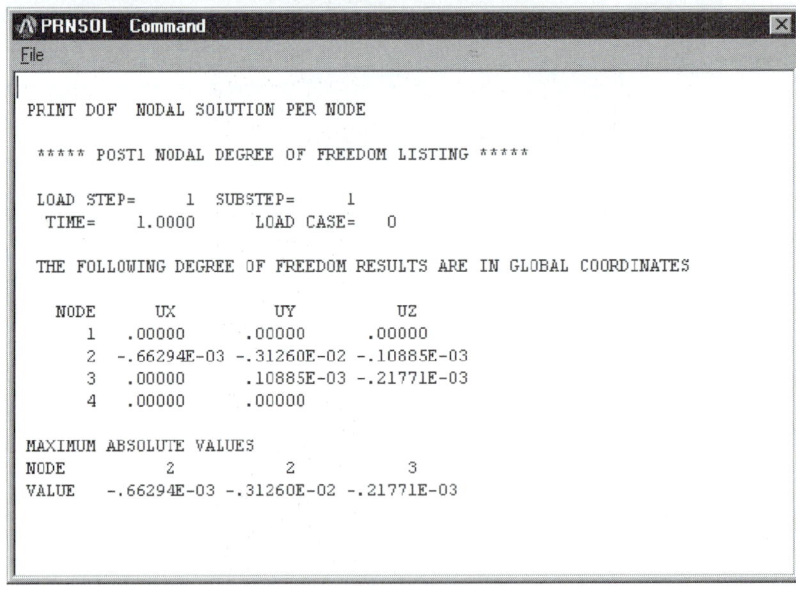

```
PRNSOL Command
File

PRINT DOF  NODAL SOLUTION PER NODE

 ***** POST1 NODAL DEGREE OF FREEDOM LISTING *****

 LOAD STEP=     1  SUBSTEP=     1
  TIME=    1.0000      LOAD CASE=   0

 THE FOLLOWING DEGREE OF FREEDOM RESULTS ARE IN GLOBAL COORDINATES

   NODE       UX          UY          UZ
      1   .00000      .00000      .00000
      2  -.66294E-03 -.31260E-02 -.10885E-03
      3   .00000      .10885E-03 -.21771E-03
      4   .00000      .00000

MAXIMUM ABSOLUTE VALUES
NODE          2           2           3
VALUE   -.66294E-03 -.31260E-02 -.21771E-03
```

To review other results, such as axial forces and axial stresses, we must copy these results into element tables. These items are obtained using *item label* and *sequence numbers*, as given in the Table 4.1–4.3 section of the ANSYS elements manual. So, we run the following sequence:

main menu: **General Postproc** → **Element Table** → **Define Table** ...

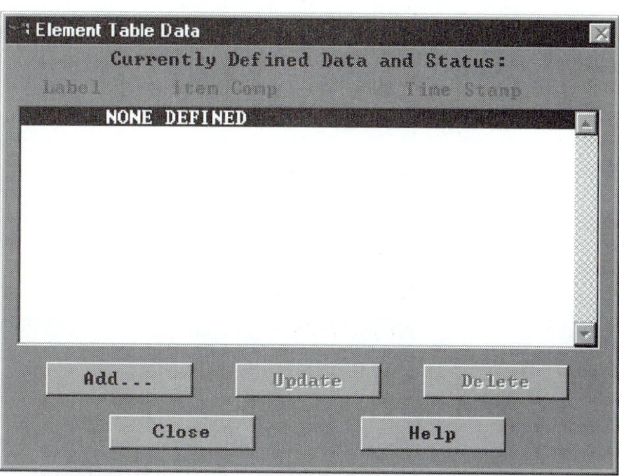


Define Additional Element Table Items
[ETABLE] Define Additional Element Table Items
Lab User label for item AXFORCE
Item,Comp Results data item
Energy
Error estimation
Geometry
Strain-elastic
Strain-thermal
Strain-plastic
Strain-creep
Strain-other
By sequence num
SMISC,
NMISC,
LS,
LEPEL,
LEPTH,
LEPPL,
LEPCR,
SMISC, 1
(For "By sequence num", enter sequence no. in Selection box. See Table 4.xx-3 in Elements Manual for seq. numbers.)
OK Apply Cancel Help

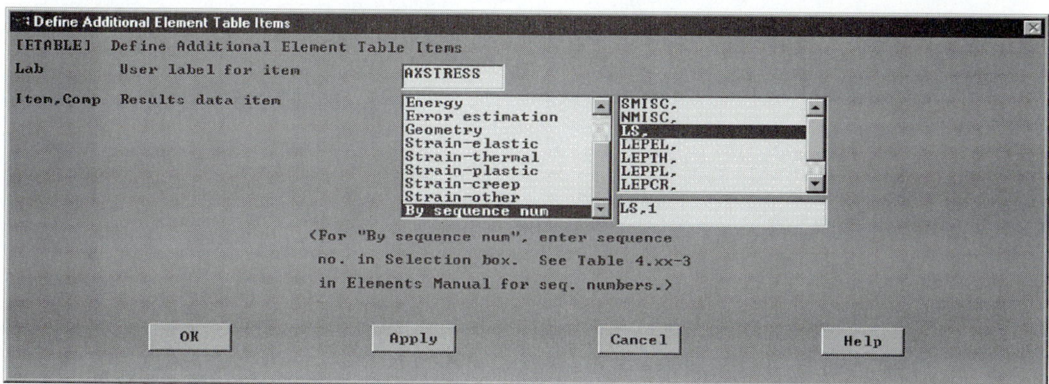

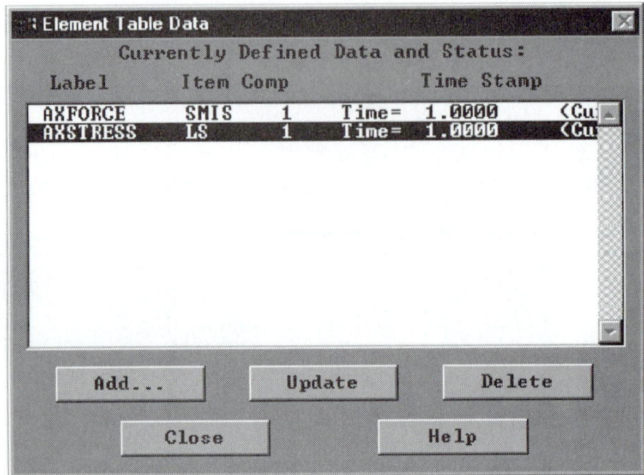


main menu: **General Postproc → Element Table → List Element Table** …

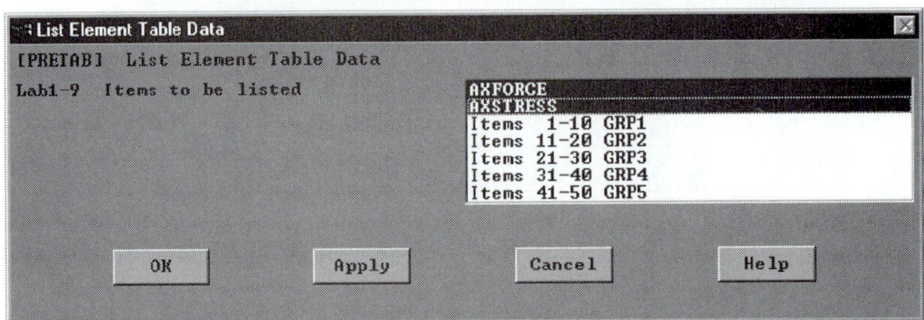


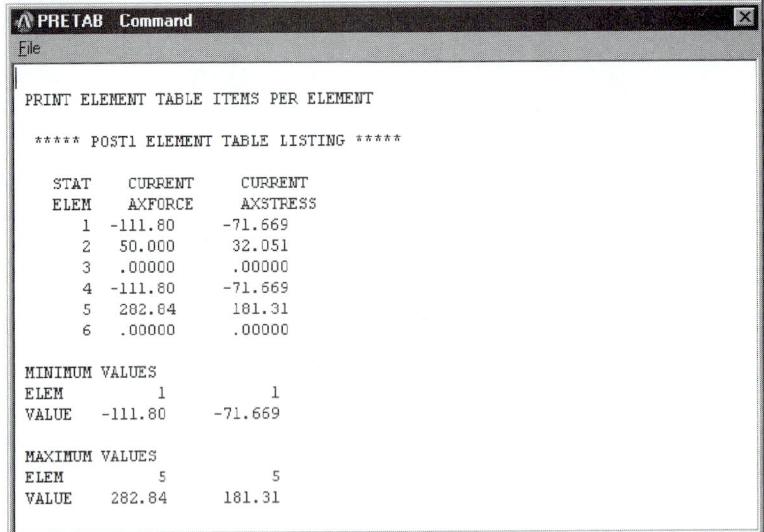


```
PRINT ELEMENT TABLE ITEMS PER ELEMENT

 ***** POST1 ELEMENT TABLE LISTING *****

   STAT    CURRENT     CURRENT
   ELEM    AXFORCE     AXSTRESS
      1  -111.80      -71.669
      2   50.000       32.051
      3   .00000       .00000
      4  -111.80      -71.669
      5   282.84       181.31
      6   .00000       .00000

MINIMUM VALUES
ELEM          1           1
VALUE   -111.80     -71.669

MAXIMUM VALUES
ELEM          5           5
VALUE    282.84      181.31
```

**Close**

List reaction solutions:

main menu: **General Postproc → List Results → Reaction Solu** …

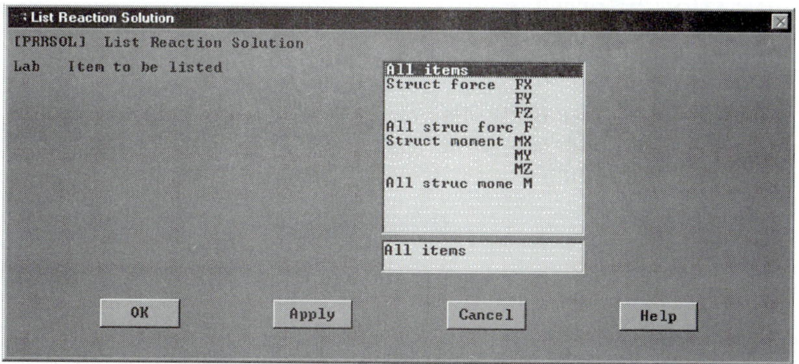


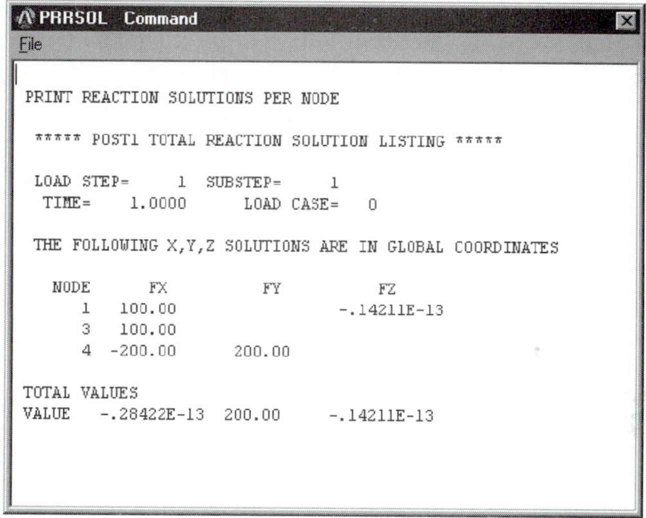


**Close**

Exit ANSYS and save everything, including element tables and reaction forces:

ANSYS Toolbar: **QUIT**

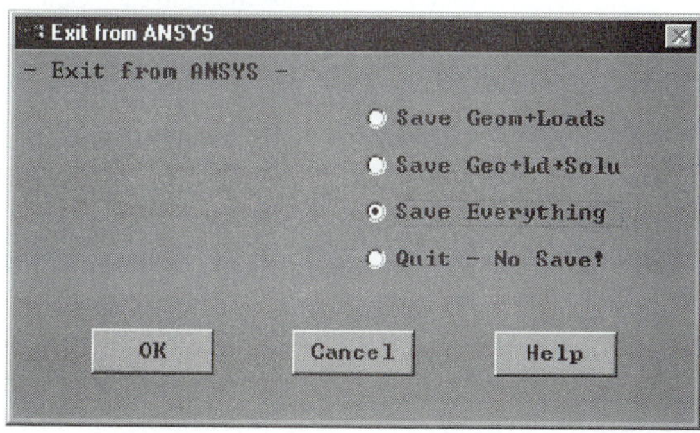

## 2.6 VERIFICATION OF RESULTS

There are various ways to verify your findings.

1. *Check the reaction forces.*
   We can use the computed reaction forces and the external forces to check for statics equilibrium:

$$\Sigma F_X = 0$$
$$\Sigma F_Y = 0$$

and

$$\Sigma M_{\text{node}} = 0$$

The reaction forces computed by ANSYS are: $F_{1X} = 1500$ lb; $F_{1Y} = 0$; $F_{3X} = -1500$ lb; and $F_{3Y} = 1000$ lb. Using the free body diagram shown in the accompanying figure and applying the static equilibrium equations, we have:

$$\Sigma F_X = 0 \qquad 1500 - 1500 = 0$$
$$\Sigma F_Y = 0 \qquad 1000 - 500 - 500 = 0$$
$$\Sigma M_{\text{node1}} = 0 \qquad (1500)(3) - (500)(3) - (500)(6) = 0$$

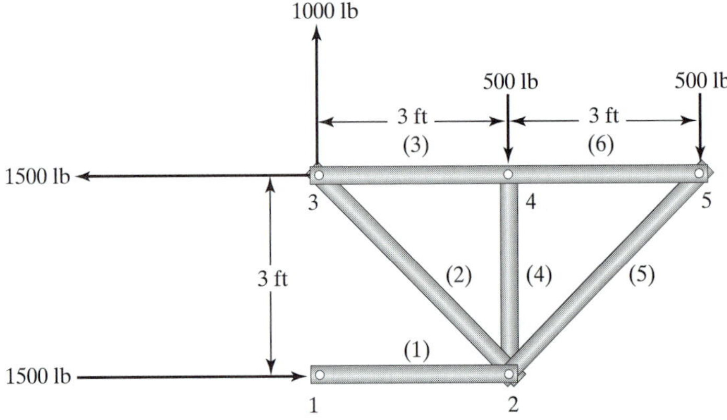


Now consider the internal forces of Example 2.1 as computed by ANSYS, shown in Table 2.2.

TABLE 2.2 Internal forces in each element as computed by ANSYS

| Element Number | Internal Forces (lb) |
|---|---|
| 1 | –1500 |
| 2 | 1414 |
| 3 | 500 |
| 4 | –500 |
| 5 | –707 |
| 6 | 500 |

2. *The sum of the forces at each node should be zero.*
Choose an arbitrary node and apply the equilibrium conditions. As an example, let us choose node 5. Using the free body diagram shown in the accompanying figure, we have:

$$\Sigma F_X = 0 \quad -500 + 707 \cos 45 = 0$$

$$\Sigma F_Y = 0 \quad -500 + 707 \sin 45 = 0$$

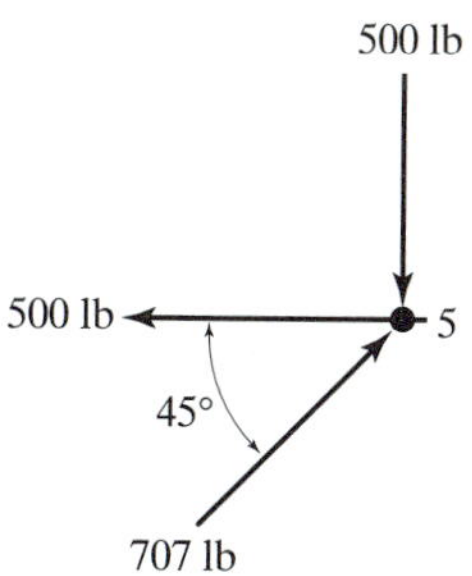


3. *Pass an arbitrary section through the truss.*
Another way of checking for the validity of your FEA findings is by arbitrarily cutting a section through the truss and applying the statics equilibrium conditions. For example, consider cutting a section through elements (1), (2), and (3), as shown in the accompanying figure.

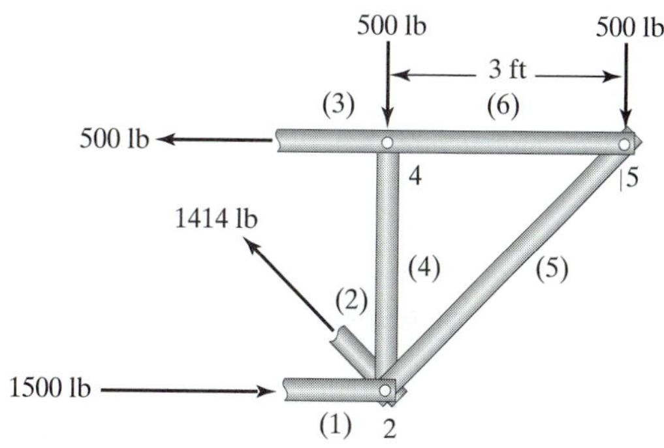


$$\Sigma F_X = 0 \qquad -500 + 1500 - 1414 \cos 45 = 0$$

$$\Sigma F_Y = 0 \qquad -500 - 500 + 1414 \cos 45 = 0$$

$$\Sigma M_{\text{node2}} = 0 \qquad -(500)(3) + (500)(3) = 0$$

Again, the validity of the computed internal forces is verified. Moreover, it is important to realize that when you analyze statics problems, statics equilibrium conditions must always be satisfied.

## SUMMARY

At this point you should:

1. have a good understanding of the underlying assumptions in truss analysis.
2. understand the significance of using global and local coordinate systems in describing a problem. You should also have a clear understanding of their role in describing nodal displacements and how information presented with respect to each frame of reference is related through the transformation matrix.
3. know the difference between the elemental stiffness matrix and the global stiffness matrix and know how to assemble elemental stiffness matrices to obtain a truss's global stiffness matrix.
4. know how to apply the boundary conditions and loads to a global matrix to obtain the nodal displacement solution.
5. know how to obtain internal forces and stresses in each member from displacement results.
6. have a good grasp of the basic concepts and commands of ANSYS. You should realize that a typical analysis using ANSYS involves: the *preprocessing phase*, where you provide data such as geometry, materials, and element type to the program; the *solution phase*, where you apply boundary conditions, apply loads, and initiate a finite element solution; and the *postprocessing phase*, where you review the results of the analysis through graphics displays and or tabular listings.
7. know how to verify the results of your truss analysis.

## REFERENCES

*ANSYS User's Manual*: *Procedures*, Vol. I, Swanson Analysis Systems, Inc.

*ANSYS User's Manual*: *Commands*, Vol. II, Swanson Analysis Systems, Inc.

*ANSYS User's Manual*: *Elements*, Vol. III, Swanson Analysis Systems, Inc.

Beer F. P., and Johnston, E. R., *Vector Mechanics for Engineers: Statics*, 5th ed., New York, McGraw-Hill, 1988.

Segrlind, L., Applied Finite Element Analysis, 2d. ed., New York, John Wiley and Sons, 1984.

## PROBLEMS

1. Starting with the transformation matrix, show that the inverse of the transformation matrix is its transpose. That is, show that

$$[\mathbf{T}]^{-1} = \begin{bmatrix} \cos\theta & \sin\theta & 0 & 0 \\ -\sin\theta & \cos\theta & 0 & 0 \\ 0 & 0 & \cos\theta & \sin\theta \\ 0 & 0 & -\sin\theta & \cos\theta \end{bmatrix}$$

2. Starting with Eq. (2.14), $\{\mathbf{F}\} = [\mathbf{T}][\mathbf{K}][\mathbf{T}]^{-1}\{\mathbf{U}\}$, and substituting for values of the $[\mathbf{T}]$, $[\mathbf{K}]$, $[\mathbf{T}]^{-1}$, and $\{\mathbf{U}\}$ matrices in Eq. (2.14), verify the elemental relationship

$$\begin{Bmatrix} F_{iX} \\ F_{iY} \\ F_{jX} \\ F_{jY} \end{Bmatrix} = k \begin{bmatrix} \cos^2\theta & \sin\theta\cos\theta & -\cos^2\theta & -\sin\theta\cos\theta \\ \sin\theta\cos\theta & \sin^2\theta & -\sin\theta\cos\theta & -\sin^2\theta \\ -\cos^2\theta & -\sin\theta\cos\theta & \cos^2\theta & \sin\theta\cos\theta \\ -\sin\theta\cos\theta & -\sin^2\theta & \sin\theta\cos\theta & \sin^2\theta \end{bmatrix} \begin{Bmatrix} U_{iX} \\ U_{iY} \\ U_{jX} \\ U_{jY} \end{Bmatrix}$$

3. The members of the truss shown in the accompanying figure have a cross-sectional area of 2.3 $\text{in}^2$ and are made of aluminum alloy ($E = 10.0 \times 10^6\ \text{lb/in}^2$). Using hand calculations, determine the deflection of joint $A$, the stress in each member, and the reaction forces. Verify your results.

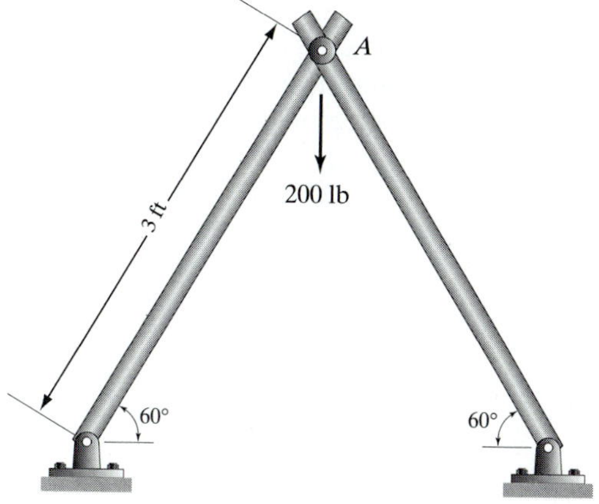


4. The members of the truss shown in the accompanying figure have a cross-sectional area of 8 $\text{cm}^2$ and are made of steel ($E = 200$ GPa). Using hand calculations, determine the deflection of each joint, the stress in each member, and the reaction forces. Verify your results.

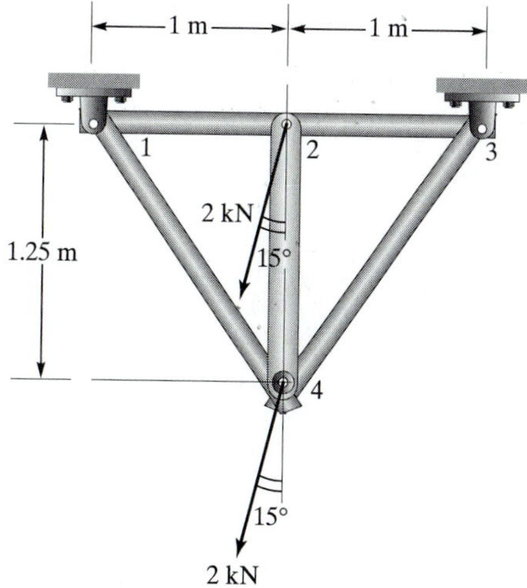


5. The members of the truss shown in the accompanying figure have a cross-sectional area of 15 $\text{cm}^2$ and are made of aluminum alloy (2024-T4). Using hand calculations, determine the deflection of each joint, the stress in each member, and the reaction forces. Verify your results.

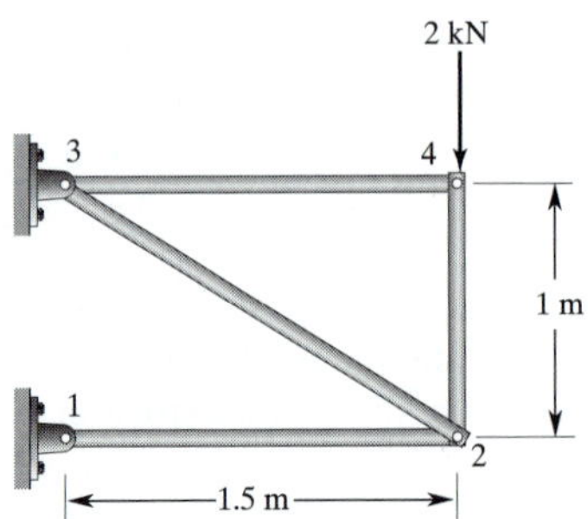


6. The members of the truss shown in the accompanying figure have a cross-sectional area of 2 $in^2$ and are made of structural steel (ASTM-A36). Using hand calculations, determine the deflection of each joint, the stress in each member, and the reaction forces. Verify your results.

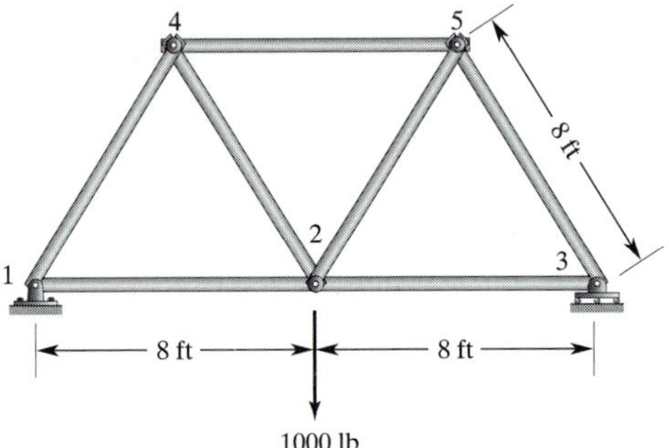


7. The members of the three-dimensional truss shown in the accompanying figure have a cross-sectional area of 2.5 $in^2$ and are made of aluminum alloy ($E = 10.0 \times 10^6$ $lb/in^2$). Using hand calculations, determine the deflection of joint *A*, the stress in each member, and the reaction forces. Verify your results.

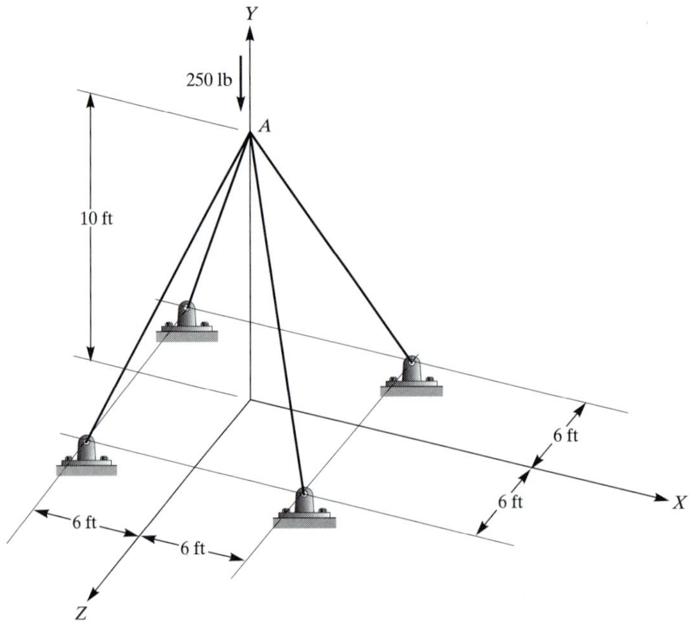

**8.** The members of the three-dimensional truss shown in the accompanying figure have a cross-sectional area of 15 $cm^2$ and are made of steel ($E = 200$ GPa). Using hand calculations, determine the deflection of joint $A$, the stress in each member, and the reaction forces. Verify your results.

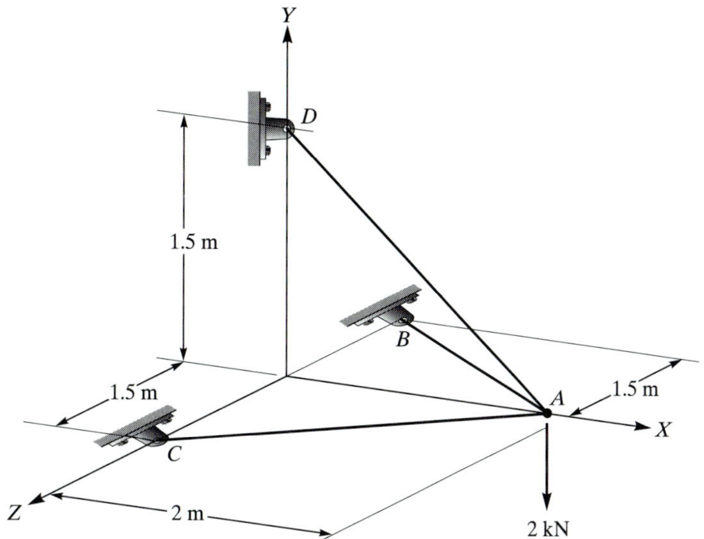


**9.** Consider the power transmission-line tower shown in the accompanying figure. The members have a cross-sectional area of 10 $in^2$ and a modulus of elasticity of $E = 29 \times 10^6$ lb/$in^2$. Using ANSYS, determine the deflection of each joint, the stress in each member, and the reaction forces at the base. Verify your results.

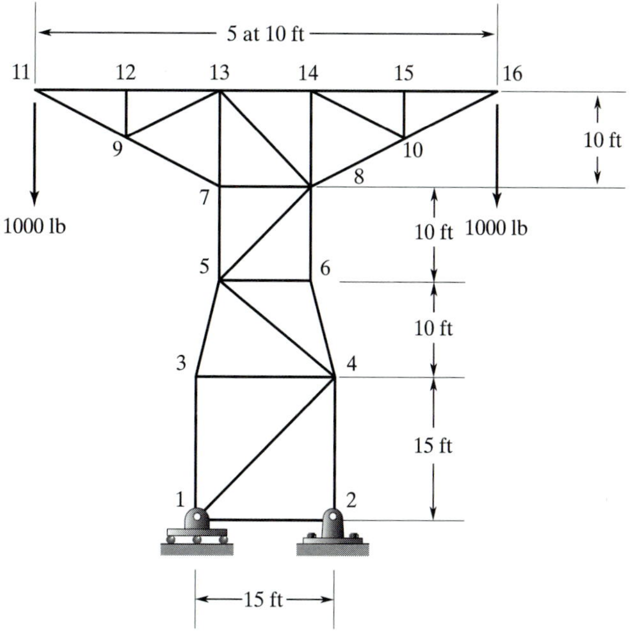

**10.** Consider the staircase truss shown in the accompanying figure. There are 14 steps, each with a rise of 8 in and a run of 12 in. The members have a cross-sectional area of 4 $in^2$ and are made of steel with a modulus of elasticity of $E = 29 \times 10^6$ lb/in$^2$. Using ANSYS, determine the deflection of each joint, the stress in each member, and the reaction forces. Verify your results.

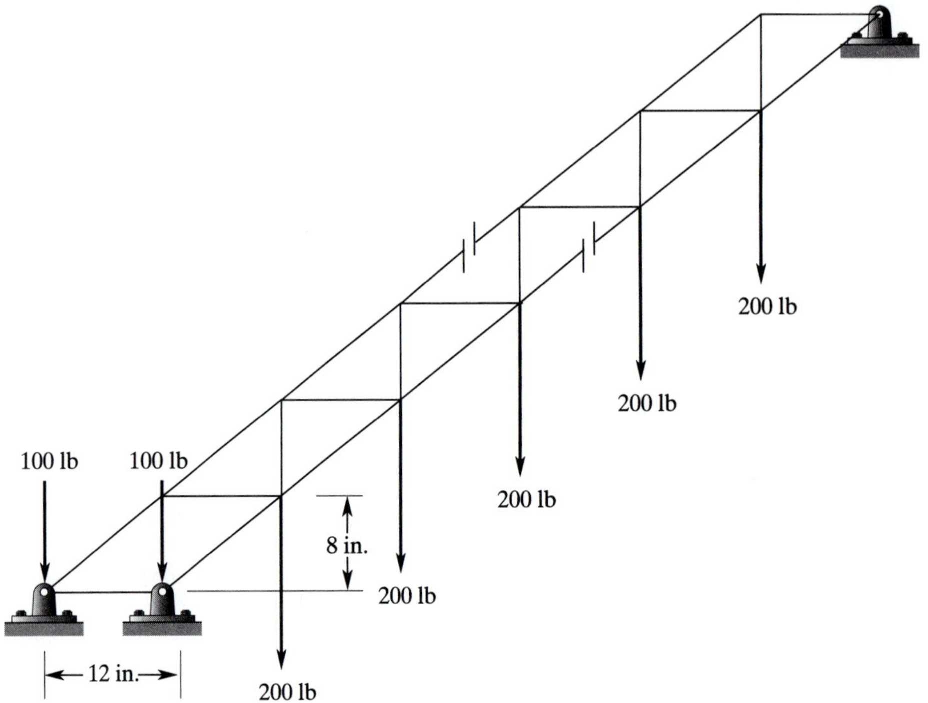


**11.** The members of the roof truss shown in the accompanying figure have a cross-sectional area of approximately 21.5 $in^2$ and are made of Douglas-fir wood with a modulus of elasticity of $E = 1.9 \times 10^6$ lb/in$^2$. Using ANSYS, determine the deflection of each joint, the stresses in each member, and the reaction forces. Verify your results. Also, replace one of the fixed boundary conditions with rollers and obtain the stresses in each member. Discuss the difference in results.

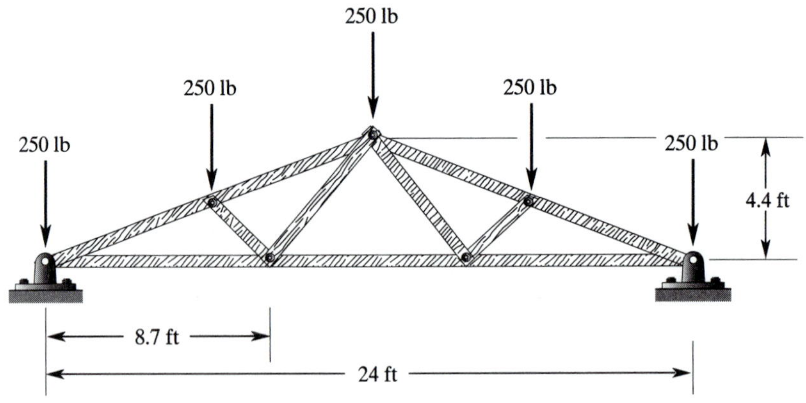

12. The members of the floor truss shown in the accompanying figure have a cross-sectional area of approximately 21.5 $in^2$ and are made of Douglas-fir wood with a modulus of elasticity of $E = 1.9 \times 10^6$ $lb/in^2$. Using ANSYS, determine the deflection of each joint, the stresses in each member, and the reaction forces. Verify your results. Also, replace one of the fixed boundary conditions with rollers and solve the problem again to obtain the stresses in each member. Discuss the difference in results.

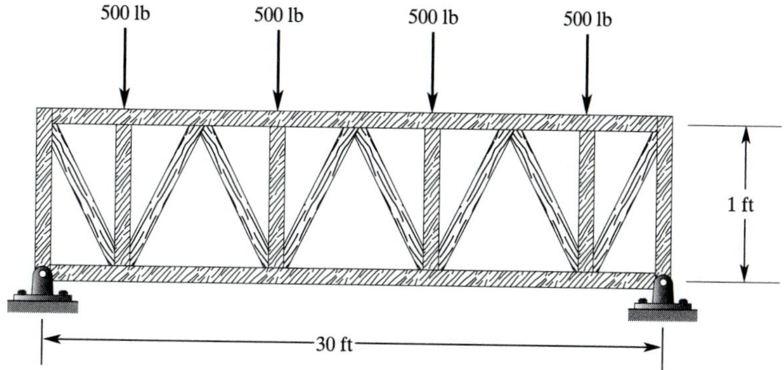


13. The three-dimensional truss shown in the accompanying figure is made of aluminum alloy ($E = 10.9 \times 10^6$ psi) and is to support a load of 500 lb. The Cartesian coordinates of the joints with respect to the coordinate system shown in the figure are given in feet. The cross-sectional area of each member is 2.246 $in^2$. Using ANSYS, determine the deflection of each joint, the stress in each member, and the reaction forces. Knowing that the second moment of area is 4.090 $in^4$, do you think that buckling is a concern for this truss? Verify your results.

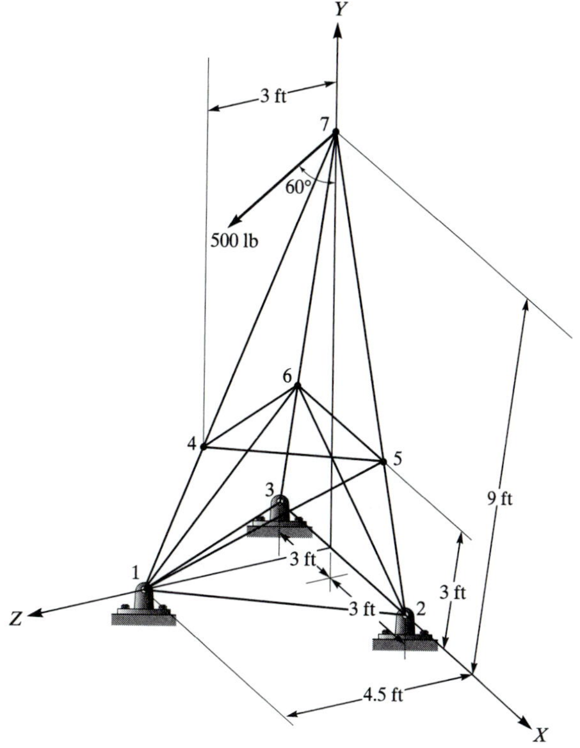

**14.** The three-dimensional truss shown in the accompanying figure is made of aluminum alloy ($E = 10.4 \times 10^6$ lb/in$^2$) and is to support a sign weighing 1000 lb. The Cartesian coordinates of the joints with respect to the coordinate system shown in the figure are given in feet. The cross-sectional area of each member is 3.14 in$^2$. Using ANSYS, determine the deflection of joint $E$, the stresses in each member, and the reaction forces. Verify your results.

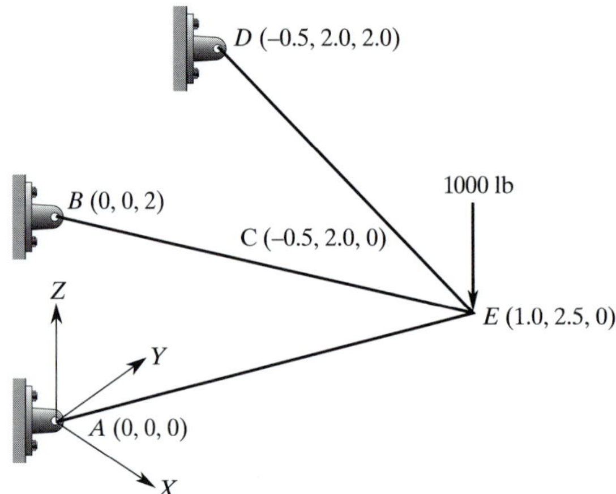


**15.** The three-dimensional truss shown in the accompanying figure is made of steel ($E = 29 \times 10^6$ psi) and is to support the load shown in the figure. The Cartesian coordinates of the joints with respect to the system shown in the figure are given in feet. The cross-sectional area of each member is 3.093 in$^2$. Using ANSYS, determine the deflection of each joint, the stresses in each member, and the reaction forces. Verify your results.

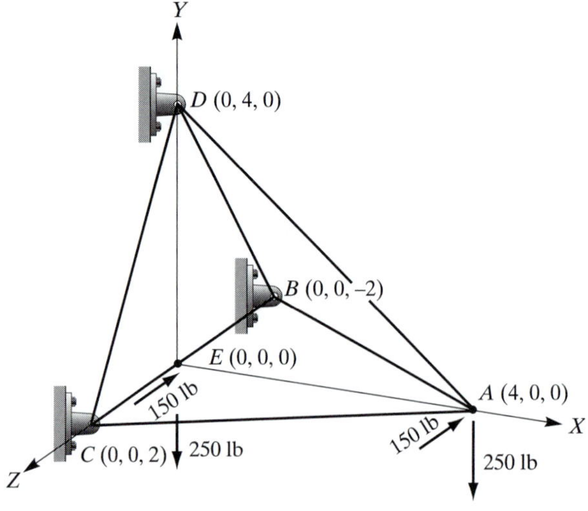

**16.** During a maintenance process on the three-dimensional truss in Problem 15, the $AB$ member is replaced with a member with the following properties: $E = 28 \times 10^6$ psi and $A = 2.246$ in$^2$. Using ANSYS, determine the deflection of each joint and the stresses in each member. Hint: you may need to ask your instructor for some help with this problem or you may want to study example 4.2 (revisited) on your own to learn about how to assign different attributes to an element in ANSYS.

**17.** During a maintenance process on the three-dimensional truss in Problem 13, members 4–5, 4–6, and 5–6 are replaced with steel members with the following properties: $E = 29 \times 10^6$ psi and $A = 1.25$ in$^2$. Member 1–5 is also replaced with a steel member with a cross-sectional area of 1.35 in$^2$. Using ANSYS, determine the deflection of each joint and the stresses in each member. See the hint given for Problem 16.

**18.** Derive the transformation matrix for an arbitrary member of a space truss, shown in the accompanying figure. The directional cosines, in terms of the difference between the coordinates of nodes $j$ and $i$ of a member and its length, are

$$\cos\theta_X = \frac{X_j - X_i}{L}; \qquad \cos\theta_Y = \frac{Y_j - Y_i}{L}; \qquad \cos\theta_Z = \frac{Z_j - Z_i}{L}$$

where $L$ is the length of the member and is

$$L = \sqrt{(X_j - X_i)^2 + (Y_j - Y_i)^2 + (Z_j - Z_i)^2}.$$

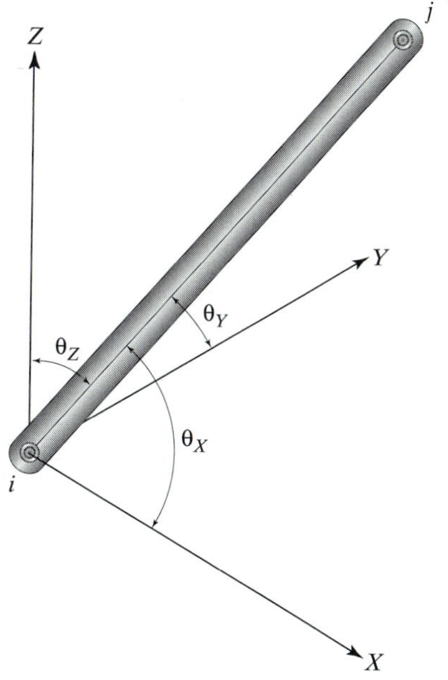

19. The three-dimensional truss shown in the accompanying figure is made of steel ($E = 29 \times 10^6$ psi) and is to support the load shown in the figure. Dimensions are given in feet. The cross-sectional area of each member is 3.25 in$^2$. Using ANSYS, determine the deflection of each joint, the stresses in each member, and the reaction forces. Verify your results.

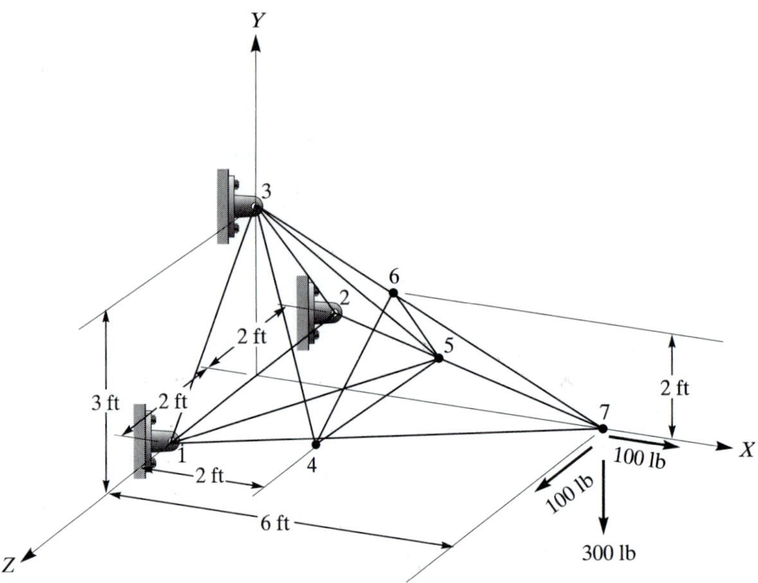


20. **Design Problem** Size the cross section of each member for the outdoor truss structure shown in the accompanying figure so that the end deflection of the truss is kept under 1 in. Select appropriate material and discuss how you arrived at your final design.

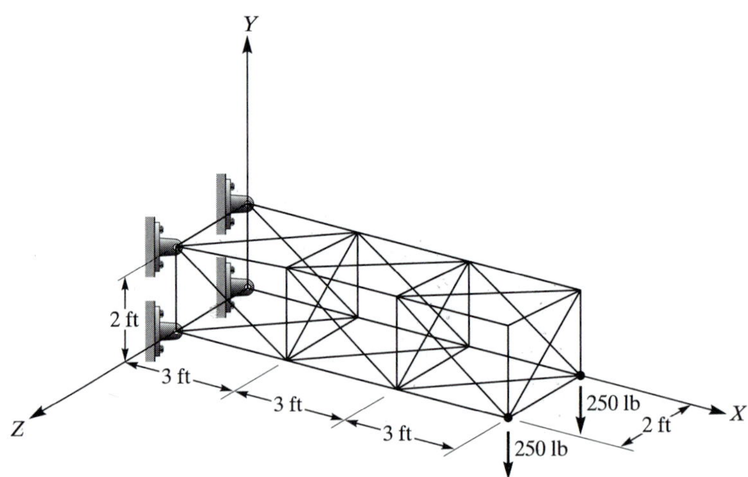

CHAPTER 3

# One-Dimensional Elements

The objectives of this chapter are to introduce the concepts of one-dimensional elements and shape functions and their properties. The idea of local and natural coordinate systems will also be presented here. In addition, one-dimensional elements used by ANSYS will be discussed. These are the main topics discussed in Chapter 3:



## 3.1 LINEAR ELEMENTS

The structural and heat transfer examples in this section are employed to introduce the basic ideas of one-dimensional elements and shape functions. Steel columns are commonly used to support loads from various floors of multi story buildings, as shown in Figure 3.1. The column shown in the figure may be divided into four elements and five nodes to generate a finite element model. The loading from the floors causes vertical displacements of various points along the column. Assuming axial central loading, we may approximate the actual *deflection* of the column by using a series of *linear functions*, describing the deflection over each element or each section of the column. In Chapter 4, we will formulate the stiffness and loading matrices for problems in which columns are subjected to axial loading.

As another example, let us consider a heat transfer problem. Fins are commonly used in a variety of engineering applications to enhance cooling. Common examples include a motorcycle engine head, a lawn mower engine head, extended surfaces (heat sinks) used in electronic equipment, and finned-tube heat exchangers. A straight fin of a uniform cross section is shown in Figure 3.2, along with a typical temperature distribution along the fin. As a first approximation, let us divide the fin into three elements and four nodes. The actual temperature distribution may be approximated by a combination of linear functions, as shown in Figure 3.2. To better approximate the actual temperature gradient near the base of the fin in our finite element model, we have placed the nodes

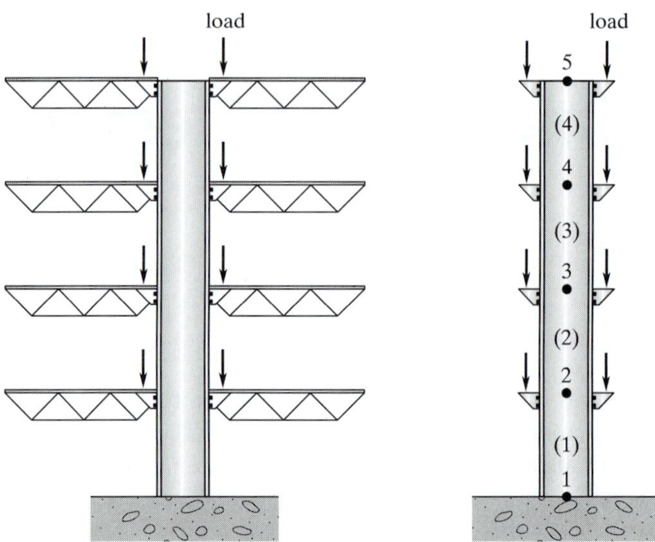


**FIGURE 3.1** Deflection of a steel column subject to floor loading.

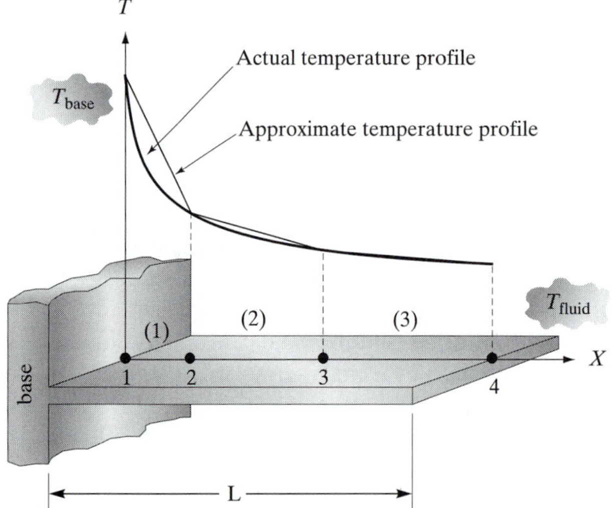


**FIGURE 3.2** Temperature distribution for a fin of uniform cross section.

closer to each other in that region. It should be clear that we can improve the accuracy of our approximation by increasing the number of elements as well. However, for now, let us be content with the three-element model and focus our attention on a typical element, as shown in Figure 3.3. The temperature distribution along the element may be interpolated (or approximated) using a linear function, as depicted in Figure 3.3.

The linear temperature distribution for a typical element may be expressed as

$$T^{(e)} = c_1 + c_2 X \tag{3.1}$$

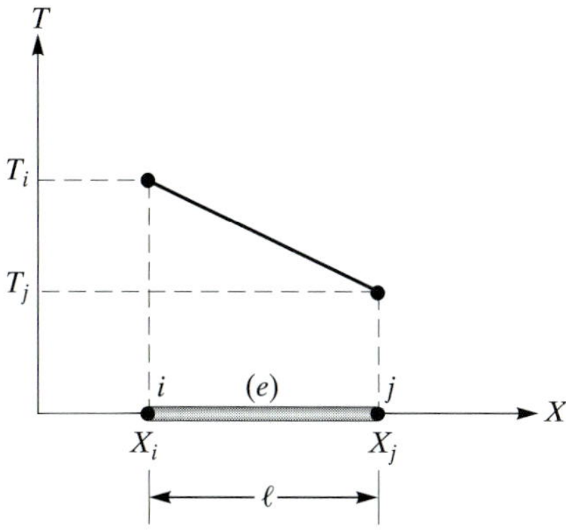


**FIGURE 3.3** Linear approximation of temperature distribution for an element.

The element's end conditions are given by the nodal temperatures $T_i$ and $T_j$, according to the conditions

$$T = T_i \quad \text{at} \quad X = X_i \qquad (3.2)$$
$$T = T_j \quad \text{at} \quad X = X_j$$

Substitution of nodal values into Eq. (3.1) results in two equations and two unknowns:

$$T_i = c_1 + c_2 X_i \qquad (3.3)$$
$$T_j = c_1 + c_2 X_j$$

Solving for the unknowns $c_1$ and $c_2$, we get

$$c_1 = \frac{T_i X_j - T_j X_i}{X_j - X_i} \qquad (3.4)$$

$$c_2 = \frac{T_j - T_i}{X_j - X_i} \qquad (3.5)$$

The element's temperature distribution in terms of its nodal values is:

$$T^{(e)} = \frac{T_i X_j - T_j X_i}{X_j - X_i} + \frac{T_j - T_i}{X_j - X_i} X \qquad (3.6)$$

Grouping the $T_i$ terms together and the $T_j$ terms together, we obtain

$$T^{(e)} = \left(\frac{X_j - X}{X_j - X_i}\right) T_i + \left(\frac{X - X_i}{X_j - X_i}\right) T_j \qquad (3.7)$$

We now define the *shape functions*, $S_i$ and $S_j$, according to the equations

$$S_i = \frac{X_j - X}{X_j - X_i} = \frac{X_j - X}{\ell} \qquad (3.8)$$

$$S_j = \frac{X - X_i}{X_j - X_i} = \frac{X - X_i}{\ell} \qquad (3.9)$$

where $\ell$ is the length of the element. Thus, the temperature distribution of an element in terms of the shape functions can be written as

$$T^{(e)} = S_i T_i + S_j T_j \tag{3.10}$$

Equation (3.10) can also be expressed in matrix form as

$$T^{(e)} = \begin{bmatrix} S_i & S_j \end{bmatrix} \begin{Bmatrix} T_i \\ T_j \end{Bmatrix} \tag{3.11}$$

For the structural example in Figure 3.1, the deflection $u^{(e)}$ for a typical column element is represented by

$$u^{(e)} = \begin{bmatrix} S_i & S_j \end{bmatrix} \begin{Bmatrix} u_i \\ u_j \end{Bmatrix} \tag{3.12}$$

where $u_i$ and $u_j$ represent the deflections of nodes $i$ and $j$ of an arbitrary element (e). It should be clear by now that we can represent the spatial variation of any unknown variable over a given element by using shape functions and the corresponding nodal values. Thus, in general, we can write

$$\Psi^{(e)} = \begin{bmatrix} S_i & S_j \end{bmatrix} \begin{Bmatrix} \Psi_i \\ \Psi_j \end{Bmatrix} \tag{3.13}$$

where $\Psi_i$ and $\Psi_j$ represent the nodal values of the unknown variable, such as temperature, or deflection, or velocity.

## Properties of Shape Functions

The shape functions possess unique properties that are important for us to understand because they simplify the evaluation of certain integrals when we are deriving the conductance or stiffness matrices. One of the inherent properties of a shape function is that it has a value of unity at its corresponding node and has a value of zero at the other adjacent node. Let us demonstrate this property by evaluating the shape functions at $X = X_i$ and $X = X_j$. Evaluating $S_i$ at $X = X_i$ and $X = X_j$, we get

$$S_i|_{X=X_i} = \left.\frac{X_j - X}{\ell}\right|_{X=X_i} = \frac{X_j - X_i}{\ell} = 1 \text{ and } S_i|_{X=X_j} = \left.\frac{X_j - X}{\ell}\right|_{X=X_j} = \frac{X_j - X_j}{\ell} = 0 \tag{3.14}$$

Also evaluating $S_j$ at $X = X_i$ and $X = X_j$, we obtain

$$S_j|_{X=X_i} = \left.\frac{X - X_i}{\ell}\right|_{X=X_i} = \frac{X_i - X_i}{\ell} = 0 \text{ and } S_j|_{X=X_j} = \left.\frac{X - X_i}{\ell}\right|_{X=X_j} = \frac{X_j - X_i}{\ell} = 1 \tag{3.15}$$

This property is also illustrated in Figure 3.4.

Another important property associated with shape functions is that the shape functions add up to a value of unity. That is,

$$S_i + S_j = \frac{X_j - X}{X_j - X_i} + \frac{X - X_i}{X_j - X_i} = 1 \tag{3.16}$$

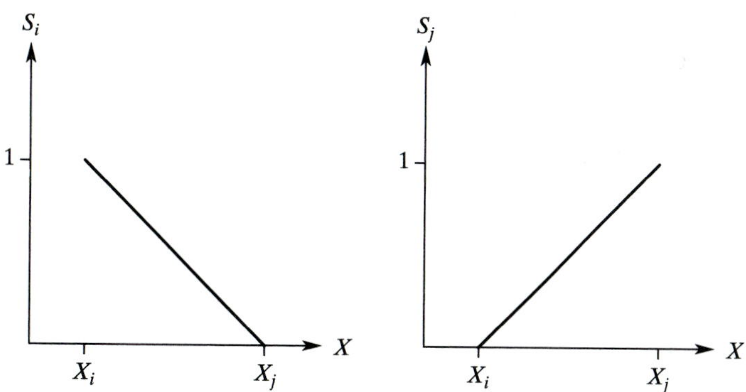


**FIGURE 3.4** Linear shape functions.

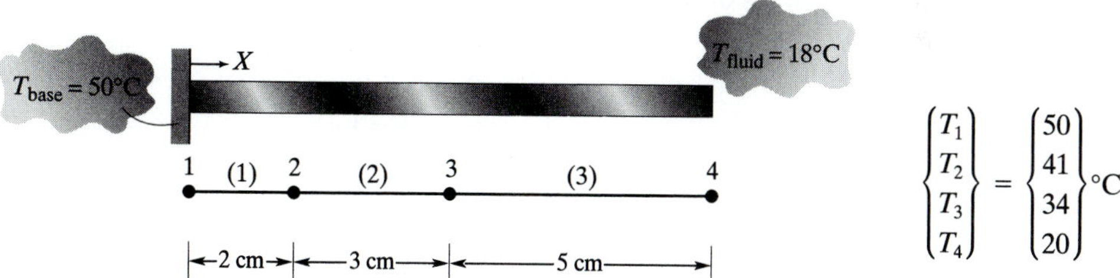


$$\begin{Bmatrix} T_1 \\ T_2 \\ T_3 \\ T_4 \end{Bmatrix} = \begin{Bmatrix} 50 \\ 41 \\ 34 \\ 20 \end{Bmatrix} °C$$

**FIGURE 3.5** The nodal temperatures and their corresponding positions along the fin in Example 3.1.

It can also be readily shown that for linear shape functions, the sum of the derivatives with respect to $X$ is zero. That is,

$$\frac{d}{dX}\left(\frac{X_j - X}{X_j - X_i}\right) + \frac{d}{dX}\left(\frac{X - X_i}{X_j - X_i}\right) = -\frac{1}{X_j - X_i} + \frac{1}{X_j - X_i} = 0 \qquad (3.17)$$

---

## EXAMPLE 3.1

We have used linear one-dimensional elements to approximate the temperature distribution along a fin. The nodal temperatures and their corresponding positions are shown in Figure 3.5. What is the temperature of the fin at a) $X = 4$ cm and b) $X = 8$ cm?

In Chapter 4, we will discuss in detail the analysis of one-dimensional fin problems, including the computation of nodal temperatures. However, for now, using the given nodal temperatures, we can proceed to answer both parts of the question:

**(a)** The temperature of the fin at $X = 4$ cm is represented by element (2);

$$T^{(2)} = S_2^{(2)} T_2 + S_3^{(2)} T_3 = \frac{X_3 - X}{\ell} T_2 + \frac{X - X_2}{\ell} T_3$$

$$T = \frac{5 - 4}{3}(41) + \frac{4 - 2}{3}(34) = 36.3 \text{ °C}$$

**(b)** The temperature of the fin at $X = 8$ cm is represented by element (3);

$$T^{(3)} = S_3^{(3)}T_3 + S_4^{(3)}T_4 = \frac{X_4 - X}{\ell}T_3 + \frac{X - X_3}{\ell}T_4$$

$$T = \frac{10 - 8}{5}(34) + \frac{8 - 5}{5}(20) = 25.6°\text{C}$$

For this example, note the difference between $S_3^{(2)}$ and $S_3^{(3)}$.

## EXAMPLE 3.2

Consider a four-story building with steel columns. One column is subjected to the loading shown in Figure 3.6. Under axial loading assumption and using linear elements, the vertical displacements of the column at various floor–column connection points were determined to be

$$\begin{Bmatrix} u_1 \\ u_2 \\ u_3 \\ u_4 \\ u_5 \end{Bmatrix} = \begin{Bmatrix} 0 \\ 0.03283 \\ 0.05784 \\ 0.07504 \\ 0.08442 \end{Bmatrix} \text{ in.}$$

The modulus of elasticity of $E = 29 \times 10^6$ lb/in$^2$, and area of $A = 39.7$ in$^2$ were used in the calculations. A detailed analysis of this problem is given in Chapter 4. For now, given

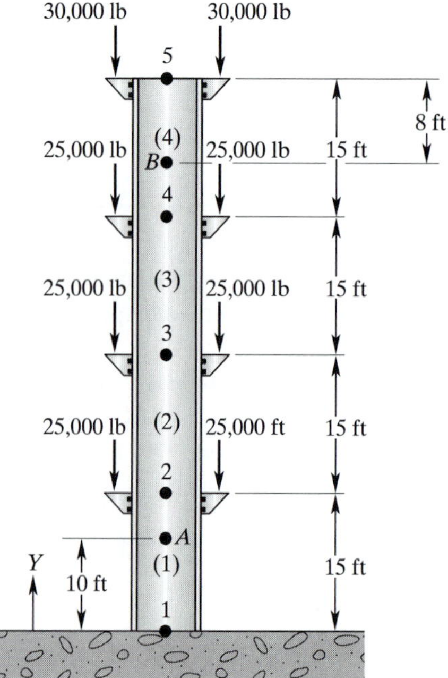


**FIGURE 3.6** The column in Example 3.2.

the nodal displacement values, we are interested in determining the deflections of points $A$ and $B$.

**(a)** Using the global coordinate $Y$, the displacement of point $A$ is represented by element (1):

$$u^{(1)} = S_1^{(1)} u_1 + S_2^{(1)} u_2 = \frac{Y_2 - Y}{\ell} u_1 + \frac{Y - Y_1}{\ell} u_2$$

$$u = \frac{15 - 10}{15}(0) + \frac{10 - 0}{15}(0.03283) = 0.02188 \text{ in.}$$

**(b)** The displacement of point $B$ is represented by element (4):

$$u^{(4)} = S_4^{(4)} u_4 + S_5^{(4)} u_5 = \frac{Y_5 - Y}{\ell} u_4 + \frac{Y - Y_4}{\ell} u_5$$

$$u = \frac{60 - 52}{15}(0.07504) + \frac{52 - 45}{15}(0.08442) = 0.07941 \text{ in.}$$

## 3.2 QUADRATIC ELEMENTS

We can increase the accuracy of our finite element findings by either increasing the number of linear elements used in the analysis or by using higher order interpolation functions. For example, we can employ a quadratic function to represent the spatial variation of an unknown variable. Utilizing a quadratic function instead of a linear function requires that we use three nodes to define an element. We need three nodes to define an element because in order to fit a quadratic function, we need three points. The third point can be created by placing a node, such as node $k$, in the middle of an element, as shown in Figure 3.7. Referring to the previous example of a fin, using quadratic approximation, the temperature distribution for a typical element can be represented by

$$T^{(e)} = c_1 + c_2 X + c_3 X^2 \tag{3.18}$$

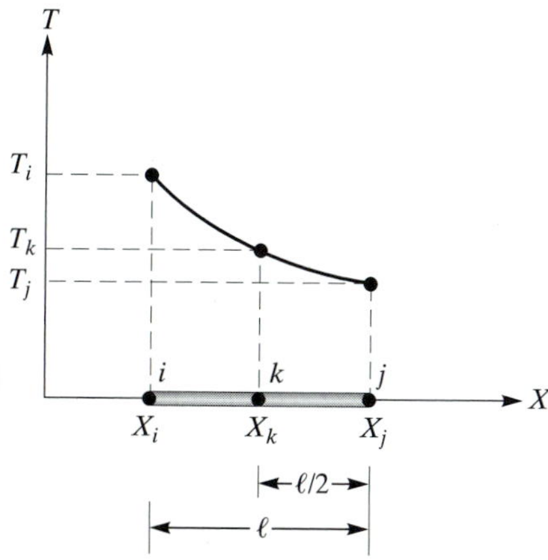


**FIGURE 3.7** Quadratic approximation of the temperature distribution for an element.

and the nodal values are

$$T = T_i \quad \text{at} \quad X = X_i \tag{3.19}$$

$$T = T_k \quad \text{at} \quad X = X_k$$

$$T = T_j \quad \text{at} \quad X = X_j$$

Three equations and three unknowns are created upon substitution of the nodal values into Eq. (3.18):

$$T_i = c_1 + c_2 X_i + c_3 X_i^2 \tag{3.20}$$

$$T_k = c_1 + c_2 X_k + c_3 X_k^2$$

$$T_j = c_1 + c_2 X_j + c_3 X_j^2$$

Solving for $c_1, c_2$, and $c_3$ and rearranging terms leads to the element's temperature distribution in terms of the nodal values and the shape functions:

$$T^{(e)} = S_i T_i + S_j T_j + S_k T_k \tag{3.21}$$

In matrix form, the above expression is

$$T^{(e)} = \begin{bmatrix} S_i & S_j & S_k \end{bmatrix} \begin{Bmatrix} T_i \\ T_j \\ T_k \end{Bmatrix} \tag{3.22}$$

where the shape functions are

$$S_i = \frac{2}{\ell^2}(X - X_j)(X - X_k) \tag{3.23}$$

$$S_j = \frac{2}{\ell^2}(X - X_i)(X - X_k)$$

$$S_k = \frac{-4}{\ell^2}(X - X_i)(X - X_j)$$

In general, for a given element the variation of any parameter, $\Psi$ in terms of its nodal values may be written as

$$\Psi^{(e)} = \begin{bmatrix} S_i & S_j & S_k \end{bmatrix} \begin{Bmatrix} \Psi_i \\ \Psi_j \\ \Psi_k \end{Bmatrix} \tag{3.24}$$

It is important to note here that the quadratic shape functions possess properties similar to those of the linear shape functions; that is: (1) a shape function has a value of unity at its corresponding node and a value of zero at the other adjacent node and (2) if we sum up the shape functions, we will again come up with a value of unity. The main difference between linear shape functions and quadratic shape functions is in their derivatives. The sum of the derivatives of the quadratic shape functions with respect to $X$ is not zero.

## 3.3 CUBIC ELEMENTS

The quadratic interpolation functions offer good results in finite element formulations. However, if additional accuracy is needed, we can resort to even higher order interpo-

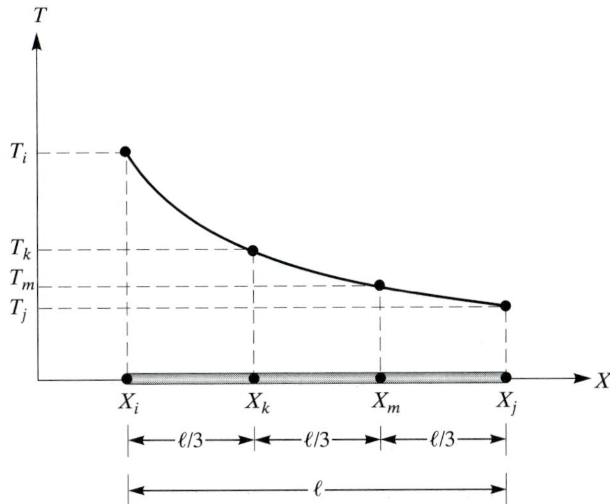


**FIGURE 3.8** Cubic approximation of the temperature distribution for an element.

lation functions, such as third-order polynomials. Thus, we can use cubic functions to represent the spatial variation of a given variable. Utilizing a cubic function instead of a quadratic function requires that we use four nodes to define an element. We need four nodes to define an element because in order to fit a third-order polynomial, we need four points. The element is divided into three equal lengths. The placement of the four nodes is depicted in Figure 3.8. Referring to the previous example of a fin, using cubic approximation, the temperature distribution for a typical element can be represented by

$$T^{(e)} = c_1 + c_2 X + c_3 X^2 + c_4 X^3 \tag{3.25}$$

and the nodal values are

$$\begin{aligned} T &= T_i \quad \text{at} \quad X = X_i \\ T &= T_k \quad \text{at} \quad X = X_k \\ T &= T_m \quad \text{at} \quad X = X_m \\ T &= T_j \quad \text{at} \quad X = X_j \end{aligned} \tag{3.26}$$

Four equations and four unknowns are created upon substitution of the nodal values into Eq. (3.25). Solving for $c_1, c_2, c_3$, and $c_4$ and rearranging terms leads to the element's temperature distribution in terms of the nodal values and the shape functions:

$$T^{(e)} = S_i T_i + S_j T_j + S_k T_k + S_m T_m \tag{3.27}$$

In matrix form, the above expression is

$$T^{(e)} = \begin{bmatrix} S_i & S_j & S_k & S_m \end{bmatrix} \begin{Bmatrix} T_i \\ T_j \\ T_k \\ T_m \end{Bmatrix} \tag{3.28}$$

where the shape functions are

$$S_i = -\frac{9}{2\ell^3}(X - X_j)(X - X_k)(X - X_m) \tag{3.29}$$

$$S_j = \frac{9}{2\ell^3}(X - X_i)(X - X_k)(X - X_m)$$

$$S_k = \frac{27}{2\ell^3}(X - X_i)(X - X_j)(X - X_m)$$

$$S_m = -\frac{27}{2\ell^3}(X - X_i)(X - X_j)(X - X_k)$$

It is worth noting here that when the order of the interpolating function increases, it is necessary to employ Lagrange interpolation functions instead of taking the above approach to obtain the shape functions. The main advantage the Lagrange method offers is that using it, we do not have to solve a set of equations simultaneously to obtain the unknown coefficients of the interpolating function. Instead, we represent the shape functions in terms of the products of linear functions. For cubic interpolating functions, the shape function associated with each node can be represented in terms of the product of three functions. For a given node—for example, $i$—we select the functions such that their product will produce a value of zero at other nodes—namely, $j$, $k$, and $m$—and a value of unity at the given node, $i$. Moreover, the product of the functions must produce linear and nonlinear terms similar to the ones given by a general third-order polynomial function.

To demonstrate this method, let us consider node $i$, with the global coordinate $X_i$. First, the functions must be selected such that when evaluated at nodes $j$, $k$, and $m$, the outcome is a value of zero. We select

$$S_i = a_1(X - X_j)(X - X_k)(X - X_m) \tag{3.30}$$

which satisfies the above condition. That is, if you substitute for $X = X_j$, or $X = X_k$, or $X = X_m$, the value of $S_i$ is zero. We then evaluate $a_1$ such that when the shape function $S_i$ is evaluated at node $i$, it will produce a value of unity:

$$1 = a_1(X_i - X_j)(X_i - X_k)(X_i - X_m) = a_1(-\ell)\left(-\frac{\ell}{3}\right)\left(-\frac{2\ell}{3}\right)$$

Solving for $a_1$, we get

$$a_1 = -\frac{9}{2\ell^3}$$

and substituting into Eq. (3.30), we have

$$S_i = -\frac{9}{2\ell^3}(X - X_j)(X - X_k)(X - X_m)$$

The other shape functions are obtained in a similar fashion. Keeping in mind the explanation offered above, we can generate shape functions of an $(n - 1)$-order polynomial directly from the Lagrange polynomial formula:

$$S_K = \prod_{M=1}^{N} \frac{X - X_M \text{ omitting } (X - X_K)}{X_K - X_M \text{ omitting } (X_K - X_K)} = \frac{(X - X_1)(X - X_2)\cdots(X - X_N)}{(X_K - X_1)(X_K - X_2)\cdots(X_K - X_N)} \quad (3.31)$$

Note that in order to accommodate any order polynomial representation in Eq. (3.31) numeral values are assigned to the nodes and the subscripts of the shape functions.

In general, using a cubic interpolation function, the variation of any parameter $\Psi$ in terms of its nodal values may be written as

$$\Psi^{(e)} = [S_i \quad S_j \quad S_k \quad S_m] \begin{Bmatrix} \Psi_i \\ \Psi_j \\ \Psi_k \\ \Psi_m \end{Bmatrix}$$

Once again, note that the cubic shape functions possess properties similar to those of the linear and the quadratic shape functions; that is, (1) a shape function has a value of unity at its corresponding node and a value of zero at the other adjacent node and (2) if we sum up the shape functions, we will come up with a value of unity. However, note that taking the spatial derivative of cubic shape functions will produce quadratic results.

## 3.4 GLOBAL, LOCAL, AND NATURAL COORDINATES

Most often, in finite element modeling, it is convenient to use several frames of reference, as we briefly discussed in Chapter 2. We need a global coordinate system to represent the location of each node, orientation of each element, and to apply boundary conditions and loads (in terms of their respective global components). Moreover, the solution, such as nodal displacements, is generally represented with respect to the global directions. On the other hand, we need to employ local and natural coordinates because they offer certain advantages when we construct the geometry or compute integrals. The advantage becomes apparent particularly when the integrals contain products of shape functions. For one-dimensional elements, the relationship between a global coordinate $X$ and a local coordinate $x$ is given by $X = X_i + x$, as shown in Figure 3.9.

Substituting for $X$ in terms of the local coordinate $x$ in Eqs. (3.8) and (3.9), we get

$$S_i = \frac{X_j - X}{\ell} = \frac{X_j - (X_i + x)}{\ell} = 1 - \frac{x}{\ell} \quad (3.32)$$

$$S_j = \frac{X - X_i}{\ell} = \frac{(X_i + x) - X_i}{\ell} = \frac{x}{\ell} \quad (3.33)$$

where the local coordinate $x$ varies from 0 to $\ell$; that is $0 \le x \le \ell$.

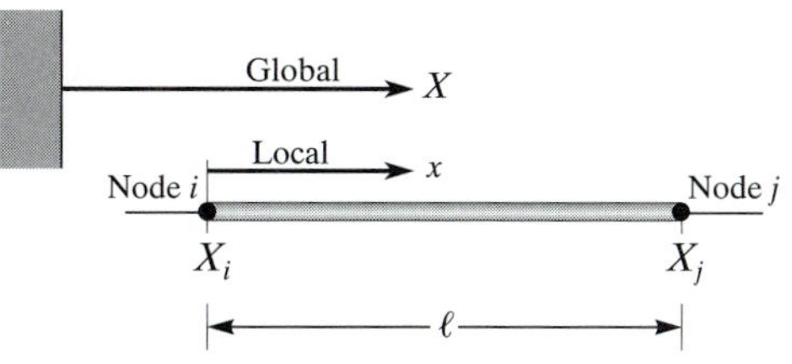


**FIGURE 3.9** The relationship between a global coordinate $X$ and a local coordinate $x$.

### One-Dimensional Linear Natural Coordinates

Natural coordinates are basically local coordinates in a dimensionless form. It is often necessary to use numerical methods to evaluate integrals for the purpose of calculating elemental stiffness or conductance matrices. Natural coordinates offer the convenience of having $-1$ and 1 for the limits of integration. For example, if we let

$$\xi = \frac{2x}{\ell} - 1$$

where $x$ is the local coordinate, then we can specify the coordinates of node $i$ as $-1$ and node $j$ by 1. This relationship is shown in Figure 3.10.

We can obtain the natural linear shape functions by substituting for $x$ in terms of $\xi$ into Eqs. (3.32) and (3.33). This substitution yields

$$S_i = \frac{1}{2}(1 - \xi) \tag{3.34}$$

$$S_j = \frac{1}{2}(1 + \xi) \tag{3.35}$$

Natural linear shape functions possess the same properties as linear shape functions; that is, a shape function has a value of unity at its corresponding node and has a value of zero at the other adjacent node in a given element. As an example, the temperature distribution over an element of a one-dimensional fin may expressed by

$$T^{(e)} = S_iT_i + S_jT_j = \frac{1}{2}(1 - \xi)T_i + \frac{1}{2}(1 + \xi)T_j \tag{3.36}$$

It is clear that at $\xi = -1, T = T_i$ and at $\xi = 1, T = T_j$. It should also be clear that we can represent other variables, such as the displacement $u$, in terms of the natural shape functions $S_i$ and $S_j$ according to the equation

$$u^{(e)} = S_iu_i + S_ju_j = \frac{1}{2}(1 - \xi)u_i + \frac{1}{2}(1 + \xi)u_j \tag{3.36a}$$

Also note that the transformation from the global coordinate $X$ $(X_i \leq X \leq X_j)$ or the local coordinate $x$ $(0 \leq x \leq \ell)$ to $\xi$ can be made using the same shape functions $S_i$ and $S_j$. That is,

$$X = S_iX_i + S_jX_j = \frac{1}{2}(1 - \xi)X_i + \frac{1}{2}(1 + \xi)X_j \tag{3.36b}$$

or

$$x = S_ix_i + S_jx_j = \frac{1}{2}(1 - \xi)x_i + \frac{1}{2}(1 + \xi)x_j$$

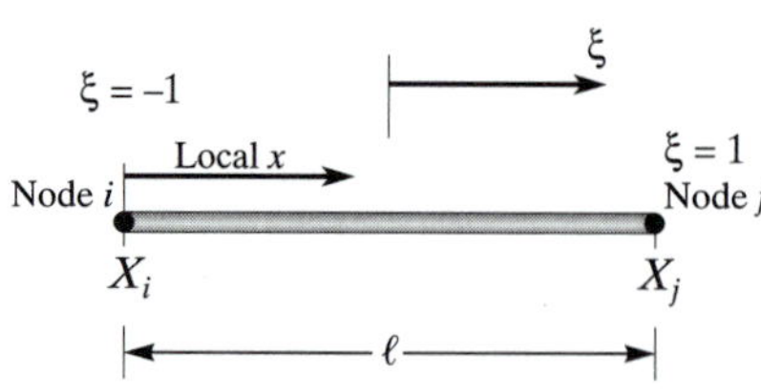


**FIGURE 3.10** The relationship between the local coordinate $x$ and the natural coordinate $\xi$.

Comparing the relationships given by Eqs. (3.36), (3.36a), and (3.36b), we note that we have used a *single* set of parameters (such as $S_i, S_j$) to define the unknown variables $u$, $T$, and so on, and we used the same parameters ($S_i, S_j$) to express the geometry. Finite element formulation that makes use of this idea is commonly referred to as *isoparametric* formulation, and an element expressed in such a manner is called an isoparametric element. We will discuss isoparametric formulation further in Chapters 5 and 8.

## EXAMPLE 3.3

Determine the temperature of the fin in Example 3.1 at the global location $X = 8$ cm using local coordinates. Also determine the temperature of the fin at the global location $X = 7.5$ cm using natural coordinates.

**(a)** Using local coordinates, we find that the temperature of the fin at $X = 8$ cm is represented by element (3) according to the equation

$$T^{(3)} = S_3^{(3)} T_3 + S_4^{(3)} T_4 = \left(1 - \frac{x}{\ell}\right) T_3 + \frac{x}{\ell} T_4$$

Note that element (3) has a length of 5 cm and the location of a point 8 cm from the base is represented by the local coordinate $x = 3$:

$$T = \left(1 - \frac{3}{5}\right)(34) + \frac{3}{5}(20) = 25.6\ °\text{C}$$

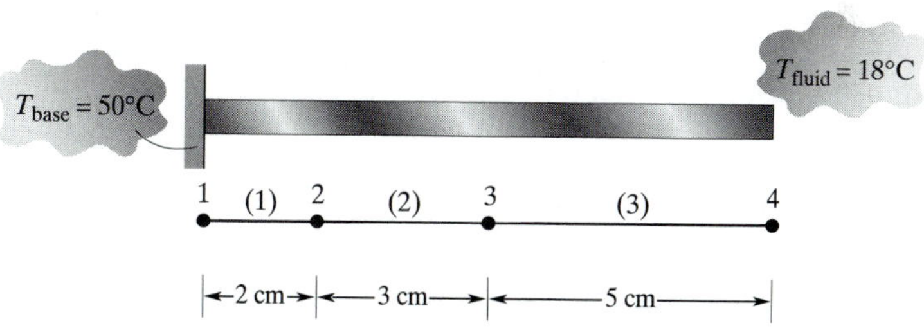


**(b)** Using natural coordinates, we find that the temperature of the fin at $X = 7.5$ cm is represented by element (3) according to the equation

$$T^{(3)} = S_3^{(3)} T_3 + S_4^{(3)} T_4 = \frac{1}{2}(1 - \xi) T_3 + \frac{1}{2}(1 + \xi) T_4$$

Because the point with the global coordinate $X = 7.5$ cm is located in the middle of element (3), the natural coordinate of this point is given by $\xi = 0$:

$$T^{(3)} = \frac{1}{2}(1 - 0)(34) + \frac{1}{2}(1 + 0)(20) = 27\ °\text{C}$$

### One-Dimensional Natural Quadratic and Cubic Shape Functions

The natural one-dimensional quadratic and cubic shape functions can be obtained in a way similar to the method discussed in the previous section. The quadratic natural shape functions are:

$$S_i = -\frac{1}{2}\xi(1 - \xi) \tag{3.37}$$

$$S_j = \frac{1}{2}\xi(1 + \xi) \tag{3.38}$$

$$S_k = (1 + \xi)(1 - \xi) \tag{3.39}$$

The natural one-dimensional cubic shape functions are:

$$S_i = \frac{1}{16}(1 - \xi)(3\xi + 1)(3\xi - 1) \tag{3.40}$$

$$S_j = \frac{1}{16}(1 + \xi)(3\xi + 1)(3\xi - 1) \tag{3.41}$$

$$S_k = \frac{9}{16}(1 + \xi)(\xi - 1)(3\xi - 1) \tag{3.42}$$

$$S_m = \frac{9}{16}(1 + \xi)(1 - \xi)(3\xi + 1) \tag{3.43}$$

For the sake of convenience, the results of Sections 3.1 to 3.4 are summarized in Table 3.1. Make sure to distinguish the differences among presentations of the shape functions using global, local, and natural coordinates.

---

**EXAMPLE 3.4**

Evaluate the integral $\int_{X_i}^{X_j} S_j^2\,dX$ using (a) global coordinates and (b) local coordinates.

**(a)** Using global coordinates, we obtain

$$\int_{X_i}^{X_j} S_j^2\,dX = \int_{X_i}^{X_j}\left(\frac{X - X_i}{\ell}\right)^2 dX = \frac{1}{3\ell^2}(X - X_i)^3\Big|_{X_i}^{X_j} = \frac{\ell}{3}$$

**(b)** Using local coordinates, we obtain

$$\int_{X_i}^{X_j} S_j^2\,dX = \int_0^{\ell}\left(\frac{x}{\ell}\right)^2 dx = \frac{x^3}{3\ell^2}\Big|_0^{\ell} = \frac{\ell}{3}$$

This simple example demonstrates that local coordinates offer a simple way to evaluate integrals containing products of shape functions.

---

TABLE 3.1 One-dimensional shape functions

| Interpolation function | In terms of global coordinate $X$<br>$X_i \le X \le X_j$ | In terms of local coordinate $x$<br>$0 \le x \le \ell$ | In terms of natural coordinate $\xi$<br>$-1 \le \xi \le 1$ |
|---|---|---|---|
| Linear | $S_i = \frac{X_j - X}{\ell}$ | $S_i = 1 - \frac{x}{\ell}$ | $S_i = \frac{1}{2}(1 - \xi)$ |
| | $S_j = \frac{X - X_i}{\ell}$ | $S_j = \frac{x}{\ell}$ | $S_j = \frac{1}{2}(1 + \xi)$ |
| Quadratic | $S_i = \frac{2}{\ell^2}(X - X_j)(X - X_k)$ | $S_i = \left(\frac{x}{\ell} - 1\right)\left(2\left(\frac{x}{\ell}\right) - 1\right)$ | $S_i = -\frac{1}{2}\xi(1 - \xi)$ |
| | $S_j = \frac{2}{\ell^2}(X - X_i)(X - X_k)$ | $S_j = \left(\frac{x}{\ell}\right)\left(2\left(\frac{x}{\ell}\right) - 1\right)$ | $S_j = \frac{1}{2}\xi(1 + \xi)$ |
| | $S_k = \frac{-4}{\ell^2}(X - X_i)(X - X_j)$ | $S_k = 4\left(\frac{x}{\ell}\right)\left(1 - \left(\frac{x}{\ell}\right)\right)$ | $S_k = (1 - \xi)(1 + \xi)$ |
| Cubic | $S_i = -\frac{9}{2\ell^3}(X - X_j)(X - X_k)(X - X_m)$ | $S_i = \frac{1}{2}\left(1 - \frac{x}{\ell}\right)\left(2 - 3\left(\frac{x}{\ell}\right)\right)\left(1 - 3\left(\frac{x}{\ell}\right)\right)$ | $S_i = \frac{1}{16}(1 - \xi)(3\xi + 1)(3\xi - 1)$ |
| | $S_j = \frac{9}{2\ell^3}(X - X_i)(X - X_k)(X - X_m)$ | $S_j = \frac{1}{2}\left(\frac{x}{\ell}\right)\left(2 - 3\left(\frac{x}{\ell}\right)\right)\left(1 - 3\left(\frac{x}{\ell}\right)\right)$ | $S_j = \frac{1}{16}(1 + \xi)(3\xi + 1)(3\xi - 1)$ |
| | $S_k = \frac{27}{2\ell^3}(X - X_i)(X - X_j)(X - X_m)$ | $S_k = \frac{9}{2}\left(\frac{x}{\ell}\right)\left(2 - 3\left(\frac{x}{\ell}\right)\right)\left(1 - \left(\frac{x}{\ell}\right)\right)$ | $S_k = \frac{9}{16}(1 + \xi)(\xi - 1)(3\xi - 1)$ |
| | $S_m = -\frac{27}{2\ell^3}(X - X_i)(X - X_j)(X - X_k)$ | $S_m = \frac{9}{2}\left(\frac{x}{\ell}\right)\left(3\left(\frac{x}{\ell}\right) - 1\right)\left(1 - \left(\frac{x}{\ell}\right)\right)$ | $S_m = \frac{9}{16}(1 + \xi)(1 - \xi)(3\xi + 1)$ |

## 3.5 NUMERICAL INTEGRATION: GAUSS–LEGENDRE QUADRATURE

As we discussed earlier, natural coordinates are basically local coordinates in a dimensionless form. Moreover, most finite element programs perform element numerical integration by Gaussian quadratures, and as the limit of integration, they use an interval from $-1$ to 1. This approach is taken because when the function being integrated is known, the Gauss–Legendre formulae offer a more efficient way of evaluating an integral as compared to other numerical integration methods such as the trapezoidal method. Whereas the trapezoidal method or Simpson's method can be used to evaluate integrals dealing with discrete data, the Gauss–Legendre method is based on the evaluation of a known function at nonuniformally spaced points to compute the integral. The two-point Gauss–Legendre formula is developed next in this section. The basic goal behind the Gauss–Legendre formulae is to represent an integral in terms of the sum of the product of certain weighting coefficients and the value of the function at some selected points. So, we begin with

$$I = \int_a^b f(x)\,dx = \sum_{i=1}^{n} w_i f(x_i) \tag{3.44}$$

Next, we must ask: (1) How do we determine the value of the weighting coefficients, represented by the $w_i$'s? (2) Where do we evaluate the function, or, in other words, how do we select these points? We begin by changing the limits of integration from $a$–$b$ to $-1$ to 1 with the introduction of the variable $\lambda$ such that

$$x = c_0 + c_1\lambda$$

Matching the limits, we get

$$a = c_0 + c_1(-1)$$

$$b = c_0 + c_1(1)$$

and solving for $c_0$ and $c_1$, we have

$$c_0 = \frac{(b + a)}{2}$$

and

$$c_1 = \frac{(b - a)}{2}$$

Therefore,

$$x = \frac{(b + a)}{2} + \frac{(b - a)}{2}\lambda \tag{3.45}$$

and

$$dx = \frac{(b - a)}{2}d\lambda \tag{3.46}$$

Thus, using Eqs. (3.45) and (3.46), we find that any integral in the form of Eq. (3.44) can be expressed in terms of an integral with its limits at $-1$ and 1:

$$I = \int_{-1}^{1} f(\lambda)\, d\lambda = \sum_{i=1}^{n} w_i f(\lambda_i) \tag{3.47}$$

The two-point Gauss–Legendre formulation requires the determination of two weighting factors $w_1$ and $w_2$ and two sampling points $\lambda_1$ and $\lambda_2$ to evaluate the function at these points. Because there are four unknowns, four equations are created using Legendre polynomials $(1, \lambda, \lambda^2, \lambda^3)$ as follows:

$$w_1 f(\lambda_1) + w_2 f(\lambda_2) = \int_{-1}^{1} 1\, d\lambda = 2$$

$$w_1 f(\lambda_1) + w_2 f(\lambda_2) = \int_{-1}^{1} \lambda\, d\lambda = 0$$

$$w_1 f(\lambda_1) + w_2 f(\lambda_2) = \int_{-1}^{1} \lambda^2\, d\lambda = \frac{2}{3}$$

$$w_1 f(\lambda_1) + w_2 f(\lambda_2) = \int_{-1}^{1} \lambda^3\, d\lambda = 0$$

The above equations lead to the equations

$$w_1(1) + w_2(1) = 2$$

$$w_1(\lambda_1) + w_2(\lambda_2) = 0$$

$$w_1(\lambda_1)^2 + w_2(\lambda_2)^2 = \frac{2}{3}$$

$$w_1(\lambda_1)^3 + w_2(\lambda_2)^3 = 0$$

Solving for $w_1$, $w_2$, $\lambda_1$, and $\lambda_2$, we have $w_1 = w_2 = 1$, $\lambda_1 = -0.577350269$, and $\lambda_2 = 0.577350269$. The weighting factors and the 2, 3, 4, and 5 sampling points for Gauss–Legendre formulae are given in Table 3.2. Note that as the number of sampling point increases, so does the accuracy of the calculations. As you will see in Chapter 5, we can readily extend the Gauss–Legendre quadrature formulation to two- or three-dimensional problems.

## EXAMPLE 3.5

Evaluate the integral $I = \int_{2}^{6} (x^2 + 5x + 3)\, dx$ using the Gauss–Legendre two-point sampling formula.

This integral is simple and can be evaluated analytically, leading to the solution $I = 161.333333333$. The purpose of this example is to demonstrate the Gauss–Legendre procedure. We begin by changing the variable $x$ to $\lambda$ by using Eq. (3.45). So, we obtain

$$x = \frac{(b + a)}{2} + \frac{(b - a)}{2}\lambda = \frac{(6 + 2)}{2} + \frac{(6 - 2)}{2}\lambda = 4 + 2\lambda$$

TABLE 3.2 Weighting factors and sampling points for Gauss–Legendre formulae

| Points ($n$) | Weighting Factors ($w_i$) | Sampling points ($\lambda_i$) |
|---|---|---|
| 2 | $w_1 = 1.00000000$<br>$w_1 = 1.00000000$ | $\lambda_1 = -0.577350269$<br>$\lambda_2 = 0.577350269$ |
| 3 | $w_1 = 0.55555556$<br>$w_2 = 0.88888889$<br>$w_3 = 0.55555556$ | $\lambda_1 = -0.774596669$<br>$\lambda_2 = 0$<br>$\lambda_3 = 0.774596669$ |
| 4 | $w_1 = 0.3478548$<br>$w_2 = 0.6521452$<br>$w_3 = 0.6521452$<br>$w_4 = 0.3478548$ | $\lambda_1 = -0.861136312$<br>$\lambda_2 = -0.339981044$<br>$\lambda_3 = 0.339981044$<br>$\lambda_4 = 0.861136312$ |
| 5 | $w_1 = 0.2369269$<br>$w_2 = 0.4786287$<br>$w_3 = 0.5688889$<br>$w_4 = 0.4786287$<br>$w_5 = 0.2369269$ | $\lambda_1 = -0.906179846$<br>$\lambda_2 = -0.538469310$<br>$\lambda_3 = 0$<br>$\lambda_4 = 0.538469310$<br>$\lambda_5 = 0.906179846$ |

and

$$dx = \frac{(b-a)}{2}\,d\lambda = \frac{(6-2)}{2}\,d\lambda = 2\,d\lambda$$

Thus, the integral $I$ can be expressed in terms of $\lambda$:

$$I = \int_2^6 \overbrace{(x^2 + 5x + 3)}^{f(x)}\,dx = \int_{-1}^{1} \overbrace{(2)[(4 + 2\lambda)^2 + 5(4 + 2\lambda) + 3)]}^{f(\lambda)}\,d\lambda$$

Using the Gauss–Legendre two-point formula and Table 3.2, we compute the value of the integral $I$ from:

$$I \cong w_1 f(\lambda_1) + w_2 f(\lambda_2)$$

From Table 3.2, we find that $w_1 = w_2 = 1$, and evaluating $f(\lambda)$ at $\lambda_1 = -0.577350269$ and $\lambda_2 = 0.577350269$, we obtain

$$f(\lambda_1) = (2)\big[[4 + 2(-0.577350269)]^2 + 5(4 + 2(-0.577350269) + 3)\big] = 50.6444526769$$

$$f(\lambda_2) = (2)\big[[4 + 2(0.577350269)]^2 + 5(4 + 2(0.577350269) + 3)\big] = 110.688880653$$

$$I = (1)(50.6444526769) + (1)110.688880653 = 161.33333333$$

## EXAMPLE 3.6

Evaluate the integral $\int_{X_i}^{X_j} S_j^2\,dX$ in Example 3.4 using the Gauss–Legendre two-point formula.

Recall from Eq. (3.35) that $S_j = \frac{1}{2}(1 + \xi)$ and $dx = \frac{\ell}{2}\,d\xi$. Also note that for this problem, $\xi = \lambda$. So,

$$I = \int_{X_i}^{X_j} S_j^2\,dX = \int_{X_i}^{X_j}\left(\frac{X - X_i}{\ell}\right)^2 dX = \int_0^{\ell}\left(\frac{x}{\ell}\right)^2 dx = \frac{\ell}{2}\int_{-1}^{1}\left[\frac{1}{2}(1 + \xi)\right]^2 d\xi$$

Using the Gauss–Legendre two-point formula and Table 3.2, we compute the value of the integral $I$ from

$$I \cong w_1 f(\lambda_1) + w_2 f(\lambda_2)$$

From Table 3.2, we find that $w_1 = w_2 = 1$, and evaluating $f(\lambda)$ at $\lambda_1 = -0.577350269$ and $\lambda_2 = 0.577350269$, we obtain

$$f(\xi_1) = \frac{\ell}{2}\left[\frac{1}{2}(1 + \xi_1)\right]^2 = \frac{\ell}{2}\left[\frac{1}{2}(1 - 0.577350269)\right]^2 = 0.022329099389\ell$$

$$f(\xi_2) = \frac{\ell}{2}\left[\frac{1}{2}(1 + \xi_2)\right]^2 = \frac{\ell}{2}\left[\frac{1}{2}(1 + 0.577350269)\right]^2 = 0.31100423389\ell$$

$$I = (1)(0.022329099389\ell) + (1)(0.31100423389\ell) = 0.333333333\ell$$

Note that the above result is identical to the results of Example 3.4.

## 3.6 EXAMPLES OF ONE-DIMENSIONAL ELEMENTS IN ANSYS

ANSYS offers uniaxial link elements that may be used to represent one-dimensional problems. These link elements include LINK31, LINK32, and LINK34. The LINK32 element is a uniaxial heat conduction element. It allows for the transfer of heat between its two nodes via conduction mode. The nodal degree of freedom associated with this element is temperature. The element is defined by its two nodes, cross-sectional area, and material properties such as thermal conductivity. The LINK34 element is a uniaxial convection link that allows for heat transfer between its nodes by convection. This element is defined by its two nodes, a convective surface area, and a convective heat transfer (film) coefficient. The LINK31 element can be used to model radiation heat transfer between two points in space. The element is defined by its two nodes, a radiation surface area, a geometric shape factor, emmissivity, and the Stefan–Boltzman constant. In Chapter 4, we will use LINK32 and LINK34 to solve a one-dimensional heat-conduction problem.

## SUMMARY

At this point you should:

1. have a good understanding of the linear one-dimensional elements and shape functions, their properties, and their limitations.
2. have a good understanding of the quadratic and cubic one-dimensional elements and shape functions, their properties, and their advantages over linear elements.
3. know why it is important to use local and natural coordinate systems.
4. know what is meant by isoparametric element and formulation.
5. have a good understanding of Gauss–Legendre quadrature.
6. know examples of one-dimensional elements in ANSYS.

## REFERENCES

*ANSYS User's Manual: Elements*, Vol. III, Swanson Analysis Systems, Inc.

Chandrupatla, T., and Belegundu, A., *Introduction to Finite Elements in Engineering*, Prentice Hall, 1991.

Incropera, F. P., and DeWitt, D. P., *Fundamentals of Heat and Mass Transfer*, 2d. ed., New York, John Wiley and Sons, 1985.

Segrlind, L., *Applied Finite Element Analysis*, 2d. ed., New York John Wiley and Sons, 1984.

## PROBLEMS

1. We have used linear one-dimensional elements to approximate the temperature distribution along a fin. The nodal temperatures and their corresponding positions are shown in the accompanying figure. (a) What is the temperature of the fin at $X = 7$ cm? (b) Evaluate the heat loss from the fin using the relationship

$$Q = -kA\frac{dT}{dX}\Big|_{X=0}$$

where $k = 180$ W/m · K and $A = 10$ mm$^2$.



$$\begin{Bmatrix} T_1 \\ T_2 \\ T_3 \\ T_4 \end{Bmatrix} = \begin{Bmatrix} 100 \\ 80 \\ 64 \\ 41 \end{Bmatrix} °C$$

2. Evaluate the integral $\int_{X_i}^{X_j} S_i^2\, dX$ for a linear shape function using a) global coordinates and b) local coordinates.
3. Starting with the equations

$$T_i = c_1 + c_2 X_i + c_3 X_i^2$$
$$T_k = c_1 + c_2 X_k + c_3 X_k^2$$
$$T_j = c_1 + c_2 X_j + c_3 X_j^2$$

solve for $c_1$, $c_2$, and $c_3$, and rearrange terms to verify the shape functions given by

$$S_i = \frac{2}{\ell^2}(X - X_j)(X - X_k)$$

$$S_j = \frac{2}{\ell^2}(X - X_i)(X - X_k)$$

$$S_k = \frac{-4}{\ell^2}(X - X_i)(X - X_j)$$

4. For Problem 3, use the Lagrange functions to derive the quadratic shape functions by the method discussed in Section 3.3.

5. Derive the expressions for quadratic shape functions in terms of the local coordinates and compare your results to the results given in Table 3.1.

6. Verify the results given for one-dimensional quadratic natural shape functions in Table 3.1 by showing that (1) a shape function has a value of unity at its corresponding node and a value of zero at the other nodes and (2) if we sum up the shape functions, we will come up with a value of unity.

7. Verify the results given for the local cubic shape functions in Table 3.1 by showing that (1) a shape function has a value of unity at its corresponding node and a value of zero at the other nodes and (2) if we sum up the shape functions, we will come up with a value of unity.

8. Verify the results given for the natural cubic shape functions in Table 3.1 by showing that (1) a shape function has a value of unity at its corresponding node and a value of zero at the other nodes and (2) if we sum up the shape functions, we will come up with a value of unity.

9. Obtain expressions for the spatial derivatives of the quadratic and cubic shape functions.

10. As previously explained, we can increase the accuracy of our finite element findings either by increasing the number of elements used in the analysis to represent a problem or by using a higher order approximation. Derive the local cubic shape functions.

11. Evaluate the integral $\int_{X_i}^{X_j} S_i\, dX$ for a quadratic shape function using a) global coordinates, b) natural coordinates, and c) local coordinates.

12. Assume that the deflection of a cantilever beam was approximated with linear one-dimensional elements. The nodal deflections and their corresponding positions are shown in the accompanying figure. a) What is the deflection of the beam at $X = 2$ ft? b) Evaluate the slope at the endpoint.

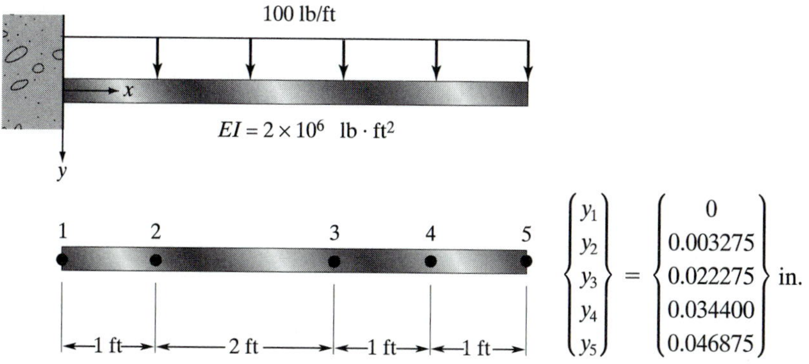


13. We have used linear one-dimensional elements to approximate the temperature distribution inside a metal plate. A heating element is embedded within a plate. The nodal temperatures and their corresponding positions are shown in the accompanying figure. What is the temperature of the plate at $X = 25$ mm? Assume that (a) linear elements were used in obtaining nodal temperatures and (b) quadratic elements were used.

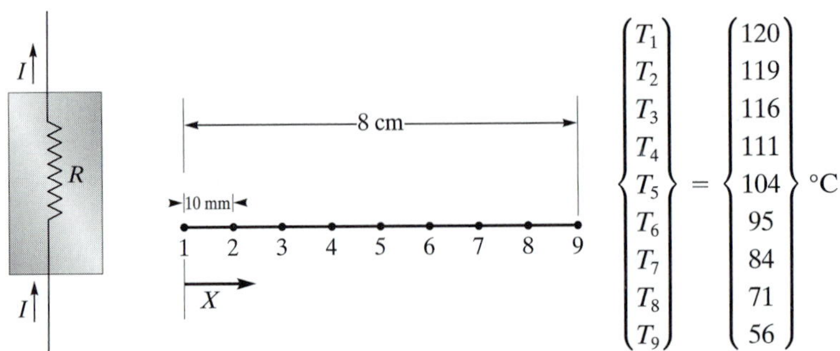


**14.** Quadratic elements are used to approximate the temperature distribution in a straight fin. The nodal temperatures and their corresponding positions are shown in the accompanying figure. What is the temperature of the fin at $X = 7$ cm?

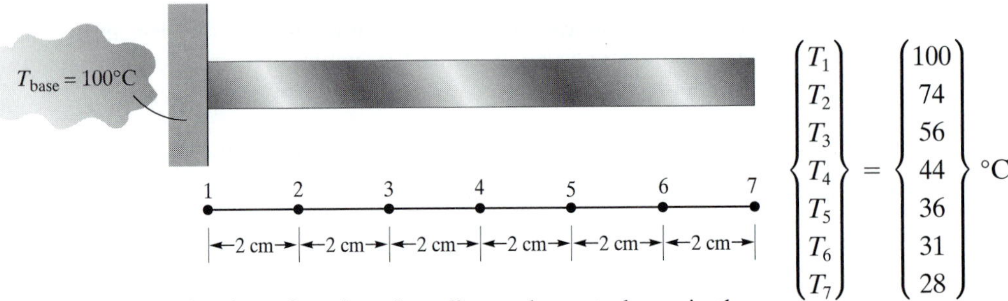


**15.** Develop the shape functions for a linear element, shown in the accompanying figure, using the local coordinate $x$ whose origin lies at the one-fourth point of the element.

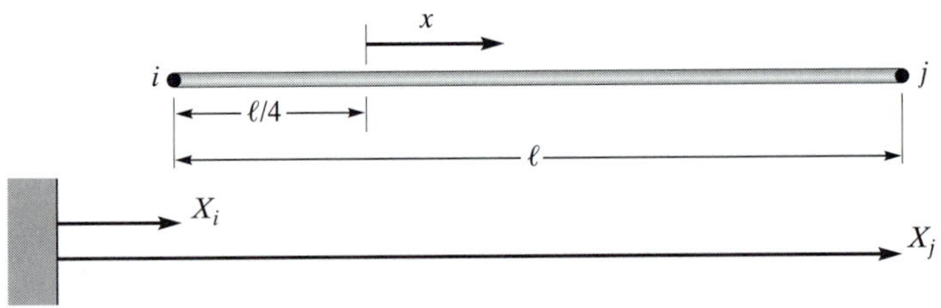


**16.** Using the natural coordinate system shown in the accompanying figure, develop the natural shape functions for a linear element.

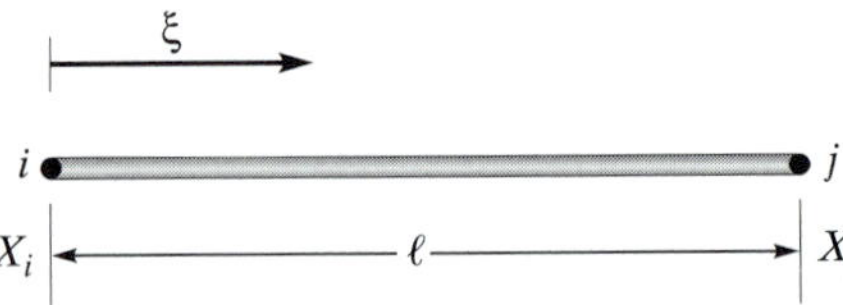


**17.** In the accompanying figure, the deflection of nodes 2 and 3 are 0.02 mm and 0.025 mm, respectively. What are the deflections at point $A$ and point $B$, provided that linear elements were used in the analysis?

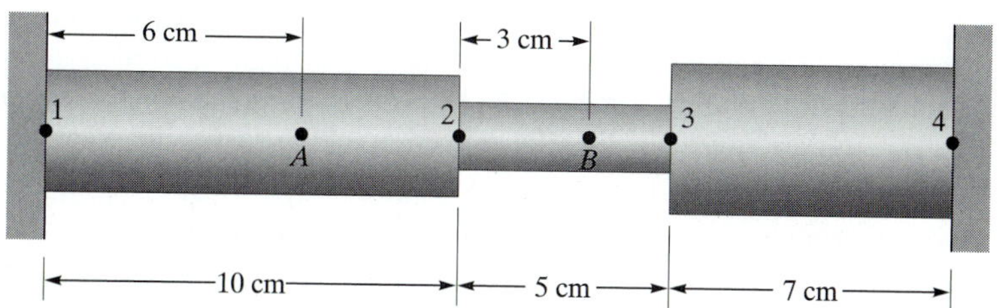


**18.** Consider the steel column in Example 3.2, depicted in the accompanying figure. Under the assumption of axial loading, and using linear elements, we determined that the vertical displacements of the column at various floor–column connection points are

$$\begin{Bmatrix} u_1 \\ u_2 \\ u_3 \\ u_4 \\ u_5 \end{Bmatrix} = \begin{Bmatrix} 0 \\ 0.03283 \\ 0.05784 \\ 0.07504 \\ 0.08442 \end{Bmatrix} \text{ in.}$$

Using local shape functions, determine the deflections of points $A$ and $B$, located in the middle of elements (3) and (4), respectively.

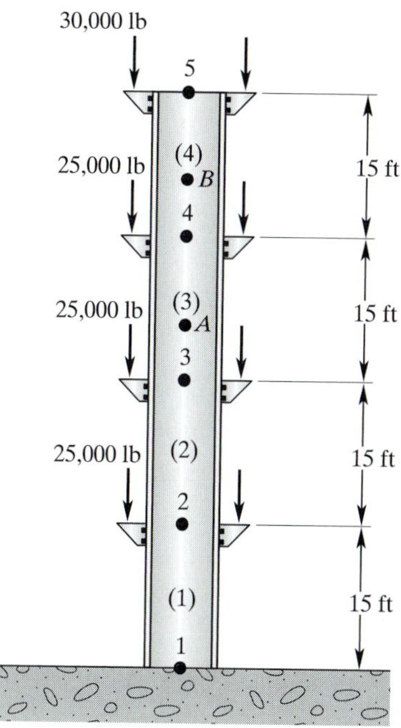


**19.** Determine the deflection of points $A$ and $B$ on the column in Problem 18 using natural coordinates.

**20.** A 20-ft-tall post is used to support advertisement signs at various locations along its height, as shown in the accompanying figure. The post is made of structural steel with a modulus of elasticity of $E = 29 \times 10^6$ lb/in$^2$. Not considering wind loading on the signs and using linear elements, we determined that the deflections of the post at the points of load application are

$$\begin{Bmatrix} u_1 \\ u_2 \\ u_3 \\ u_4 \end{Bmatrix} = \begin{Bmatrix} 0 \\ 6.312 \times 10^{-4} \\ 8.718 \times 10^{-4} \\ 11.470 \times 10^{-4} \end{Bmatrix} \text{ in.}$$

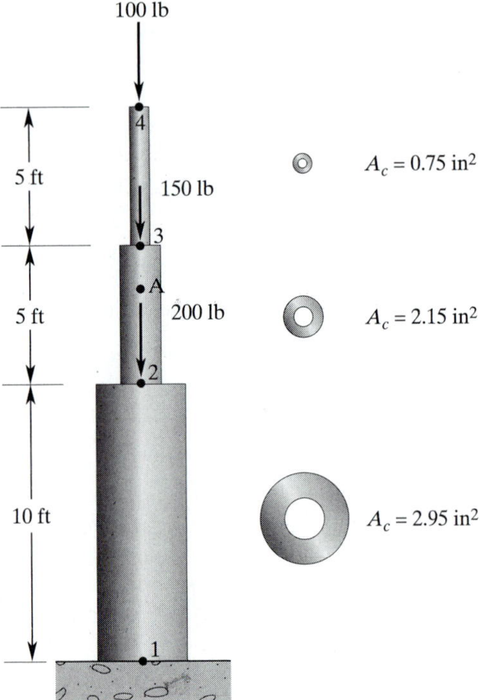


Determine the deflection of point $A$, located at the midpoint of the middle member, using (a) global shape functions, (b) local shape functions, and (c) natural shape functions.

**21.** Evaluate the integral in problem 11 using Gauss–Legendre two-point formula.

CHAPTER 4

# Analysis of One-Dimensional Problems

The main objective of this chapter is to introduce the analysis of one-dimensional problems. Most often, a physical problem is not truly one-dimensional in nature; however, as a starting point, we may model the behavior of a system using one-dimensional approximation. This approach can usually provide some basic insight into a more complex problem. If necessary, as a next step we can always analyze the problem using a two- or three-dimensional approach. This chapter first presents the one-dimensional Galerkin formulation used for heat transfer problems. This presentation will be followed by a discussion of the minimum potential energy formulation of one-dimensional solid mechanics problems. The main topics discussed in Chapter 4 are:

**4.1** Heat Transfer Problems
**4.2** Solid Mechanics Problems
**4.3** An Example Using ANSYS
**4.4** Verification of Results

## 4.1 HEAT TRANSFER PROBLEMS

Recall that in Chapter 1 we discussed the basic steps involved in any finite element analysis; to refresh your memory, these steps are repeated here:

### PREPROCESSING PHASE

1. Create and discretize the solution domain into finite elements; that is, subdivide the problem into nodes and elements.
2. Assume shape functions to represent the behavior of an element; that is, assume an approximate continuous function to represent the solution for a element. The one-dimensional linear and quadratic shape functions were discussed in Chapter 3.
3. Develop equations for an element. This step is the main focus of the current chapter. We will use the Galerkin approach, as well as the minimum potential energy theorem, to formulate elemental descriptions.
4. Assemble the elements to represent the entire problem. Construct the global stiffness or conductance matrix.
5. Apply boundary conditions and loading.

**SOLUTION PHASE**

**6.** Solve a set of linear algebraic equations simultaneously to obtain nodal results, such as the temperature at different nodes or displacements.

**POSTPROCESSING PHASE**

**7.** Obtain other important information. We may be interested in determining the heat loss or stress in each element.

We will now focus our attention on step 3 of the preprocessing phase. We will formulate the conductance and the thermal load matrices for a typical one-dimensional fin element. We considered a straight fin of a uniform cross section in Chapter 3. For the sake of convenience, the fin is shown here again in Figure 4.1. The fin is modeled using three elements and four nodes. The temperature distribution along the element is interpolated using linear functions. The actual and the approximate piecewise linear temperature distribution along the fin are shown in Figure 4.1. We will concentrate on a typical element belonging to the fin and formulate the conductance matrix and the thermal load matrix for such an element.

One-dimensional heat transfer in a straight fin is governed by the following heat equation, as given in any introductory text on heat transfer:

$$kA\frac{d^2T}{dX^2} - hpT + hpT_f = 0 \tag{4.1}$$

Equation (4.1) is derived by applying the conservation of energy to a differential section of a fin, as shown in Figure 4.2. The heat transfer in the fin is accomplished by conduction in the longitudinal direction ($x$-direction) and convection to the surrounding fluid. In Eq. (4.1), $k$ is the thermal conductivity, and $A$ denotes the cross-sectional area of the fin. The convective heat transfer coefficient is represented by $h$, the perimeter of the fin is denoted by $p$, and $T_f$ is the temperature of the surrounding fluid. Equation

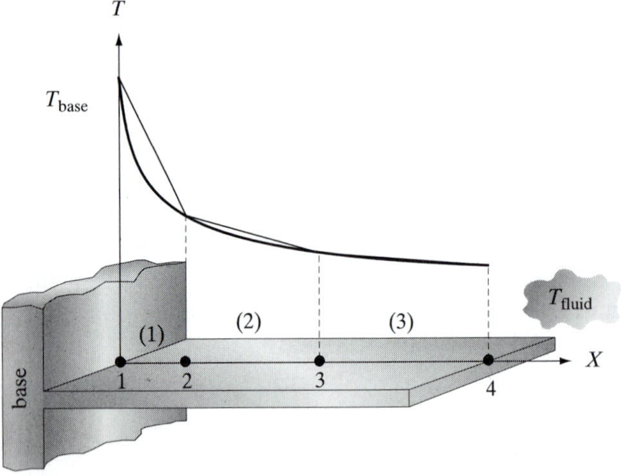


**FIGURE 4.1** The actual and approximate temperature distribution for a fin of uniform cross section.

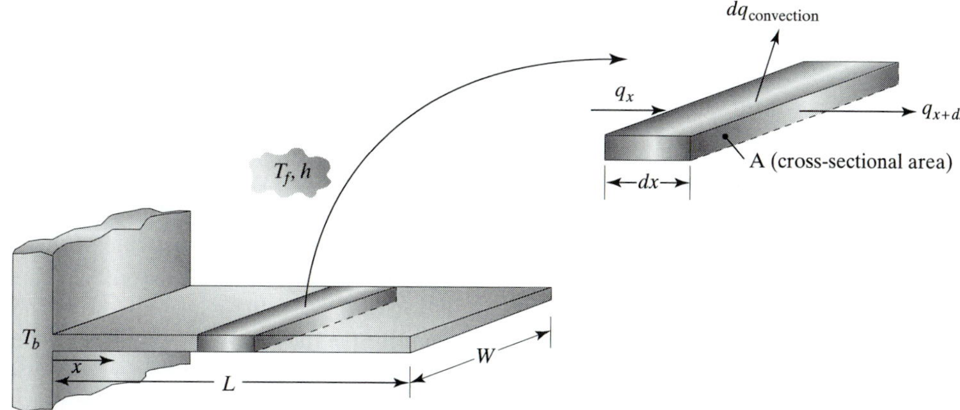


We start by applying the energy balance to a differential element

$$q_x = q_{x+dx} + dq_{\text{convection}}$$

$$q_x = q_x + \frac{dq_x}{dx}dx + dq_{\text{convection}}$$

Next we use Fourier's Law

$$q_x = -kA\frac{dT}{dx}$$

and use Newton's Law of Cooling,

$$dq_{\text{convection}} = h(dA_s)(T - T_f)$$

$$0 = \frac{dq_x}{dx}dx + dq_{\text{convection}} = \frac{d}{dx}\left(-kA\frac{dT}{dx}\right)dx + h(dA_s)(T - T_f)$$

Writing $dA_s$ (differential surface area) in terms of the perimeter of the fin and $dx$ and simplifying, we are left with

$$-kA\frac{d^2T}{dx^2} + hp(T - T_f) = 0$$

**FIGURE 4.2** Derivation of the heat equation for a fin.

(4.1) is subjected to a set of boundary conditions. First, the temperature of the base is generally known; that is;

$$T(0) = T_b \tag{4.2}$$

The other boundary condition deals with the heat loss at the tip of the fin. In general, there are three possibilities. One possibility is that the tip is long enough so that the temperature of the tip is equal to the temperature of the surrounding fluid temperature. This situation is represented by the condition

$$T(L) = T_f \tag{4.3}$$

The situation in which the heat loss from the tip of the fin may be neglected is represented by the condition

$$-kA\frac{dT}{dX}\bigg|_{X=L} = 0 \tag{4.4}$$

If the heat loss from the tip of the fin should be included in the analysis, then we have the condition

$$-kA\frac{dT}{dX}\bigg|_{X=L} = hA(T_L - T_f) \tag{4.5}$$

Equation (4.5) is obtained by applying the energy balance to the cross-sectional area of the tip. Equation (4.5) simply states that the heat conducted to the tip's surface is convected away by the surrounding fluid. Therefore, we can use one of the boundary conditions given by Eqs. (4.3)–(4.5) and the base temperature to model an actual problem. Before we proceed with the formulation of the conductance matrix and the thermal load matrix for a typical element, let us emphasize the following points: (1) The governing differential equation of the fin represents the balance of energy at any point along the fin and, thus, governs the balance of energy at all nodes of a finite element model as well and (2) the exact solution of the governing differential equation (if possible) subject to two appropriate boundary conditions renders the detailed temperature distribution along the fin, and the finite element solution represents an approximation of this solution. We will now focus on a typical element and proceed with the formulation of the conductance matrix, recalling that the temperature distribution for a typical element may be approximated using linear shape functions, as discussed in Chapter 3. That is,

$$T^{(e)} = \begin{bmatrix} S_i & S_j \end{bmatrix}\begin{Bmatrix} T_i \\ T_j \end{Bmatrix} \tag{4.6}$$

where the shape functions are given by:

$$S_i = \frac{X_j - X}{\ell} \quad \text{and} \quad S_j = \frac{X - X_i}{\ell} \tag{4.7}$$

In order to make this derivation as general as possible and applicable to other type of problems with the same form of differential equations, let $c_1 = kA$, $c_2 = -hp$, $c_3 = hpT_f$, and $\Psi = T$. Thus, Eq. (4.1) can be written as

$$c_1\frac{d^2\Psi}{dX^2} + c_2\Psi + c_3 = 0 \tag{4.8}$$

Recall from our introductory discussion of weighted residual methods in Chapter 1 that when we substitute an approximate solution into the governing differential equation, the approximate solution does not satisfy the differential equation exactly, and thus, an error, or a residual, is produced. Also recall that the Galerkin formulation requires the error to be orthogonal to some weighting functions. Furthermore, the weighting functions are chosen to be members of the approximate solution. Here we will use the shape functions as the weighting functions because they are members of the approximate solution. The Galerkin residuals for an arbitrary element with nodes $i$ and $j$ are forced to be zero according to the equations

$$R_i^{(e)} = \int_{X_i}^{X_j} S_i\left(c_1\frac{d^2\Psi}{dX^2} + c_2\Psi + c_3\right)dX = 0 \tag{4.9}$$

$$R_j^{(e)} = \int_{X_i}^{X_j} S_j\left( c_1 \frac{d^2\Psi}{dX^2} + c_2\Psi + c_3 \right) dX = 0 \tag{4.10}$$

Evaluation of the integrals given by Eqs. (4.9) and (4.10) will result in the elemental formulation. But first, because the second derivative of a linear function equals zero, we need to manipulate the second-order terms into first-order terms. This manipulation is accomplished by using the chain rule in the following manner:

$$\frac{d}{dX}\left( S_i \frac{d\Psi}{dX} \right) = S_i \frac{d^2\Psi}{dX^2} + \frac{dS_i}{dX}\frac{d\Psi}{dX} \tag{4.11}$$

$$S_i \frac{d^2\Psi}{dX^2} = \frac{d}{dX}\left( S_i \frac{d\Psi}{dX} \right) - \frac{dS_i}{dX}\frac{d\Psi}{dX} \tag{4.12}$$

Substituting Eq. (4.12) into Eq. (4.9), we obtain

$$R_i^{(e)} = \int_{X_i}^{X_j} \left( c_1\left( \frac{d}{dX}\left( S_i \frac{d\Psi}{dX} \right) - \frac{dS_i}{dX}\frac{d\Psi}{dX} \right) + S_i(c_2\Psi + c_3) \right) dX = 0 \tag{4.13}$$

We eventually need to follow the same procedure for Eq. (4.10) as well, but for now let us focus only on one of the residual equations. There are four terms in Eq. (4.13) that need to be evaluated:

$$R_i^{(e)} = \int_{X_i}^{X_j} c_1\left( \frac{d}{dX}\left( S_i \frac{d\Psi}{dX} \right) \right) dX + \int_{X_i}^{X_j} c_1\left( -\frac{dS_i}{dX}\frac{d\Psi}{dX} \right) dX$$
$$+ \int_{X_i}^{X_j} S_i(c_2\Psi)\, dX + \int_{X_i}^{X_j} S_i c_3\, dX = 0 \tag{4.14}$$

Considering and evaluating the first term, we have:

$$\int_{X_i}^{X_j} c_1\left( \frac{d}{dX}\left( S_i \frac{d\Psi}{dX} \right) \right) dX = -c_1 \left. \frac{d\Psi}{dX} \right|_{X=X_i} \tag{4.15}$$

It is important to realize that in order for us to obtain the result given by Eq. (4.15), $S_i$ is zero at $X = X_j$ and $S_i = 1$ at $X = X_i$. The second integral in Eq. (4.14) is evaluated as

$$\int_{X_i}^{X_j} c_1\left( -\frac{dS_i}{dX}\frac{d\Psi}{dX} \right) dX = -\frac{c_1}{\ell}(\Psi_i - \Psi_j) \tag{4.16}$$

Evaluation of the third and the fourth integrals in Eq. (4.14) yields:

$$\int_{X_i}^{X_j} S_i(c_2\Psi)\, dX = \frac{c_2\ell}{3}\Psi_i + \frac{c_2\ell}{6}\Psi_j \tag{4.17}$$

$$\int_{X_i}^{X_j} S_i c_3\, dX = c_3\frac{\ell}{2} \tag{4.18}$$

In exactly the same manner, we can evaluate the second residual equation for node $j$, as given by Eq. (4.10). This evaluation results in the following equations:

$$\int_{X_i}^{X_j} c_1\left(\frac{d}{dX}\left(S_j\frac{d\Psi}{dX}\right)\right)dX = c_1\left.\frac{d\Psi}{dX}\right|_{X=X_j} \tag{4.19}$$

$$\int_{X_i}^{X_j} c_1\left(-\frac{dS_j}{dX}\frac{d\Psi}{dX}\right)dX = -\frac{c_1}{\ell}(-\Psi_i + \Psi_j) \tag{4.20}$$

$$\int_{X_i}^{X_j} S_j(c_2\Psi)\,dX = \frac{c_2\ell}{6}\Psi_i + \frac{c_2\ell}{3}\Psi_j \tag{4.21}$$

$$\int_{X_i}^{X_j} S_jc_3\,dX = c_3\frac{\ell}{2} \tag{4.22}$$

It should be clear by now that evaluation of Eqs. (4.9) and (4.10) results in two sets of linear equations, as given by:

$$\begin{Bmatrix} R_i \\ R_j \end{Bmatrix} = \begin{Bmatrix} -c_1\left.\dfrac{d\Psi}{dX}\right|_{X=X_i} \\ c_1\left.\dfrac{d\Psi}{dX}\right|_{X=X_j} \end{Bmatrix} - \frac{c_1}{\ell}\begin{Bmatrix} 1 & -1 \\ -1 & 1 \end{Bmatrix}\begin{Bmatrix} \Psi_i \\ \Psi_j \end{Bmatrix}$$
$$+\frac{c_2\ell}{6}\begin{bmatrix} 2 & 1 \\ 1 & 2 \end{bmatrix}\begin{Bmatrix} \Psi_i \\ \Psi_j \end{Bmatrix} + \frac{c_3\ell}{2}\begin{Bmatrix} 1 \\ 1 \end{Bmatrix} = \begin{Bmatrix} 0 \\ 0 \end{Bmatrix} \tag{4.23}$$

We can rewrite the Eq. (4.23) as

$$\begin{Bmatrix} c_1\left.\dfrac{d\Psi}{dX}\right|_{X=X_i} \\ -c_1\left.\dfrac{d\Psi}{dX}\right|_{X=X_j} \end{Bmatrix} + \frac{c_1}{\ell}\begin{Bmatrix} 1 & -1 \\ -1 & 1 \end{Bmatrix}\begin{Bmatrix} \Psi_i \\ \Psi_j \end{Bmatrix} + \frac{-c_2\ell}{6}\begin{bmatrix} 2 & 1 \\ 1 & 2 \end{bmatrix}\begin{Bmatrix} \Psi_i \\ \Psi_j \end{Bmatrix} = \frac{c_3\ell}{2}\begin{Bmatrix} 1 \\ 1 \end{Bmatrix} \tag{4.24}$$

Combining the unknown nodal parameters, we obtain

$$\begin{Bmatrix} c_1\left.\dfrac{d\Psi}{dX}\right|_{X=X_i} \\ -c_1\left.\dfrac{d\Psi}{dX}\right|_{X=X_j} \end{Bmatrix} + \{[\mathbf{K}]_{c_1}^{(e)} + [\mathbf{K}]_{c_2}^{(e)}\}\begin{Bmatrix} \Psi_i \\ \Psi_j \end{Bmatrix} = \{\mathbf{F}\}^{(e)} \tag{4.25}$$

where

$$[\mathbf{K}]_{c_1}^{(e)} = \frac{c_1}{\ell}\begin{bmatrix} 1 & -1 \\ -1 & 1 \end{bmatrix}$$

is the elemental conductance for a heat transfer problem (or, it could represent the stiffness for solid mechanics problems) due to the $c_1$ coefficient,

$$[\mathbf{K}]_{c_2}^{(e)} = \frac{-c_2\ell}{6}\begin{bmatrix} 2 & 1 \\ 1 & 2 \end{bmatrix}$$

is the elemental conductance (or, for a solid mechanics problem, the stiffness) due to the $c_2$ coefficient, and

$$\{\mathbf{F}\}^{(e)} = \frac{c_3 \ell}{2}\begin{Bmatrix} 1 \\ 1 \end{Bmatrix}$$

is the load matrix for a given element. The terms

$$\begin{Bmatrix} c_1 \left.\dfrac{d\Psi}{dX}\right|_{X=X_i} \\ -c_1 \left.\dfrac{d\Psi}{dX}\right|_{X=X_j} \end{Bmatrix}$$

contribute to both the conductance (or, for a solid-mechanics problem, the stiffness) matrix and the load matrix. They need to be evaluated for specific boundary conditions. We shall undertake this task shortly. However, let us first write down the conductance matrix for a typical one-dimensional fin in terms of its parameters. The conductance matrices are given by:

$$[\mathbf{K}]_{c_1}^{(e)} = \frac{c_1}{\ell}\begin{bmatrix} 1 & -1 \\ -1 & 1 \end{bmatrix} = \frac{kA}{\ell}\begin{bmatrix} 1 & -1 \\ -1 & 1 \end{bmatrix} \tag{4.26}$$

and

$$[\mathbf{K}]_{c_2}^{(e)} = \frac{-c_2 \ell}{6}\begin{bmatrix} 2 & 1 \\ 1 & 2 \end{bmatrix} = \frac{hp\ell}{6}\begin{bmatrix} 2 & 1 \\ 1 & 2 \end{bmatrix} \tag{4.27}$$

In general, the elemental conductance matrix may consist of three terms: The $[\mathbf{K}]_{c_1}^{(e)}$ term is due to conduction loss along the fin (through the cross-sectional area); the $[\mathbf{K}]_{c_2}^{(e)}$ term represents the heat loss through the top, bottom, and side surfaces (periphery) of an element of a fin; and, depending on the boundary condition of the tip, an additional elemental conductance matrix $[\mathbf{K}]_{\text{B.C.}}^{(e)}$ can exist. For the very last element containing the tip surface, and referring to the boundary condition given by Eq. (4.5), the heat loss through the tip surface can be evaluated as

$$\begin{Bmatrix} c_1 \left.\dfrac{d\Psi}{dX}\right|_{X=X_i} \\ -c_1 \left.\dfrac{d\Psi}{dX}\right|_{X=X_j} \end{Bmatrix} = \begin{Bmatrix} kA \left.\dfrac{dT}{dX}\right|_{X=X_i} \\ -kA \left.\dfrac{dT}{dX}\right|_{X=X_j} \end{Bmatrix} = \begin{Bmatrix} 0 \\ hA(T_j - T_f) \end{Bmatrix} \tag{4.28}$$

By rearranging and simplifying, we have:

$$\begin{Bmatrix} kA \left.\dfrac{dT}{dX}\right|_{X=X_i} \\ -kA \left.\dfrac{dT}{dX}\right|_{X=X_j} \end{Bmatrix} = \begin{Bmatrix} 0 \\ hA(T_j - T_f) \end{Bmatrix} = \begin{bmatrix} 0 & 0 \\ 0 & hA \end{bmatrix}\begin{Bmatrix} T_i \\ T_j \end{Bmatrix} - \begin{Bmatrix} 0 \\ hAT_f \end{Bmatrix} \tag{4.29}$$

$$[\mathbf{K}]_{\text{B.C.}}^{(e)} = \begin{bmatrix} 0 & 0 \\ 0 & hA \end{bmatrix} \tag{4.30}$$

The term $-\begin{Bmatrix} 0 \\ hAT_f \end{Bmatrix}$ belongs to the right side of Eq. (4.25) with the thermal load matrix. It shows the contribution of the boundary condition of the tip to the load matrix:

$$\{\mathbf{F}\}^{(e)}_{\text{B.C.}} = \begin{Bmatrix} 0 \\ hAT_f \end{Bmatrix} \tag{4.31}$$

To summarize, the conductance matrix for all elements, excluding the last element, is given by

$$[\mathbf{K}]^{(e)} = \left\{ \frac{kA}{\ell} \begin{bmatrix} 1 & -1 \\ -1 & 1 \end{bmatrix} + \frac{hp\ell}{6} \begin{bmatrix} 2 & 1 \\ 1 & 2 \end{bmatrix} \right\} \tag{4.32}$$

If the heat loss through the tip of the fin must be accounted for, the conductance matrix for the very last element must be computed from the equation

$$[\mathbf{K}]^{(e)} = \left\{ \frac{kA}{\ell} \begin{bmatrix} 1 & -1 \\ -1 & 1 \end{bmatrix} + \frac{hp\ell}{6} \begin{bmatrix} 2 & 1 \\ 1 & 2 \end{bmatrix} + \begin{bmatrix} 0 & 0 \\ 0 & hA \end{bmatrix} \right\} \tag{4.33}$$

The thermal load matrix for all elements, excluding the last element, is given by

$$\{\mathbf{F}\}^{(e)} = \frac{hp\ell T_f}{2} \begin{Bmatrix} 1 \\ 1 \end{Bmatrix} \tag{4.34}$$

If the heat loss through the tip of the fin must be included in the analysis, the thermal load matrix for the very last element must be computed from the relation

$$\{\mathbf{F}\}^{(e)} = \frac{hp\ell T_f}{2} \begin{Bmatrix} 1 \\ 1 \end{Bmatrix} + \begin{Bmatrix} 0 \\ hAT_f \end{Bmatrix} \tag{4.35}$$

The next set of examples demonstrates the assembly of elements to present the entire problem and the treatment of other boundary conditions.

---

**EXAMPLE 4.1: A Fin Problem**

Aluminum fins of a rectangular profile, shown in Figure 4.3, are used to remove heat from a surface whose temperature is 100°C. The temperature of the ambient air is 20°C. The thermal conductivity of aluminum is 168 W/m·K (W/m·°C). The natural convective heat transfer coefficient associated with the surrounding air is 30 W/m$^2$·K (W/m$^2$·°C). The fins are 80 mm long, 5 mm wide, and 1 mm thick. (a) Determine the temperature distribution along the fin using the finite element model shown in Figure 4.3. (b) Compute the heat loss per fin.

We will solve this problem using two boundary conditions for the tip. First, let us include the heat transfer from the tip's surface in the analysis. For elements (1), (2), and (3) in the situation, the conductance and thermal load matrices are given by

$$[\mathbf{K}]^{(e)} = \left\{ \frac{kA}{\ell} \begin{bmatrix} 1 & -1 \\ -1 & 1 \end{bmatrix} + \frac{hp\ell}{6} \begin{bmatrix} 2 & 1 \\ 1 & 2 \end{bmatrix} \right\}$$

$$\{\mathbf{F}\}^{(e)} = \frac{hp\ell T_f}{2} \begin{Bmatrix} 1 \\ 1 \end{Bmatrix}$$

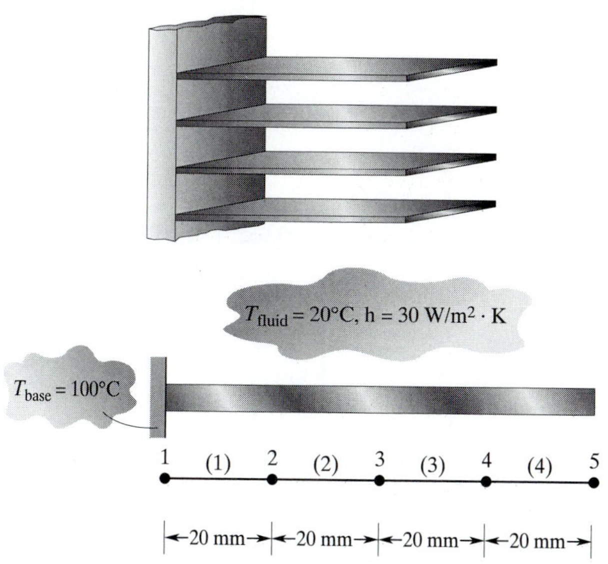


**FIGURE 4.3** Finite element model of a straight fin.

Substituting for the properties, we obtain

$$[\mathbf{K}]^{(e)} = \left\{ \frac{(168)(5 \times 1 \times 10^{-6})}{20 \times 10^{-3}} \begin{bmatrix} 1 & -1 \\ -1 & 1 \end{bmatrix} + \frac{30 \times 12 \times 20 \times 10^{-6}}{6} \begin{bmatrix} 2 & 1 \\ 1 & 2 \end{bmatrix} \right\}$$

$$\{\mathbf{F}\}^{(e)} = \frac{30 \times 12 \times 20 \times 10^{-6} \times 20}{2} \begin{Bmatrix} 1 \\ 1 \end{Bmatrix} = \begin{Bmatrix} 0.072 \\ 0.072 \end{Bmatrix}$$

The conductance matrix for elements (1), (2), and (3) is

$$[\mathbf{K}]^{(1)} = [\mathbf{K}]^{(2)} = [\mathbf{K}]^{(3)} = \begin{bmatrix} 0.0444 & -0.0408 \\ -0.0408 & 0.0444 \end{bmatrix} \frac{\text{W}}{^{\circ}\text{C}}$$

and the thermal-load matrix for elements (1), (2), and (3) is

$$\{\mathbf{F}\}^{(1)} = \{\mathbf{F}\}^{(2)} = \{\mathbf{F}\}^{(3)} = \begin{Bmatrix} 0.072 \\ 0.072 \end{Bmatrix} \text{W}$$

Including the boundary condition of the tip, the conductance and load matrices for element (4) are obtained in the following manner:

$$[\mathbf{K}]^{(e)} = \left\{ \frac{kA}{\ell} \begin{bmatrix} 1 & -1 \\ -1 & 1 \end{bmatrix} + \frac{hp\,\ell}{6} \begin{bmatrix} 2 & 1 \\ 1 & 2 \end{bmatrix} + \begin{bmatrix} 0 & 0 \\ 0 & hA \end{bmatrix} \right\}$$

$$[\mathbf{K}]^{(4)} = \begin{bmatrix} 0.0444 & -0.0408 \\ -0.0408 & 0.0444 \end{bmatrix} + \begin{bmatrix} 0 & 0 \\ 0 & (30 \times 5 \times 1 \times 10^{-6}) \end{bmatrix} = \begin{bmatrix} 0.0444 & -0.0408 \\ -0.0408 & 0.04455 \end{bmatrix} \frac{\text{W}}{^{\circ}\text{C}}$$

$$\{\mathbf{F}\}^{(e)} = \frac{hp\ell T_f}{2} \begin{Bmatrix} 1 \\ 1 \end{Bmatrix} + \begin{Bmatrix} 0 \\ hAT_f \end{Bmatrix}$$

$$\{\mathbf{F}\}^{(4)} = \begin{Bmatrix} 0.072 \\ 0.072 \end{Bmatrix} + \begin{Bmatrix} 0 \\ (30 \times 5 \times 1 \times 10^{-6} \times 20) \end{Bmatrix} = \begin{Bmatrix} 0.072 \\ 0.075 \end{Bmatrix} \text{W}$$

Assembly of the elements leads to the global conductance matrix $[\mathbf{K}]^{(G)}$ and the global load matrix $\{\mathbf{F}\}^{(G)}$:

$$[\mathbf{K}]^{(G)} = \begin{bmatrix} 0.0444 & -0.0408 & 0 & 0 & 0 \\ -0.0408 & 0.0444 + 0.0444 & -0.0408 & 0 & 0 \\ 0 & -0.0408 & 0.0444 + 0.0444 & -0.0408 & 0 \\ 0 & 0 & -0.0408 & 0.0444 + 0.0444 & -0.0408 \\ 0 & 0 & 0 & -0.0408 & 0.04455 \end{bmatrix}$$

$$\{\mathbf{F}\}^{(G)} = \begin{Bmatrix} 0.072 \\ 0.072 + 0.072 \\ 0.072 + 0.072 \\ 0.072 + 0.072 \\ 0.075 \end{Bmatrix}$$

Applying the base boundary condition $T_1 = 100°\text{C}$, we find that the final set of linear equations becomes

$$\begin{bmatrix} 1 & 0 & 0 & 0 & 0 \\ -0.0408 & 0.0888 & -0.0408 & 0 & 0 \\ 0 & -0.0408 & 0.0888 & -0.0408 & 0 \\ 0 & 0 & -0.0408 & 0.0888 & -0.0408 \\ 0 & 0 & 0 & -0.0408 & 0.04455 \end{bmatrix} \begin{Bmatrix} T_1 \\ T_2 \\ T_3 \\ T_4 \\ T_5 \end{Bmatrix} = \begin{Bmatrix} 100 \\ 0.144 \\ 0.144 \\ 0.144 \\ 0.075 \end{Bmatrix}$$

We can obtain the nodal temperatures from the solution of the above equation. The nodal solutions are:

$$\begin{Bmatrix} T_1 \\ T_2 \\ T_3 \\ T_4 \\ T_5 \end{Bmatrix} = \begin{Bmatrix} 100 \\ 75.03 \\ 59.79 \\ 51.56 \\ 48.90 \end{Bmatrix} °\text{C}$$

Note that the nodal temperatures are given in °C and not in °K.

Because the cross-sectional area of the given fin is relatively small, we could have neglected the heat loss from the tip. Under this assumption, the elemental conductance and forcing matrices for all elements are given by:

$$[\mathbf{K}]^{(1)} = [\mathbf{K}]^{(2)} = [\mathbf{K}]^{(3)} = [\mathbf{K}]^{(4)} = \begin{bmatrix} 0.0444 & -0.0408 \\ -0.0408 & 0.0444 \end{bmatrix} \frac{\text{W}}{°\text{C}}$$

$$\{\mathbf{F}\}^{(1)} = \{\mathbf{F}\}^{(2)} = \{\mathbf{F}\}^{(3)} = \{\mathbf{F}\}^{(4)} = \begin{Bmatrix} 0.072 \\ 0.072 \end{Bmatrix} \text{W}$$

Assembly of the elements leads to the global conductance matrix $[\mathbf{K}]^G$ and the global load matrix $\{\mathbf{F}\}^G$:

$$[\mathbf{K}]^{(G)} = \begin{bmatrix} 0.0444 & -0.0408 & 0 & 0 & 0 \\ -0.0408 & 0.0444 + 0.0444 & -0.0408 & 0 & 0 \\ 0 & -0.0408 & 0.0444 + 0.0444 & -0.0408 & 0 \\ 0 & 0 & -0.0408 & 0.0444 + 0.0444 & -0.0408 \\ 0 & 0 & 0 & -0.0408 & 0.0444 \end{bmatrix}$$

$$\{\mathbf{F}\}^{(G)} = \begin{Bmatrix} 0.072 \\ 0.072 + 0.072 \\ 0.072 + 0.072 \\ 0.072 + 0.072 \\ 0.072 \end{Bmatrix}$$

Applying the base boundary condition $T_1 = 100°\text{C}$, we find that the final set of linear equations becomes

$$\begin{bmatrix} 1 & 0 & 0 & 0 & 0 \\ -0.0408 & 0.0888 & -0.0408 & 0 & 0 \\ 0 & -0.0408 & 0.0888 & -0.0408 & 0 \\ 0 & 0 & -0.0408 & 0.0888 & -0.0408 \\ 0 & 0 & 0 & -0.0408 & 0.0444 \end{bmatrix} \begin{Bmatrix} T_1 \\ T_2 \\ T_3 \\ T_4 \\ T_5 \end{Bmatrix} = \begin{Bmatrix} 100 \\ 0.144 \\ 0.144 \\ 0.144 \\ 0.072 \end{Bmatrix}$$

which is approximately the same solution as that calculated previously:

$$\begin{Bmatrix} T_1 \\ T_2 \\ T_3 \\ T_4 \\ T_5 \end{Bmatrix} = \begin{Bmatrix} 100 \\ 75.08 \\ 59.89 \\ 51.74 \\ 49.19 \end{Bmatrix} °\text{C}$$

Compared to the previous results, the nodal temperatures are slightly higher because we neglected the heat loss through the end surface of the tip.

The total heat loss $Q$ from the fin can be determined by summing the heat loss through individual elements:

$$Q_{\text{total}} = \Sigma Q^{(e)} \tag{4.36}$$

$$Q^{(e)} = \int_{X_i}^{X_j} hp(T - T_f)\, dX$$

$$= \int_{X_i}^{X_j} hp((S_i T_i + S_j T_j) - T_f)\, dX = hp\ell\left(\left(\frac{T_i + T_j}{2}\right) - T_f\right) \tag{4.37}$$

Applying the temperature results to Eqs. (4.36) and (4.37), we have:

$$Q_{\text{total}} = Q^{(1)} + Q^{(2)} + Q^{(3)} + Q^{(4)}$$

$$Q^{(1)} = hp\,\ell\left(\left(\frac{T_i + T_j}{2}\right) - T_f\right) = 30 \times 12 \times 20 \times 10^{-6}\left(\left(\frac{100 + 75.08}{2}\right) - 20\right) = 0.4862 \text{ W}$$

$$Q^{(2)} = 30 \times 12 \times 20 \times 10^{-6}\left(\left(\frac{75.08 + 59.89}{2}\right) - 20\right) = 0.3418 \text{ W}$$

$$Q^{(3)} = 30 \times 12 \times 20 \times 10^{-6}\left(\left(\frac{59.89 + 51.74}{2}\right) - 20\right) = 0.2578 \text{ W}$$

$$Q^{(4)} = 30 \times 12 \times 20 \times 10^{-6}\left(\left(\frac{51.74 + 49.19}{2}\right) - 20\right) = 0.2193 \text{ W}$$

$$Q_{\text{total}} = 1.3051 \text{ W}$$

---

### EXAMPLE 4.2: A Composite Wall Problem

A wall of an industrial oven consists of three different materials, as depicted in Figure 4.4. The first layer is composed of 5 cm of insulating cement with a clay binder that has a thermal conductivity of 0.08 W/m · K. The second layer is made from 15 cm of 6-ply

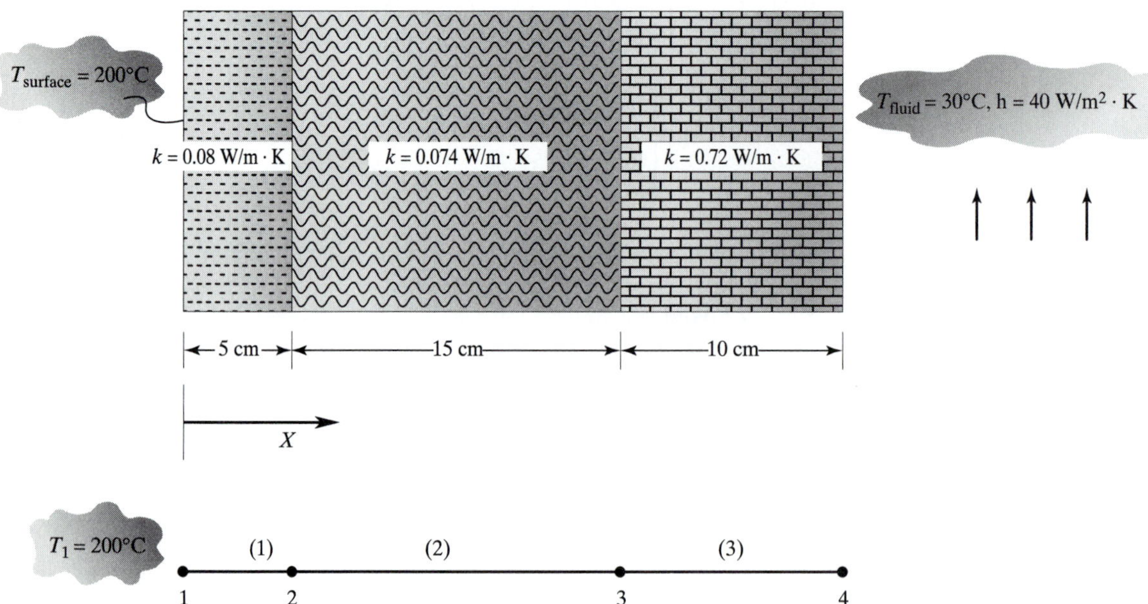


**FIGURE 4.4** A composite wall of an industrial oven.

asbestos board with a thermal conductivity of 0.074 W/m · K. The exterior consists of 10-cm common brick with a thermal conductivity of 0.72 W/m$^2$ · K. The inside wall temperature of the oven is 200°C, and the outside air is 30°C with a convection coefficient of 40 W/m$^2$ · K. Determine the temperature distribution along the composite wall.

This heat conduction problem is governed by the equation

$$kA\frac{d^2T}{dX^2} = 0 \tag{4.38}$$

and is subjected to the boundary conditions $T_1 = 200°\text{C}$ and $-kA\left.\frac{dT}{dX}\right|_{X=30\text{ cm}} = hA(T_4 - T_f)$. For this example, we compare Eq. (4.38) to Eq. (4.8), finding that $c_1 = kA$, $c_2 = 0, c_3 = 0$, and $\Psi = T$. Thus, for element (1), we have:

$$[\mathbf{K}]^{(1)} = \frac{kA}{\ell}\begin{bmatrix} 1 & -1 \\ -1 & 1 \end{bmatrix} = \frac{0.08 \times 1}{0.05}\begin{bmatrix} 1 & -1 \\ -1 & 1 \end{bmatrix} = \begin{bmatrix} 1.6 & -1.6 \\ -1.6 & 1.6 \end{bmatrix}\frac{\text{W}}{°\text{C}}$$

$$\{\mathbf{F}\}^{(1)} = \begin{Bmatrix} 0 \\ 0 \end{Bmatrix}\text{W}$$

For element (2), we have:

$$[\mathbf{K}]^{(2)} = \frac{kA}{\ell}\begin{bmatrix} 1 & -1 \\ -1 & 1 \end{bmatrix} = \frac{0.074 \times 1}{0.15}\begin{bmatrix} 1 & -1 \\ -1 & 1 \end{bmatrix} = \begin{bmatrix} 0.493 & -0.493 \\ -0.493 & 0.493 \end{bmatrix}\frac{\text{W}}{°\text{C}}$$

$$\{\mathbf{F}\}^{(2)} = \begin{Bmatrix} 0 \\ 0 \end{Bmatrix}\text{W}$$

For element (3), including the boundary condition, we have

$$[\mathbf{K}]^{(3)} = \frac{kA}{\ell}\begin{bmatrix} 1 & -1 \\ -1 & 1 \end{bmatrix} + \begin{bmatrix} 0 & 0 \\ 0 & hA \end{bmatrix} = \frac{0.72 \times 1}{0.1}\begin{bmatrix} 1 & -1 \\ -1 & 1 \end{bmatrix} + \begin{bmatrix} 0 & 0 \\ 0 & (40 \times 1) \end{bmatrix}$$

$$= \begin{bmatrix} 7.2 & -7.2 \\ -7.2 & 47.2 \end{bmatrix}\frac{\text{W}}{°\text{C}}$$

$$\{\mathbf{F}\}^{(3)} = \begin{Bmatrix} 0 \\ hAT_f \end{Bmatrix} = \begin{Bmatrix} 0 \\ (40 \times 1 \times 30) \end{Bmatrix} = \begin{Bmatrix} 0 \\ 1200 \end{Bmatrix}\text{W}$$

Assembling elements, we obtain

$$[\mathbf{K}]^{(G)} = \begin{bmatrix} 1.6 & -1.6 & 0 & 0 \\ -1.6 & 1.6 + 0.493 & -0.493 & 0 \\ 0 & -0.493 & 0.493 + 7.2 & -7.2 \\ 0 & 0 & -7.2 & 47.2 \end{bmatrix}$$

$$\{\mathbf{F}\}^{(G)} = \begin{Bmatrix} 0 \\ 0 \\ 0 \\ 1200 \end{Bmatrix}$$

Applying the boundary condition at the inside furnace wall, we get

$$\begin{bmatrix} 1 & 0 & 0 & 0 \\ -1.6 & 2.093 & -0.493 & 0 \\ 0 & -0.493 & 7.693 & -7.2 \\ 0 & 0 & -7.2 & 47.2 \end{bmatrix} \begin{Bmatrix} T_1 \\ T_2 \\ T_3 \\ T_4 \end{Bmatrix} = \begin{Bmatrix} 200 \\ 0 \\ 0 \\ 1200 \end{Bmatrix}$$

and solving the set of linear equations, we have the following results:

$$\begin{Bmatrix} T_1 \\ T_2 \\ T_3 \\ T_4 \end{Bmatrix} = \begin{Bmatrix} 200 \\ 162.3 \\ 39.9 \\ 31.5 \end{Bmatrix} {}^\circ\mathrm{C}$$

Note that this type of heat conduction problem can be solved just as easily using fundamental concepts of heat transfer without resorting to finite element formulation. The point of this exercise was to demonstrate the steps involved in finite element analysis using a simple problem.

---

**EXAMPLE 4.3: A Fluid Mechanics Problem**

In a chemical processing plant, aqueous glycerin solution flows in a narrow channel, as shown in Figure 4.5. The pressure drop along the channel is continuously monitored. The upper wall of the channel is maintained at 50°C, while the lower wall is kept at 20°C. The variation of viscosity and density of the glycerin with the temperature is given in Table 4.1. For a relatively low flow, the pressure drop along the channel is measured to be 120 Pa/m. The channel is 3 m long, 9 cm high, and 40 cm wide. Determine the velocity profile and the mass flow rate of the fluid through the channel.

The laminar flow of a fluid with a constant viscosity inside a channel is governed by the balance between the net shear forces and the net pressure forces acting on a parcel of fluid. The equation of motion is

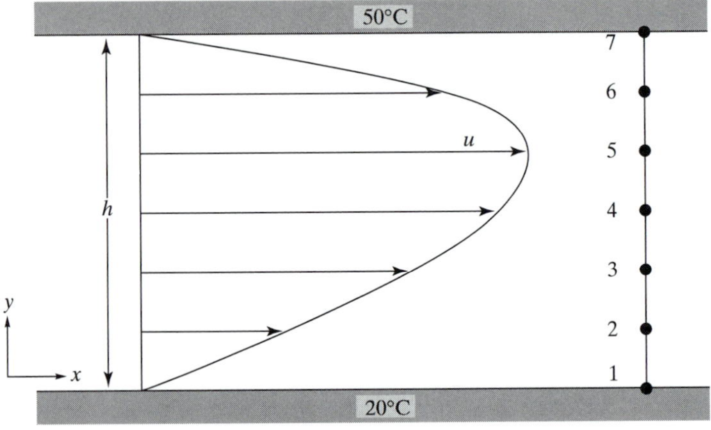


**FIGURE 4.5** Laminar flow of aqueous glycerin solution through a channel.

TABLE 4.1 Properties of aqueous glycerin solution as a function of temperature

| Temperature (°C) | Viscosity (kg/m • S) | Density (kg/m³) |
|---|---|---|
| 20 | 0.90 | 1255 |
| 25 | 0.65 | 1253 |
| 30 | 0.40 | 1250 |
| 35 | 0.28 | 1247 |
| 40 | 0.20 | 1243 |
| 45 | 0.12 | 1238 |
| 50 | 0.10 | 1233 |

$$\mu \frac{d^2u}{dy^2} - \frac{dp}{dx} = 0 \tag{4.39}$$

subject to the boundary conditions $u(0) = 0$ and $u(h) = 0$. Here, $u$ represents fluid velocity, $\mu$ is the dynamic viscosity of the fluid, and $\frac{dp}{dx}$ is the pressure drop in the direction of the flow. For this problem, when comparing Eq. (4.39) to Eq. (4.8), we find that $c_1 = \mu$, $c_2 = 0, c_3 = -\frac{dp}{dx}$, and $\Psi = u$.

Here, the viscosity of the aqueous glycerin solution varies with the height of the channel. We will use an average value of viscosity over each element when computing the elemental resistance matrices. The average values of viscosity and density associated with each element are given in Table 4.2.

TABLE 4.2 Properties of each element

| Element | Average Viscosity (kg/m·s) | Average Density (kg/m³) |
|---|---|---|
| 1 | 0.775 | 1254 |
| 2 | 0.525 | 1252 |
| 3 | 0.34 | 1249 |
| 4 | 0.24 | 1245 |
| 5 | 0.16 | 1241 |
| 6 | 0.11 | 1236 |

Using the properties from Table 4.2, we can compute the elemental flow-resistance matrices as:

$$[\mathbf{K}]^{(1)} = \frac{\mu}{\ell}\begin{bmatrix} 1 & -1 \\ -1 & 1 \end{bmatrix} = \frac{0.775}{1.5 \times 10^{-2}}\begin{bmatrix} 1 & -1 \\ -1 & 1 \end{bmatrix} = \begin{bmatrix} 51.67 & -51.67 \\ -51.67 & 51.67 \end{bmatrix} \frac{kg}{m^2 \cdot s}$$

$$[\mathbf{K}]^{(2)} = \frac{\mu}{\ell}\begin{bmatrix} 1 & -1 \\ -1 & 1 \end{bmatrix} = \frac{0.525}{1.5 \times 10^{-2}}\begin{bmatrix} 1 & -1 \\ -1 & 1 \end{bmatrix} = \begin{bmatrix} 35 & -35 \\ -35 & 35 \end{bmatrix} \frac{kg}{m^2 \cdot s}$$

$$[\mathbf{K}]^{(3)} = \frac{\mu}{\ell}\begin{bmatrix} 1 & -1 \\ -1 & 1 \end{bmatrix} = \frac{0.340}{1.5 \times 10^{-2}}\begin{bmatrix} 1 & -1 \\ -1 & 1 \end{bmatrix} = \begin{bmatrix} 22.67 & -22.67 \\ -22.67 & 22.67 \end{bmatrix}\frac{kg}{m^2 \cdot s}$$

$$[\mathbf{K}]^{(4)} = \frac{\mu}{\ell}\begin{bmatrix} 1 & -1 \\ -1 & 1 \end{bmatrix} = \frac{0.240}{1.5 \times 10^{-2}}\begin{bmatrix} 1 & -1 \\ -1 & 1 \end{bmatrix} = \begin{bmatrix} 16 & -16 \\ -16 & 16 \end{bmatrix}\frac{kg}{m^2 \cdot s}$$

$$[\mathbf{K}]^{(5)} = \frac{\mu}{\ell}\begin{bmatrix} 1 & -1 \\ -1 & 1 \end{bmatrix} = \frac{0.160}{1.5 \times 10^{-2}}\begin{bmatrix} 1 & -1 \\ -1 & 1 \end{bmatrix} = \begin{bmatrix} 10.67 & -10.67 \\ -10.67 & 10.67 \end{bmatrix}\frac{kg}{m^2 \cdot s}$$

$$[\mathbf{K}]^{(6)} = \frac{\mu}{\ell}\begin{bmatrix} 1 & -1 \\ -1 & 1 \end{bmatrix} = \frac{0.110}{1.5 \times 10^{-2}}\begin{bmatrix} 1 & -1 \\ -1 & 1 \end{bmatrix} = \begin{bmatrix} 7.33 & -7.33 \\ -7.33 & 7.33 \end{bmatrix}\frac{kg}{m^2 \cdot s}$$

Since $[\mathbf{K}]$ represents resistance to flow, we have opted to use the term elemental flow-resistance matrix instead of elemental stiffness matrix. Because the flow is fully developed, $\frac{dp}{dx}$ is constant; thus, the forcing matrix has the same value for all elements:

$$\{\mathbf{F}\}^{(1)} = \{\mathbf{F}\}^{(2)} = \ldots = \{\mathbf{F}\}^{(5)} = \{\mathbf{F}\}^{(6)} = \frac{-\frac{dp}{dx}\ell}{2}\begin{Bmatrix} 1 \\ 1 \end{Bmatrix} = \frac{-(-120)(1.5 \times 10^{-2})}{2}\begin{Bmatrix} 1 \\ 1 \end{Bmatrix} = \begin{Bmatrix} 0.9 \\ 0.9 \end{Bmatrix}\frac{N}{m^2}$$

The negative value associated with the pressure drop represents the decreasing nature of the pressure along the direction of flow in the channel. The global resistance matrix is obtained by assembling the elemental resistance matrices:

$$[\mathbf{K}]^{(G)} = \begin{bmatrix} 51.67 & -51.67 & 0 & 0 & 0 & 0 & 0 \\ -51.67 & 51.67 + 35 & -35 & 0 & 0 & 0 & 0 \\ 0 & -35 & 35 + 22.67 & -22.67 & 0 & 0 & 0 \\ 0 & 0 & -22.67 & 22.67 + 16 & -16 & 0 & 0 \\ 0 & 0 & 0 & -16 & 16 + 10.67 & -10.67 & 0 \\ 0 & 0 & 0 & 0 & -10.67 & 10.67 + 7.33 & -7.33 \\ 0 & 0 & 0 & 0 & 0 & -7.33 & 7.33 \end{bmatrix}$$

and the global forcing matrix is

$$\{\mathbf{F}\}^{(G)} = \begin{Bmatrix} 0.9 \\ 0.9 + 0.9 \\ 0.9 + 0.9 \\ 0.9 + 0.9 \\ 0.9 + 0.9 \\ 0.9 + 0.9 \\ 0.9 \end{Bmatrix}$$

Applying the no-slip boundary conditions at the walls leads to the matrix

$$\begin{bmatrix} 1 & 0 & 0 & 0 & 0 & 0 & 0 \\ -51.67 & 86.67 & -35 & 0 & 0 & 0 & 0 \\ 0 & -35 & 57.67 & -22.67 & 0 & 0 & 0 \\ 0 & 0 & -22.67 & 38.67 & -16 & 0 & 0 \\ 0 & 0 & 0 & -16 & 26.67 & -10.67 & 0 \\ 0 & 0 & 0 & 0 & -10.67 & 18 & -7.33 \\ 0 & 0 & 0 & 0 & 0 & 0 & 1 \end{bmatrix} \begin{Bmatrix} u_1 \\ u_2 \\ u_3 \\ u_4 \\ u_5 \\ u_6 \\ u_7 \end{Bmatrix} = \begin{Bmatrix} 0 \\ 1.8 \\ 1.8 \\ 1.8 \\ 1.8 \\ 1.8 \\ 0 \end{Bmatrix}$$

The solution provides the fluid velocities at each node:

$$\begin{Bmatrix} u_1 \\ u_2 \\ u_3 \\ u_4 \\ u_5 \\ u_6 \\ u_7 \end{Bmatrix} = \begin{Bmatrix} 0 \\ 0.1233 \\ 0.2538 \\ 0.3760 \\ 0.4366 \\ 0.3588 \\ 0 \end{Bmatrix} \text{ m/s}$$

The mass flow rate through the channel can be determined from

$$\dot{m}_{\text{total}} = \Sigma \dot{m}^{(e)} \tag{4.40}$$

$$\dot{m}^{(e)} = \int_{y_i}^{y_j} \rho u W \, dy = \int_{y_i}^{y_j} \rho W (S_i u_i + S_j u_j) \, dy = \rho W \ell \left( \frac{u_i + u_j}{2} \right) \tag{4.41}$$

In Eq. (4.41), $W$ represents the width of the channel. The elemental and total mass flow rates are given by:

$$\dot{m}^{(1)} = \rho W \ell \left( \frac{u_i + u_j}{2} \right) = 1254 \times 0.4 \times 1.5 \times 10^{-2} \times \frac{0 + 0.1233}{2} = 0.4638 \text{ kg/s}$$

$$\dot{m}^{(2)} = 1252 \times 0.4 \times 1.5 \times 10^{-2} \times \frac{0.1233 + 0.2538}{2} = 1.4164 \text{ kg/s}$$

$$\dot{m}^{(3)} = 1249 \times 0.4 \times 1.5 \times 10^{-2} \times \frac{0.2538 + 0.3760}{2} = 2.3598 \text{ kg/s}$$

$$\dot{m}^{(4)} = 1245 \times 0.4 \times 1.5 \times 10^{-2} \times \frac{0.3760 + 0.4366}{2} = 3.0350 \text{ kg/s}$$

$$\dot{m}^{(5)} = 1241 \times 0.4 \times 1.5 \times 10^{-2} \times \frac{0.4366 + 0.3588}{2} = 2.9612 \text{ kg/s}$$

$$m^{\cdot(6)} = 1236 \times 0.4 \times 1.5 \times 10^{-2} \times \frac{0.3588 + 0}{2} = 1.3304 \text{ kg/s}$$

$$m^{\cdot}_{\text{total}} = 11.566 \text{ kg/s}$$

## 4.2 SOLID MECHANICS PROBLEMS

In this section, we will use the minimum total potential energy formulation to generate finite element models for members under axial loading. Consider a column supporting several floors, as shown in Figure 4.6. Assuming axial loading, we can approximate the exact deflection of the column by a series of linear functions.

As discussed in Section 1.6, applied external loads cause a body to deform. During the deformation, the work done by the external forces is stored in the material in the form of elastic energy, called strain energy. For a member (element) under axial loading, the strain energy $\Lambda^{(e)}$ is given by

$$\Lambda^{(e)} = \int_V \frac{\sigma\varepsilon}{2} dV = \int_V \frac{E\varepsilon^2}{2} dV \tag{4.42}$$

The total potential energy $\Pi$ for a body consisting of $n$ elements and $m$ nodes is the difference between the total strain energy and the work done by the external forces:

$$\Pi = \sum_{e=1}^{n} \Lambda^{(e)} - \sum_{i=1}^{m} F_i u_i \tag{4.43}$$

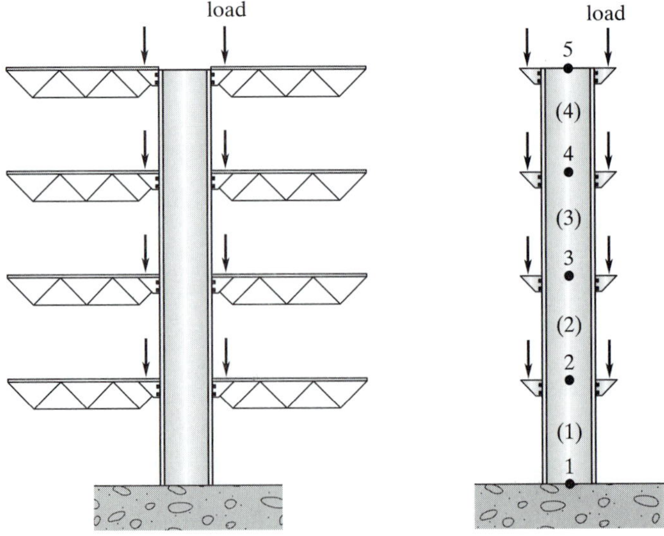


**FIGURE 4.6** Deflection of a steel column supporting several floors.

The minimum total potential energy principle states that for a stable system, the displacement at the equilibrium position occurs such that the value of the system's total potential energy is a minimum. That is,

$$\frac{\partial \Pi}{\partial u_i} = \frac{\partial}{\partial u_i}\sum_{e=1}^{n} \Lambda^{(e)} - \frac{\partial}{\partial u_i}\sum_{i=1}^{m} F_i u_i = 0 \quad \text{for } i = 1, 2, 3, \ldots, m \tag{4.44}$$

where $i$ takes on different values of node numbers. The deflection for an arbitrary element with nodes $i$ and $j$ in terms of local shape functions is given by

$$u^{(e)} = S_i u_i + S_j u_j \tag{4.45}$$

where $S_i = 1 - \frac{y}{\ell}$ and $S_j = \frac{y}{\ell}$ and $y$ is the element's local coordinate, with its origin at node $i$. The strain in each member can be computed using the relation $\varepsilon = \frac{du}{dy}$ as:

$$\varepsilon = \frac{du}{dy} = \frac{d}{dy}[S_i u_i + S_j u_j] = \frac{d}{dy}\left[\left(1 - \frac{y}{\ell}\right)u_i + \frac{y}{\ell}u_j\right] = \frac{-u_i + u_j}{\ell} \tag{4.46}$$

Incorporating Eq. (4.46) into Eq. (4.42) yields the strain energy for an arbitrary element $(e)$:

$$\Lambda^{(e)} = \int_V \frac{E\varepsilon^2}{2}\, dV = \frac{AE}{2\,\ell}\left(u_j^2 + u_i^2 - 2u_j u_i\right) \tag{4.47}$$

Minimizing the strain energy with respect to $u_i$ and $u_j$ leads to

$$\frac{\partial \Lambda^{(e)}}{\partial u_i} = \frac{AE}{\ell}(u_i - u_j) \tag{4.48}$$

$$\frac{\partial \Lambda^{(e)}}{\partial u_j} = \frac{AE}{\ell}(u_j - u_i)$$

or, in matrix form,

$$\begin{Bmatrix} \dfrac{\partial \Lambda^{(e)}}{\partial u_i} \\ \dfrac{\partial \Lambda^{(e)}}{\partial u_j} \end{Bmatrix} = \begin{bmatrix} k & -k \\ -k & k \end{bmatrix} \begin{Bmatrix} u_i \\ u_j \end{Bmatrix} \tag{4.49}$$

where $k = \dfrac{(AE)}{\ell}$. Minimizing the work done by external forces, the second term on the right-hand side of Eq. (4.44), results in the load matrix:

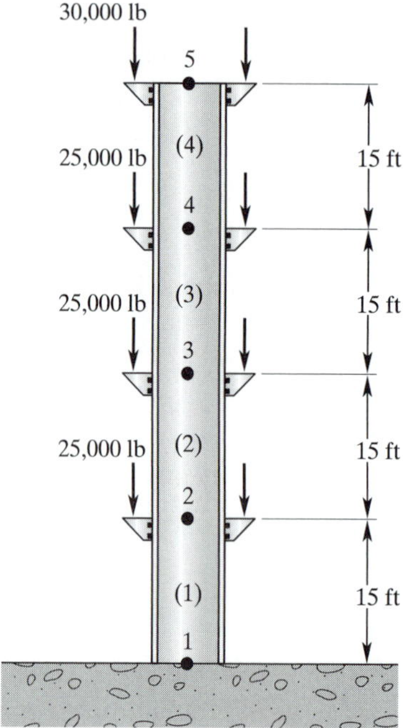


**FIGURE 4.7** A schematic of the column in Example 4.4.

$$\{\mathbf{F}\}^{(e)} = \begin{Bmatrix} F_i \\ F_j \end{Bmatrix}$$

Computing individual elemental stiffness and load matrices and connecting them leads to a global stiffness matrix. This step is demonstrated by the next example.

## EXAMPLE 4.4: A Column Problem

Consider a four-story building with steel columns. One column is subjected to the loading shown in Figure 4.7. Assuming axial loading, determine: (a) vertical displacements of the column at various floor–column connection points and (b) the stresses in each portion of the column. $E = 29 \times 10^6$ lb/in$^2$, $A = 39.7$ in$^2$.

Because all elements have the same length, cross-sectional area, and physical properties, the elemental stiffness for elements (1), (2), (3), and (4) is given by:

$$[\mathbf{K}]^{(e)} = \frac{AE}{\ell}\begin{bmatrix} 1 & -1 \\ -1 & 1 \end{bmatrix} = \frac{39.7 \times 29 \times 10^6}{15 \times 12}\begin{bmatrix} 1 & -1 \\ -1 & 1 \end{bmatrix} = 6.396 \times 10^6 \begin{bmatrix} 1 & -1 \\ -1 & 1 \end{bmatrix}$$

$$[\mathbf{K}]^{(1)} = [\mathbf{K}]^{(2)} = [\mathbf{K}]^{(3)} = [\mathbf{K}]^{(4)} = 6.396 \times 10^6 \begin{bmatrix} 1 & -1 \\ -1 & 1 \end{bmatrix} \frac{\text{lb}}{\text{in}}$$

The global stiffness matrix is obtained by assembling the elemental matrices:

$$[\mathbf{K}]^{(e)} = 6.396 \times 10^6 \begin{bmatrix} 1 & -1 & 0 & 0 & 0 \\ -1 & 1+1 & -1 & 0 & 0 \\ 0 & -1 & 1+1 & -1 & 0 \\ 0 & 0 & -1 & 1+1 & -1 \\ 0 & 0 & 0 & -1 & 1 \end{bmatrix}$$

The global forcing matrix is obtained from

$$\{\mathbf{F}\}^{(G)} = \left\{ \frac{\partial F_i u_i}{\partial u_i} \right\}_{i=1,5} = \begin{Bmatrix} F_1 \\ F_2 \\ F_3 \\ F_4 \\ F_5 \end{Bmatrix} = \begin{Bmatrix} 0 \\ 50000 \\ 50000 \\ 50000 \\ 60000 \end{Bmatrix} \text{lb}$$

Application of the boundary condition and loads results in

$$6.396 \times 10^6 \begin{bmatrix} 1 & 0 & 0 & 0 & 0 \\ -1 & 2 & -1 & 0 & 0 \\ 0 & -1 & 2 & -1 & 0 \\ 0 & 0 & -1 & 2 & -1 \\ 0 & 0 & 0 & -1 & 1 \end{bmatrix} \begin{Bmatrix} u_1 \\ u_2 \\ u_3 \\ u_4 \\ u_5 \end{Bmatrix} = \begin{Bmatrix} 0 \\ 50000 \\ 50000 \\ 50000 \\ 60000 \end{Bmatrix}$$

Solving for displacements, we have

$$\begin{Bmatrix} u_1 \\ u_2 \\ u_3 \\ u_4 \\ u_5 \end{Bmatrix} = \begin{Bmatrix} 0 \\ 0.03283 \\ 0.05784 \\ 0.07504 \\ 0.08442 \end{Bmatrix} \text{in}$$

The axial stresses in each element are determined from:

$$\sigma^{(1)} = \frac{E(u_i - u_j)}{\ell} = \frac{29 \times 10^6(0 - 0.03283)}{15 \times 12} = -5289 \text{ lb/in}^2$$

$$\sigma^{(2)} = \frac{29 \times 10^6(0.03283 - 0.05784)}{15 \times 12} = -4029 \text{ lb/in}^2$$

$$\sigma^{(3)} = \frac{29 \times 10^6(0.05784 - 0.07504)}{15 \times 12} = -2771 \text{ lb/in}^2$$

$$\sigma^{(4)} = \frac{29 \times 10^6(0.07504 - 0.08442)}{15 \times 12} = -1511 \text{ lb/in}^2$$

## 4.3 AN EXAMPLE USING ANSYS

**EXAMPLE 4.2 (Revisited)**

A wall of an industrial oven consists of three different materials, as shown in Figure 4.4, repeated here as Figure 4.8. The first layer is composed of 5 cm of insulating cement with a clay binder that has a thermal conductivity of 0.08 W/m · K. The second layer is made from 15 cm of 6-ply asbestos board with a thermal conductivity of 0.074 W/m · K. The exterior consists of 10-cm common brick with a thermal conductivity of 0.72 W/m$^2$ · K. The inside wall temperature of the oven is 200°C, and the outside air is 30°C with a convection coefficient of 40 W/m$^2$ · K. Determine the temperature distribution along the composite wall.

The following steps demonstrate how to create one-dimensional conduction problems with convective boundary conditions in ANSYS. This task includes choosing ap-

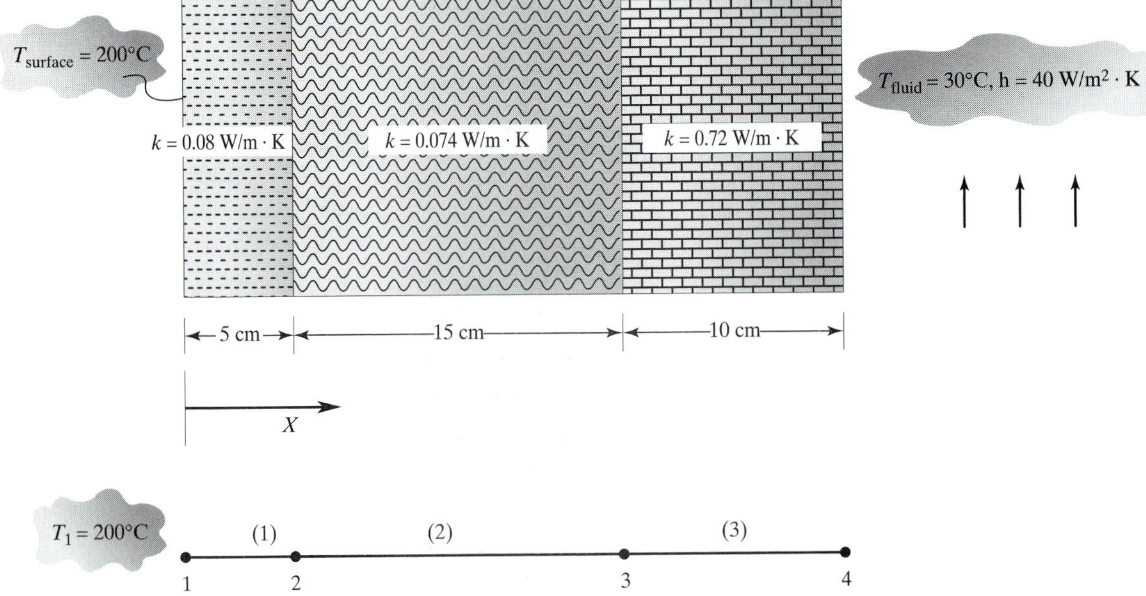


**FIGURE 4.8** A composite wall of an industrial oven.

propriate element types, assigning attributes, applying boundary conditions, and obtaining results.

To solve this problem using ANSYS, we employ the following steps:

Enter the **ANSYS** program by using the Launcher.

Type **xansys54** on the command line if you are running ANSYS on a UNIX platform, or consult your system administrator for information on how to run ANSYS from your computer system's platform.

Pick **Interactive** from the Launcher menu.

Type **HeatTran** (or a file name of your choice) in the **Initial Jobname** entry field of the dialog box.

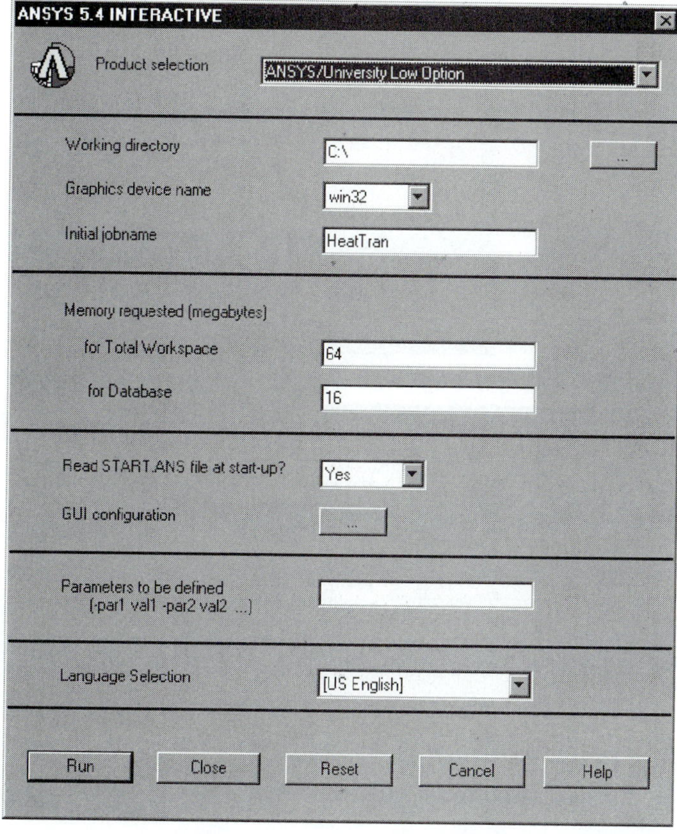


Pick **Run** to start the Graphic User Interface (GUI). A window will open with some disclaimer information. You will eventually be asked to press the **Return** key to start the graphics window and the main menu. Do so in order to proceed.

Create a title for the problem. This title will appear on ANSYS display windows to provide a simple way of identifying the displays:

utility menu: **File** → **Change Title** ...

Change Title
[/TITLE] Enter new title HeatTran
OK Cancel Help

Define the element type and material properties:

main menu: **Preprocessor** → **Element Type** → **Add/Edit/Delete** ...

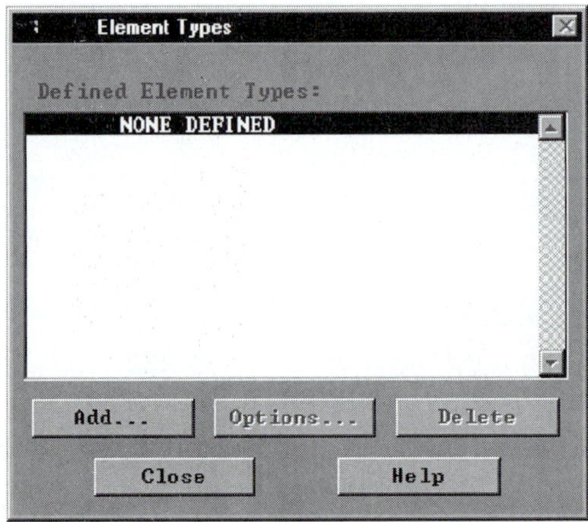


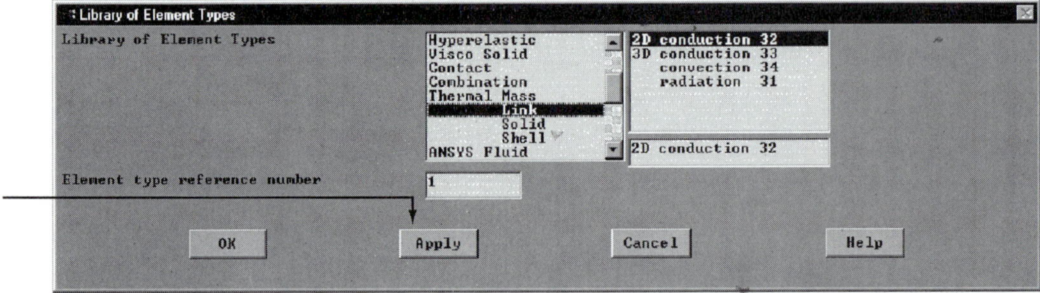


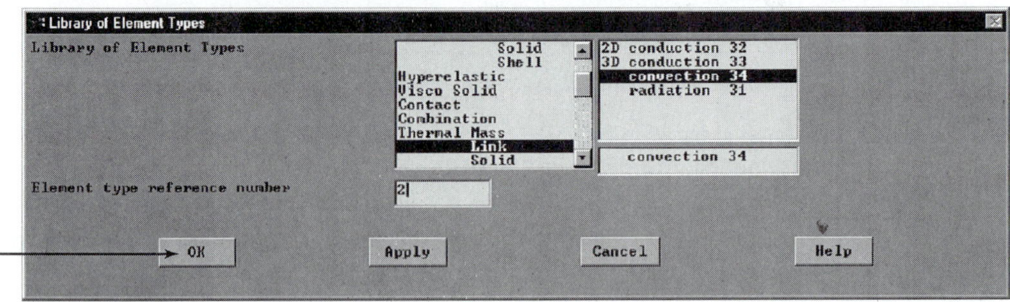

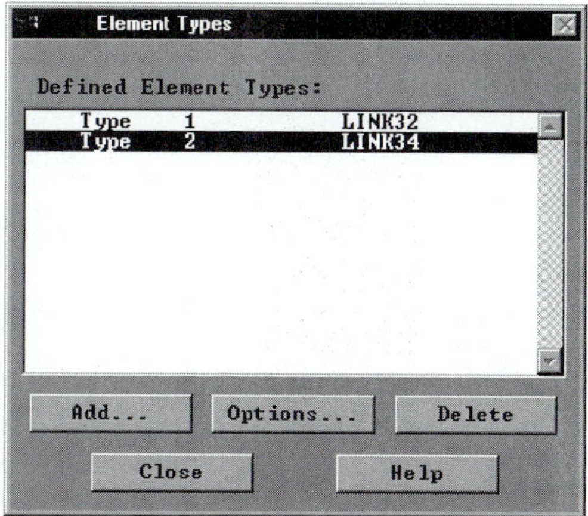


Assign the cross-sectional area of the wall.

main menu: **Preprocessor** → **Real Constants** ...

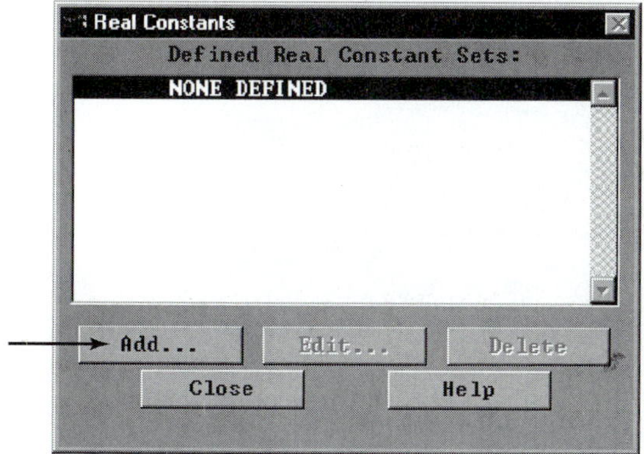


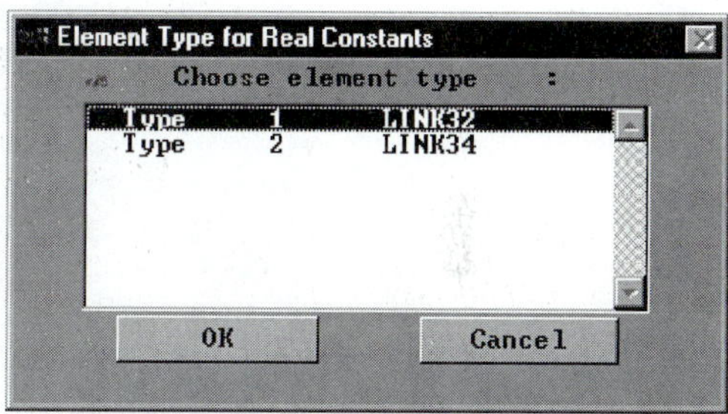


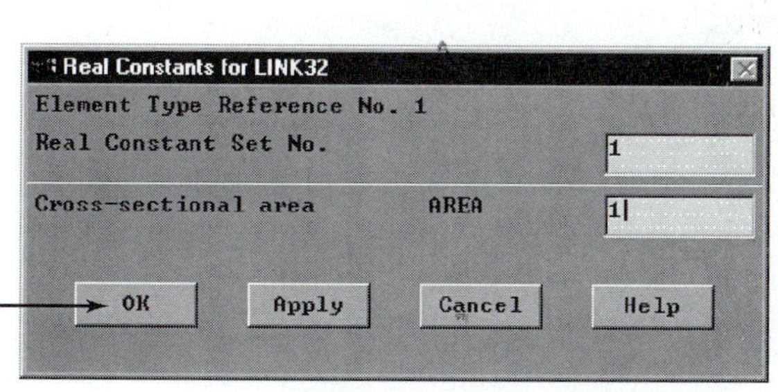


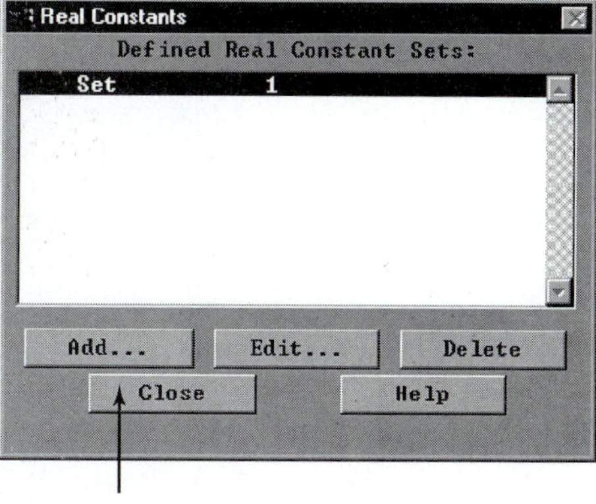


Assign the thermal conductivity values.

main menu: **Preprocessor** → **Material Props** → **-Constant-Isotropic** ...

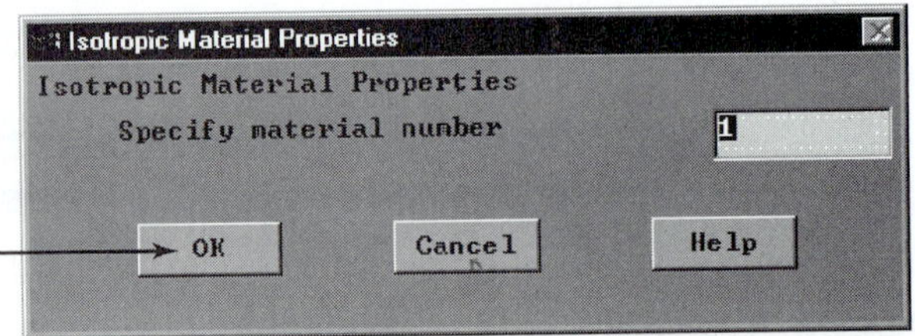
Isotropic Material Properties
Isotropic Material Properties
Specify material number
1
OK
Cancel
Help

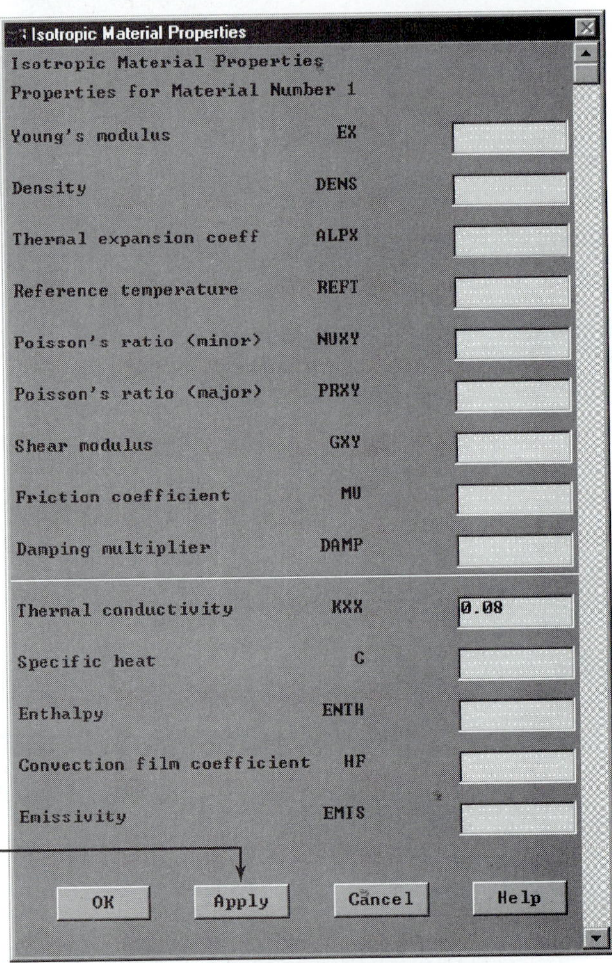
Isotropic Material Properties
Isotropic Material Properties
Properties for Material Number 1
Young's modulus EX
Density DENS
Thermal expansion coeff ALPX
Reference temperature REFT
Poisson's ratio (minor) NUXY
Poisson's ratio (major) PRXY
Shear modulus GXY
Friction coefficient MU
Damping multiplier DAMP
Thermal conductivity KXX
0.08
Specific heat C
Enthalpy ENTH
Convection film coefficient HF
Emissivity EMIS
OK
Apply
Cancel
Help

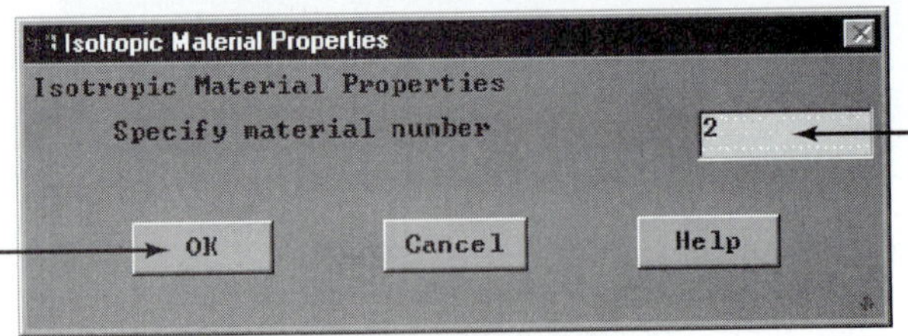
Isotropic Material Properties
Isotropic Material Properties
Specify material number
2
OK
Cancel
Help

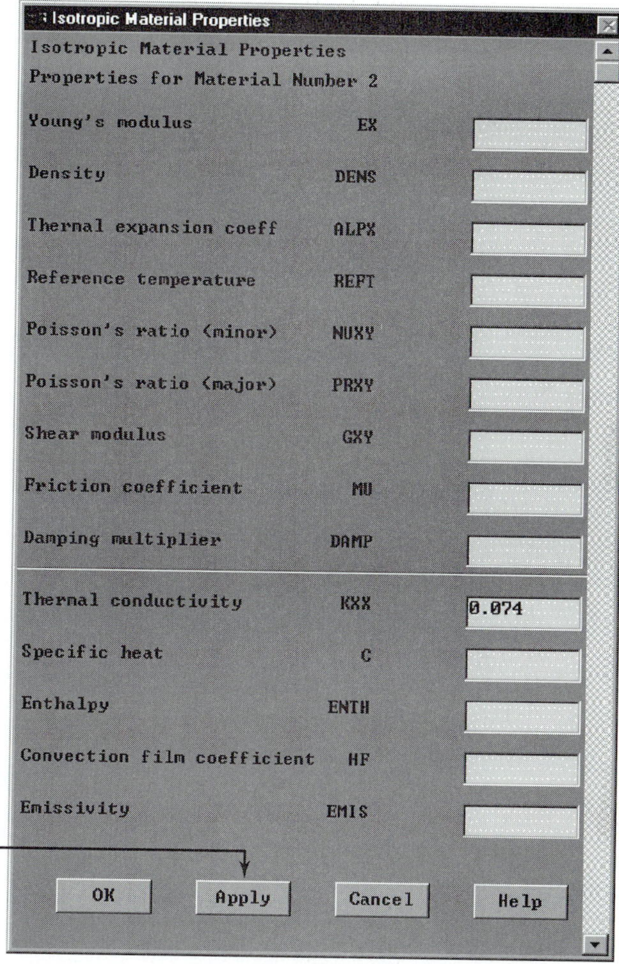
Isotropic Material Properties
Isotropic Material Properties
Properties for Material Number 2
Young's modulus EX
Density DENS
Thermal expansion coeff ALPX
Reference temperature REFT
Poisson's ratio (minor) NUXY
Poisson's ratio (major) PRXY
Shear modulus GXY
Friction coefficient MU
Damping multiplier DAMP
Thermal conductivity KXX
0.074
Specific heat C
Enthalpy ENTH
Convection film coefficient HF
Emissivity EMIS
OK
Apply
Cancel
Help

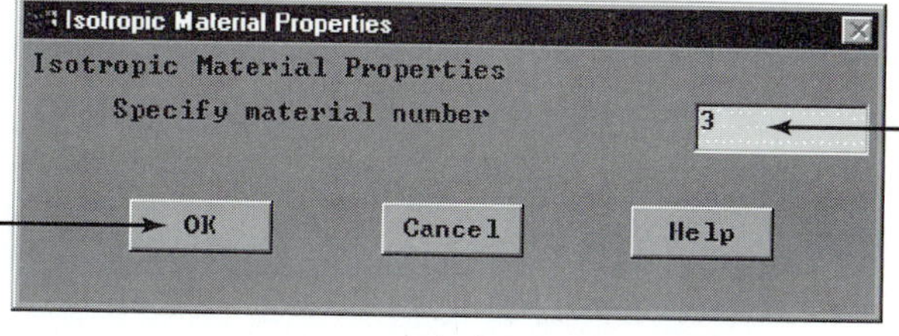
Isotropic Material Properties
Isotropic Material Properties
Specify material number
3
OK
Cancel
Help

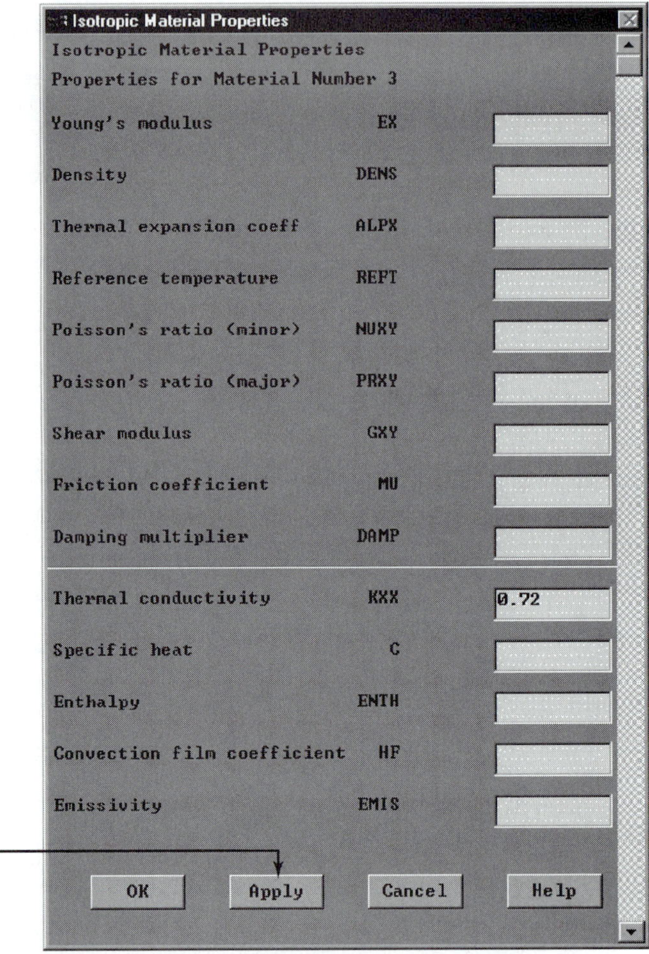
Isotropic Material Properties
Isotropic Material Properties
Properties for Material Number 3
Young's modulus EX
Density DENS
Thermal expansion coeff ALPX
Reference temperature REFT
Poisson's ratio (minor) NUXY
Poisson's ratio (major) PRXY
Shear modulus GXY
Friction coefficient MU
Damping multiplier DAMP
Thermal conductivity KXX
0.72
Specific heat C
Enthalpy ENTH
Convection film coefficient HF
Emissivity EMIS
OK
Apply
Cancel
Help

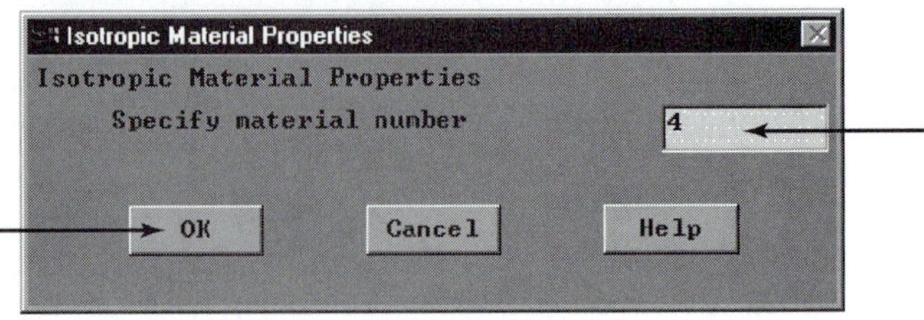
Isotropic Material Properties
Isotropic Material Properties
Specify material number
4
OK
Cancel
Help

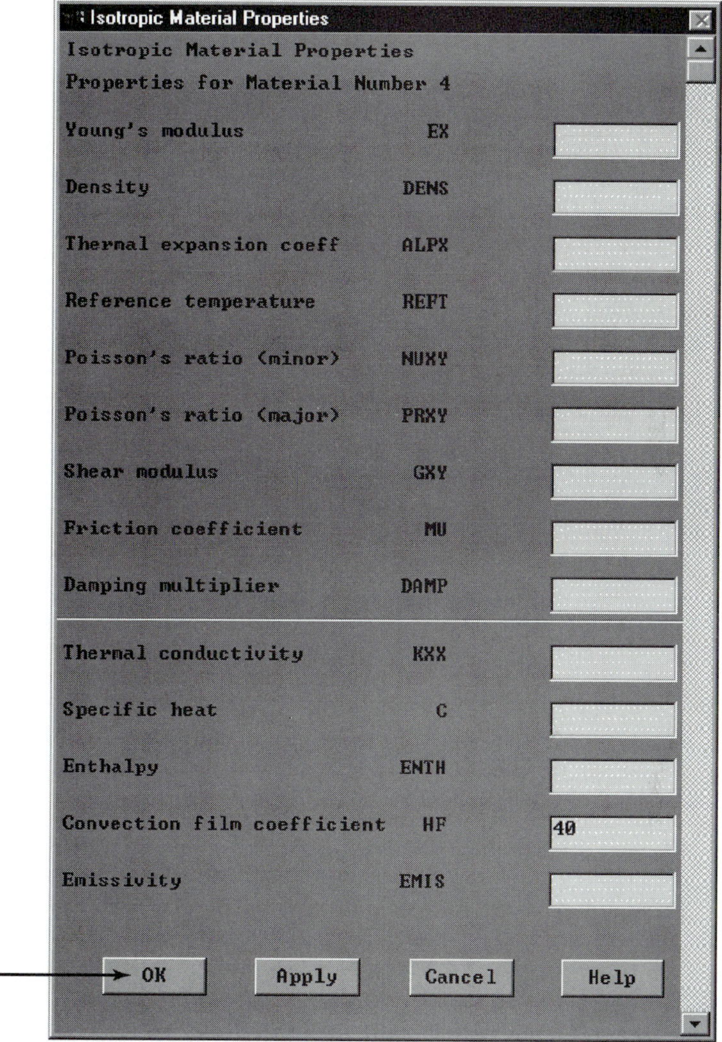


ANSYS Toolbar: **SAVE_DB**

Set up the graphics area (i.e., workplane, zoom, etc.):

utility menu: **Workplane** → **WP Settings** …

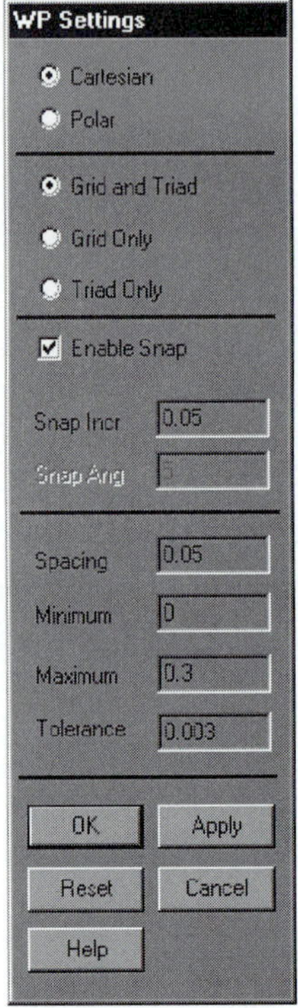


Toggle on the workplane by the following sequence:

utility menu: **Workplane** → **Display working plane**

Bring the workplane to view using the following sequence:

utility menu: **PlotCtrls** → **Pan, Zoom, Rotate** ...

Create nodes by picking points on the workplane:

main menu: **Preprocessor** → **-Modeling-Create** → **Nodes** → **On Working Plane** +

On the workplane, pick the location of nodes and apply them:

**[WP = 0,0]**

**[WP = 0.05,0]**

**[WP = 0.2,0]**

**[WP = 0.3,0]**

Create the node for the convection element:

**[WP = 0.3,0]**

**OK**

You may want to turn off the workplane now and turn on node numbering:

utility menu: **Workplane → Display Working plane**

utility menu: **PlotCtrls → Numbering** ...

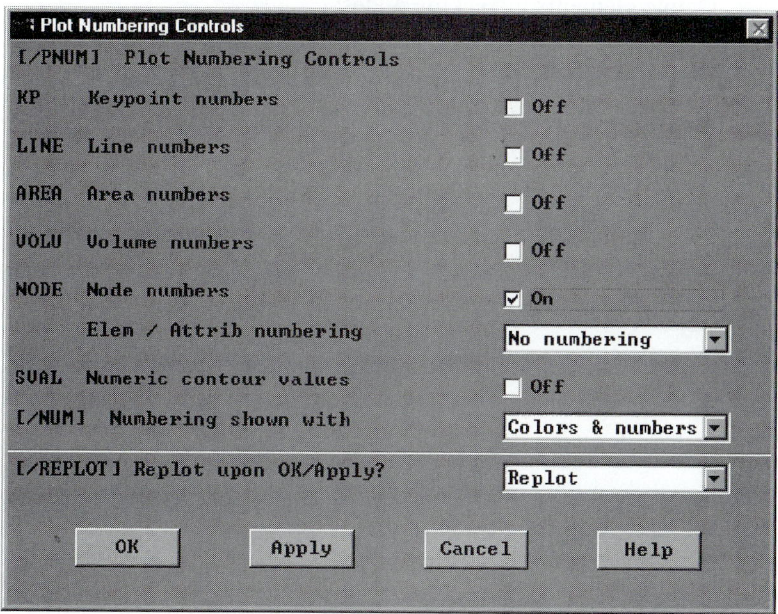


You may want to list nodes at this point in order to check your work:

utility menu: **List → Nodes** ...

NLIST Command

File

LIST ALL SELECTED NODES. DSYS= 0
SORT TABLE ON NODE NODE NODE

| NODE | X | Y | Z | THXY | THYZ | THZX |
|---|---|---|---|---|---|---|
| 1 | .00000 | .00000 | .00000 | .00 | .00 | .00 |
| 2 | .50000E-01 | .00000 | .00000 | .00 | .00 | .00 |
| 3 | .20000 | .00000 | .00000 | .00 | .00 | .00 |
| 4 | .30000 | .00000 | .00000 | .00 | .00 | .00 |
| 5 | .30000 | .00000 | .00000 | .00 | .00 | .00 |

**Close**

ANSYS Toolbar**: SAVE_DB**

Define elements by picking nodes:

main menu: **Preprocessor** → **-Modeling-Create** → **Elements** → **-Auto Numbered-Thru Nodes** +

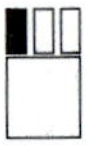

[node 1 and then node 2]

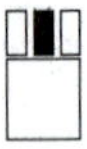

[use the middle button anywhere in the ANSYS graphics window to apply.]

**OK**

Assign the thermal conductivity of the second layer (element), and then connect the nodes to define the element:

main menu: **Preprocessor** → **-Modeling-Create** → **Elements** → **Element Attributes** ...

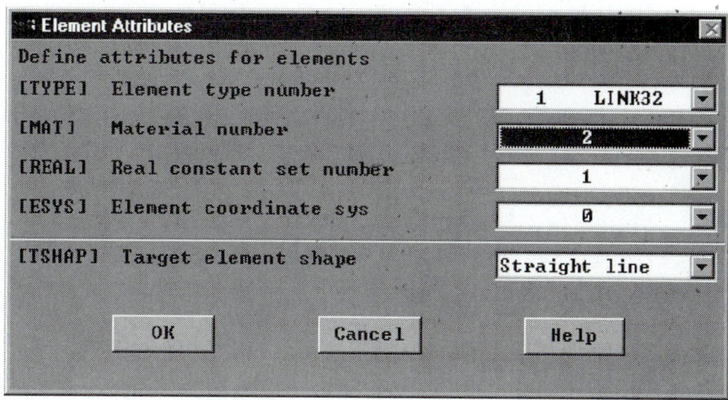


main menu: **Preprocessor** → **-Modeling-Create** → **Elements** → **-Auto Numbered-Thru Nodes** +

[node 2 and then node 3]

[anywhere in the ANSYS graphics window]

**OK**

Assign the thermal conductivity of third layer (element), and then connect the nodes to define the element:

main menu: **Preprocessor** → **-Modeling-Create** → **Elements** → **Element Attributes** ...

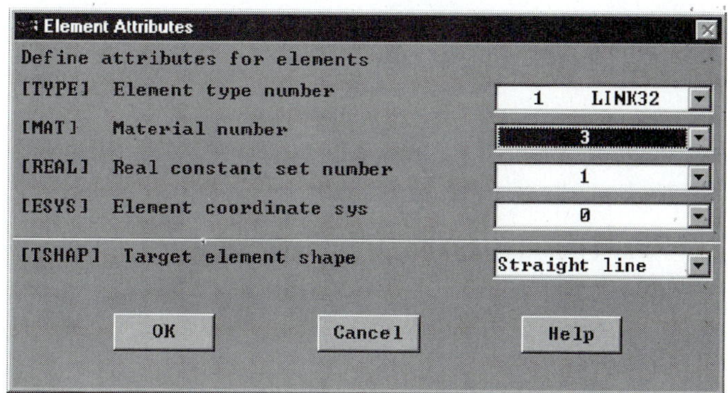


main menu: **Preprocessor** → **-Modeling-Create** → **Elements** → **-Auto Numbered** → **Thru Nodes** +

[node 3 and then node 4*]

[anywhere in the ANSYS graphics window]

**OK**

Create the convection link:

main menu: **Preprocessor** → **-Modeling-Create** → **Elements** → **Element Attributes** ...

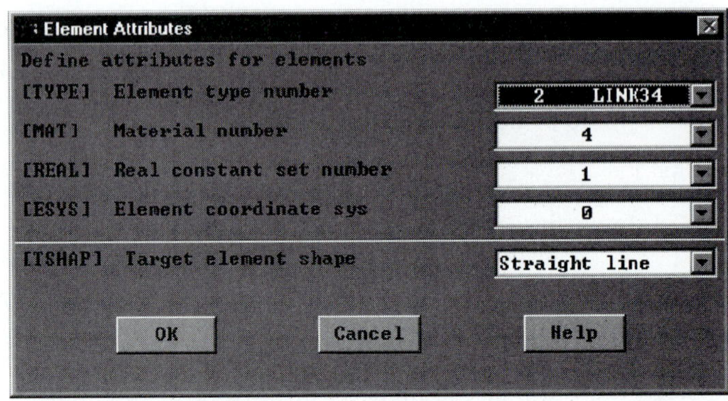


main menu: **Preprocessor** → **-Modeling-Create** → **Elements** → **-Auto Numbered-Thru Nodes** +

On the command line type **4** and press the <**Return**> key. Then type **5** and press the <**Return**> key.

**OK**

ANSYS Toolbar: **SAVE_DB**

Apply boundary conditions:

main menu: **Solution** → **-Loads-Apply** → **-Thermal-Temperature** → **On Nodes** +

[node 1]

[anywhere in the ANSYS graphics window]

*Press the **OK** key of the Multiple-Entities window and proceed.

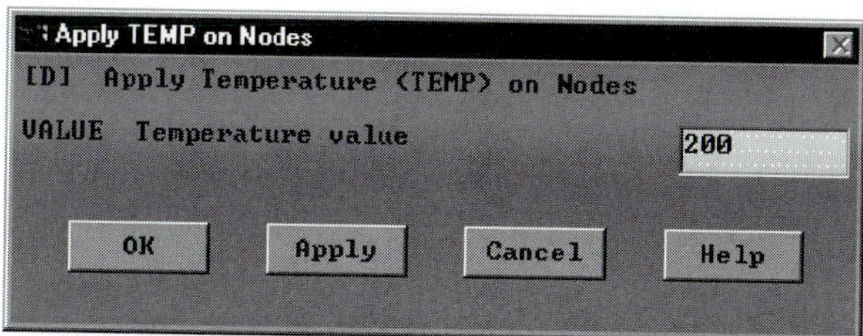


main menu: **Solution** → **-Loads-Apply** → **-Thermal-Temperature** → **On Nodes** +

[node 5*]

[anywhere in the ANSYS graphics window]

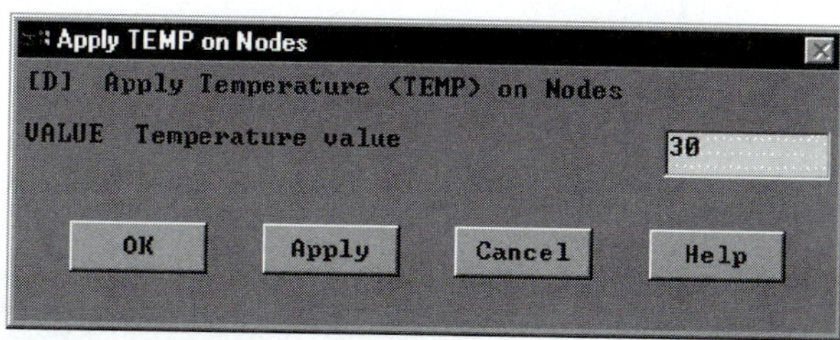


ANSYS Toolbar: **SAVE_DB**

Solve the problem:

main menu: **Solution** → **-Solve-Current LS**

**OK**

**Close** (the solution is done!) window.

**Close** (the /STAT Command) window.

For the postprocessing phase, obtain information such as nodal temperatures:

main menu: **General Postproc** → **List Results** → **Nodal Solution**

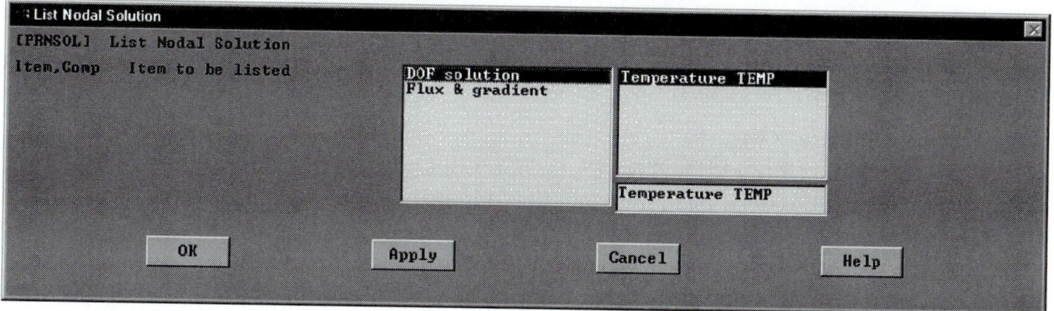


*Press the **Next** key and then the **OK** key of the Multiple-Entities window and proceed.

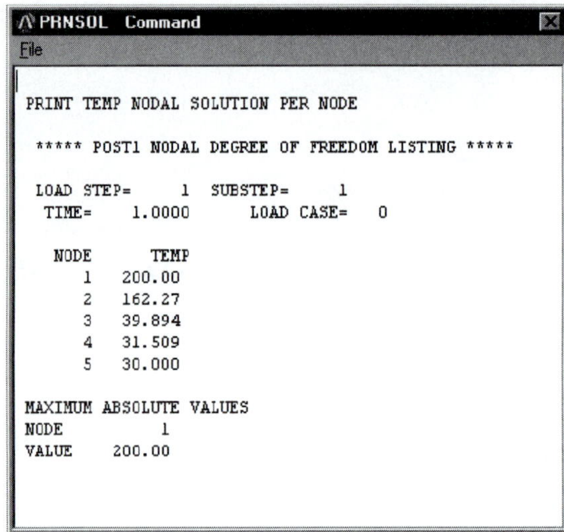


**Close**

Exit ANSYS and save everything.

Toolbar: **QUIT**

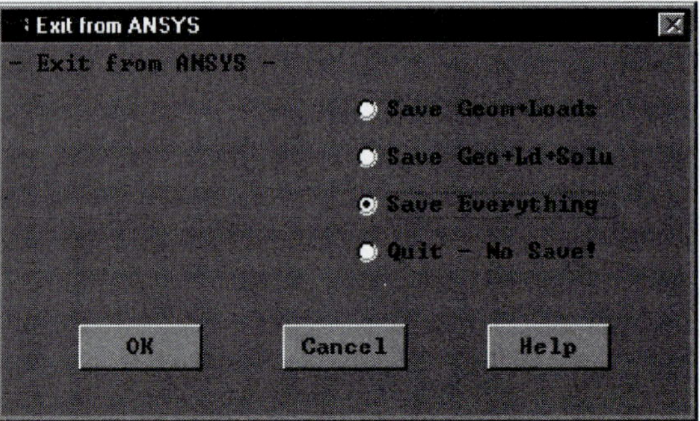


## 4.4 VERIFICATION OF RESULTS

There are various ways to verify your findings. Consider the nodal temperatures of Example 4.2, as computed by ANSYS and diplayed in Table 4.3.

In general, for a heat transfer problem under steady state conditions, conservation of energy applied to a control volume surrounding an arbitrary node must be satisfied. Are the energies flowing into and out of a node balanced out? Let us use Example 4.2 to demonstrate this important concept. The heat loss through each layer of the composite wall must be equal. Furthermore, heat loss from the last layer must equal the heat removed by the surrounding air. So,

TABLE 4.3 Nodal temperatures

| Node Number | Temperature (°C) |
|---|---|
| 1 | 200 |
| 2 | 162.3 |
| 3 | 39.9 |
| 4 | 31.5 |
| 5 | 30 |

$$Q^{(1)} = Q^{(2)} = Q^{(3)} = Q^{(4)}$$

$$Q^{(1)} = kA\frac{\Delta T}{\ell} = (0.08)(1)\left(\frac{200 - 162.3}{0.05}\right) = 60 \text{ W}$$

$$Q^{(2)} = (0.074)(1)\left(\frac{162.3 - 39.9}{0.15}\right) = 60 \text{ W}$$

$$Q^{(3)} = (0.72)(1)\left(\frac{39.9 - 31.5}{0.1}\right) = 60 \text{ W}$$

and the heat removal by the fluid is given by

$$Q^{(4)} = hA\Delta T = (40)(1)(31.5 - 30) = 60 \text{ W}$$

For thermal elements, ANSYS provides information such as heat flow through each element. Therefore, we could have compared those values with the one we calculated above.

Another check on the validity of your results could have come from examining the slopes of temperatures in each layer. The first layer has a temperature slope of 754°C/m. For the second layer, the slope of the temperature is 816°C/m. This layer consists of a material with relatively low thermal conductivity and, therefore, a relatively large temperature drop. The slope of the temperature in the exterior wall is 84°C/m. Because the exterior wall is made of a material with relatively high thermal conductivity, we expect the temperature drop through this layer not to be as significant as the other layers.

Now consider the fin problem in Example 4.1. For this problem, recall that all elements have the same length. We determined the temperature distribution and the heat loss from each element. Comparing heat loss results, it is important to realize that element (1) has the highest value because the greatest thermal potential exists at the base of the fin, and as the temperature of the fin drops, so does the rate of heat loss for each element. This outcome is certainly consistent with the results we obtained previously.

Refer now to Example 4.4. One way of checking for the validity of our FEA findings of Example 4.4 is to arbitrarily cut a section through the column and apply the static equilibrium conditions. As an example, consider cutting a section through the column containing element (2), as shown in the accompanying illustration.

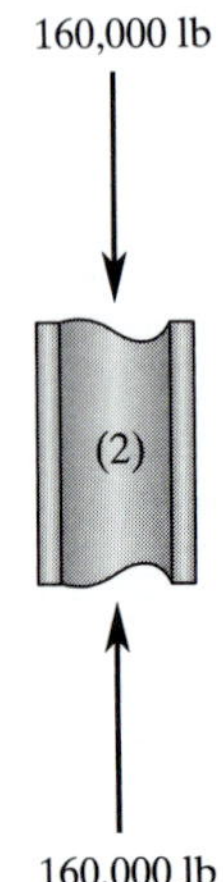


The average normal stress in that section of the column is

$$\sigma^{(2)} = \frac{f_{\text{internal}}}{A} = \frac{160{,}000}{39.7} = 4030 \text{ lb/in}^2$$

In a similar way, the average stress in element (4) can be checked by:

$$\sigma^{(4)} = \frac{f_{\text{internal}}}{A} = \frac{60{,}000}{39.7} = 1511 \text{ lb/in}^2$$

The stresses computed in this manner are identical to the results obtained earlier using the energy method. These simple problems illustrate the importance of checking for equilibrium conditions when verifying results.

## SUMMARY

At this point you should:

1. know how to formulate stiffness (conductance or resistance) matrices, and be able to formulate load matrices for various one-dimensional problems.
2. know how to apply appropriate boundary conditions.
3. have a good understanding of the Galerkin and energy formulations of one-dimensional problems.
4. know how to verify your results.

## REFERENCES

*ANSYS User's Manual: Procedures*, Vol. I, Swanson Analysis Systems, Inc.

*ANSYS User's Manual: Commands*, Vol. II, Swanson Analysis Systems, Inc.

*ANSYS User's Manual: Elements*, Vol. III, Swanson Analysis Systems, Inc.

Incropera, F., and Dewitt, D., *Fundamentals of Heat and Mass Transfer*, 2d. ed., New York, John Wiley and Sons, 1985.

Glycerin Producers' Association, "Physical Properties of Glycerin And Its Solutions," New York.

Segrlind, L., *Applied Finite Element Analysis*, 2d. ed., New York, John Wiley and Sons, 1984.

## PROBLEMS

1. Aluminum fins, similar to the ones in Example 4.1, with rectangular profiles are used to remove heat from a surface whose temperature is 150°C. The temperature of the ambient air is 20°C. The thermal conductivity of aluminum is 168 W/m • K. The natural convective coefficient associated with the surrounding air is 35 W/m$^2$ • K. The fins are 150 mm long, 5 mm wide, and 1 mm thick. (a) Determine the temperature distribution along a fin using five equally spaced elements. (b) Approximate the total heat loss for an array of 50 fins.

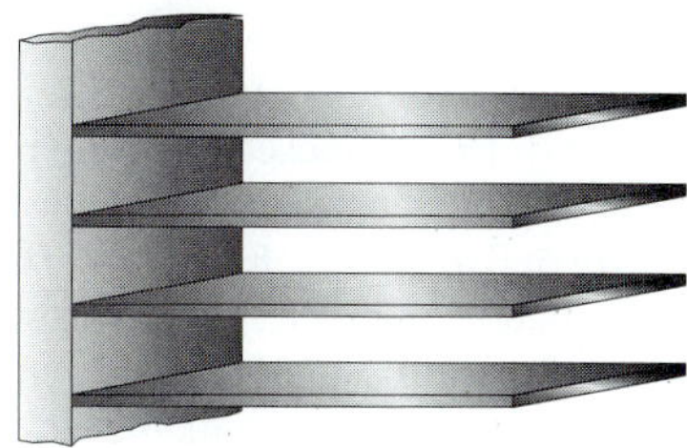

2. For the aluminum fins in Problem 1, determine the temperature of a point on the fin 45 mm from the base. Also compute the fraction of the total heat that is lost through this section of the fin.
3. A pin fin, or spine, is a fin with a circular cross section. An array of aluminum pin fins are used to remove heat from a surface whose temperature is 120°C. The temperature of the ambient air is 25°C. The thermal conductivity of aluminum is 168 W/m • K. The natural convective coefficient associated with the surrounding air is 30 W/m$^2$ • K. The fins are 100 mm long and have respective diameters of 4 mm. (a) Determine the temperature distribution along a fin using five equally spaced elements. (b) Approximate the total heat loss for an array of 100 fins.
4. A rectangular aluminum fin is used to remove heat from a surface whose temperature is 80°C. The temperature of the ambient air varies between 18°C and 25°C. The thermal conductivity of aluminum is 168 W/m • K. The natural convective coefficient associated with the surrounding air is 25 W/m$^2$ • K. The fin is 100 mm long, 5 mm wide, and 1 mm thick. (a) Determine the temperature distribution along a fin using five equally spaced elements for both ambient conditions. (b) Approximate the total heat loss for an array of 50 fins for each ambient condition. (c) The exact temperature distribution and heat loss for a fin with a negligible heat loss at its tip is given by

$$\frac{T(x) - T_f}{T_b - T_f} = \frac{\cosh\left[\sqrt{\frac{hp}{kA_c}}(L - x)\right]}{\cosh\left[\sqrt{\frac{hp}{kA_c}}(L)\right]}$$

$$Q = \sqrt{hpkA_c}\left(\tanh\left[\sqrt{\frac{hp}{kA_c}}(L)\right]\right)(T_b - T_f)$$

Compare your finite element results to the exact results.

5. Evaluate the integral $\int_{X_i}^{X_j} S_i hpT_f \, dX$ for a situation in which the heat transfer coefficient $h$ varies linearly over a given element.

6. The front window of a car is defogged by supplying warm air at 90°F to its inner surface. The glass has a thermal conductivity of $k = 0.8$ W/m · °C with a thickness of approximately 1/4 in. With the supply fan running at moderate speed, the heat transfer coefficient associated with the air is 50 W/m$^2$ · K. The outside air is at a temperature of 20°F with an associated heat transfer coefficient of 110 W/m$^2$ · K. Determine (a) the temperatures of the inner and outer surfaces of the glass and (b) the heat loss through the glass if the area of the glass is approximately 10 ft$^2$.

7. A wall of an industrial oven consists of three different materials, as shown in the accompanying figure. The first layer is composed of 10 cm of insulating cement with a thermal conductivity of 0.12 W/m · K. The second layer is made from 20 cm of 8-ply asbestos board with a thermal conductivity of 0.071 W/m · K. The exterior consists of 12-cm cement mortar with a thermal conductivity of 0.72 W/m$^2$ · K. The inside wall temperature of the oven is 250°C, and the outside air is at a temperature of 35°C with a convection coefficient of 45 W/m$^2$ · K. Determine the temperature distribution along the composite wall.

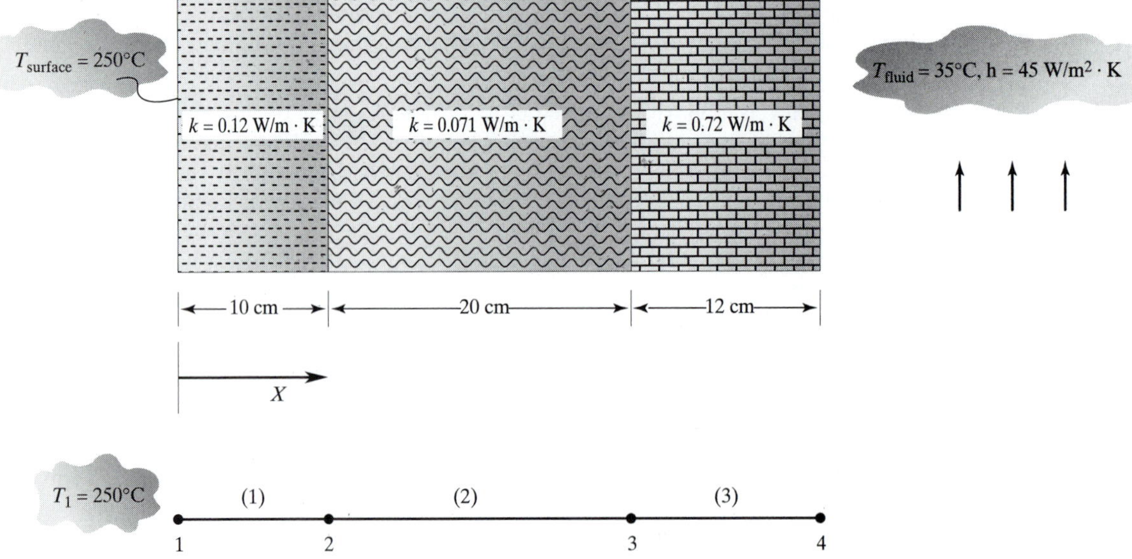


8. Replace the temperature boundary condition of the inside wall of the oven in Problem 7 with air temperature of 400°C and an associated convection coefficient of 100 W/m$^2$ · K. Show the contribution of this boundary condition to the conductance matrix and the forcing matrix of element (1). Also determine the temperature distribution along the composite wall.

9. Determine the deflections of point $D$ and point $F$ and the axial stress in each member of the system shown in the accompanying figure. ($E = 29 \times 10^3$ ksi.)

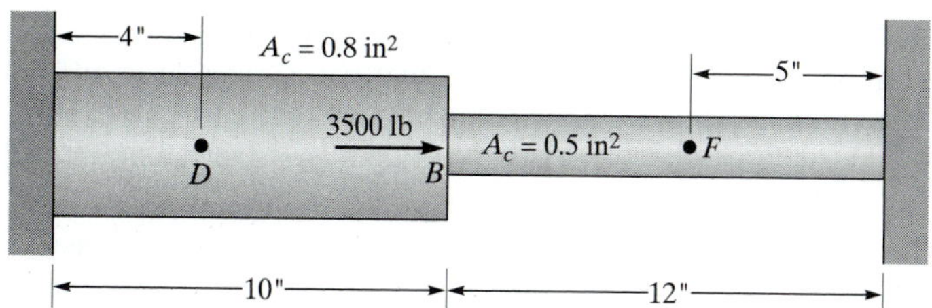


**10.** Consider a four-story building with steel columns similar to the one presented in Example 4.4. The column is subjected to the loading shown in the accompanying figure. Assuming axial loading, (a) determine vertical displacements of the column at various floor–column connection points and (b) determine the stresses in each portion of the column. $E = 29 \times 10^6$ lb/in², $A = 59.1$ in².

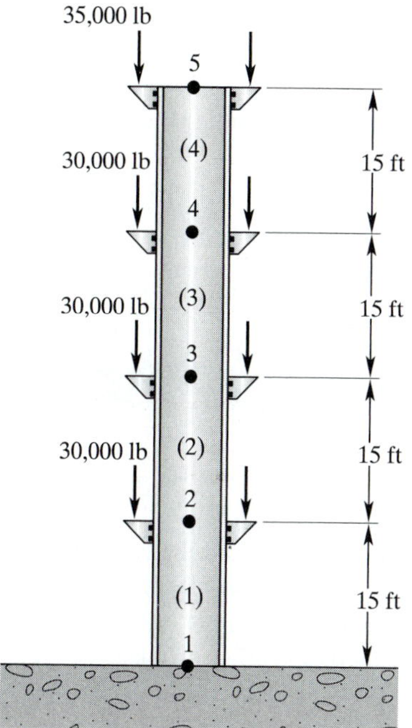


**11.** The equation for the heat diffusion of a one-dimensional system with heat generation in a Cartesian coordinate system is

$$k\frac{\partial^2 T}{\partial X^2} + q^{\cdot} = 0$$

The rate of thermal energy generation $q^{\cdot}$ represents the conversion of energy from electrical, chemical, nuclear, or electromagnetic forms to thermal energy within the volume of a given system. Derive the contribution of $q^{\cdot}$ to the load matrix. Consider a strip of heating elements embedded within the rear glass of a car producing a uniform heat generation at a rate of approximately 7000 W/m³. The glass has a thermal conductivity of $k = 0.8$ W/m · °C with a thickness of approximately 6 mm. The heat transfer coefficient associated with the

20°C air inside the back of the car is approximately 20 W/m$^2$ · K. The outside air is at a temperature of −5 °C with an associated heat transfer coefficient of 50 W/m$^2$ · K. Determine the temperatures of the inner and outer surfaces of the glass.

**12.** Verify the evaluation of the integral given by Eq. (4.17):

$$\int_{X_i}^{X_j} S_i(c_2 \Psi)\, dX = \frac{c_2 \ell}{3} \Psi_i + \frac{c_2 \ell}{6} \Psi_j$$

**13.** Verify the evaluation of the integral given by Eq. (4.20):

$$\int_{X_i}^{X_j} c_1 \left( -\frac{dS_j}{dX} \frac{d\Psi}{dX} \right) dX = -\frac{c_1}{\ell} (-\Psi_i + \Psi_j)$$

**14.** The deformation of a simply supported beam under distributed load, shown in the accompanying figure, is governed by the equation

$$\frac{d^2 Y}{dX^2} = \frac{M(X)}{EI}$$

where $M(X)$ is the internal bending moment and is given by:

$$M(X) = \frac{wX(L - X)}{2}$$

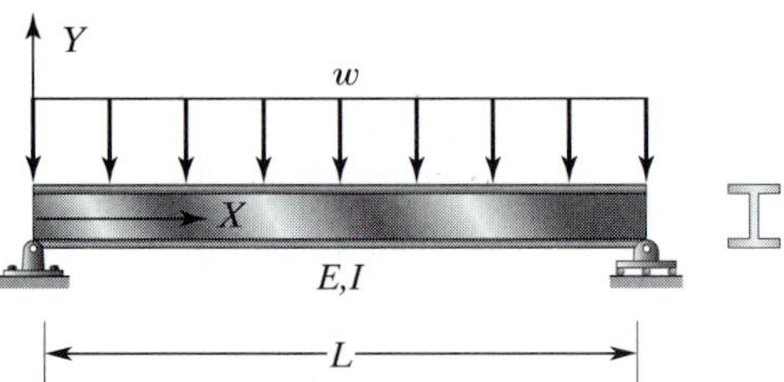


Formulate the stiffness matrix and the loading matrix for an arbitrary element using the approach discussed in this chapter. In Chapter 8, we will discuss the general formulation for beams.

**15.** Determine the deflection of point $D$ and the axial stress in each member in the system shown in the accompanying figure. ($E$ = 10.6 ksi.)

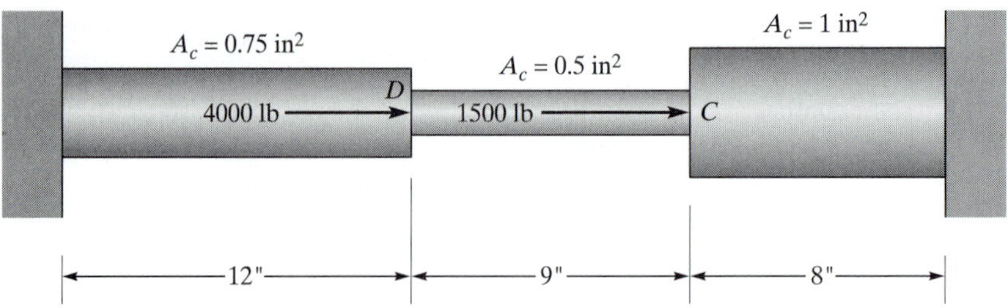


**16.** A 20-ft-tall post is used to support advertisement signs, at various locations along its height, as shown in the accompanying figure. The post is made of structural steel with a modulus of elasticity of $E = 29 \times 10^6$ lb/in$^2$. Not considering wind loading on the signs, (a) determine displacements of the post at the points of load application and (b) determine stresses in the post.

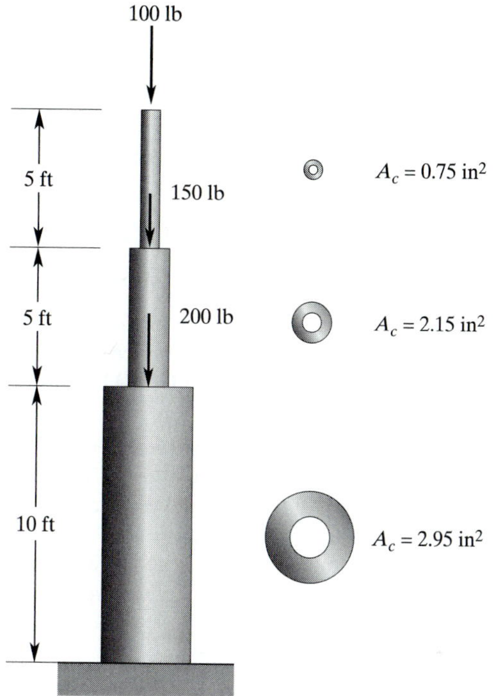


**17.** Determine the deflections of point $D$ and point $F$ in the system in the accompanying figure. Also compute the axial force and stress in each member. ($E = 29 \times 10^3$ ksi.)

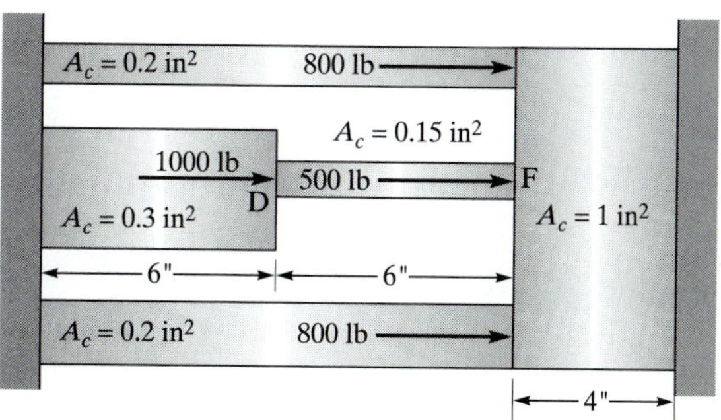


**18.** Determine the deflections of point $D$ and point $F$ in the system in the accompanying figure. Also compute the axial force and stress in each member.

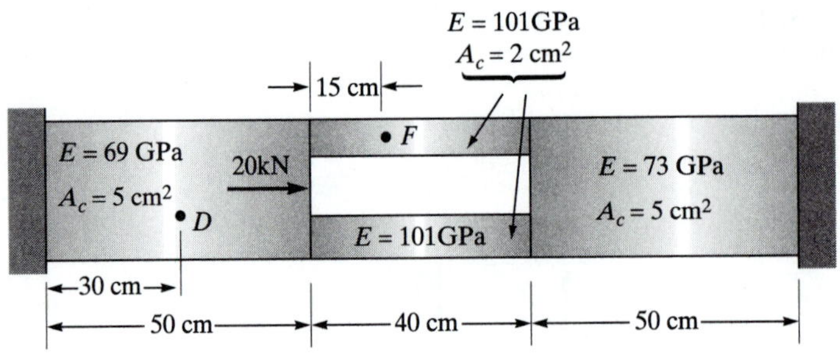


**19.** The deformation of an axial element of length $\ell$ due to the change in its temperature is given by

$$\delta_T = \alpha \ell \Delta T,$$

where $\delta_T$ is the change in the length of the element, $\alpha$ is the thermal expansion coefficient of the material, and $\Delta T$ represents the temperature change. Formulate the contribution of thermal strains to the strain energy of an element. Also formulate the stiffness and the load matrices for such an element.

**20.** You are to size fins of a rectangular cross section to remove a total of 200 W from a 400-cm$^2$ surface whose temperature is to be kept at 80°C. The temperature of the surrounding air is 25°C, and you may assume that the natural convection coefficient value is 25 W/m$^2$ · K. Because of restrictions on the amount of space, the fins cannot be extended more than 100 mm from the hot surface. You may select from the following materials: aluminum, copper, or steel. In a brief report, explain how you came up with your final design.

CHAPTER 5

# Two-Dimensional Elements

The objective of this chapter is to introduce the concept of two-dimensional shape functions, along with two-dimensional elements and their properties. Natural coordinates associated with quadrilateral and triangular elements will also be presented. We will derive the shape functions for rectangular elements, quadratic quadrilateral elements, and triangular elements. Examples of two-dimensional thermal and structural elements in ANSYS will also be presented here. The main topics discussed in Chapter 5 include the following:

- **5.1** Rectangular Elements
- **5.2** Quadratic Quadrilateral Elements
- **5.3** Linear Triangular Elements
- **5.4** Quadratic Triangular Elements
- **5.5** Isoparametric Elements
- **5.6** Two-Dimensional Integrals: Gauss–Legendre Quadrature
- **5.7** Examples of Two-Dimensional Elements in ANSYS

## 5.1 RECTANGULAR ELEMENTS

In Chapter 4, we studied the analysis of one-dimensional problems. We investigated heat transfer in a straight fin. We used one-dimensional linear shape functions to approximate temperature distributions along elements and formulated the conductance matrix and the thermal load matrix. The resulting systems of equations, once solved, yielded the nodal temperatures. In this chapter, we will lay the groundwork for the analysis of two-dimensional problems by first studying two-dimensional shape functions and elements. To aid us in presenting this material, let us consider the straight fin shown in Figure 5.1. Here, the dimensions of the fin and thermal boundary conditions are such that we cannot accurately approximate temperature distribution along the fin by a one-dimensional function. The temperature varies in both the $X$-direction and the $Y$-direction.

At this point, it is important to understand that the one-dimensional solutions are approximated by line segments, whereas the two-dimensional solutions are represented by plane segments. This point is illustrated in Figure 5.1. A close-up look at a typical rectangular element and its nodal values is shown in Figure 5.2.

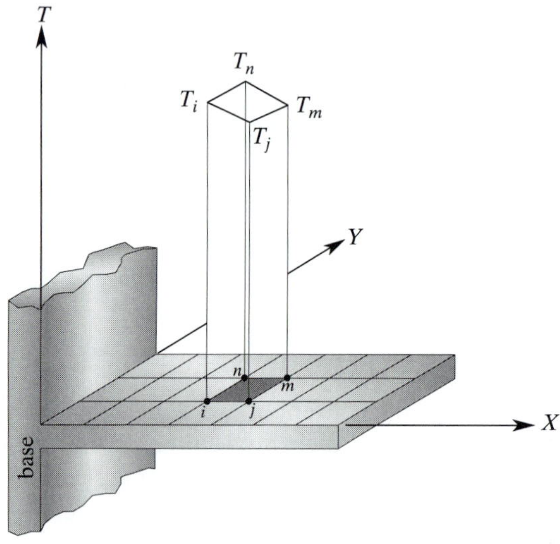


FIGURE 5.1 Using rectangular elements to describe a two-dimensional temperature distribution.

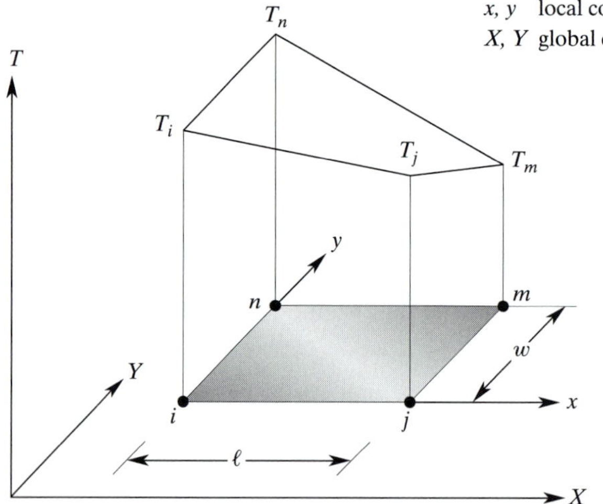


FIGURE 5.2 A typical rectangular element.

It is clear from examining Figure 5.2 that the temperature distribution over the element is a function of both $X$- and $Y$-coordinates. We can approximate the temperature distribution for an arbitrary rectangular element by

$$T^{(e)} = b_1 + b_2 x + b_3 y + b_4 xy \tag{5.1}$$

Note that there are four unknowns in Eq. (5.1), because a rectangular element is defined by four nodes: $i, j, m, n$. Also note that the function varies linearly along the edges of the element, and it becomes nonlinear inside the element. Elements with these types of characteristics are commonly referred to as bilinear elements. The procedure for deriving two-dimensional shape functions is essentially the same as that for one-dimensional elements. To obtain $b_1, b_2, b_3$, and $b_4$, we will use the local coordinates $x$ and $y$. Considering nodal temperatures, we must satisfy the following conditions:

$$\begin{aligned}
T &= T_i \quad \text{at} \quad x = 0 \quad \text{and} \quad y = 0 \\
T &= T_j \quad \text{at} \quad x = \ell \quad \text{and} \quad y = 0 \\
T &= T_m \quad \text{at} \quad x = \ell \quad \text{and} \quad y = w \\
T &= T_n \quad \text{at} \quad x = 0 \quad \text{and} \quad y = w
\end{aligned} \tag{5.2}$$

Applying the nodal conditions given by Eq. (5.2) to Eq. (5.1) and solving for $b_1, b_2, b_3$, and $b_4$, we have:

$$b_1 = T_i \qquad b_2 = \frac{1}{\ell}(T_j - T_i)$$

$$b_3 = \frac{1}{w}(T_n - T_i) \qquad b_4 = \frac{1}{\ell w}(T_i - T_j + T_m - T_n) \tag{5.3}$$

Substituting expressions given for $b_1, b_2, b_3$, and $b_4$, into Eq. (5.1) and regrouping parameters will result in the temperature distribution for a typical element in terms of shape functions:

$$T^{(e)} = \begin{bmatrix} S_i & S_j & S_m & S_n \end{bmatrix} \begin{Bmatrix} T_i \\ T_j \\ T_m \\ T_n \end{Bmatrix} \tag{5.4}$$

The shape functions in the above expression are

$$S_i = \left(1 - \frac{x}{\ell}\right)\left(1 - \frac{y}{w}\right) \tag{5.5}$$

$$S_j = \frac{x}{\ell}\left(1 - \frac{y}{w}\right)$$

$$S_m = \frac{xy}{\ell w}$$

$$S_n = \frac{y}{w}\left(1 - \frac{x}{\ell}\right)$$

It should be clear that we can use these shape functions to represent the variation of any unknown variable $\Psi$ over a rectangular region in terms of its nodal values $\Psi_i$, $\Psi_j$, $\Psi_m$, and $\Psi_n$. Thus, in general, we can write

$$\Psi^{(e)} = \begin{bmatrix} S_i & S_j & S_m & S_n \end{bmatrix} \begin{Bmatrix} \Psi_i \\ \Psi_j \\ \Psi_m \\ \Psi_n \end{Bmatrix} \tag{5.6}$$

For example, $\Psi$ could represent a solid element displacement field.

### Natural Coordinates

As was discussed in Chapter 3, natural coordinates are basically local coordinates in a dimensionless form. Moreover, most finite element programs perform element numerical integration by Gaussian quadratures. As the limits of integration, they use an interval from $-1$ to 1. The origin of the local coordinate system $x, y$ used earlier coincides with the natural coordinates $\xi = -1$ and $\eta = -1$, as shown in Figure 5.3.

If we let $\xi = \dfrac{2x}{\ell} - 1$ and $\eta = \dfrac{2y}{w} - 1$, then the shape functions in terms of the natural coordinates $\xi$ and $\eta$ are

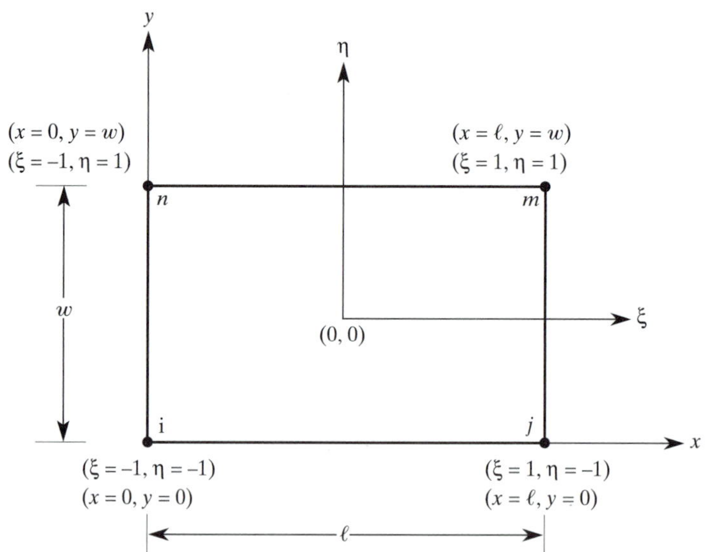


**FIGURE 5.3** Natural coordinates used to describe a quadrilateral element.

$$S_i = \frac{1}{4}(1 - \xi)(1 - \eta) \tag{5.7}$$

$$S_j = \frac{1}{4}(1 + \xi)(1 - \eta)$$

$$S_m = \frac{1}{4}(1 + \xi)(1 + \eta)$$

$$S_n = \frac{1}{4}(1 - \xi)(1 + \eta)$$

These shape functions have the same general basic properties as their one-dimensional counterparts. For example, $S_i$ has a value of 1 when evaluated at the coordinates of node $i$, but has a value of zero at all other nodes.

## 5.2 QUADRATIC QUADRILATERAL ELEMENTS

The eight-node quadratic quadrilateral element is basically a higher order version of the two-dimensional four-node quadrilateral element. This type of element is better suited for modeling problems with curved boundaries. A typical eight-node quadratic element is shown in Figure 5.4. When compared to the linear elements, for the same number of elements, quadratic elements offer better nodal results. In terms of the natural coordinates $\xi$ and $\eta$, the eight-node quadratic element has the general form of

$$\Psi^{(e)} = b_1 + b_2\xi + b_3\eta + b_4\xi\eta + b_5\xi^2 + b_6\eta^2 + b_7\xi^2\eta + b_8\xi\eta^2 \tag{5.8}$$

To solve for $b_1, b_2, b_3, \ldots, b_8$, we must first apply the nodal conditions and create eight equations from which we can solve for these coefficients. Instead of using this laborious and difficult method, we will follow an alternative approach, which is demonstrated next.

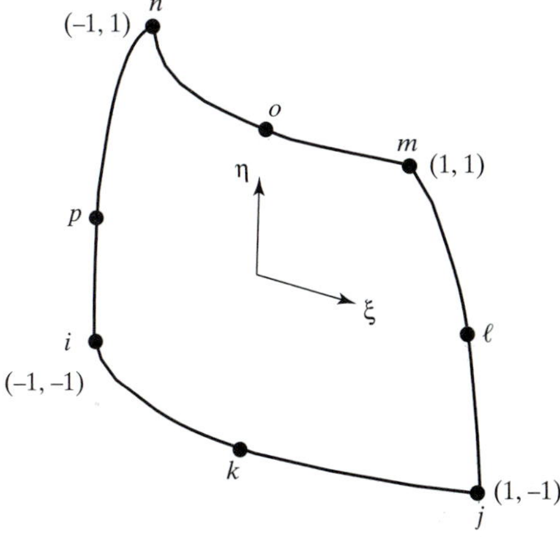


**FIGURE 5.4** Eight-node quadratic quadrilateral element.

In general, the shape function associated with each node can be represented in terms of the product of two functions $F_1$ and $F_2$:

$$S = F_1(\xi, \eta)F_2(\xi, \eta) \tag{5.9}$$

For a given node, we select the first function $F_1$ such that it will produce a value of zero when evaluated along the sides of the element that the given node does not contact. Moreover, the second function $F_2$ is selected such that when multiplied by $F_1$, it will produce a value of unity at the given node and a value of zero at other neighboring nodes. The product of functions $F_1$ and $F_2$ must also produce linear and nonlinear terms similar to the ones given by Eq. (5.8). To demonstrate this method, let us consider corner node $m$, with natural coordinates $\xi = 1$ and $\eta = 1$. First, $F_1$ must be selected such that along the $ij$-side ($\eta = -1$) and $in$-side ($\xi = -1$), the function will produce a value of zero. We select

$$F_1(\xi, \eta) = (1 + \xi)(1 + \eta)$$

which satisfies the condition. We then select

$$F_2(\xi, \eta) = c_1 + c_2\xi + c_3\eta$$

The coefficients in $F_2$ should be selected such that when multiplied by $F_1$, they will produce a value of unity at the given node $m$, and a value of zero at the adjacent neighboring nodes $\ell$ and $o$. Evaluating $S_m$ at node $m$ should give $S_m = 1$ for $\xi = 1$ and $\eta = 1$; evaluating $S_m$ at node $\ell$ should give $S_m = 0$ for $\xi = 1$ and $\eta = 0$; and evaluating $S_m$ at node $o$ should give $S_m = 0$ for $\xi = 0$ and $\eta = 1$. Applying these conditions to Eq. (5.9), we obtain

$$1 = \overbrace{(1 + 1)(1 + 1)}^{F_1(\xi, \eta)}\overbrace{\big(c_1 + c_2(1) + c_3(1)\big)}^{F_2(\xi, \eta)} = 4c_1 + 4c_2 + 4c_3$$
$$0 = (1 + 1)(1 + 0)\big(c_1 + c_2(1) + c_3(0)\big) = 2c_1 + 2c_2$$
$$0 = (1 + 0)(1 + 1)\big(c_1 + c_2(0) + c_3(1)\big) = 2c_1 + 2c_3$$

which results in $c_1 = -\frac{1}{4}$, $c_2 = \frac{1}{4}$, and $c_3 = \frac{1}{4}$ with $S_m = (1 + \xi)(1 + \eta)\big(-\frac{1}{4} + \frac{1}{4}\xi + \frac{1}{4}\eta\big)$. The shape functions associated with the other corner nodes are determined in a similar fashion. The corner node shape functions are:

$$S_i = -\frac{1}{4}(1 - \xi)(1 - \eta)(1 + \xi + \eta) \tag{5.10}$$

$$S_j = \frac{1}{4}(1 + \xi)(1 - \eta)(-1 + \xi - \eta)$$

$$S_m = \frac{1}{4}(1 + \xi)(1 + \eta)(-1 + \xi + \eta)$$

$$S_n = -\frac{1}{4}(1 - \xi)(1 + \eta)(1 + \xi - \eta)$$

Let us turn our attention to the shape functions for the middle nodes. As an example, we will develop the shape function associated with node $o$. First, $F_1$ must be selected such that along the *ij*-side ($\eta = -1$), *in*-side ($\xi = -1$), and *jm*-side ($\xi = 1$), the function will produce a value of zero. We select

$$F_1(\xi, \eta) = (1 - \xi)(1 + \eta)(1 + \xi)$$

Note that the product of the terms given by $F_1$ will produce linear and nonlinear terms, as required by Eq. (5.8). Therefore, the second function $F_2$ must be a constant; otherwise, the product of $F_1$ and $F_2$ will produce third-order polynomial terms, which we certainly do not want! So,

$$F_2(\xi, \eta) = c_1$$

Applying the nodal condition

$$S_o = 1 \text{ for } \xi = 0 \text{ and } \eta = 1$$

leads to

$$1 = \overbrace{(1 - 0)(1 + 1)(1 + 0)}^{F_1(\xi, \eta)} \overbrace{c_1}^{F_2(\xi, \eta)} = 2c_1$$

resulting in $c_1 = \frac{1}{2}$ with $S_o = \frac{1}{2}(1 - \xi)(1 + \eta)(1 + \xi) = S_o = \frac{1}{2}(1 + \eta)(1 - \xi^2)$. Using a similar procedure, we can obtain the shape functions for the midpoint nodes $k$, $\ell$, and $p$. Thus the midpoint shape functions are:

$$S_k = \frac{1}{2}(1 - \eta)(1 - \xi^2) \tag{5.11}$$

$$S_\ell = \frac{1}{2}(1 + \xi)(1 - \eta^2)$$

$$S_o = \frac{1}{2}(1 + \eta)(1 - \xi^2)$$

$$S_p = \frac{1}{2}(1 - \xi)(1 - \eta^2)$$

## EXAMPLE 5.1

We have used two-dimensional rectangular elements to model the stress distribution in a thin plate. The nodal stresses for an element belonging to the plate are shown in Figure 5.5. What is the value of stress at the center of this element?

The stress distribution for the element is

$$\sigma^{(e)} = \begin{bmatrix} S_i & S_j & S_m & S_n \end{bmatrix} \begin{Bmatrix} \sigma_i \\ \sigma_j \\ \sigma_m \\ \sigma_n \end{Bmatrix}$$

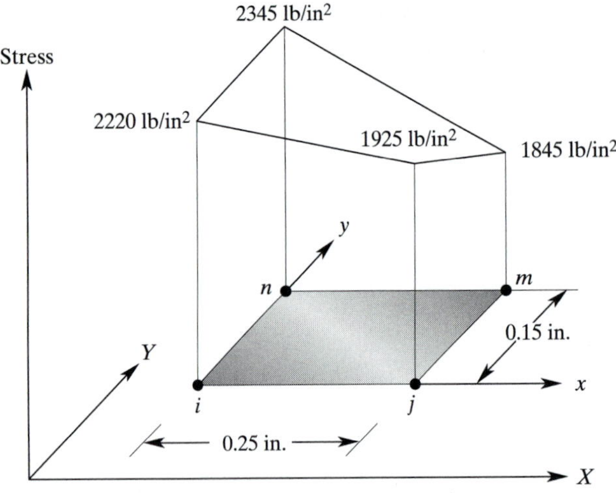


**FIGURE 5.5** Nodal stresses for Example 5.1.

where $\sigma_i, \sigma_j, \sigma_m$, and $\sigma_n$ are stresses at nodes $i, j, m$, and $n$, respectively, and the shape functions are given by Eq. (5.5):

$$S_i = \left(1 - \frac{x}{\ell}\right)\left(1 - \frac{y}{w}\right) = \left(1 - \frac{x}{0.25}\right)\left(1 - \frac{y}{0.15}\right)$$

$$S_j = \frac{x}{\ell}\left(1 - \frac{y}{w}\right) = \frac{x}{0.25}\left(1 - \frac{y}{0.15}\right)$$

$$S_m = \frac{xy}{\ell w} = \frac{xy}{(0.25)(0.15)}$$

$$S_n = \frac{y}{w}\left(1 - \frac{x}{\ell}\right) = \frac{y}{0.15}\left(1 - \frac{x}{0.25}\right)$$

For the given element, the stress distribution in terms of the local coordinates $x$ and $y$ is given by

$$\sigma^{(e)} = \overbrace{\left(1 - \frac{x}{0.25}\right)\left(1 - \frac{y}{0.15}\right)}^{S_i}\overbrace{(2220)}^{\sigma_i} + \overbrace{\frac{x}{0.25}\left(1 - \frac{y}{0.15}\right)}^{S_j}\overbrace{(1925)}^{\sigma_j} + \overbrace{\frac{xy}{(0.25)(0.15)}}^{S_m}\overbrace{(1845)}^{\sigma_m}$$

$$+ \overbrace{\frac{y}{0.15}\left(1 - \frac{x}{0.25}\right)}^{S_n}\overbrace{(2345)}^{\sigma_n}$$

We can compute the stress at any point within this element from the aforementioned equation. Here, we are interested in the value of the stress at the midpoint. Substituting $x = 0.125$ and $y = 0.075$ into the equation, we have

$$\sigma(0.125,0.075) = 555 + 481 + 461 + 586 = 2083 \quad \text{lb/in}^2$$

Note that we could have solved this problem using natural coordinates. This approach may be easier because the point of interest is located at the center of the element $\xi = 0$ and $\eta = 0$. The quadrilateral natural shape functions are given by Eq. (5.7):

$$S_i = \frac{1}{4}(1 - \xi)(1 - \eta) = \frac{1}{4}(1 - 0)(1 - 0) = \frac{1}{4}$$

$$S_j = \frac{1}{4}(1 + \xi)(1 - \eta) = \frac{1}{4}(1 + 0)(1 - 0) = \frac{1}{4}$$

$$S_m = \frac{1}{4}(1 + \xi)(1 + \eta) = \frac{1}{4}(1 + 0)(1 + 0) = \frac{1}{4}$$

$$S_n = \frac{1}{4}(1 - \xi)(1 + \eta) = \frac{1}{4}(1 - 0)(1 + 0) = \frac{1}{4}$$

$$\sigma(0.125, 0.075) = \frac{1}{4}(2220) + \frac{1}{4}(1925) + \frac{1}{4}(1845) + \frac{1}{4}(2345) = 2083 \quad \text{lb/in}^2$$

Thus, the stress at the midpoint of the rectangular element is the average of the nodal stresses.

---

## EXAMPLE 5.2

Confirm the expression given for the quadratic quadrilateral shape function $S_n$.

Referring to the procedure discussed in Section 5.2, we can represent $S_n$ by

$$S_n = F_1(\xi, \eta)F_2(\xi, \eta)$$

For the shape function $S_n$, $F_1$ should be selected such that it will have a value of zero along the *ij*-side ($\eta = -1$) and the *jm*-side ($\xi = 1$) So, we choose

$$F_1(\xi, \eta) = (1 - \xi)(1 + \eta)$$

Furthermore, $F_2$ is given by:

$$F_2(\xi, \eta) = c_1 + c_2\xi + c_3\eta$$

and the coefficients $c_1$, $c_2$, and $c_3$ are determined by applying the following conditions:

$$S_n = 1 \quad \text{for} \quad \xi = -1 \quad \text{and} \quad \eta = 1$$

$$S_n = 0 \quad \text{for} \quad \xi = 0 \quad \text{and} \quad \eta = 1$$

$$S_n = 0 \quad \text{for} \quad \xi = -1 \quad \text{and} \quad \eta = 0$$

So, we get

$$1 = 4c_1 - 4c_2 + 4c_3$$

$$0 = 2c_1 + 2c_3$$

$$0 = 2c_1 - 2c_2$$

resulting in $c_1 = -\frac{1}{4}$, $c_2 = -\frac{1}{4}$, and $c_3 = \frac{1}{4}$, which is identical to the expression previously given for $S_n$. That is,

$$S_n = -\frac{1}{4}(1 - \xi)(1 + \eta)(1 + \xi - \eta)$$

## 5.3 LINEAR TRIANGULAR ELEMENTS

A major disadvantage associated with using bilinear rectangular elements is that they do not conform to a curved boundary very well. In contrast, triangular elements, shown describing a two-dimensional temperature distribution in Figure 5.6, are better suited to approximate curved boundaries. A triangular element is defined by three nodes, as shown in Figure 5.7. Therefore, we can represent the variation of a dependent variable, such as temperature, over the triangular region by

$$T^{(e)} = a_1 + a_2 X + a_3 Y \tag{5.12}$$

Considering the nodal temperatures as shown in Figure 5.7, we must satisfy the following conditions:

$$\begin{aligned} T &= T_i \quad \text{at} \quad X = X_i \quad \text{and} \quad Y = Y_i \\ T &= T_j \quad \text{at} \quad X = X_j \quad \text{and} \quad Y = Y_j \\ T &= T_k \quad \text{at} \quad X = X_k \quad \text{and} \quad Y = Y_k \end{aligned} \tag{5.13}$$

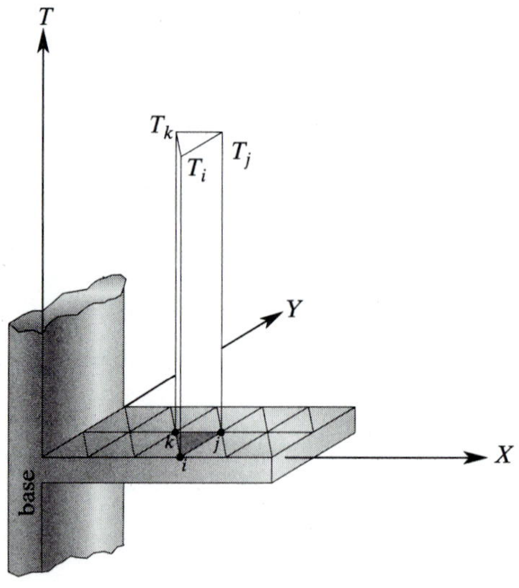


**FIGURE 5.6** Using triangular elements to describe a two-dimensional temperature distribution.

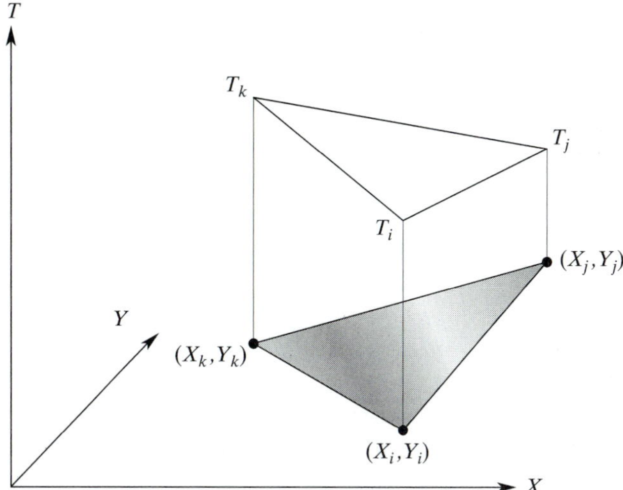


**FIGURE 5.7** A triangular element.

Substituting nodal values into Eq. (5.12), we have:

$$T_i = a_1 + a_2 X_i + a_3 Y_i \tag{5.14}$$

$$T_j = a_1 + a_2 X_j + a_3 Y_j$$

$$T_k = a_1 + a_2 X_k + a_3 Y_k$$

Solving for $a_1, a_2$, and $a_3$, we obtain

$$a_1 = \frac{1}{2A}\left[(X_j Y_k - X_k Y_j)T_i + (X_k Y_i - X_i Y_k)T_j + (X_i Y_j - X_j Y_i)T_k\right] \tag{5.15}$$

$$a_2 = \frac{1}{2A}\left[(Y_j - Y_k)T_i + (Y_k - Y_i)T_j + (Y_i - Y_j)T_k\right]$$

$$a_3 = \frac{1}{2A}\left[(X_k - X_j)T_i + (X_i - X_k)T_j + (X_j - X_i)T_k\right]$$

where $A$ is the area of the triangular element and is computed from the equation

$$2A = X_i(Y_j - Y_k) + X_j(Y_k - Y_i) + X_k(Y_i - Y_j) \tag{5.16}$$

Substituting for $a_1, a_2$, and $a_3$ into Eq. (5.12) and grouping $T_i, T_j$, and $T_k$ terms yields

$$T^{(e)} = \begin{bmatrix} S_i & S_j & S_k \end{bmatrix} \begin{Bmatrix} T_i \\ T_j \\ T_k \end{Bmatrix} \tag{5.17}$$

where the shape functions $S_i$, $S_j$, and $S_k$ are

$$S_i = \frac{1}{2A}(\alpha_i + \beta_i X + \delta_i Y) \tag{5.18}$$

$$S_j = \frac{1}{2A}(\alpha_j + \beta_j X + \delta_j Y)$$

$$S_k = \frac{1}{2A}(\alpha_k + \beta_k X + \delta_k Y)$$

and

$$\alpha_i = X_j Y_k - X_k Y_j \qquad \beta_i = Y_j - Y_k \qquad \delta_i = X_k - X_j$$

$$\alpha_j = X_k Y_i - X_i Y_k \qquad \beta_j = Y_k - Y_i \qquad \delta_j = X_i - X_k$$

$$\alpha_k = X_i Y_j - X_j Y_i \qquad \beta_k = Y_i - Y_j \qquad \delta_k = X_j - X_i$$

Again, keep in mind that triangular shape functions have some basic properties, like other shape functions defined previously. For example, $S_i$ has a value of unity when evaluated at the coordinates of node $i$ and has a value of zero at all other nodes. Or, as another example, the sum of the shape functions has a value of unity. That property is demonstrated by the equation

$$S_i + S_j + S_k = 1 \tag{5.19}$$

## Natural (Area) Coordinates for Triangular Elements

Consider point $P$ with coordinates $(x, y)$ inside the triangular region. Connecting this point to nodes, $i$, $j$, and $k$ results in dividing the area of the triangle into three smaller areas $A_1$, $A_2$, and $A_3$, as shown in Figure 5.8.

Let us now perform an experiment. We move point $P$ from its inside position to coincide with point $Q$ along the $kj$-edge of the element. In the process, the value of area $A_1$ becomes zero. Moving point $P$ to coincide with node $i$ stretches $A_1$ to fill in the entire area

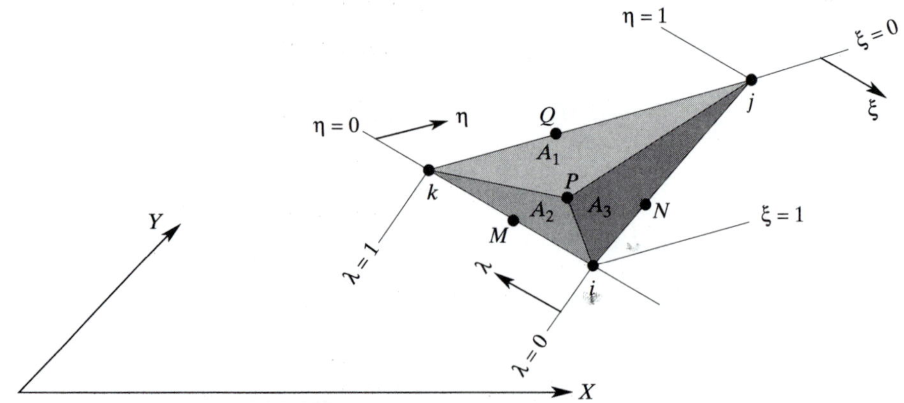


**FIGURE 5.8** Natural (area) coordinates for a triangular element.

$A$ of the element. Based on the results of our experiment, we can define a natural, or area, coordinate $\xi$ as the ratio of $A_1$ to the area $A$ of the element so that its values vary from 0 to 1. Similarly, moving point $P$ from its inside position to coincide with point $M$, along the $ki$-edge, results in $A_2 = 0$. Moving point $P$ to coincide with node $j$ stretches $A_2$ such that it fills the entire area of the element; that is, $A_2 = A$. We can define another area coordinate $\eta$ as the ratio of $A_2$ to $A$, and its magnitude varies from 0 to 1. Formally, for a triangular element, the natural (area) coordinates $\xi, \eta$, and $\lambda$ are defined by

$$\xi = \frac{A_1}{A} \tag{5.20}$$

$$\eta = \frac{A_2}{A}$$

$$\lambda = \frac{A_3}{A}$$

It is important to realize that only two of the natural coordinates are linearly independent, because

$$\frac{A_1}{A} + \frac{A_2}{A} + \frac{A_3}{A} = \frac{A}{A} = 1 = \xi + \eta + \lambda$$

For example, the $\lambda$-coordinate can be defined in terms of $\xi$ and $\eta$ by

$$\lambda = 1 - \xi - \eta \tag{5.21}$$

We can show that the triangular natural (area) coordinates are exactly identical to the shape functions $S_i, S_j$, and $S_k$. That is,

$$\xi = S_i \tag{5.22}$$

$$\eta = S_j$$

$$\lambda = S_k$$

We now offer a proof of the above relationships. We can describe all of the triangular areas in terms of the coordinates of their vertexes. As an example, consider $\xi$, which is the ratio of $A_1$ to $A$:

$$\xi = \frac{A_1}{A} = \frac{\frac{1}{2}[(X_jY_k - X_kY_j) + X(Y_j - Y_k) + Y(X_k - X_j)]}{\frac{1}{2}[X_i(Y_j - Y_k) + X_j(Y_k - Y_i) + X_k(Y_i - Y_j)]} \tag{5.23}$$

Comparison of Eq. (5.23) to Eq. (5.18)* shows that $\xi$ and $S_i$ are identical.

---

*Substitute for $A_1, \alpha_i, \beta_i, S_i$, in terms of nodal coordinates.

## 5.4 QUADRATIC TRIANGULAR ELEMENTS

The spatial variation of a dependent variable, such as temperature, over a region may be approximated more accurately by a quadratic function, such as:

$$T^{(e)} = a_1 + a_2X + a_3Y + a_4X^2 + a_5XY + a_6Y^2 \tag{5.24}$$

By now, you should understand how to develop shape functions. Therefore, the shape functions for a quadratic triangular element, shown in Figure 5.9, are given below without proof. The shape functions in terms of natural coordinates are:

$$\begin{aligned} S_i &= \xi(2\xi - 1) \\ S_j &= \eta(2\eta - 1) \\ S_k &= \lambda(2\lambda - 1) = 1 - 3(\xi + \eta) + 2(\xi + \eta)^2 \\ S_\ell &= 4\xi\eta \\ S_m &= 4\eta\lambda = 4\eta(1 - \xi - \eta) \\ S_n &= 4\xi\lambda = 4\xi(1 - \xi - \eta) \end{aligned} \tag{5.25}$$

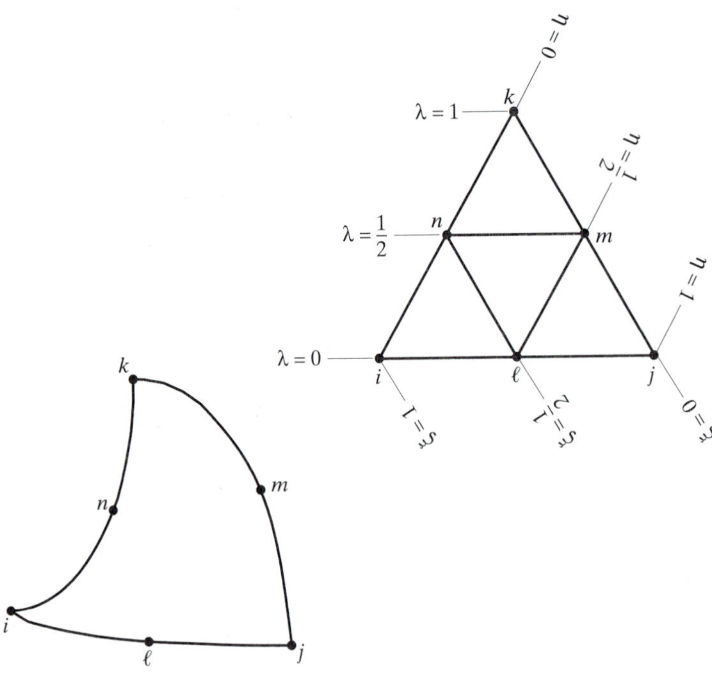


**FIGURE 5.9** A quadratic triangular element.

---

### EXAMPLE 5.3

We have used two-dimensional triangular elements to model the temperature distribution in a fin. The nodal temperatures and their corresponding positions for an element are shown in Figure 5.10. (a) What is the value of temperature at $X = 2.15$ cm and $Y = 1.1$ cm? (b) Determine the components of temperature gradients for this element. (c) Determine the location of 70°C and 75°C isotherms.

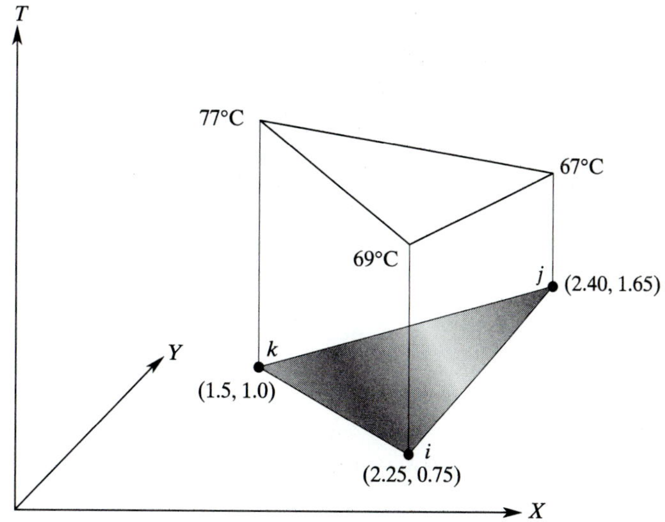


**FIGURE 5.10** Nodal temperatures and coordinates for the element in Example 5.3.

(a) The temperature distribution inside the element is

$$T^{(e)} = \begin{bmatrix} S_i & S_j & S_k \end{bmatrix} \begin{Bmatrix} T_i \\ T_j \\ T_k \end{Bmatrix}$$

where the shape functions $S_i$, $S_j$, and $S_k$ are

$$S_i = \frac{1}{2A}(\alpha_i + \beta_i X + \delta_i Y)$$

$$S_j = \frac{1}{2A}(\alpha_j + \beta_j X + \delta_j Y)$$

$$S_k = \frac{1}{2A}(\alpha_k + \beta_k X + \delta_k Y)$$

and

$$\alpha_i = X_j Y_k - X_k Y_j = (2.4)(1.0) - (1.5)(1.65) = -0.075$$
$$\alpha_j = X_k Y_i - X_i Y_k = (1.5)(0.75) - (2.25)(1.0) = -1.125$$
$$\alpha_k = X_i Y_j - X_j Y_i = (2.25)(1.65) - (2.40)(0.75) = 1.9125$$

$$\beta_i = Y_j - Y_k = 1.65 - 1.0 = 0.65$$
$$\beta_j = Y_k - Y_i = 1.0 - 0.75 = 0.25$$
$$\beta_k = Y_i - Y_j = 0.75 - 1.65 = -0.9$$

$$\delta_i = X_k - X_j = 1.50 - 2.40 = -0.9$$
$$\delta_j = X_i - X_k = 2.25 - 1.5 = 0.75$$
$$\delta_k = X_j - X_i = 2.40 - 2.25 = 0.15$$

and

$$2A = X_i(Y_j - Y_k) + X_j(Y_k - Y_i) + X_k(Y_i - Y_j)$$

$$2A = 2.25(1.65 - 1.0) + 2.40(1.0 - 0.75) + 1.5(0.75 - 1.65) = 0.7125$$

$$S_i = \frac{1}{2A}\left(\alpha_i + \beta_i X + \delta_i Y\right) = \frac{1}{0.7125}(-0.075 + 0.65X - 0.9Y)$$

$$S_j = \frac{1}{2A}\left(\alpha_j + \beta_j X + \delta_j Y\right) = \frac{1}{0.7125}(-1.125 + 0.25X + 0.75Y)$$

$$S_k = \frac{1}{2A}\left(\alpha_k + \beta_k X + \delta_k Y\right) = \frac{1}{0.7125}(1.9125 - 0.9X + 0.15Y)$$

The temperature distribution for the element is

$$T = \frac{69}{0.7125}(-0.075 + 0.65X - 0.9Y) + \frac{67}{0.7125}(-1.125 + 0.25X + 0.75Y) + \frac{77}{0.7125}(1.9125 - 0.9X + 0.15Y)$$

After simplifying, we have

$$T = 93.632 - 10.808X - 0.421Y$$

Substituting for coordinates of the point $X = 2.15$ and $Y = 1.1$ leads to $T = 69.93°C$.

(b) In general, the gradient components of a dependent variable $\Psi^{(e)}$ are computed from:

$$\frac{\partial \Psi^{(e)}}{\partial X} = \frac{\partial}{\partial X}\left[S_i\Psi_i + S_j\Psi_j + S_k\Psi_k\right]$$

$$\frac{\partial \Psi^{(e)}}{\partial Y} = \frac{\partial}{\partial Y}\left[S_i\Psi_i + S_j\Psi_j + S_k\Psi_k\right]$$

$$\begin{Bmatrix} \dfrac{\partial \Psi^{(e)}}{\partial X} \\ \dfrac{\partial \Psi^{(e)}}{\partial Y} \end{Bmatrix} = \frac{1}{2A}\begin{bmatrix} \beta_i & \beta_j & \beta_k \\ \delta_i & \delta_j & \delta_k \end{bmatrix}\begin{Bmatrix} \Psi_i \\ \Psi_j \\ \Psi_k \end{Bmatrix} \tag{5.26}$$

It should be clear from examining Eq. (5.26) that the gradients have constant values. This property is always true for linear triangular elements. The temperature gradients are computed from:

$$\begin{Bmatrix} \dfrac{\partial T^{(e)}}{\partial X} \\ \dfrac{\partial T^{(e)}}{\partial Y} \end{Bmatrix} = \frac{1}{2A}\begin{bmatrix} \beta_i & \beta_j & \beta_k \\ \delta_i & \delta_j & \delta_k \end{bmatrix}\begin{Bmatrix} T_i \\ T_j \\ T_k \end{Bmatrix} = \frac{1}{0.7125}\begin{bmatrix} 0.65 & 0.25 & -0.9 \\ -0.9 & 0.75 & 0.15 \end{bmatrix}\begin{Bmatrix} 69 \\ 67 \\ 77 \end{Bmatrix} = \begin{Bmatrix} -10.808 \\ -0.421 \end{Bmatrix}$$

Note that differentiation of the simplified temperature equation ($T = 93.632 - 10.808X - 0.421Y$) directly, would have resulted in exactly the same values.

(c) The location of 70°C and 75°C isotherms can be determined from the fact that over a triangular element, temperature varies linearly in both $X$- and $Y$- directions. Thus, we can use linear interpolation to calculate coordinates of isotherms. First, let us focus on the 70°C constant temperature line. This isotherm will intersect the 77°C–69°C-edge according to the relations

$$\frac{77 - 70}{77 - 69} = \frac{1.5 - X}{1.5 - 2.25} = \frac{1.0 - Y}{1.0 - 0.75}$$

which results in the coordinates $X = 2.16$ cm and $Y = 0.78$ cm. The 70°C isotherm also intersects the 77°C–67°C-edge:

$$\frac{77 - 70}{77 - 67} = \frac{1.5 - X}{1.5 - 2.4} = \frac{1.0 - Y}{1.0 - 1.65}$$

These relations result in the coordinates $X = 2.13$ cm and $Y = 1.45$ cm. Similarly, the location of the 75°C isotherm is determined using the 77°C–69°C-edge:

$$\frac{77 - 75}{77 - 69} = \frac{1.5 - X}{1.5 - 2.25} = \frac{1.0 - Y}{1.0 - 0.75}$$

which results in the coordinates $X = 1.69$ and $Y = 0.94$. Finally, along the 77°C–67°C-edge, we have:

$$\frac{77 - 75}{77 - 67} = \frac{1.5 - X}{1.5 - 2.4} = \frac{1.0 - Y}{1.0 - 1.65}$$

These equations result in the coordinates $X = 1.68$ and $Y = 1.13$. The isotherms and their corresponding locations are shown in Figure 5.11.

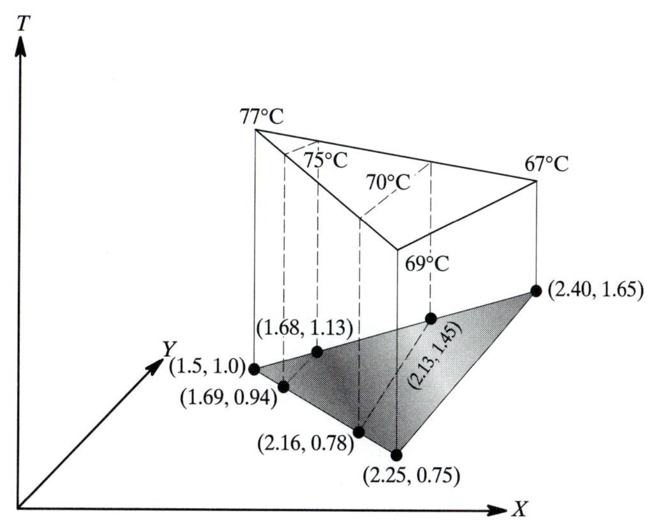


**FIGURE 5.11** The isotherms of the element in Example 5.3.

## 5.5 ISOPARAMETRIC ELEMENTS

As we discussed in Chapter 3, when we use a single set of parameters (a set of shape functions) to define the unknown variables $u$, $v$, $T$, and so on, and use the same parameters (the same shape functions) to express the geometry, we are using an *isoparametric* formulation. An element expressed in such a manner is called an isoparametric element. Let us turn our attention to the quadrilateral element shown in Figure 5.12. Let us also consider a solid mechanics problem, in which a body undergoes a deformation. Using a quadrilateral element, the displacement field within an element belonging to this solid body can be expressed in terms of its nodal values as:

$$u = S_i U_{ix} + S_j U_{jx} + S_m U_{mx} + S_n U_{nx} \tag{5.27}$$
$$v = S_i U_{iy} + S_j U_{jy} + S_m U_{my} + S_n U_{ny}$$

We can write the relations given by Eq. (5.27) in matrix form:

$$\begin{Bmatrix} u \\ v \end{Bmatrix} = \begin{bmatrix} S_i & 0 & S_j & 0 & S_m & 0 & S_n & 0 \\ 0 & S_i & 0 & S_j & 0 & S_m & 0 & S_n \end{bmatrix} \begin{Bmatrix} U_{ix} \\ U_{iy} \\ U_{jx} \\ U_{jy} \\ U_{mx} \\ U_{my} \\ U_{nx} \\ U_{ny} \end{Bmatrix} \tag{5.28}$$

Note that using isoparametric formulation, we can use the same shape functions to describe the position of any point within the element by the equations

$$x = S_i x_i + S_j x_j + S_m x_m + S_n x_n \tag{5.29}$$
$$y = S_i y_i + S_j y_j + S_m y_m + S_n y_n$$

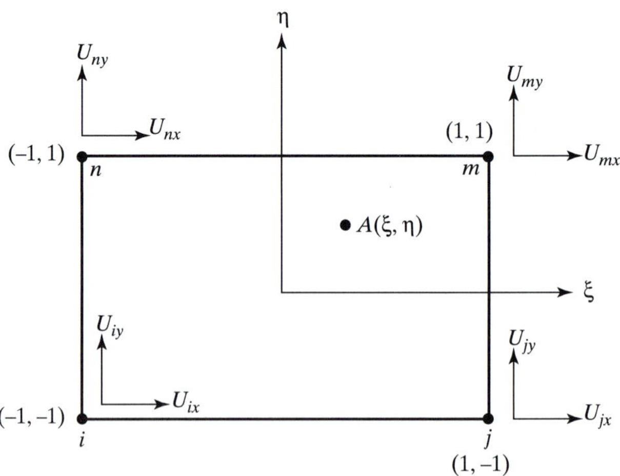


**FIGURE 5.12** A quadrilateral element used in formulating plane-stress problems.

As you will see in Chapter 8, the displacement field is related to the components of strains $\left(\varepsilon_{xx} = \frac{\partial u}{\partial x}, \varepsilon_{yy} = \frac{\partial v}{\partial y}, \text{ and } \gamma_{xy} = \frac{\partial u}{\partial y} + \frac{\partial v}{\partial x}\right)$ and, subsequently, to the nodal displacements using shape functions. In deriving the elemental stiffness matrix from strain energy, we need to take the derivatives of the components of the displacement field with respect to the $x$- and $y$- coordinates, which in turn means taking the derivatives of the appropriate shape functions with respect to $x$ and $y$. At this point, keep in mind that the shape functions are expressed in terms of $\xi$ and $\eta$. Thus, in general, it is necessary to establish relationships that allow the derivatives of a function $f(x,y)$ to be taken with respect to $x$ and $y$ and to express them in terms of derivatives of the function $f(x,y)$ with respect to $\xi$ and $\eta$. This point will become clear soon. Using the chain rule, we can write:

$$\begin{aligned}\frac{\partial f(x,y)}{\partial \xi} &= \frac{\partial f(x,y)}{\partial x}\frac{\partial x}{\partial \xi} + \frac{\partial f(x,y)}{\partial y}\frac{\partial y}{\partial \xi}\\ \frac{\partial f(x,y)}{\partial \eta} &= \frac{\partial f(x,y)}{\partial x}\frac{\partial x}{\partial \eta} + \frac{\partial f(x,y)}{\partial y}\frac{\partial y}{\partial \eta}\end{aligned} \tag{5.30}$$

Expressing Eq. (5.30) in matrix form, we have

$$\begin{Bmatrix}\dfrac{\partial f(x,y)}{\partial \xi}\\ \dfrac{\partial f(x,y)}{\partial \eta}\end{Bmatrix} = \overbrace{\begin{bmatrix}\dfrac{\partial x}{\partial \xi} & \dfrac{\partial y}{\partial \xi}\\ \dfrac{\partial x}{\partial \eta} & \dfrac{\partial y}{\partial \eta}\end{bmatrix}}^{[\mathbf{J}]}\begin{Bmatrix}\dfrac{\partial f(x,y)}{\partial x}\\ \dfrac{\partial f(x,y)}{\partial y}\end{Bmatrix} \tag{5.31}$$

where the **J** matrix is referred to as the Jacobian of the coordinate transformation. The relationships of Eq. (5.31) can be also presented as

$$\begin{Bmatrix}\dfrac{\partial f(x,y)}{\partial x}\\ \dfrac{\partial f(x,y)}{\partial y}\end{Bmatrix} = [\mathbf{J}]^{-1}\begin{Bmatrix}\dfrac{\partial f(x,y)}{\partial \xi}\\ \dfrac{\partial f(x,y)}{\partial \eta}\end{Bmatrix} \tag{5.32}$$

For a quadrilateral element, the **J** matrix can be evaluated using Eq. (5.29) and (5.7). This evalutation is left as an exercise for you; see Problem 5.24. We will discuss the derivation of the element stiffness matrix using the isoparametric formulation in Chapter 8.

## 5.6 TWO-DIMENSIONAL INTEGRALS: GAUSS–LEGENDRE QUADRATURE

As we discussed in Chapter 3, most finite element programs perform numerical integration for elements by Gaussian quadratures, and as the limits of integration, they use an interval from −1 to 1. We now extend the Gauss–Legendre quadrature formulation to two-dimensional problems as follows:

$$I = \int_{-1}^{1}\int_{-1}^{1} f(\xi,\eta)\,d\xi\,d\eta \cong \int_{-1}^{1}\left[\sum_{i=1}^{n} w_i f(\xi_i,\eta)\right]d\eta \cong \sum_{i=1}^{n}\sum_{j=1}^{n} w_i w_j f(\xi_i,\eta_j) \tag{5.33}$$

The relationships of Eq. (5.33) should be self-evident. Recall that the weighting factors and the sampling points are given in Table 3.2.

## EXAMPLE 5.4

To demonstrate the steps involved in Gauss–Legendre quadrature computation, let us consider evaluating the integral

$$I = \int_0^2 \int_0^2 (3y^2 + 2x)dx\,dy$$

As you know, the given integral can be evaluated analytically as

$$I = \int_0^2 \int_0^2 (3y^2 + 2x)dxdy = \int_0^2 \left[ \int_0^2 (3y^2 + 2x)dx \right] dy = \int_0^2 [(3y^2x + x^2)]_0^2 dy = \int_0^2 (6y^2 + 4)dy = 24$$

We now evaluate the integral using Gauss–Legendre quadrature. We begin by changing $y$- and $x$- variables into $\xi$ and $\eta$, using the relationships of Eq. (3.45):

$$x = 1 + \xi \quad \text{and} \quad dx = d\xi$$

$$y = 1 + \eta \quad \text{and} \quad dy = d\eta$$

Thus, the integral $I$ can be expressed by:

$$I = \int_0^2 \int_0^2 (3y^2 + 2x)dxdy = \int_{-1}^1 \int_{-1}^1 [3(1 + \eta)^2 + 2(1 + \xi)]d\xi d\eta$$

Using the two-point sampling formula, we have

$$I \cong \sum_{i=1}^{n} \sum_{j=1}^{n} w_i w_j f(\xi_i, \eta_j)$$

$$I \cong \sum_{i=1}^{2} \sum_{j=1}^{2} w_i w_j [3(1 + \eta_j)^2 + 2(1 + \xi_i)]$$

To evaluate the summation, we start with $i = 1$, while changing $j$ from 1 to 2, and we repeat the process with $i = 2$ while changing $j$ from 1 to 2:

$$\begin{aligned} I \cong\ & [(1)(1)[3(1 + (-0.577350269))^2 + 2(1 + (-0.577350269))] \\ & + (1)(1)[3(1 + (0.577350269))^2 + 2(1 + (-0.577350269))]] \\ & + [(1)(1)[3(1 + (-0.577350269))^2 + 2(1 + (0.577350269))] \\ & + (1)(1)[3(1 + (0.577350269))^2 + 2(1 + (0.577350269))]] = 24.000000000 \end{aligned}$$

## 5.7 EXAMPLES OF TWO-DIMENSIONAL ELEMENTS IN ANSYS

ANSYS offers many two-dimensional elements that are based on linear and quadratic quadrilateral and triangular shape functions. We will discuss the formulation of two-dimensional thermal- and solid-structural problems in detail in Chapters 7 and 8. For now, consider some examples of two-dimensional structural-solid and thermal-solid elements.

### Plane2

is a six-node triangular structural-solid element. The element has quadratic displacement behavior with two degrees of freedom at each node, translation in the nodal $x$- and $y$-directions. The element input data can include thickness if KEYOPTION 3 (plane stress with thickness input) is selected. Surface pressure loads may be applied to element faces. Output data include nodal displacements and element data, such as directional stresses and principal stresses.

### Plane35

is a six-node triangular thermal solid element. The element has one degree of freedom at each node, the temperature. Convection and heat fluxes may be input as surface loads at the element faces. The output data for this element include nodal temperatures and element data, such as thermal gradients and thermal fluxes.

### Plane42

is a four-node quadrilateral element used in modeling solid problems. The element is defined by four nodes, with two degrees of freedom at each node, the translation in the $x$- and $y$-directions. The element input data can include thickness if KEYOPTION 3 (plane stress with thickness input) is selected. Surface pressure loads may be applied to element faces. Output data include nodal displacements and element data, such as directional stresses and principal stresses.

### Plane55

is a four-node quadrilateral element used in modeling two-dimensional conduction heat transfer problems. The element has a single degree of freedom, the temperature. Convection or heat fluxes may be input at the element faces. Output data include nodal temperatures and element data, such as thermal gradient and thermal flux components.

### Plane77

is an eight-node quadrilateral element used in modeling two-dimensional heat conduction problems. It is basically a higher order version of the two-dimensional, four-node quadrilateral element PLANE55. This element is more capable of modeling problems with curved boundaries. At each node, the element has a single degree of freedom, the temperature. Output data include nodal temperatures and element data, such as thermal gradient and thermal flux components.

### Plane82

is an eight-node quadrilateral element used in modeling two-dimensional structural solid problems. It is a higher order version of two-dimensional, four-node quadrilateral element PLANE42. This element offers more accuracy when modeling problems with curved boundaries. At each node, there are two degrees of freedom, the translation in the $x$- and $y$- directions. The element input data can include thickness if KEYOPTION 3 (plane stress with thickness input) is selected. Surface pressure loads may be applied to element faces. Output data include nodal displacements and element data, such as directional stresses and principal stresses.

Finally, it may be worth noting that although you generally achieve better results and greater accuracy with higher order elements, these elements require more computational time. This time requirement is because numerical integration of elemental matrices is more involved.

## SUMMARY

At this point you should:

1. have a good understanding of the linear two-dimensional rectangular and triangular shape functions and of elements, along with their properties and limitations.
2. have a good understanding of the quadratic two-dimensional triangular and quadrilateral elements, as well as shape functions, along with their properties and their advantages over linear elements.
3. know why it is important to use natural coordinate systems.
4. know what is meant by isoparametric element and formulation.
5. know how to use Gauss–Legendre quadrature to evaluate two-dimensional integrals.
6. know examples of two-dimensional elements in ANSYS.

## REFERENCES

*ANSYS User's Manual: Elements*, Vol. III, Swanson Analysis Systems, Inc.

Chandrupatla, T., and Belegundu, A., *Introduction to Finite Elements in Engineering*, Englewood Cliffs, NJ, Prentice Hall, 1991.

CRC Standard Mathematical Tables, 25th ed., Boca Raton, FL, CRC Press, 1979.

Segrlind, L., *Applied Finite Element Analysis*, 2d. ed., New York, John Wiley and Sons, 1984.

## PROBLEMS

1. We have used two-dimensional rectangular elements to model temperature distribution in a thin plate. The values of nodal temperatures for an element belonging to such a plate are given in the accompanying figure. Using local shape functions, what is the temperature at the center of this element?

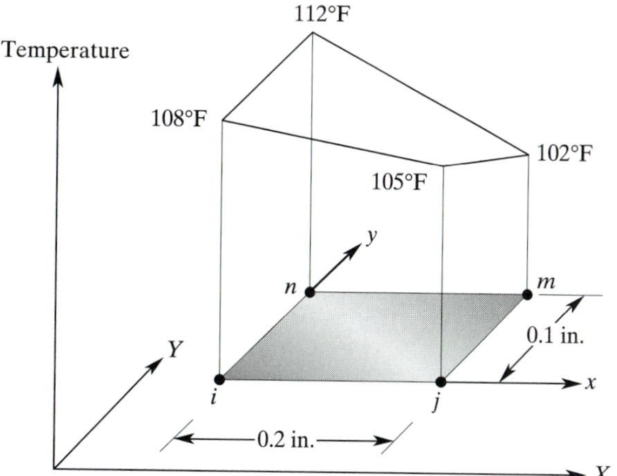


2. Determine the temperature at the center of the element in Problem 1 using natural shape functions.

3. For a rectangular element, derive the $x$- and $y$-components of the gradients of a dependent variable $\Psi$.

4. Determine the components of temperature gradients at the midpoint of the element in Problem 1. Knowing that the element has a thermal conductivity of $k = 92$ Btu/hr · ft · °F, compute the $x$- and $y$-components of the heat flux.

5. Compute the location of the 103°F and 107°F isotherms for the element in Problem 1. Also, plot these isotherms.

6. Two-dimensional triangular elements have been used to determine the stress distribution in a machine part. The nodal stresses and their corresponding positions for a triangular element are shown in the accompanying figure. What is the value of stress at $x = 2.15$ cm and $y = 1.1$ cm?

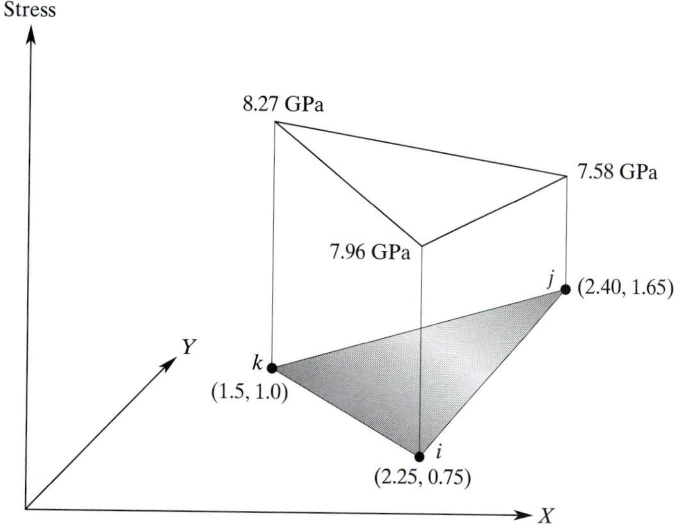

**7.** Plot the 8.0 GPa and 7.86 GPa stress contour lines for an element of the machine part in Problem 6.

**8.** For a quadratic quadrilateral element, confirm the expressions given for the shape functions $S_i$ and $S_j$.

**9.** For a quadratic quadrilateral element, confirm the expressions given for the shape functions $S_k$ and $S_\ell$.

**10.** For triangular elements, the integral that includes products of area coordinates may be evaluated using the factorial relationship shown below:

$$\int_A \xi^a \eta^b \lambda^c \, dA = \frac{a!b!c!}{(a+b+c+2)!} 2A$$

Using the above relationship, evaluate the integral $\int_A (S_i^2 + S_j S_k) dA$

**11.** Show that the area $A$ of a triangular element can be computed from the determinant of

$$\begin{vmatrix} 1 & X_i & Y_i \\ 1 & X_j & Y_j \\ 1 & X_k & Y_k \end{vmatrix} = 2A$$

**12.** In the formulation of two-dimensional heat transfer problems, the need to evaluate the integral $\int_A [\mathbf{S}]^T hT \, dA$ arises; $h$ is the heat transfer coefficient, and $T$ represents the temperature. Using a linear triangular element, evaluate the aforementioned integral, provided that temperature variation is given by

$$T^{(e)} = \begin{bmatrix} S_i & S_j & S_k \end{bmatrix} \begin{Bmatrix} T_i \\ T_j \\ T_k \end{Bmatrix}$$

and $h$ is a constant. Also, note that for triangular elements, the integral that includes products of area coordinates may be evaluated using the factorial relationship shown below:

$$\int_A \xi^a \eta^b \lambda^c \, dA = \frac{a!b!c!}{(a+b+c+2)!} 2A$$

**13.** In the formulation of two-dimensional heat transfer problems, the need to evaluate the integral

$$\int_A k\left(\frac{\partial[\mathbf{S}]^T}{\partial X}\frac{\partial T}{\partial X}\right) dA$$

arises. Using a bilinear rectangular element, evaluate the aforementioned integral, provided temperature is given by

$$T^{(e)} = \begin{bmatrix} S_i & S_j & S_m & S_n \end{bmatrix} \begin{Bmatrix} T_i \\ T_j \\ T_m \\ T_n \end{Bmatrix}$$

and $k$ is the thermal conductivity of the element and is a constant.

**14.** Look up the expressions for the nine-node quadratic quadrilateral element (Lagrangian element). Discuss its properties and compare it to the eight-node quadratic quadrilateral element. What is the basic difference between the Lagrangian element and its eight-node quadratic quadrilateral counterpart?

**15.** For triangular elements, show that the area coordinate $\eta = S_j$ and the area coordinate $\lambda = S_k$.

**16.** Verify the results given for natural quadrilateral shape functions in Eq. (5.7) by showing that (1) a shape function has a value of unity at its corresponding node and a value of zero at the other nodes and (2) if we sum up the shape functions, we will come up with a value of unity.

**17.** Verify the results given for natural quadratic triangular shape functions in Eq. (5.25) by showing that a shape function has a value of unity at its corresponding node and a value of zero at the other nodes.

**18.** For plane stress problems, using triangular elements, we can represent the displacements $u$ and $v$ using a linear triangular element similar to the one shown in the accompanying figure.

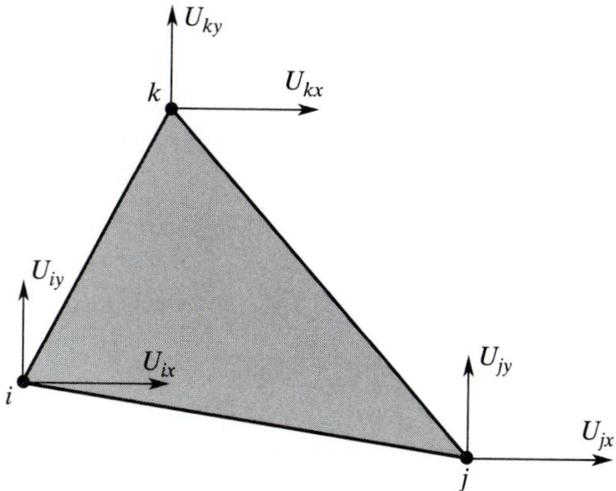


The displacement variables, in terms of linear triangular shape functions and the nodal displacements, are

$$u = S_i U_{ix} + S_j U_{jx} + S_k U_{kx}$$

$$v = S_i U_{iy} + S_j U_{jy} + S_k U_{ky}$$

Moreover, for plane stress situations, the strain displacement relationships are

$$\varepsilon_{xx} = \frac{\partial u}{\partial x} \qquad \varepsilon_{yy} = \frac{\partial v}{\partial y} \qquad \gamma_{xy} = \frac{\partial u}{\partial y} + \frac{\partial v}{\partial x}$$

Show that for a triangular element, strain components are related to the nodal displacements according to the relation

$$\begin{Bmatrix} \varepsilon_{xx} \\ \varepsilon_{yy} \\ \gamma_{xy} \end{Bmatrix} = \frac{1}{2A}\begin{bmatrix} \beta_i & 0 & \beta_j & 0 & \beta_k & 0 \\ 0 & \delta_i & 0 & \delta_j & 0 & \delta_k \\ \delta_i & \beta_i & \delta_j & \beta_j & \delta_k & \beta_k \end{bmatrix}\begin{Bmatrix} U_{ix} \\ U_{iy} \\ U_{jx} \\ U_{jy} \\ U_{kx} \\ U_{ky} \end{Bmatrix}$$

**19.** Consider point $Q$ along the $kj$-side of the triangular element shown in the accompanying figure. Connecting this point to node $i$ results in dividing the area of the triangle into two smaller areas $A_2$ and $A_3$, as shown.

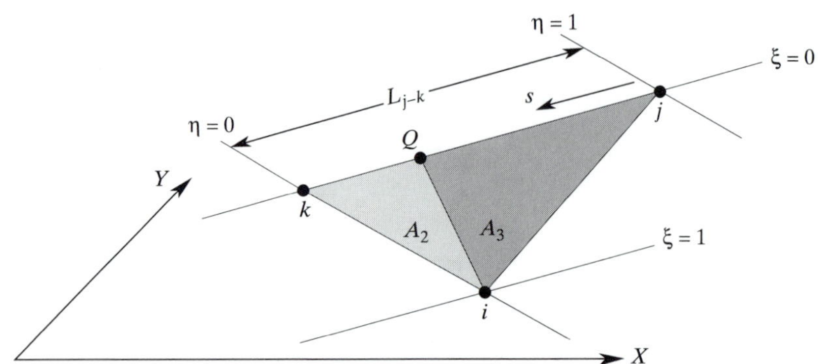


Along the $kj$-edge, the natural, or area, coordinate $\xi$ has a value of zero. Show that along the $kj$-edge, the other natural (area) coordinates $\eta$ and $\lambda$ reduce to one-dimensional natural coordinates that can be expressed in terms of the local coordinate $s$ according to the equations

$$\eta = \frac{A_2}{A} = 1 - \frac{s}{L_{j-k}}$$

$$\lambda = \frac{A_3}{A} = \frac{s}{L_{j-k}}$$

**20.** For the element in Problem 19, derive the simplified area coordinates along the $ij$ and $ki$-edges using the one-dimensional coordinate $s$.

**21.** As you will see in Chapters 7 and 8, we need to evaluate integrals along the edges of a triangular element to develop the load matrix in terms of surface loads or derivative boundary conditions. Referring to Problem 19 and making use of the relations

$$\int_0^1 (x)^{m-1}(1 - x)^{n-1}dx = \frac{\Gamma(m)\Gamma(n)}{\Gamma(m + n)}$$

$$\Gamma(n) = (n - 1)! \quad \text{and} \quad \Gamma(m) = (m - 1)!$$

show that

$$L\int_0^1 \left(1 - \frac{s}{L}\right)^a \left(\frac{s}{L}\right)^b d\left(\frac{s}{L}\right) = \frac{a!b!}{(a + b + 1)!}L$$

and

$$\int_0^{L_{k-j}} (\eta)^a(\lambda)^b ds = L_{j-k}\int_0^1 \left(1 - \frac{s}{L_{j-k}}\right)^a \left(\frac{s}{L_{j-k}}\right)^b d\left(\frac{s}{L_{j-k}}\right) = \frac{a!b!}{(a + b + 1)!}L_{j-k}$$

**22.** Consider a triangular element subjected to a distributed load along its $ki$-edge, as shown in the accompanying figure.

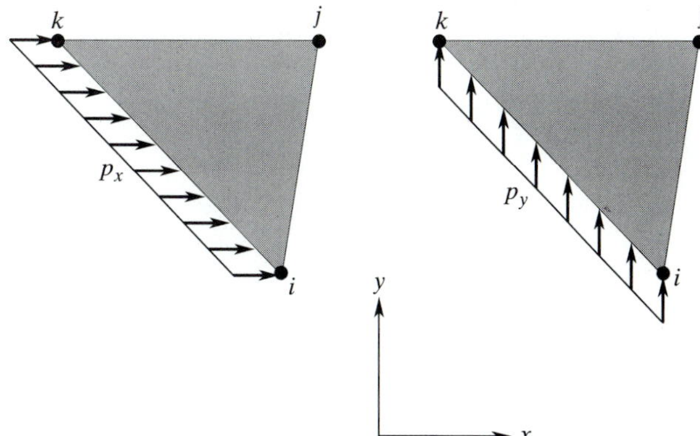


Using the minimum total potential-energy method, the differentiation of the work done by these distributed loads with respect to the nodal displacements gives the load matrix, which is computed from

$$\{\mathbf{F}\}^{(e)} = \int_A [\mathbf{S}]^T \{\mathbf{p}\} dA,$$

where

$$[\mathbf{S}]^T = \begin{bmatrix} S_i & 0 \\ 0 & S_i \\ S_j & 0 \\ 0 & S_j \\ S_k & 0 \\ 0 & S_k \end{bmatrix} \quad \text{and} \quad \{\mathbf{p}\} = \begin{Bmatrix} p_x \\ p_y \end{Bmatrix}$$

Realizing that along the *ki*-edge, $S_j = 0$, evaluate the load matrix for a situation in which the load is applied along the *ki*-edge. Use the results of Problem 21 to help you. Note, in this problem, A is equal to the product of the element thickness and the edge length.

**23.** For the element in Problem 22, evaluate the load matrices for a situation in which the distributed load is acting along the *ij*-edge and the *jk*-edge.

**24.** For a quadrilateral element, evaluate the Jacobian matrix J and its inverse $\mathbf{J}^{-1}$ using Eqs. (5.29) and (5.7).

CHAPTER 6

# More ANSYS*

The main objective of this chapter is to introduce the essential capabilities and the organization of the ANSYS program. The basic steps in creating and analyzing a model with ANSYS are discussed here, along with an example used to demonstrate these steps at the end of the chapter. The main topics discussed in Chapter 6 include the following:

**6.1** ANSYS Program
**6.2** ANSYS Database and Files
**6.3** Creating a Finite Element Model With ANSYS: Preprocessing
**6.4** Applying Boundary Conditions, Loads, and the Solution
**6.5** Results of Your Finite Element Model: Postprocessing
**6.6** Selection Options
**6.7** Graphics Capabilities
**6.8** An Example Problem

## 6.1 ANSYS PROGRAM

The ANSYS program has two basic levels: the *Begin level* and the *Processor level*. When you first enter ANSYS, you are at the Begin level. From the Begin level, you can enter one of the ANSYS processors, as shown in Figure 6.1. A processor is a collection of functions and routines to serve specific purposes. You can clear the database or change a file assignment from the Begin level.

There are three processors that are used most frequently: (1) the *preprocessor* **(PREP7)**, (2) the *processor* **(SOLUTION)**, and (3) the general *postprocessor* **(POST1)**. The preprocessor **(PREP7)** contains the commands needed to build a model:

- define element types and options
- define element real constants
- define material properties
- create model geometry
- define meshing controls
- mesh the object created.

*Materials were adapted with permission from ANSYS documents.

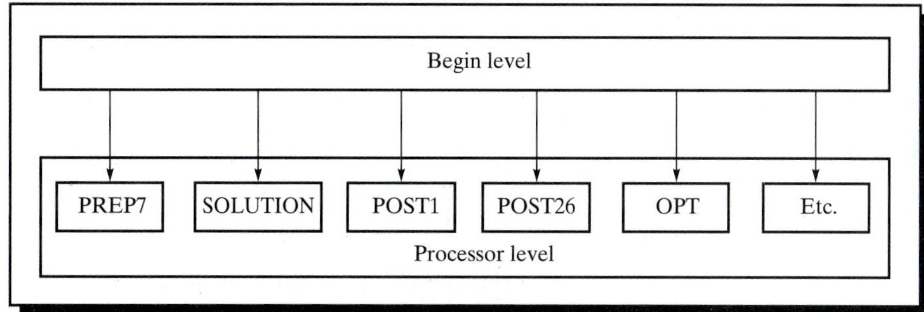


**FIGURE 6.1** Organization of ANSYS program.

The solution *processor* **(SOLUTION)** has the commands that allow you to apply boundary conditions and loads. For example, for structural problems, you can define displacement boundary conditions and forces, or for heat transfer problems, you can define boundary temperatures or convective surfaces. Once all the information is made available to the solution *processor* **(SOLUTION)**, it solves for the nodal solutions. The general *postprocessor* **(POST1)** contains the commands that allow you to list and display results of an analysis:

- read results data from results file
- read element results data
- plot results
- list results.

There are other processors that allow you to perform additional tasks. For example, the *time-history postprocessor* **(POST26)** contains the commands that allow you to review results over time in a transient analysis at a certain point in the model. The *design optimization processor* **(OPT)** allows the user to perform a design optimization analysis.

## 6.2 ANSYS DATABASE AND FILES

The previous section explained how the ANSYS program is organized. This section discusses the ANSYS database. ANSYS writes and reads many files during a typical analysis. The information you input when modeling a problem (e.g., element type, material property, dimensions, geometry, etc.) is stored as *input data*. During the solution phase, ANSYS computes various results, such as displacements, temperatures, stresses, etc. This information is stored as *results data*. The input data and the results data are stored in the ANSYS database. The database can be accessed from anywhere in the ANSYS program. The database resides in the memory until the user saves the database to a database file **Jobname.DB**. Jobname is a name that the user specifies upon entering the ANSYS program; this feature will be explained in more detail later. The database can be saved and resumed at any time. When you issue the RESUME command, the database is read into the memory from the database file that was saved most recently. In other words, the database becomes what you saved most recently. When you are uncertain about the next step you should take in your analysis, or if you want to test something, you should issue the SAVE database command before proceeding with your test. That way, if you are unhappy with the results of your test, you can issue the RESUME database command, which will allow you to go back to the place in your analysis where you started testing. The SAVE, RESUME, and EXIT commands are located in the utility menu. In addition, the "clear" option, located on the utility menu, allows the user to clear the database. This option is useful when you want to start anew, but do not wish to leave and reenter ANSYS.

When you are ready to exit the ANSYS program, you will be given four options: (1) Save all model data; (2) Save all model data and solution data; (3) Save all model data, solution data, and postprocessing data; or (4) Save nothing.

As previously explained, ANSYS writes and reads many files during a typical analysis. The files are of the form of *Jobname.Ext*. Recall that Jobname is a name you specify when you enter the ANSYS program at the beginning of an analysis. The default jobname is *file*. Files are also given unique extensions to identify their content. Typical files include the following:

- The *log file* (**Jobname.LOG**): This file is opened when ANSYS is first entered. Every command you issue in ANSYS is copied to the log file. Jobname.LOG is closed when you exit ANSYS. Jobname.LOG can be used to recover from a system crash or a serious user error by reading in the file with the **/INPUT** command.
- The *error file* (**Jobname.ERR**): This file is opened when you first enter ANSYS. Every warning and error message given by ANSYS is captured by this file. If Jobname.ERR already exists when you begin a new ANSYS session, all new warnings and error messages will be appended to the bottom of this file.
- The *database file* (**Jobname.DB**): This file is one of the most important ANSYS files because it contains all of your input data and possibly some results. The model portion of the database is automatically saved when you exit the ANSYS program.
- The *output file* (**Jobname.OUT**): This file is opened when you first enter ANSYS. Jobname.OUT is available if you are using the GUI; otherwise, your computer monitor is your output file. Jobname.OUT captures responses given by ANSYS to every command executed by the user. It also records warning and error messages and some results. If you change the Jobname while in a given ANSYS session, the output file name is not changed to the new Jobname.

Other ANSYS files include the *structural analysis results file* (**Jobname.RST**); the *thermal results file* (**Jobname.RTH**); the *magnetic results file* (**Jobname.RMG**); the *graphics file* (**Jobname.GRPH**); and the *element matrices file* (**Jobname.EMAT**).

## 6.3 CREATING A FINITE ELEMENT MODEL WITH ANSYS: PREPROCESSING

The preprocessor **(PREP7)** contains the commands needed to create a finite element model:

1. define element types and options
2. define element real constants if required for the chosen element type

3. define material properties
4. create model geometry
5. define meshing controls
6. mesh the object created.

1. **Define element types and options.**
ANSYS provides more than one hundred various elements to be used to analyze different problems. Selecting the correct element type is a very important part of the analysis process. A good understanding of finite element theory will benefit you the most in this respect, helping you choose the correct element for your analysis. In ANSYS, each element type is identified by a *category name* followed by a *number*. For example, two-dimensional solid elements have the category name **PLANE**. Furthermore, **PLANE42** is a four-node quadrilateral element used to model structural solid problems. The element is defined by four nodes having two degrees of freedom at each node, translation in the $x$- and $y$-directions. **PLANE 82** is an eight-node (four corner points and four midside nodes) quadrilateral element used to model two-dimensional structural solid problems. It is a higher order version of the two-dimensional, four-node quadrilateral element type, PLANE42. Therefore, the PLANE82 element type offers more accuracy when modeling problems with curved boundaries. At each node, there are two degrees of freedom, translation in the $x$- and $y$-directions. Many of the elements used by ANSYS have options that allow you to specify additional information for your analysis. These options are known in ANSYS as *keyoptions* (**KEOPTs**). For example, for PLANE 82, with **KEOPT (3)** you can choose plane stress, axisymmetric, plane strain, or plane stress with the thickness analysis option. A complete list of elements used by ANSYS is shown in Table 6.1 at the end of this chapter. You can define element types and options by choosing:

main menu: **Preprocessor** → **Element Type** → **Add/Edit/Delete** ...

You will see the Element Type dialog box next, shown in Figure 6.2.

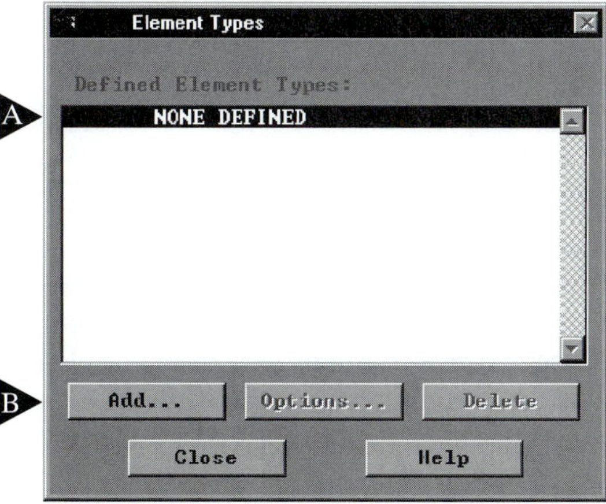


**FIGURE 6.2** Element Types dialog box.

**List:** A list of currently defined element types will be shown here. If you have not defined any elements yet, then you need to use the **Add** button to add an element. The Library of Element Types dialog box will appear next (see Figure 6.3). Then you choose the type of element you desire from the Library.

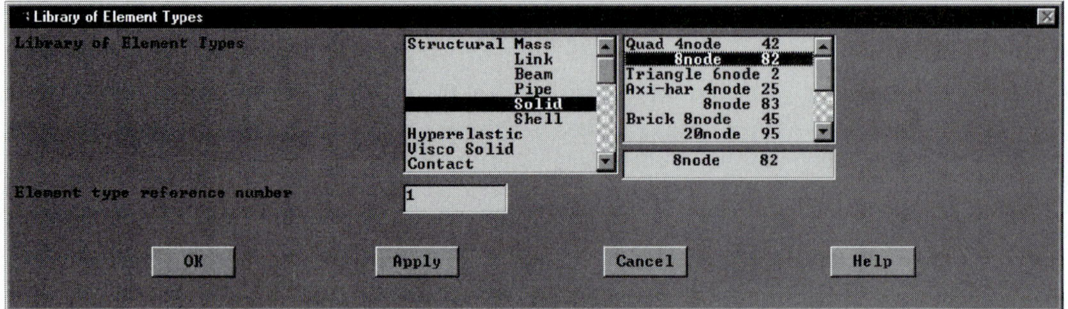


**FIGURE 6.3** Library of Element Types dialog box.

**Action Buttons:** The purpose of the **Add** button is to add an element, as we just discussed. The **Delete** button deletes the selected (highlighted) element type. The **Options** button opens the element type options dialog box. You can then choose one of the desired element options for a selected element. For example, if you had selected the element PLANE 82 with KEOPT (3) you could choose plane stress, axisymmetric, plane strain, or plane stress with the thickness analysis option, as shown in Figure 6.4.

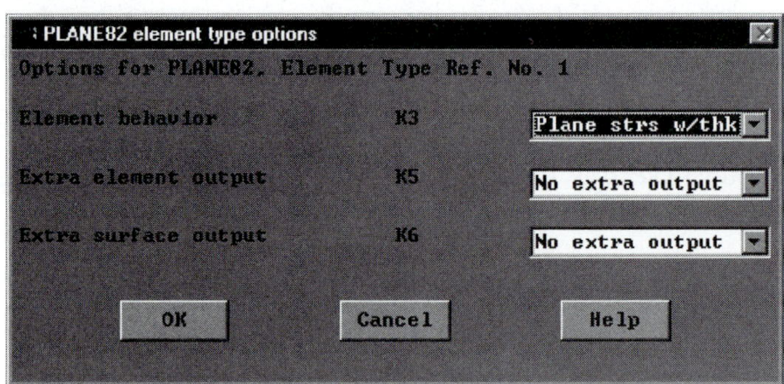


**FIGURE 6.4** The element type options dialog box.

2. **Define element real constants.**
   Element *real constants* are quantities that are specific to a particular element. For example, a beam element requires cross-sectional area, second moment of area, and so on. It is important to realize that real constants vary from one element type to another; furthermore, not all elements require real constants. Real constants can be defined by the command

   main menu: **Preprocessor** → **Real Constants** ...

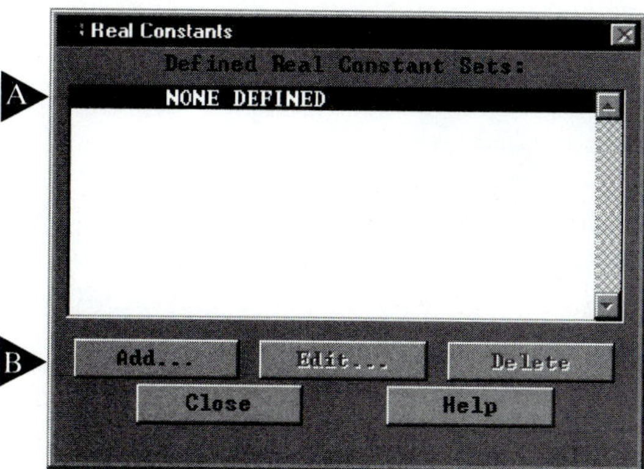


**FIGURE 6.5** Real Constants dialog box.

You will then see the Real Constants dialog box, as shown in Figure 6.5.

**List:** A list of currently defined real constants will be shown here. If you have not defined any elements at this point, you need to use the **Add** button to add real constants. An example of a dialog box for a PLANE 82 element's real constants is shown in Figure 6.6.

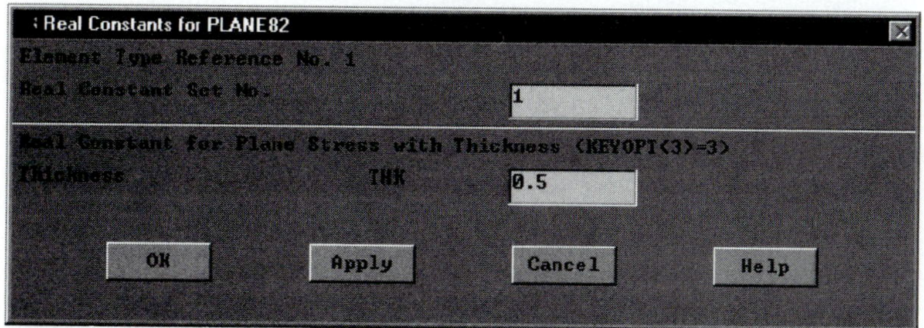


**FIGURE 6.6** An example of the dialog box for a PLANE 82 (with options) element's real constants.

**Action Buttons:** The purpose of the **Add** button has already been explained. The **Delete** Button deletes the selected (highlighted) real constants. The **Edit** button opens a new dialog box that allows you to change the values of existing real constants.

3. **Define material properties.**

At this point, you define the physical properties of your material. For example, for solid structural problems, you may need to define the modulus of elasticity, Poisson's ratio, or the density of the material, whereas for thermal problems, you may need to define thermal conductivity, specific heat, or the density of the material. You can define material properties by the command

main menu: **Preprocessor** → **Material Props** → **-Constant-Isotropic** ...

You will then see the Isotropic Material Properties reference dialog box, as shown in Figure 6.7.

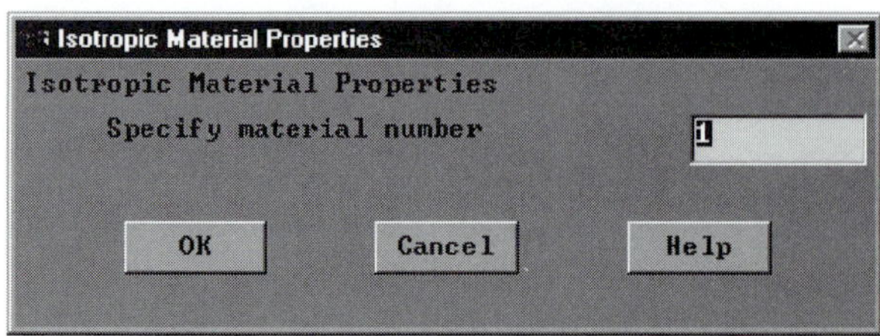


**FIGURE 6.7** Isotropic Material Properties reference dialog box.

You can use multiple materials in your model if the object you are analyzing is made of different materials. Specify the material reference number, starting with 1, and pick **OK**. The next dialog box allows you to define the appropriate properties for your analysis, as shown in Figure 6.8. You need to scroll down to access all available properties.

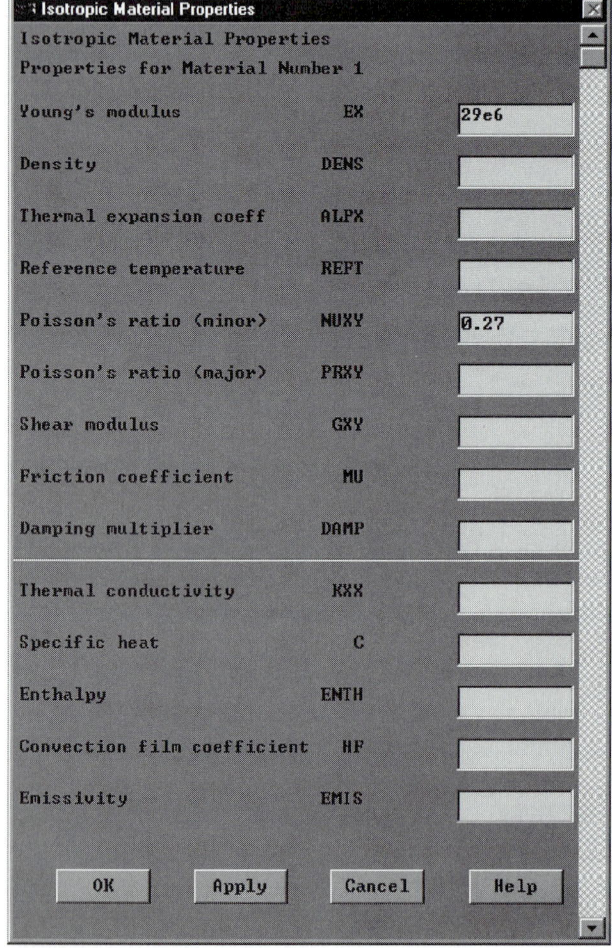


**FIGURE 6.8** Isotropic material properties dialog box.

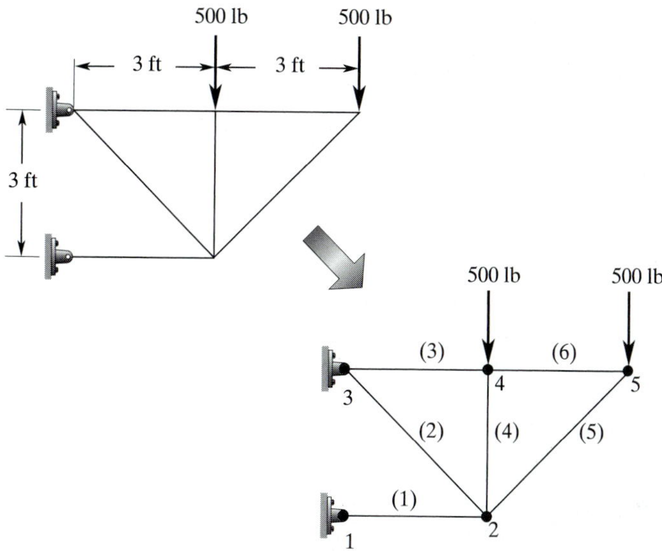


**FIGURE 6.9** A truss problem: First nodes 1–5 are created, then nodes are connected to form elements (1)–(6).

**4. Create model geometry.**

There are two approaches to constructing a finite element model's geometry: (1) *direct (manual) generation* and (2) the *solid-modeling* approach. Direct generation, or manual generation, is a simple method by which you specify the location of nodes and manually define which nodes make up an element. This approach is generally applied to simple problems that can be modeled with line elements, such as links, beams, and pipes, or if the object is made of simple geometry, such as rectangles. This approach is illustrated in Figure 6.9. Refer back to the truss problem of Example 2.1 in Chapter 2 to refresh your memory about the manual approach, if necessary.

With the solid-modeling approach, you use simple *primitives* (simple geometric shapes), such as rectangles, circles, polygons, blocks, cylinders, and spheres, to construct the model. Boolean operations are then used to combine the primitives. Examples of boolean operations include addition, subtraction, and intersection. You then specify the desired element size and shape, and ANSYS will automatically generate all the nodes and the elements. This approach is depicted in Figure 6.10.

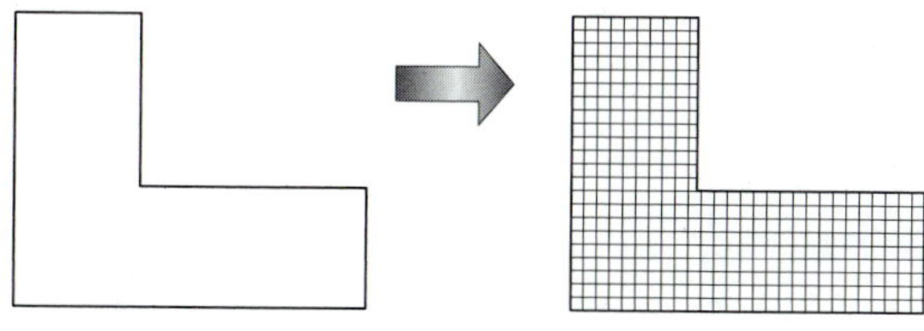

**FIGURE 6.10** An example of the solid-modeling approach.

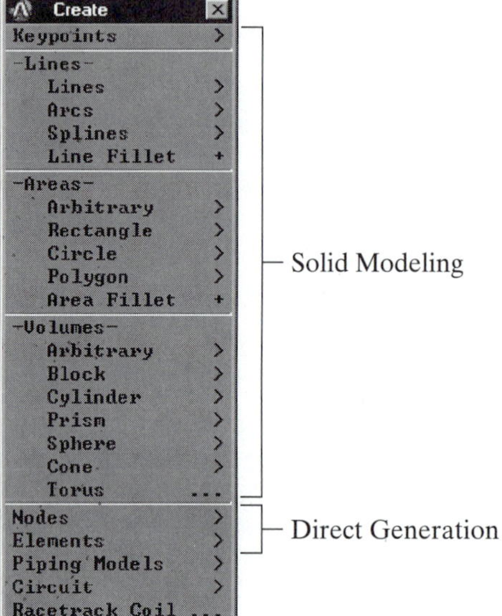


**FIGURE 6.11** The Create dialog box.

The construction of the model in ANSYS begins when you choose the **Create** option, as shown in Figure 6.11. You choose this option with the command sequence

main menu: **Preprocessor** → **-Modeling-Create**

When you create entities such as keypoints, lines, areas, or volumes, they are automatically numbered by the ANSYS program. You use *keypoints* to define the vertices of an object. *Lines* are used to represent the edges of an object. *Areas* are used to represent two-dimensional solid objects. They are also used to define the surfaces of three-dimensional objects. When using primitives to build a model, you need to pay special notice to the hierarchy of the entities. Volumes are bounded by areas, areas are bounded by lines, and lines are bounded by keypoints. Therefore, volumes are considered to be the highest entity, and the keypoints are the lowest entity in solid modeling hierarchy. Remembering this concept is particularly important if you need to delete a primitive. For example, when you define one rectangle, ANSYS automatically creates nine entities: four keypoints, four lines, and one area. The relationship among keypoints, lines, and areas is depicted in Figure 6.12.

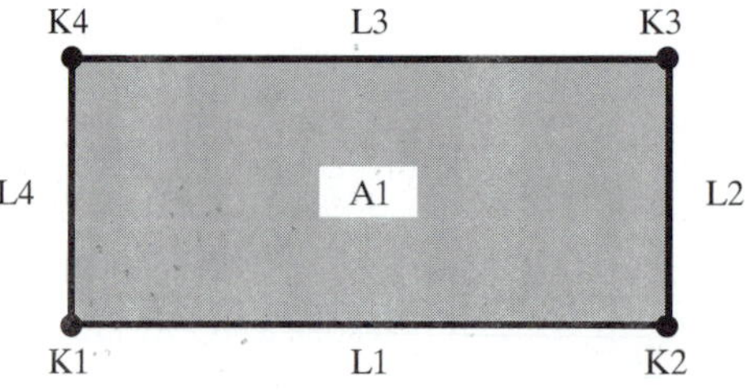


**FIGURE 6.12** The relationship among the keypoints, lines, and areas.

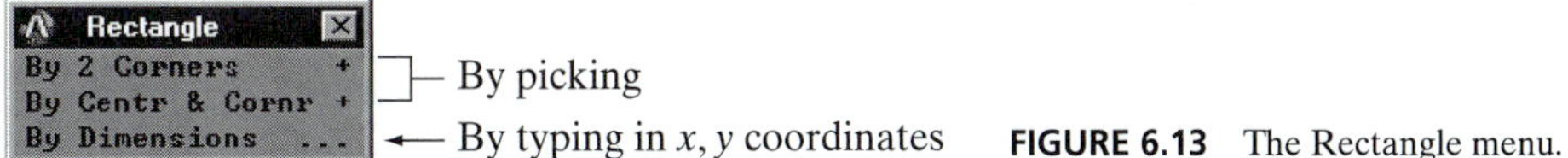


**FIGURE 6.13** The Rectangle menu.

Area primitives and volume primitives are grouped under the Areas and Volumes categories in the Create menu. Now let us consider the Rectangle and the Circle menus, because they are commonly used to build two-dimensional models. The Rectangle menu offers three methods for defining a rectangle, as shown in Figure 6.13. The command for accessing the Rectangle menu is

main menu: **Preprocessor → -Modeling-Create → Rectangle**

The Circle menu offers several methods for defining a solid circle or annulus, as shown in Figure 6.14.

The Partial Annulus option is limited to circular areas spanning 180° or less. In order to create a partial circle that spans more than 180°, you need to use the By Dimension option. An example of creating a partial annulus spanning from $\theta = 45°$ to $\theta = 315°$ is shown in Figure 6.15. Note that you can create a solid circle by setting RAD1 = 0.

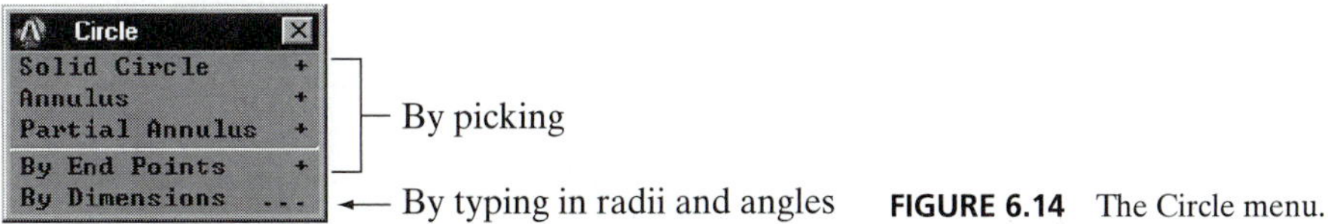


**FIGURE 6.14** The Circle menu.

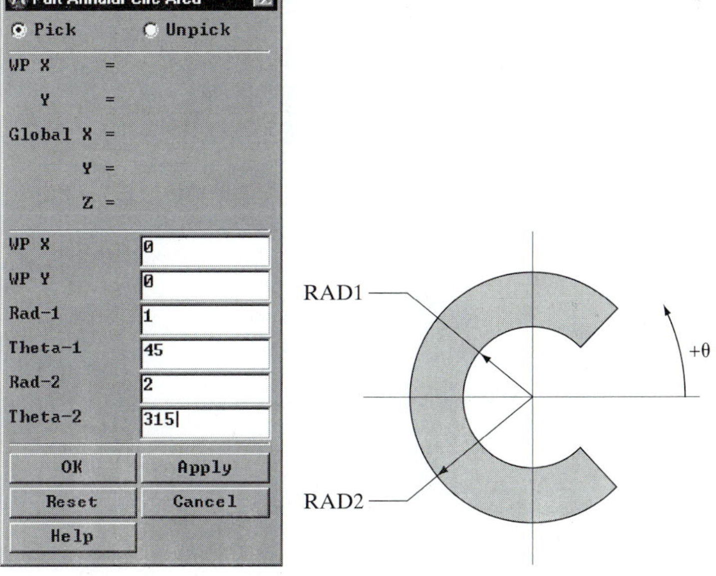


**FIGURE 6.15** An example of creating a partial annulus spanning from $\theta = 45°$ to $\theta = 315°$.

## The Working Plane (WP)

In ANSYS, you will use a *working plane (WP)* to create and orient the geometry of the object you are planning to model. All primitives and other modeling entities are defined with respect to this plane. The working plane is basically an infinite plane with a two-dimensional coordinate system. The dimensions of the geometric shapes are defined with respect to the WP. By default, the working plane is a Cartesian plane. You can change the coordinate system to a polar system, if so desired. Other attributes of the working plane may be set by opening the WP settings dialog submenu, as shown in Figure 6.16. To access this dialog box, issue the following sequence of commands:

utility menu: **Work Plane** → **WP Settings** ...

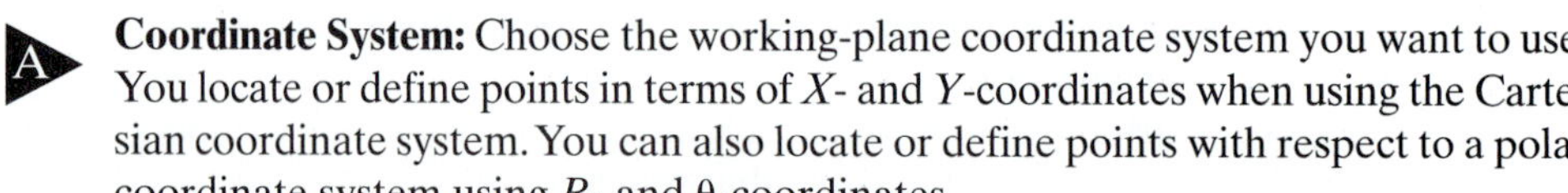

A **Coordinate System:** Choose the working-plane coordinate system you want to use. You locate or define points in terms of $X$- and $Y$-coordinates when using the Cartesian coordinate system. You can also locate or define points with respect to a polar coordinate system using $R$- and $\theta$-coordinates.

B **Display Options:** This section is where you turn on the grid or grid and triad. The triad appears in the center (0,0 coordinates) of your working plane.

C **Snap Options:** These options control the locations of points that are picked. When activated, these options allow you to pick locations nearest to the snap point. For example, in a Cartesian working plane, **Snap Incr** controls the $X$- and $Y$-increments

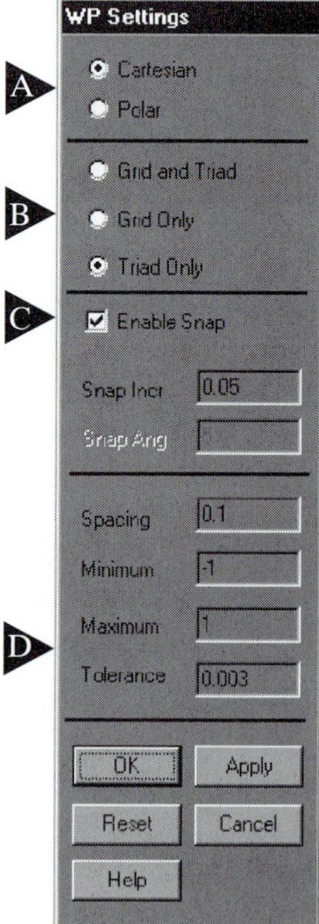


**FIGURE 6.16** The WP settings dialog box.

within the spacing grid. If you have set a spacing of 1.0 and a snap increment of 0.5, then within the $X,Y$ grid you can pick coordinates with 0.5 increments. For example, you cannot pick the coordinates 1.25 or 1.75.

**Grid Control**

*Spacing:* This number defines the spacing between the grid lines.

*Minimum:* This number is the minimum $X$-location at which you want the grid to be displayed with respect to the Cartesian coordinate system.

*Maximum:* This number is the maximum $X$-location at which you want the grid to be displayed with respect to the Cartesian coordinate system.

*Radius:* This number is the outside radius that you want the grid to be displayed with respect to the polar coordinate system.

*Tolerance:* This number is the amount that an entity can be off of the current working plane and still be considered as on the plane.

The working plane is always active and, by default, not displayed. To display the working plane, you need to issue the following command:

utility menu: **WorkPlane → Display Working Plane**

You can move the WP origin to a different location on the working plane. This feature is useful when you are defining primitives at a location other than the global location. You can move the WP origin by choosing the commands

utility menu: **WorkPlane → Offset WP to → XYZ Locations** +

You can relocate the working plane by offsetting it from its current location, as shown in Figure 6.17. To do so, issue the command

utility menu: **WorkPlane → Offset WP by Increments** ...

A **Offset buttons:** Picking these buttons will cause an immediate offset of your working plane in the direction shown on the buttons. The amount of offset is controlled by the Offset slider and the Snap-Incr value on the WP setting dialog box.

B **Offset Slider:** This number controls the amount of offset that occurs with each pick of the offset buttons. If the slider is set to 1, the offset will be one times the Snap-Incr value on the WP setting dialog box.

C **Offset Dialog Input:** This feature allows you to input the exact $X$, $Y$, and $Z$ offset values for the working plane. For instance, typing 1,2,2 into this field and pressing the Apply or OK buttons will move the working plane one unit in the positive $X$-direction and two units each in the positive $Y$- and $Z$-directions.

E **Location Status:** This section displays the current location of the working plane in global Cartesian coordinates. This status is updated each time the working plane is translated.

You can also relocate the working plane by aligning it with specified keypoints, nodes, coordinate locations, etc., as shown in Figure 6.18. To align the working plane, issue the command

utility menu: **Work Plane → Align WP with**

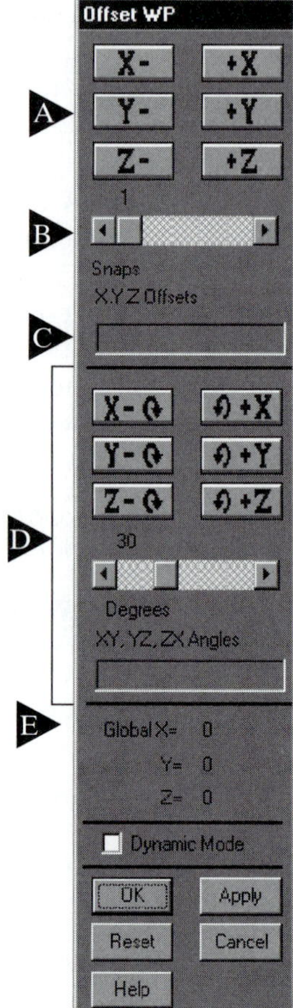


**FIGURE 6.17** The dialog box for offsetting the WP.

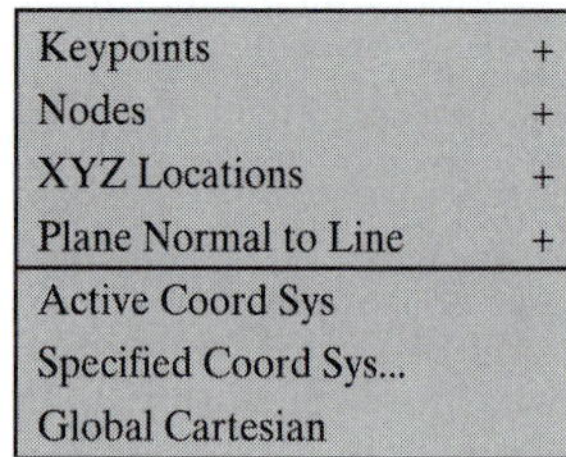


**FIGURE 6.18** Working plane-relocation using the Align command.

## Plotting Model Entities

You can plot various entities, such as keypoints, lines, areas, volumes, nodes, and elements, using the Plot menu. From the utility menu, you can issue one of the following commands to plot:

utility menu: **Plot** → **Keypoints**

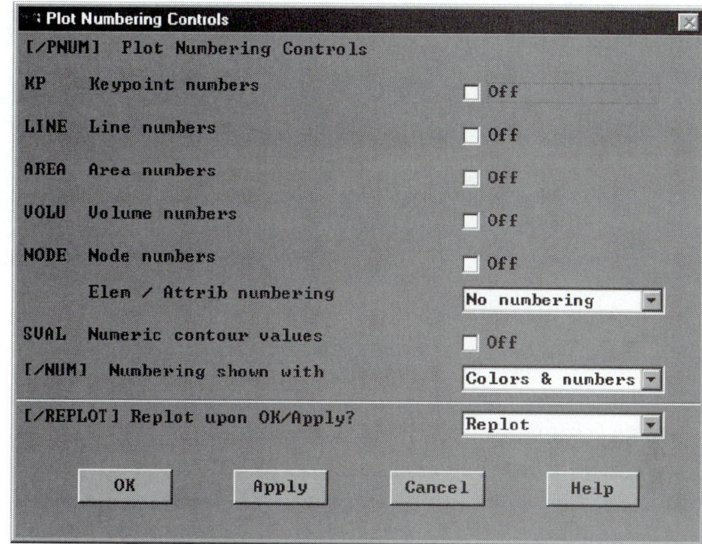


**FIGURE 6.19** The Plot Numbering Controls dialog box.

utility menu: **Plot** → **Lines**
utility menu: **Plot** → **Areas**
utility menu: **Plot** → **Volumes**
utility menu: **Plot** → **Nodes**
utility menu: **Plot** → **Elements**

The PlotCtrls menu, shown in Figure 6.19, contains a useful graphics option that allows you to turn on keypoint numbers, line numbers, area numbers, and so on to check your model. To access this option, use the command

utility menu: **PlotCtrls** → **Numbering** ...

You may need to replot to see the effects of the numbering command you issue.

5. **Define meshing controls.**

The next step in creating a finite element model is dividing the geometry into nodes and elements. This process is called meshing. The ANSYS program can automatically generate the nodes and elements, provided that you specify the *element attributes* and the *element size:*

1. The *element attributes* include element type(s), real constants, and material properties.
2. The *element size* controls the fineness of the mesh. The smaller the element size, the finer the mesh. The simplest way to define the element size is by defining a global element size. For example, if you specify an element edge length of 0.1 units, then ANSYS will generate a mesh in which no element edge is larger than 0.1 units. Another way to control the mesh size is by specifying the number of element divisions along a boundary line. The Global Element Size

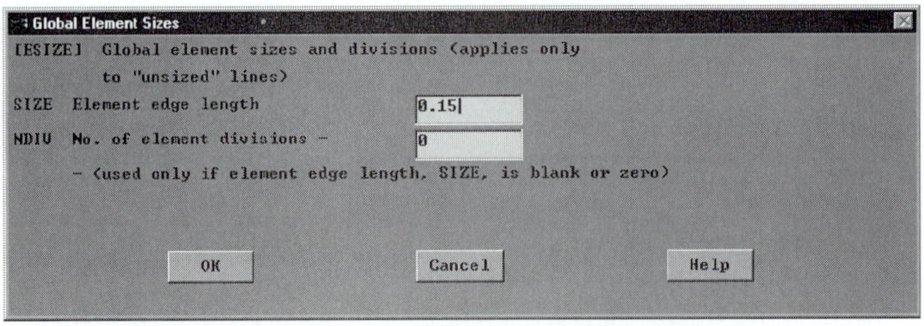

**FIGURE 6.20** The Global Element Size dialog box.

dialog box is shown in Figure 6.20. To access this dialog box, issue the following commands:

main menu: **Preprocessor** → **-Meshing-Size Cntrls**→ **-Global-Size** ...

6. **Mesh the object.**

You should get into the habit of saving the database before you initiate meshing. This way, if you are not happy with the mesh generated, you can resume the database and change the element size and remesh the model. To initiate meshing, invoke the commands

main menu: **Preprocessor** → **-Meshing-Mesh** → **-Areas-Free** +

Once a picking menu appears, you can pick individual areas or use the Pick All button to select all areas for meshing. Upon selection of the desired areas, Pick the Apply or OK buttons to mesh. The meshing process can take some time, depending on the model complexity and the speed of your computer. During the meshing process, ANSYS periodically writes a meshing status to the output window. Therefore, it is useful to bring the output window to the front to see the meshing status messages.

*Free meshing* uses either mixed-area element shapes or all-triangular area elements, whereas the *mapped meshing* option uses all quadrilateral area elements and all hexahedral (brick) volume elements. Mapped area mesh requirements include three or four sides, equal numbers of elements on opposite sides, and even numbers of elements for three-sided areas. If you want to mesh an area that is bounded by more than four lines, you can use the concatenate command to combine some of the lines to reduce the total number of lines. Concatenation is usually the last step you take before you start meshing the model. To concatenate, issue the following series of commands:

main menu: **Preprocessing** → **-Meshing-Concatenate** → **Lines** or **Areas**

### Modifying Your Meshed Model

If you want to modify your model, you must keep in mind certain rules enforced by ANSYS:

1. Meshed lines, areas, or volumes may not be deleted or moved.

2. You can delete the nodes and the elements with the meshing Clear command.

Also, areas contained in volumes may not be deleted or changed. Lines contained in areas may not be deleted. Lines can be combined or divided into smaller segments with line operation commands. Keypoints contained in lines may not be deleted. You start the clearing process by issuing the commands

main menu: **Preprocessor** → **-Meshing-Clear**

The clearing process will delete nodes and elements associated with a selected model entity. Then you can use deleting operations to remove all entities associated with an entity. The "... and below" options delete all lower entities associated with the specified entity, as well as the entity itself. For example, deleting "Area and below" will automatically remove the area, the lines, and the keypoints associated with the area.

## 6.4 APPLYING BOUNDARY CONDITIONS, LOADS, AND THE SOLUTION

The next step of finite element analysis involves applying appropriate boundary conditions and the proper loading. There are two ways to apply the boundary conditions and loading to your model in ANSYS. You can either apply the conditions to the solid model (keypoints, lines, and areas), or the conditions can be directly imposed on the nodes and elements. The first approach may be preferable because should you decide to change the meshing, you will not need to reapply the boundary conditions and the loads to the new finite element model. It is important to note that if you decide to apply the conditions to keypoints, lines, or areas during the solution phase, ANSYS automatically transfers the information to nodes. The solution *processor* **(SOLUTION)** has the commands that allow you to apply boundary conditions and loads. It includes the following options:

***for structural problems:*** displacements, forces, distributed loads (pressures), temperatures for thermal expansion, gravity
***for thermal problems:*** temperatures, heat transfer rates, convection surfaces, internal heat generation
***for fluid flow problems:*** velocities, pressures, temperatures
***for electrical problems:*** voltages, currents
***for magnetic problems:*** potentials, magnetic flux, current density

### Degrees of Freedom (DOF) Constraints

In order to constrain a model with fixed (zero displacements) boundary conditions, you need to choose the command sequence

main menu: **Solution** → **-Loads-Apply** → **-Structural-Displacement**

You can specify the given condition on the keypoints, lines, areas, or nodes. For example, if you choose to constrain certain keypoints, then you need to invoke the commands

main menu: **Solution** → **-Loads-Apply** → **-Structural-Displacement** → **On Keypoints** +

Apply U,ROT on KPs

[DK] Apply Displacements (U,ROT) on Keypoints

Lab2 DOFs to be constrained: All DOF / UX / UY

VALUE Displacement value: 0

KEXPND Expand disp to nodes? No

OK | Apply | Cancel | Help

**FIGURE 6.21** The dialog box for applying displacements on keypoints.

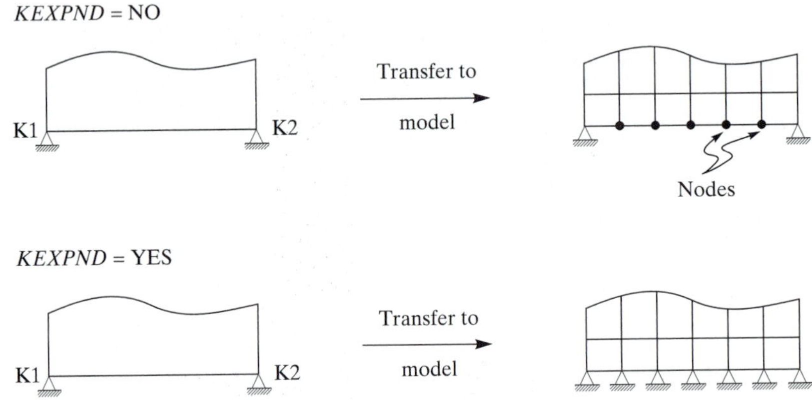


**FIGURE 6.22** KEXPND options.

A picking menu will appear. You then pick the keypoints to be constrained and press the OK button. An example of a dialog box for applying displacement constraints on keypoints is shown in Figure 6.21.

The KEXPND field in the dialog box of Figure 6.21 is used to expand the constraint specification to all nodes between the keypoints, as shown in Figure 6.22.

Once you have applied the constraints, you may want to display the constraint symbols graphically. To turn on the boundary condition symbols, open the Symbols dialog box, as shown in Figure 6.23, by choosing the commands

utility menu: **PlotCtrls** → **Symbols** ...

### Line or Surface Loads

In order to specify distributed loads on a line or surface of a model, you need to issue the following commands:

main menu: **Solution** → **-Loads-Apply** → **-Structural-Pressures** → **On Lines** or **On Surfaces**

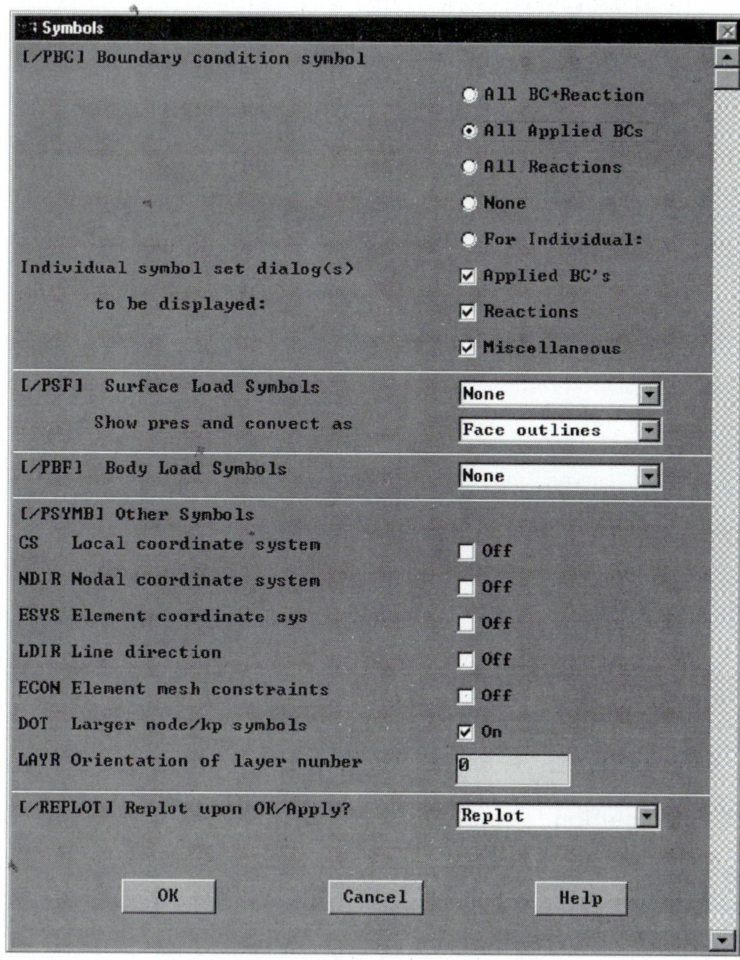


**FIGURE 6.23** The symbols dialog box.

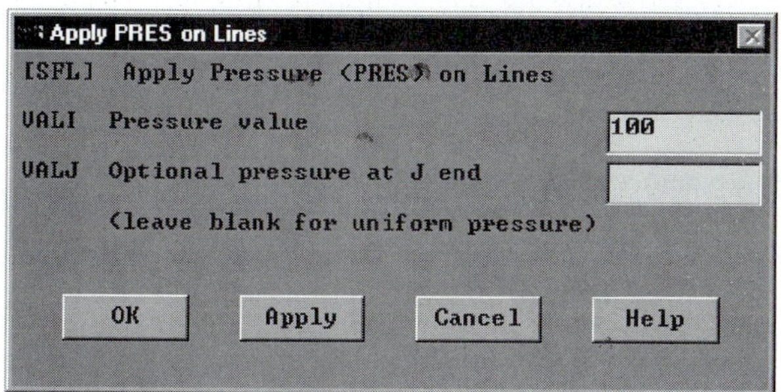


**FIGURE 6.24** The dialog box for applying pressure loads on lines.

A picking menu will appear. You then pick the line(s) or surfaces that require a pressure load and press the OK button. An example of a dialog box for applying pressure loads on line(s) is shown in Figure 6.24.

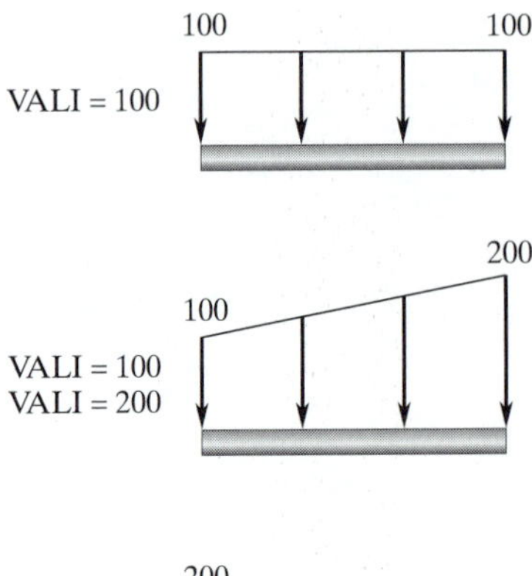


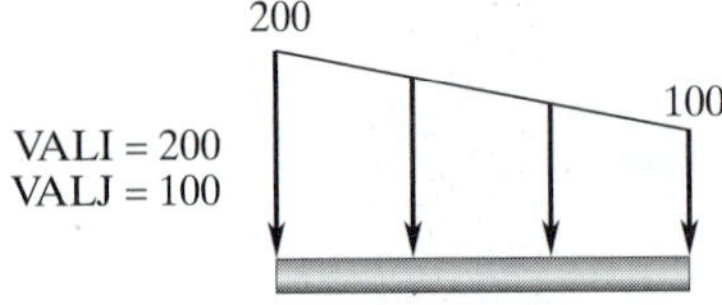


**FIGURE 6.25** An example illustrating how to apply uniform and nonuniform loads.

For uniformly distributed loads, you need to specify only VALI. For a linear distribution, you need to specify both VALI and VALJ, as shown in Figure 6.25. It is important to note that in ANSYS, a positive VALI represents pressure into the surface.

### Obtain a Solution

Once you have created the model and have applied the boundary conditions and appropriate loads, then you need to instruct ANSYS to solve the set of equations generated by your model. But first save the database. To initiate the solution, pick the commands

main menu: **Solution** $\rightarrow$ **-Solve-Current LS**

The next section is about reviewing the results of your analysis.

## 6.5 RESULTS OF YOUR FINITE ELEMENT MODEL: POSTPROCESSING

There are two postprocessors available for review of your results: (1) **POST1** and (2) **POST26.** The general postprocessor **(POST1)** contains the commands that allow you to list and display results of an analysis:

- Deformed shape displays and contour displays
- Tabular listings of the results data of the analysis
- Calculations for the results data and path operations
- Error estimations.

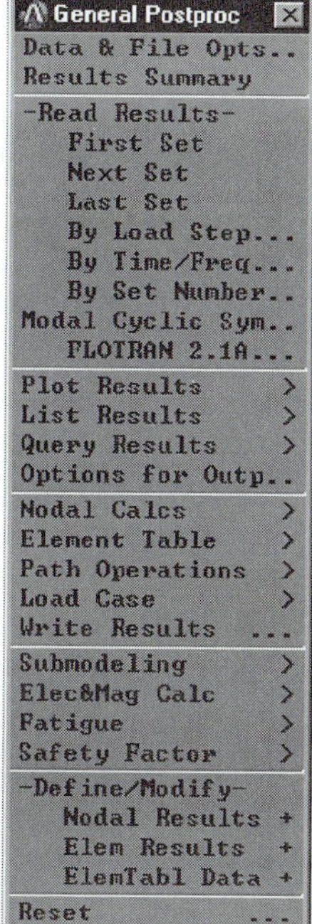


**FIGURE 6.26** The General Postprocessing dialog box.

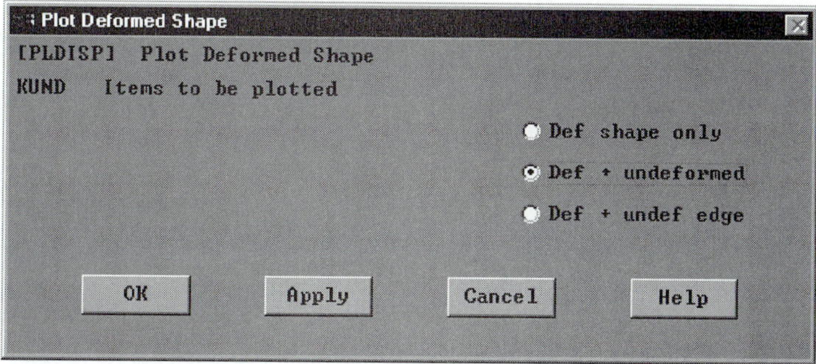


**FIGURE 6.27** The Plot Deformed Shape dialog Box.

You can read results data from the results file by using one of the choices from the dialog box shown in Figure 6.26. This dialog box may be accessed via the following command:

main menu: **General Postproc**

For example, if you are interested in viewing the deformed shape of a structure under a given loading, you choose the Plot Deformed Shape dialog box, as shown in Figure 6.27. To access this dialog box, issue the following sequence of commands:

main menu: **General Postproc** → **Plot Results** → **Deformed Shape** ...

You can also use contour displays to see the distribution of certain variables, such as a component of stress or temperature over the entire model. For example, issue the following command to access the dialog box shown in Figure 6.28.

main menu: **General Postproc** → **Plot Results** → **Nodal Solution**

Contour Nodal Solution Data

[PLNSOL] Contour Nodal Solution Data

Item,Comp Item to be contoured

DOF solution
Stress
Strain-total
Energy
Strain-elastic
Strain-thermal
Strain-plastic
Strain-creep
Strain-other

2nd principal S2
3rd principal S3
Intensity SINT
von Mises SEQV
PlasEqvStrs SEPL
StressRatio SRAT
HydrostPres HPRE

von Mises SEQV

KUND Items to be plotted

Def shape only
Def + undeformed
Def + undef edge

Fact Optional scale factor 1

[AVPRIN] Eff NU for EQV strain 0

OK Apply Cancel Help

**FIGURE 6.28** The Contour Nodal Solution Data dialog box.

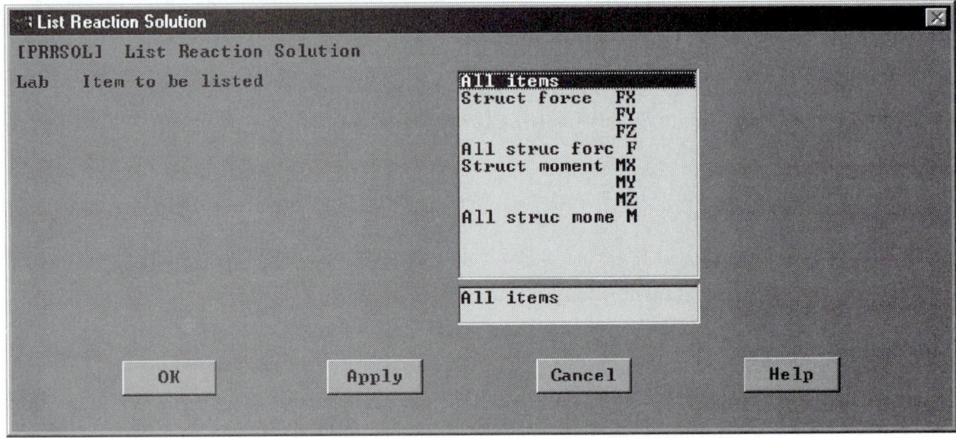


**FIGURE 6.29** The List Reaction Solution dialog box.

As already mentioned, you can list the results in a tabular form as well. For example, to list the reaction forces, you issue the following command, which gives you a dialog box similar to the one shown in Figure 6.29:

main menu: **General Postproc** → **List Results** → **Reaction Solu** ...

Select the component(s) of your choice and press the OK button.

The *time-history postprocessor* **(POST26)** contains the commands that allow you to review results over time in a transient analysis. These commands will not be discussed here, but you may consult the ANSYS on-line help for further information about how to use the time-history postprocessor.

Once you have finished reviewing the results and wish to exit the ANSYS program, choose the Quit button from the ANSYS toolbar and pick the option you want. Press the OK button.

If, for any reason, you need to come back to modify a model, first launch ANSYS, and then type the file-name in the **Initial Jobname** entry field of the interactive dialog box. Then press the Run button. From the file menu, choose **Resume Jobname.DB.** Now you have complete access to your model. You can plot keypoints, nodes, elements, and the like to make certain that you have chosen the right problem.

## 6.6 SELECTION OPTIONS

The ANSYS program uses a database to store all of the data that you define during an analysis. ANSYS also offers the user the capability to select information about only a portion of the model, such as certain nodes, elements, lines, areas, and volumes for further processing. You can select functions anywhere within ANSYS. To start selecting, issue the following command to bring up the dialogue box shown in Figure 6.30.

utility menu: **Select** → **Entities** ...

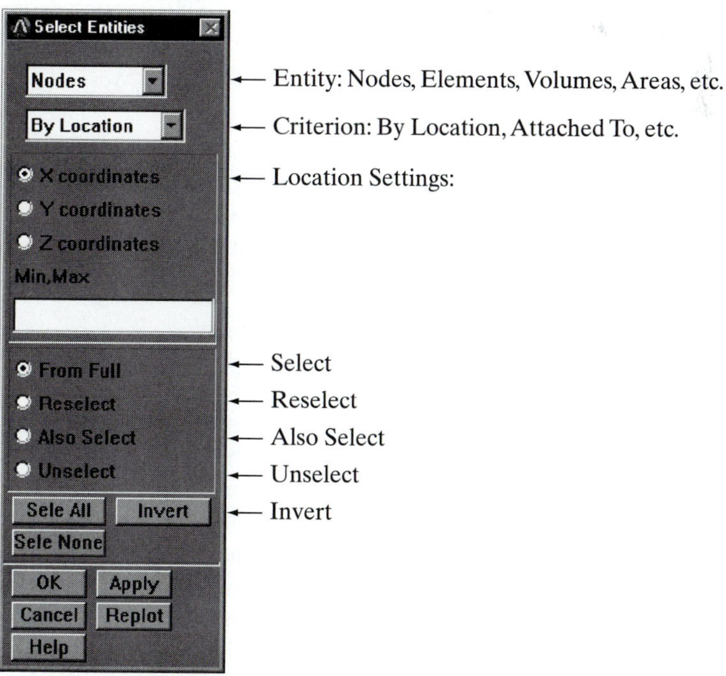


**FIGURE 6.30** The select Entities dialog box.

The various selection commands and their respective uses are as follows:

| | |
|---|---|
| **Select:** | To select a subset of active items from the full set. |
| **Reselect:** | To select again from the currently selected subset. |
| **Also Select:** | To add a different subset to the current subset. |
| **Unselect:** | To deactivate a portion of the current subset. |
| **Select All:** | To restore the full set. |
| **Select None:** | To deactivate the full set (opposite of the Select All command). |
| **Invert:** | To switch between the active and inactive portions of the set. |

The select dialog box can be used to select or unselect entities of your solid or finite element model. You can make selections based on the location of your entities in space, or you can select entities that are attached to other selected entities, such as nodes that are attached to selected elements. Be aware, however, that you must reactivate all entities before solving your model. Unselected entities will not be included in a solution. For example, if you select a subset of nodes on which to apply constraints, you should reactivate all nodes before solving. ANSYS allows the user to activate all entities with one simple operation by the command

utility menu: **Select** → **Everything**

You can also select a set of related entities in a hierarchical fashion. For example, given a subset of areas, you can select (a) all lines defining the areas, (b) all keypoints defining those lines, (c) all elements belonging to the areas, and so on. To select in this fashion, use the command

utility menu: **Select** → **Everything Below**

ANSYS also provides the capability to group some selected entities into a component. You can group one type of entity—such as nodes, elements, keypoints, or lines—into a component to be identified by a user-defined name (up to eight characters long).

## 6.7 GRAPHICS CAPABILITIES

Good graphics are especially important for visualizing and understanding a problem being analyzed. The ANSYS program provides numerous features that allow you to enhance the visual information presented to you. Some examples of the graphics capabilities of ANSYS include deformed shapes, result contours, sectional views, and animation. Consult the ANSYS procedure manual for additional information about more than 100 different graphics functions available to the user.

Up to five ANSYS windows can be opened simultaneously within one graphics window. You can display different information in different windows. ANSYS windows are defined in screen coordinates (−1 to +1 in the $x$-direction and −1 to +1 in the $y$-direction). By default, ANSYS directs all graphics information to one window (window 1).

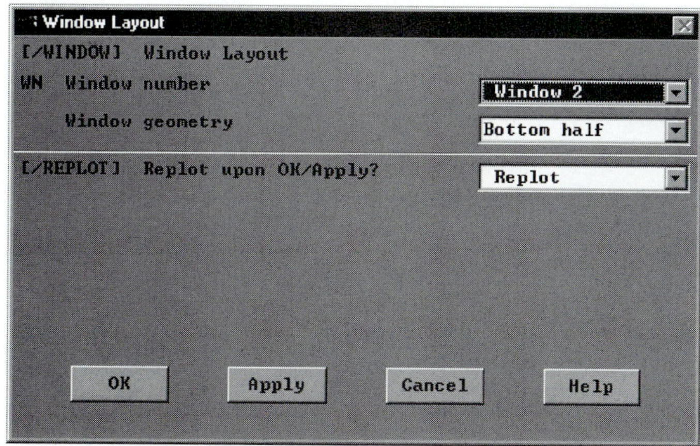


**FIGURE 6.31** The Window-layout dialog box.

In order to define additional windows, you need to access the window-layout dialog box, as shown in Figure 6.31. To do so, issue the following commands:

utility menu: **PlotCtrls** → **Window Controls** → **Window Layout** ...

There are three important concepts that you need to know with respect to window layout: (1) focus point, (2) distance, and (3) viewpoint. The focus point, with coordinates XF, YF, ZF, is the point on the model that appears at the center of the window. By changing the coordinates of the focus point, you can make a different point on the model appear at the center of the window. Distance determines the magnification of an image. As the distance approaches infinity, the image becomes a point on the screen. As the distance is decreased, the image size increases until the image fills the window. Viewpoint determines the direction from which the object is viewed. A vector is established from the viewpoint to the origin of the display coordinate. The line of sight is parallel to this vector and is directed at the focus point.

Next, the Pan–Zoom–Rotate dialog box allows you to change viewing directions, zoom in and out, or rotate your model. You can access this dialog box, shown in Figure 6.32 by the following commands:

utility menu: **PlotCtrls** → **Pan, Zoom, Rotate** ...

The various commands within the Pan, Zoom, Rotate dialog box and their respective functions are:

| | |
|---|---|
| **Zoom:** | Pick the center and the corner of the zoom rectangle. |
| **Box Zoom:** | Pick the two corners of the zoom rectangle. |
| **Win Zoom:** | Same as Box Zoom, except the zoom rectangle has the same proportions as the window. |
| • | Zoom out. |
| ● | Zoom in. |

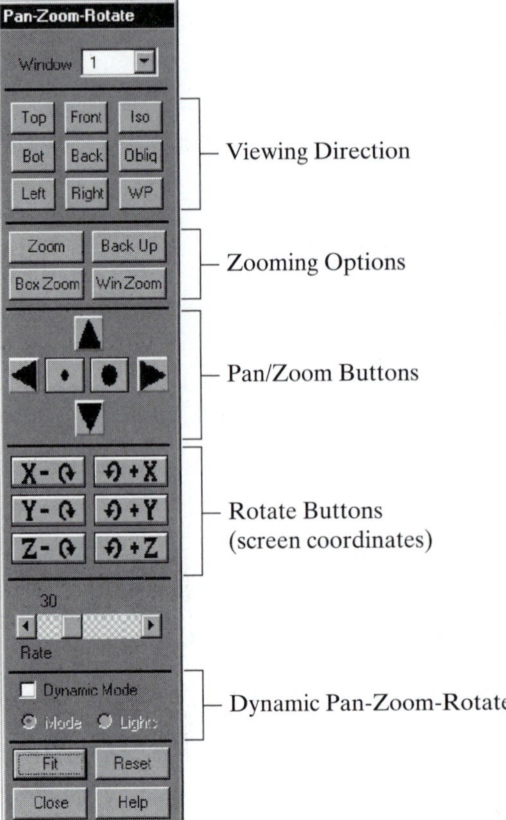


**FIGURE 6.32** The Pan, Zoom, Rotate dialog box.

**Dynamic Mode:** Allows you to pan, zoom, and rotate the image dynamically.

Pan model in *X*- and *Y*-directions.

Move the mouse right and left to rotate the model about the *Z*-axis of the screen. Move the mouse up and down to zoom in and out.

Move the mouse right and left to rotate the model about the *Y*-axis of the screen. Move the mouse up and down to rotate the model about the *X*-axis of the screen.

**Fit:** Changes the graphics specifications such that the image fits the window exactly.
**Reset:** Resets the graphics specifications to their default values.

In the next section, an example problem will demonstrate the basic steps in creating and analyzing a model with ANSYS.

## 6.8 AN EXAMPLE PROBLEM

Consider one of the many steel brackets $(E = 29 \times 10^6 \text{ lb/in}^2, \nu = 0.3)$ used to support bookshelves. The dimensions of the bracket are shown in Figure 6.33. The bracket is loaded uniformly along its top surface, and it is fixed along its left edge. Under the given loading and the constraints, plot the deformed shape; also determine the principal stresses and the von Mises stresses in the bracket.

The following steps demonstrate how to solve this problem using ANSYS:

Enter the **ANSYS** program by using the Launcher.

Type **xansys54** on the command line if you are running ANSYS on a UNIX platform, or consult your system administrator for information on how to run ANSYS from your computer system's platform.

Pick **Interactive** from the Launcher menu.

Type **Bracket** (or a file name of your choice) in the **Initial Jobname** entry field of the dialog box.

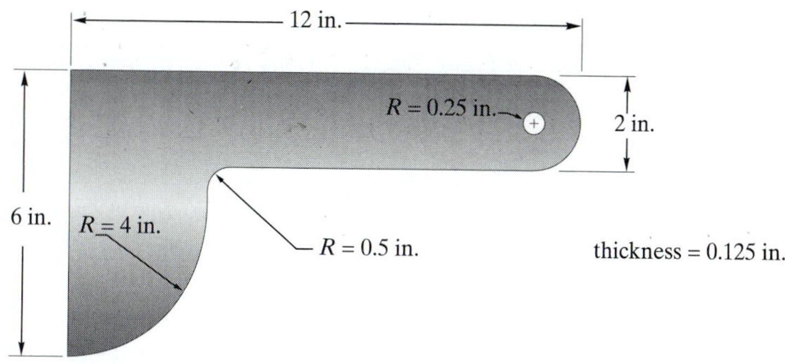


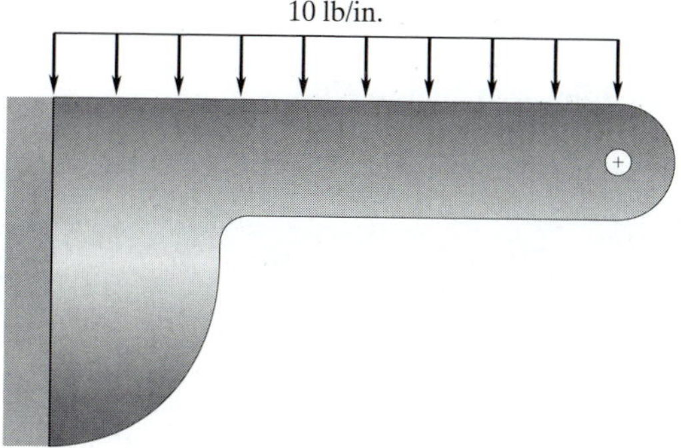


**FIGURE 6.33** A schematic of the steel bracket in the example problem.

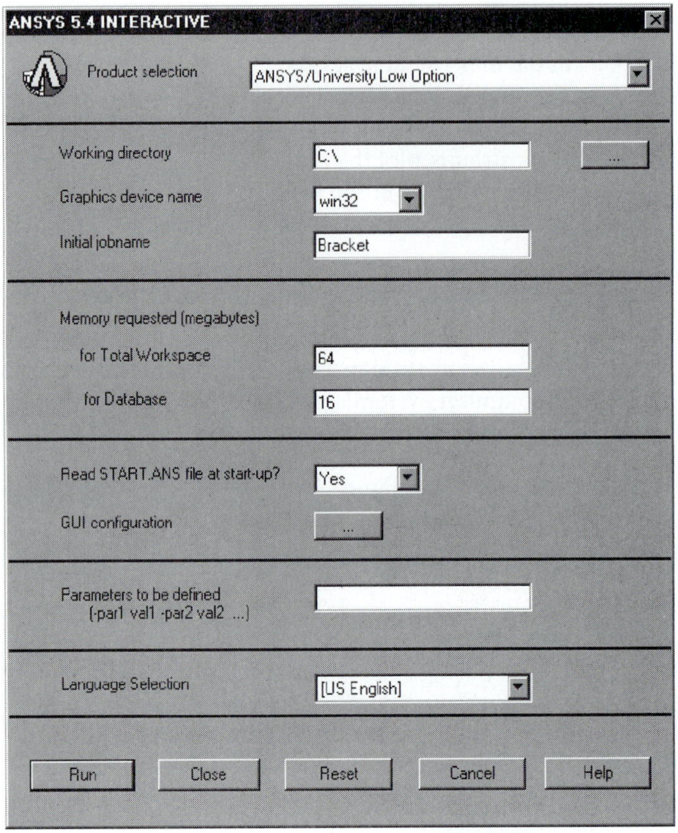

Pick **Run** to start the Graphic User Interface (GUI). A window will open with some disclaimer information. You will eventually be asked to press the <Return> key to start the graphics window and the main menu. Do so in order to proceed.

Create a title for the problem. This title will appear on ANSYS display windows to provide a simple way of identifying the displays. To create a title, issue the command

utility menu: **File** → **Change Title** ...

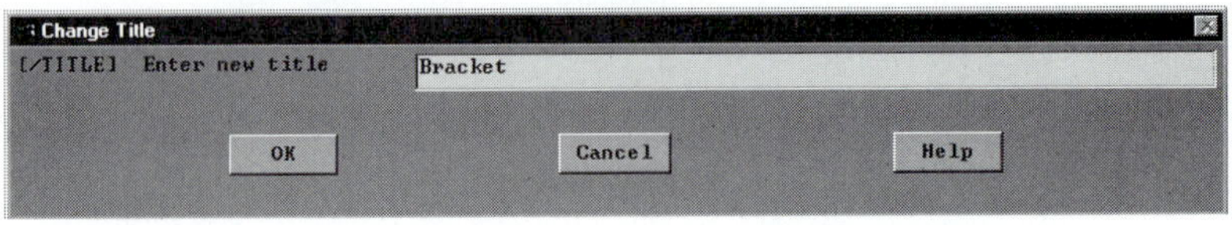

Define the element type and material properties:

main menu: **Preprocessor** → **Element Type** → **Add/Edit/Delete** ...

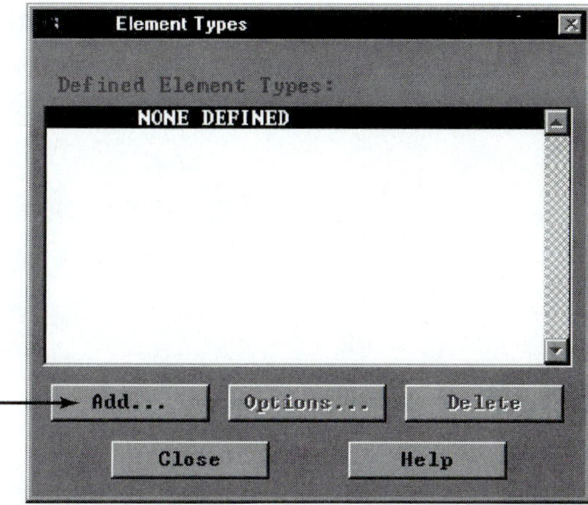
Element Types
Defined Element Types:
NONE DEFINED
Add...
Options...
Delete
Close
Help

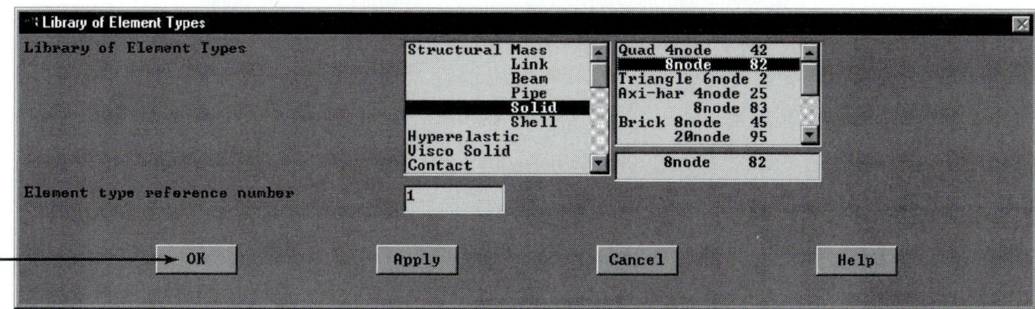
Library of Element Types
Library of Element Types
Structural Mass
Link
Beam
Pipe
Solid
Shell
Hyperelastic
Visco Solid
Contact
Quad 4node 42
8node 82
Triangle 6node 2
Axi-har 4node 25
8node 83
Brick 8node 45
20node 95
8node 82
Element type reference number
1
OK
Apply
Cancel
Help

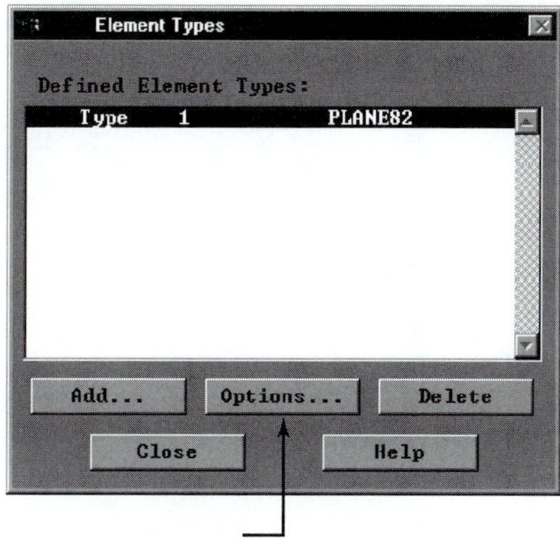
Element Types
Defined Element Types:
Type 1 PLANE82
Add...
Options...
Delete
Close
Help

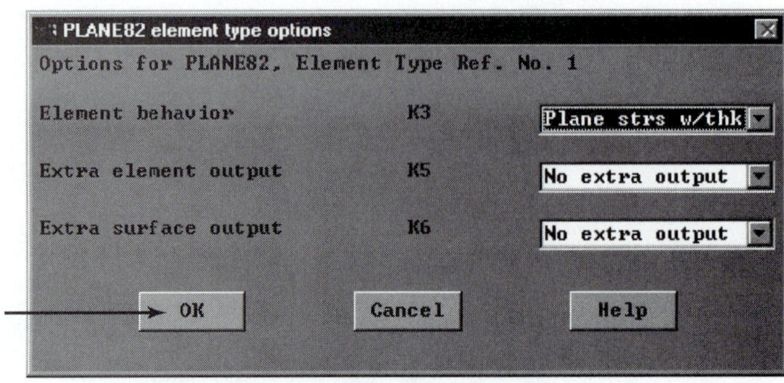


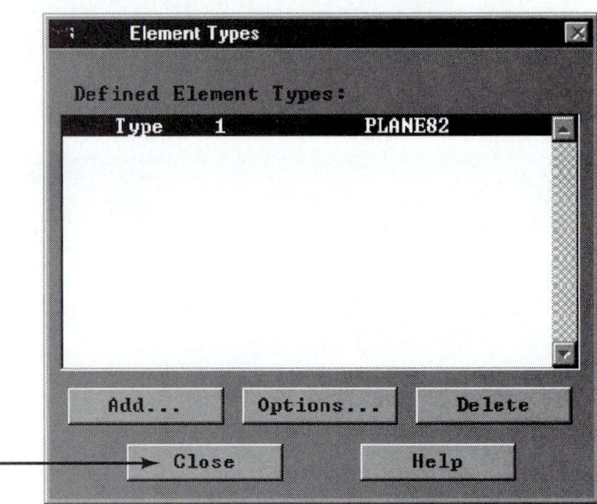


Assign the thickness of the bracket:

main menu: **Preprocessor** → **Real Constants** …

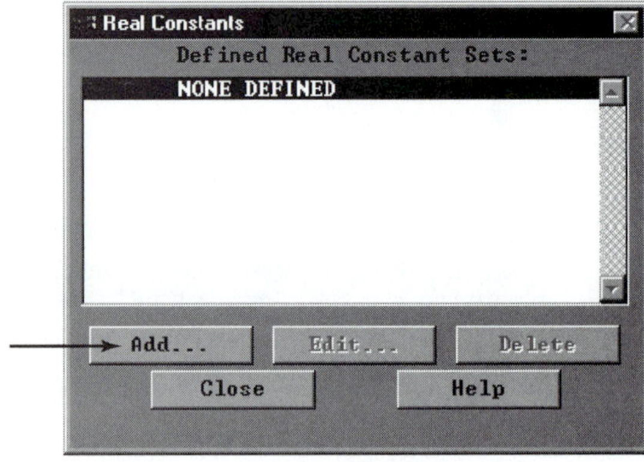

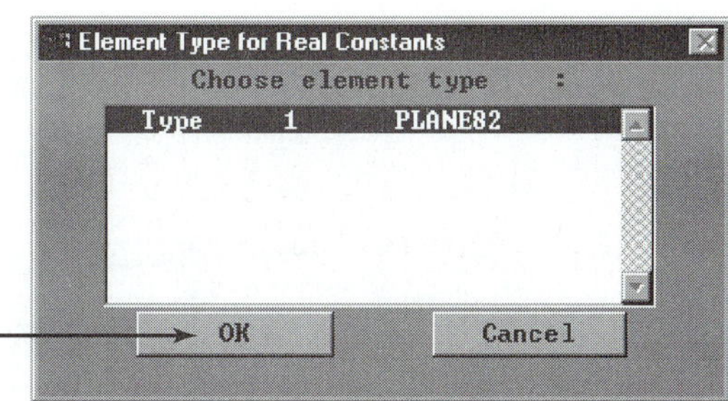


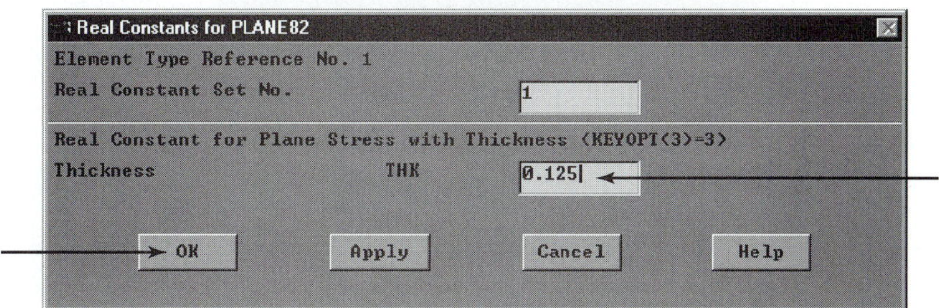


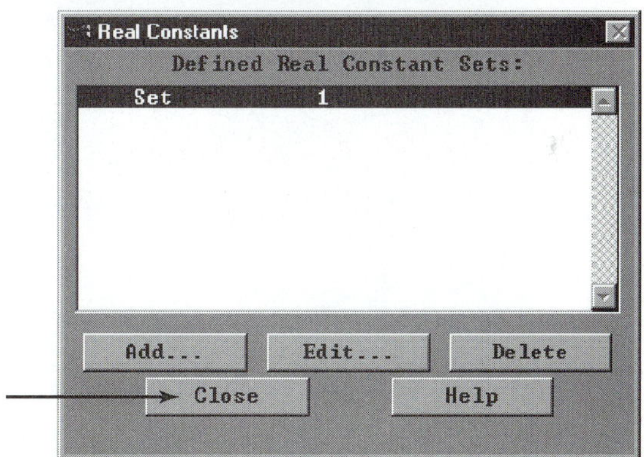


Assign the modulus of elasticity and the Poisson's-ratio values:

main menu: **Preprocessor** → **Material Props** → **-Constant-Isotropic** …

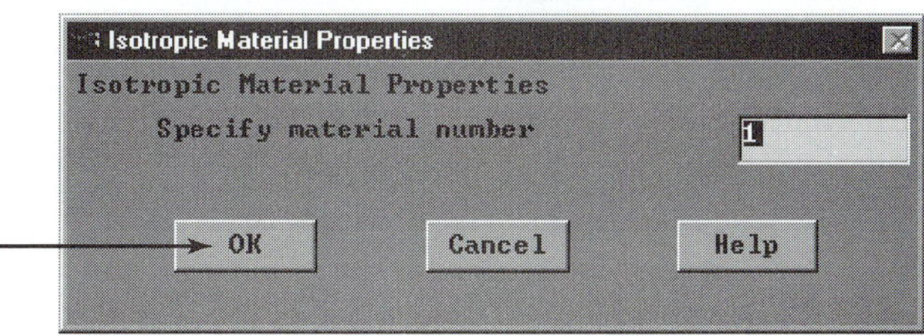

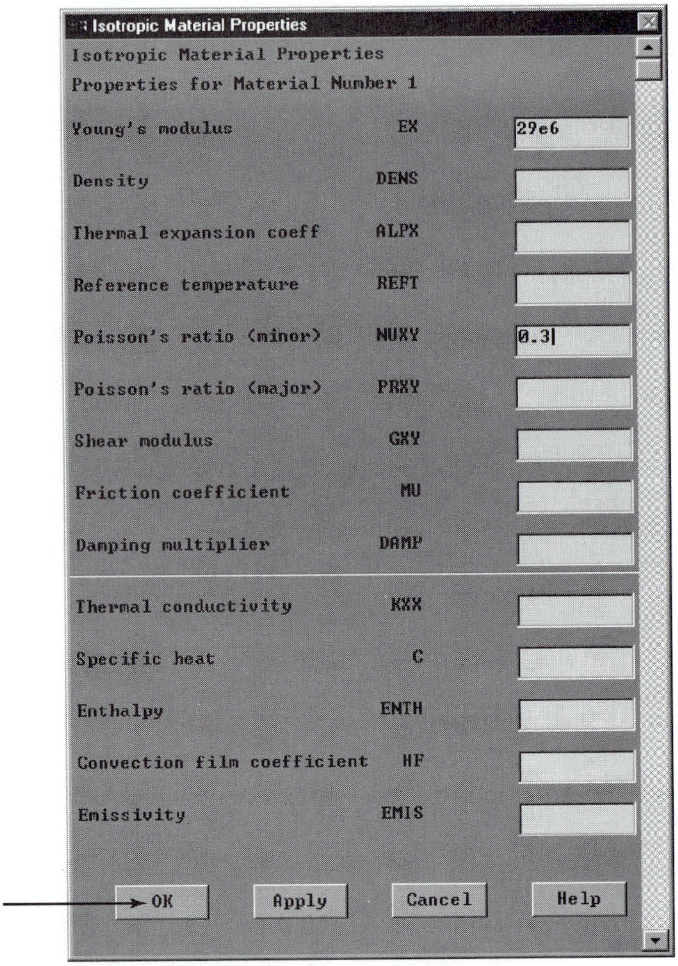

Isotropic Material Properties

Isotropic Material Properties

Properties for Material Number 1

| | | |
|---|---|---|
| Young's modulus | EX | 29e6 |
| Density | DENS | |
| Thermal expansion coeff | ALPX | |
| Reference temperature | REFT | |
| Poisson's ratio (minor) | NUXY | 0.3 |
| Poisson's ratio (major) | PRXY | |
| Shear modulus | GXY | |
| Friction coefficient | MU | |
| Damping multiplier | DAMP | |
| Thermal conductivity | KXX | |
| Specific heat | C | |
| Enthalpy | ENTH | |
| Convection film coefficient | HF | |
| Emissivity | EMIS | |

OK Apply Cancel Help

ANSYS Toolbar: **SAVE_DB**

Set up the graphics area—that is, the work plane, zoom, and so on:

utility menu: **Workplane** → **Wp Settings** ...

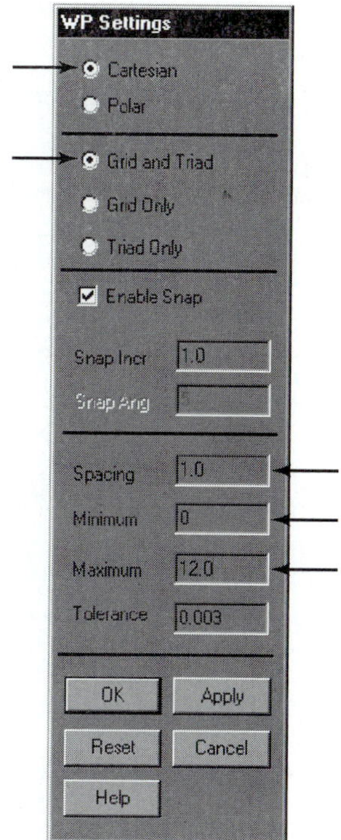


Toggle on the workplane by the command sequence

utility menu: **Workplane** → **Display Working Plane**

Bring the workplane to view by the command sequence

utility menu: **PlotCtrls** → **Pan, Zoom, Rotate** …

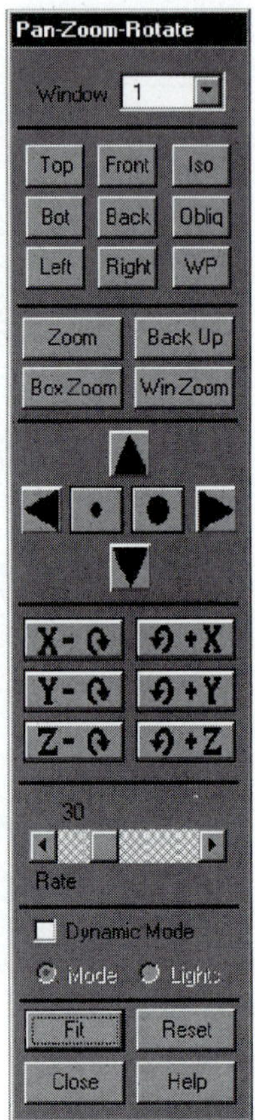


Click on the **small circle** and the **arrows** until you bring the workplane to view, and then create the geometry:

main menu: **Preprocessor** → **-Modeling-Create** → **-Areas-Rectangle** → **By 2 Corners** +

**a)** On the workplane, pick the location of the corners of Areas 1 and 2, as shown in Figure 6.34, and apply:

**[At WP = 0,12 in the upper left corner of the workplane, press the left button]**

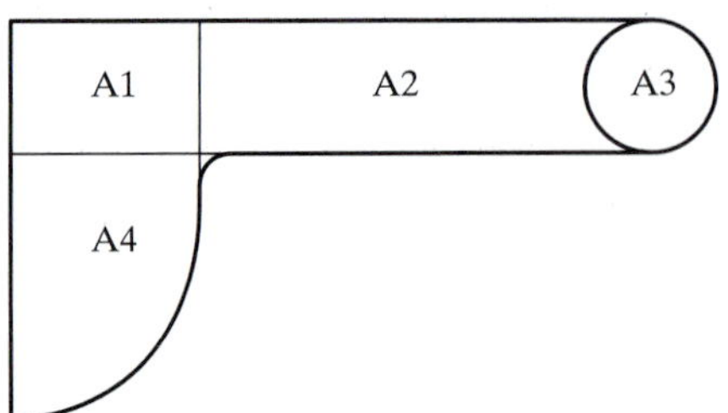


**FIGURE 6.34** The Areas making up the bracket.

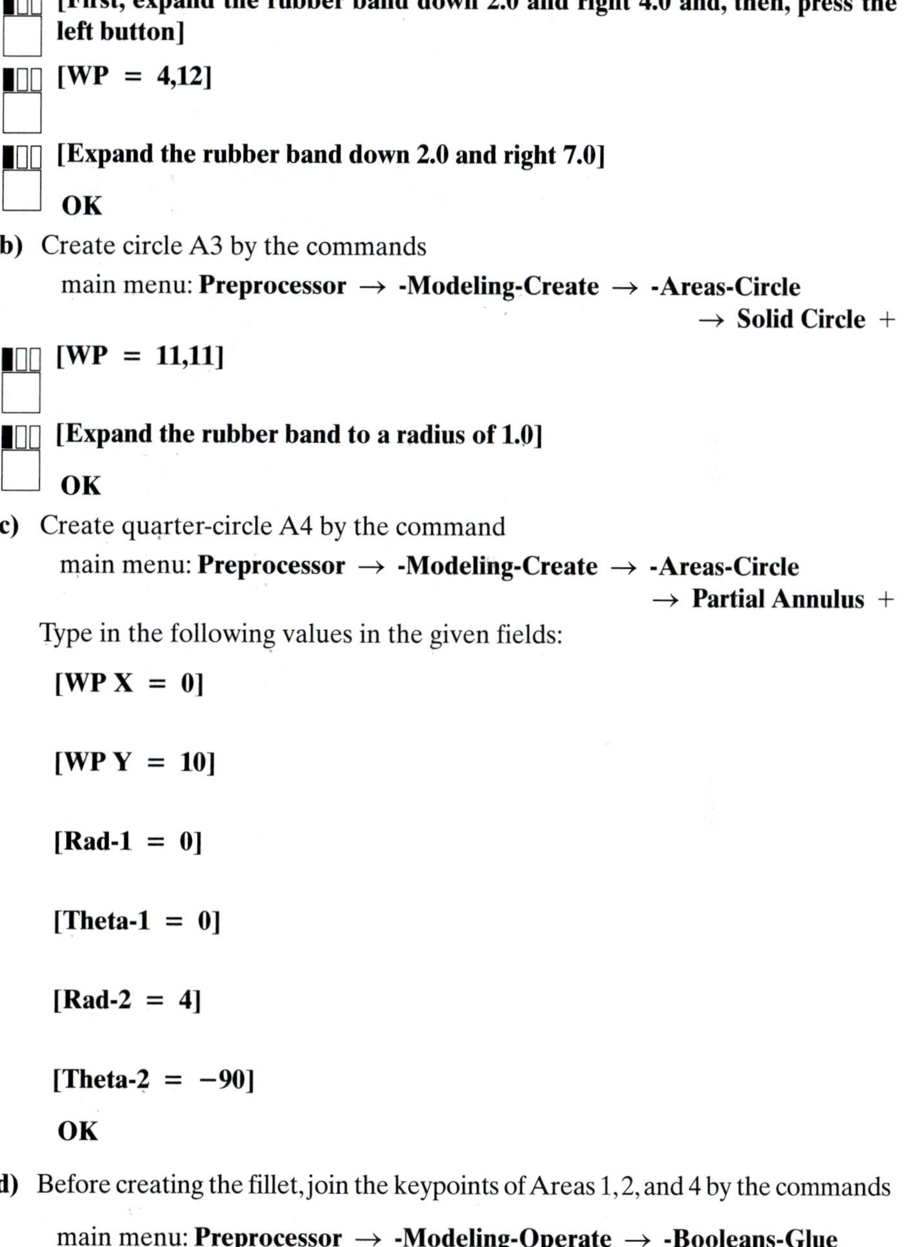

**[First, expand the rubber band down 2.0 and right 4.0 and, then, press the left button]**

**[WP = 4,12]**

**[Expand the rubber band down 2.0 and right 7.0]**

**OK**

**b)** Create circle A3 by the commands

main menu: **Preprocessor** → **-Modeling-Create** → **-Areas-Circle** → **Solid Circle** +

**[WP = 11,11]**

**[Expand the rubber band to a radius of 1.0]**

**OK**

**c)** Create quarter-circle A4 by the command

main menu: **Preprocessor** → **-Modeling-Create** → **-Areas-Circle** → **Partial Annulus** +

Type in the following values in the given fields:

**[WP X = 0]**

**[WP Y = 10]**

**[Rad-1 = 0]**

**[Theta-1 = 0]**

**[Rad-2 = 4]**

**[Theta-2 = −90]**

**OK**

**d)** Before creating the fillet, join the keypoints of Areas 1, 2, and 4 by the commands

main menu: **Preprocessor** → **-Modeling-Operate** → **-Booleans-Glue** → **Areas** +

Pick Areas 1, 2, and 4.

**OK**

**e)** Create the fillet by the commands

main menu: **Preprocessor** → **-Modeling-Create** → **-Lines-Line Fillet** +

**[Pick the bottom edge of rectangular Area 2]**

**[Pick the curved edge of quarter-circle Area 4]**

**APPLY**

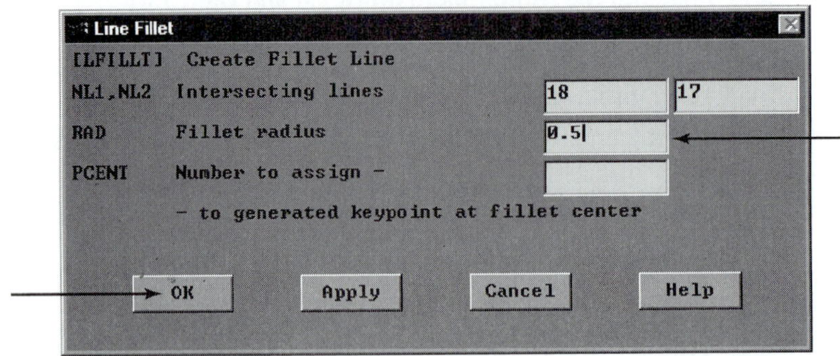


Then, issue the command

utility menu: **PlotCtrls** → **Pan, Zoom, Rotate** ...

Use the Box Zoom button to zoom about the fillet region, and issue the command

utility menu: **Plot** → **Lines**

**f)** Create an area for the fillet with the commands

main menu: **Preprocessor** → **-Modeling-Create** → **-Areas-Arbitrary** → **By Lines** +

Pick the fillet line and the two intersecting smaller lines.

**OK**

**g)** Add the areas together with the commands

main menu: **Preprocessor** → **-Modeling-Operate** → **Add** → **Areas** +

Click on the **Pick All** button and issue the command

utility menu: **PlotCtrls** → **Pan, Zoom, Rotate** ...

Click on the **Fit** button and then **Close.**

**h)** Create the area of the small hole, but first change the **Snap Incr** value in the WP Settings dialog box to 0.25.

**OK**

Then issue the commands

main menu: **Preprocessor** → **-Modeling-create** → **-Areas-Circle** → **Solid Circle**

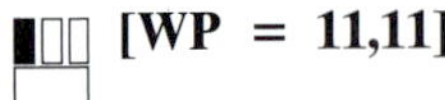
**[WP = 11,11]**

**[Expand the rubber band to a radius of 0.25]**

**OK**

**i)** Subtract the area of the small hole with the commands

main menu: **Preprocessor** → **-Modeling-Operate** → **-Booleans-Subtract** → **Areas** +

**[Pick the bracket area]**

**[anywhere in the ANSYS graphics area, apply]**

**[Pick the small circular area (*r* = 0.25)]**

**[anywhere in the ANSYS graphics area, apply]**

**OK**

Now you can toggle off the workplane grids with the command

utility menu: **Workplane** → **Display Working plane**

ANSYS Toolbar: **SAVE_DB**

You are now ready to mesh the area of the bracket to create elements and nodes. Issue the commands

main menu: **Preprocessor** → **-Meshing-Size Cntrls** → **-Manual Size-Global-Size** …

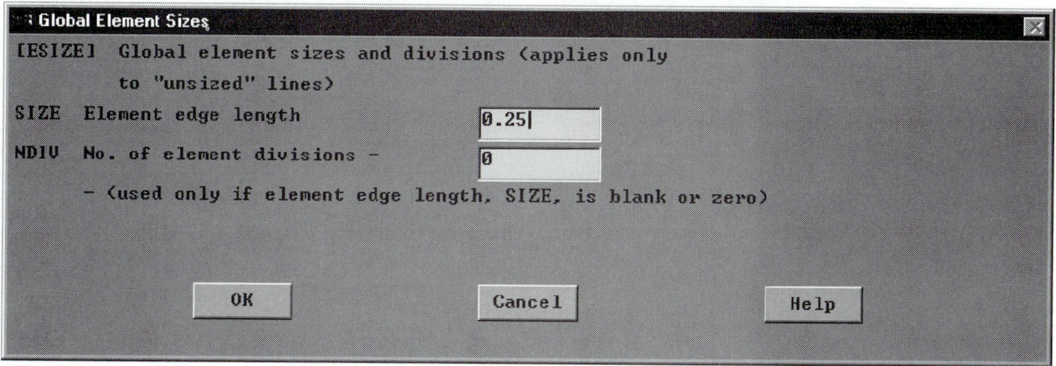
Global Element Sizes
[ESIZE] Global element sizes and divisions (applies only to "unsized" lines)
SIZE Element edge length 0.25
NDIV No. of element divisions - 0
- (used only if element edge length, SIZE, is blank or zero)
OK Cancel Help

ANSYS Toolbar: **SAVE_DB**

main menu: **Preprocessor** → **-Meshing-Mesh** → **Areas-Free** +

Click on the **Pick All** button.

Apply boundary conditions:

main menu: **Solution** → **-Loads-Apply** → **-Structural-Displacement** → **On Keypoints**

Pick the three keypoints: (1) upper left corner of Area 1, (2) two inches below the keypoint you just picked (i.e., the upper left corner of Area 4), and (3) the lower left corner of Area 4.

**OK**

Apply U,ROT on KPs
[DK] Apply Displacements (U,ROT) on Keypoints
Lab2 DOFs to be constrained All DOF UX UY
VALUE Displacement value 0
KEXPND Expand disp to nodes? Yes
OK Apply Cancel Help

main menu: **Solution** → **-Loads-Apply** → **-Structural-Pressure** → **On Lines** +

Pick the upper two horizontal lines associated with Area 1 and Area 2 (on the upper edge of the bracket).

**OK**

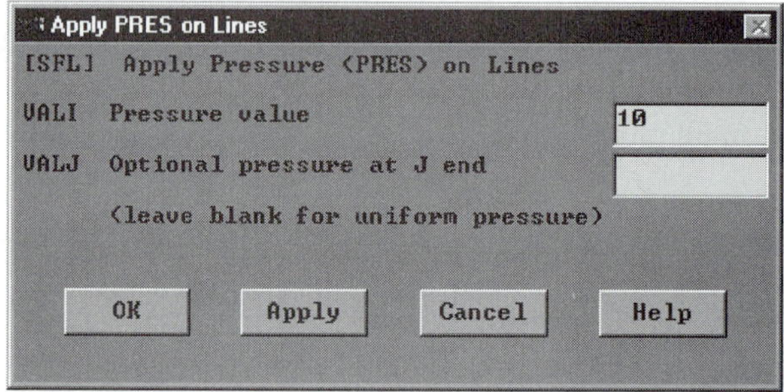


Solve the problem:

main menu: **Solution** → **-Solve-Current LS**

**OK**

**Close** (the solution is done!) window.

**Close** (the /STAT Command) window.

For the postprocessing phase, first plot the deformed shape by using the commands

main menu: **General Postproc** → **Plot Results** → **Deformed Shape** ...

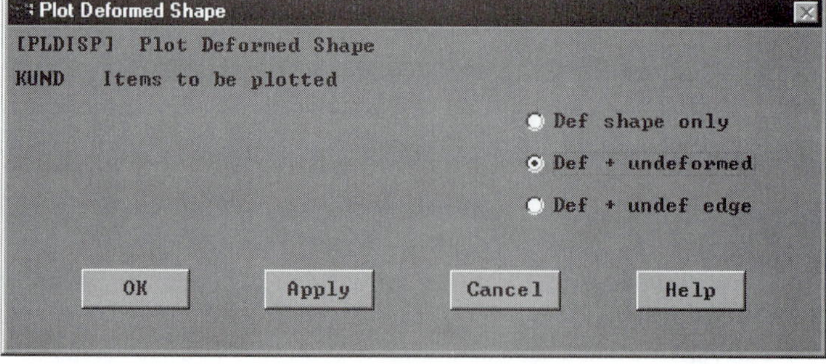

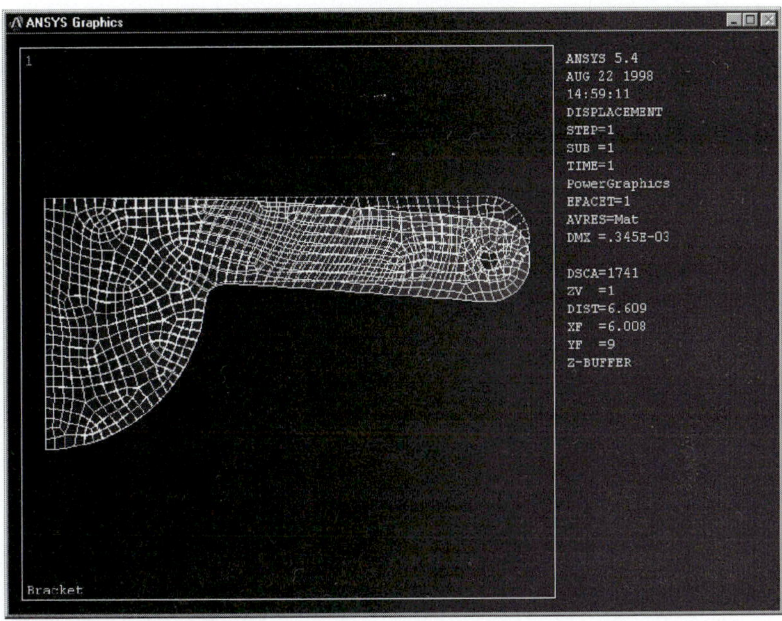


Plot the von Mises stresses with the commands

main menu: **General Postproc** → **Plot Results**
→ **-Contour Plot-Nodal Solu** ...

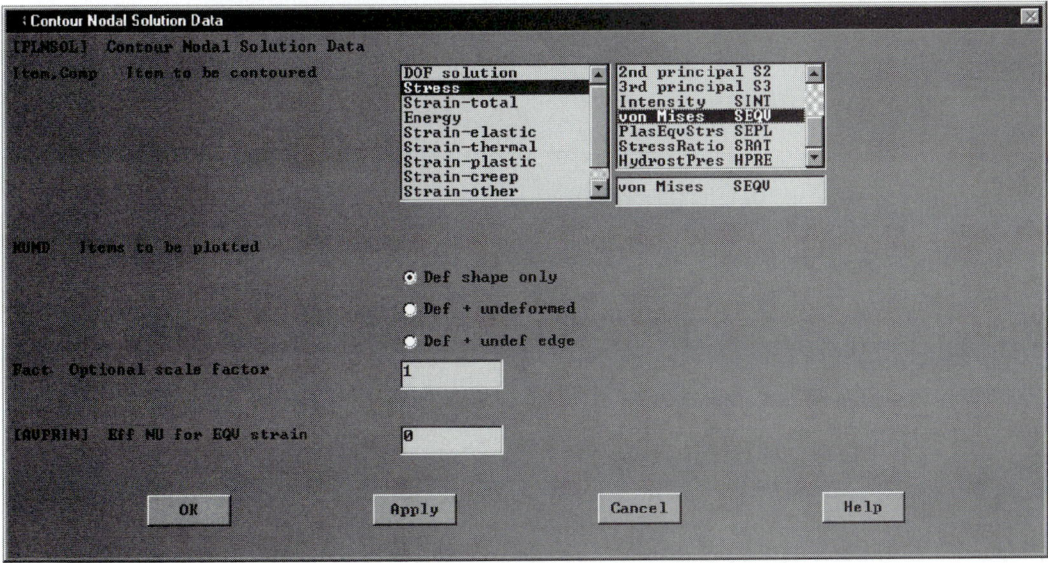

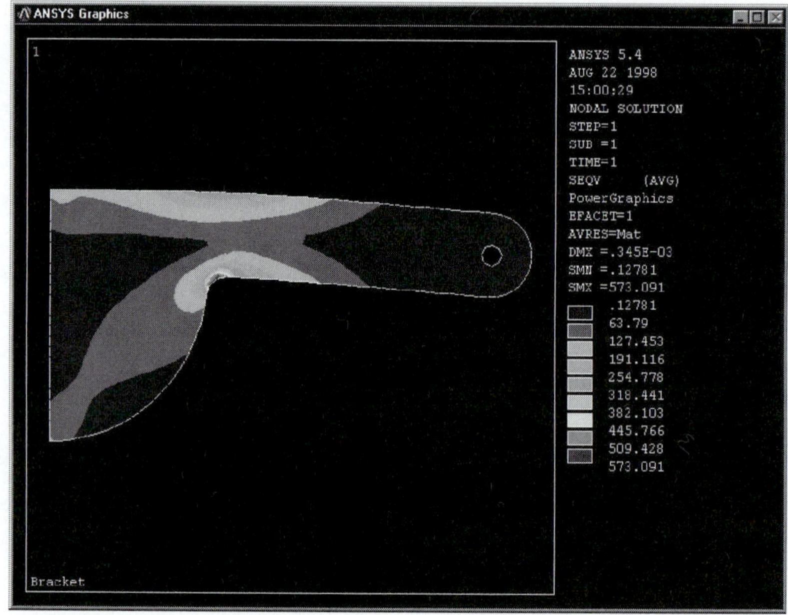


Repeat the previous step and pick the principal stresses to be plotted. Then, exit ANSYS and save everything:

Toolbar: **QUIT**

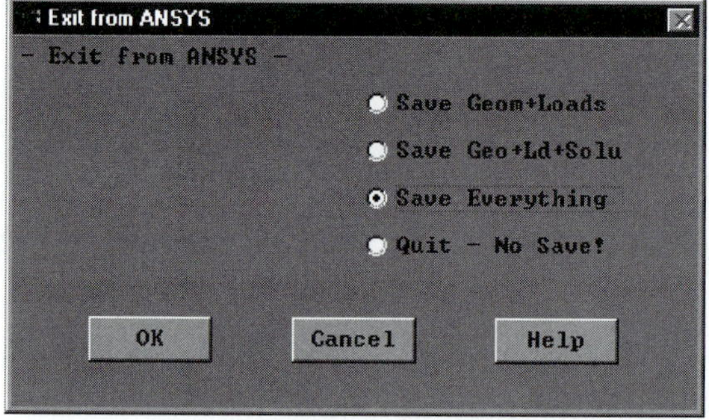


## SUMMARY

At this point you should know:

1. the basic organization of the ANSYS program. There are three processors that you will use frequently: (1) the *preprocessor* **(PREP7)**, (2) the *processor* **(SOLUTION)**, and (3) the general *postprocessor* **(POST1)**.
2. the commands the preprocessor **(PREP7)** contains that you need to use to build a model:
   - define element types and options

- define element real constants
- define material properties
- create model geometry
- define meshing controls
- mesh the object created.

3. the commands the Solution *processor* **(SOLUTION)** has that allow you to apply boundary conditions and loads. The solution processor also solves for the nodal solutions and calculates other elemental information.
4. the commands the general *postprocessor* **(POST1)** contains that allow you to list and display results of an analysis:
   - read results data from results file
   - read element results data
   - plot results
   - list results.
5. that ANSYS writes and reads many files during a typical analysis.
6. that ANSYS also offers the user the capability to select information about a portion of the model, such as certain nodes, elements, lines, areas, and volumes, for further processing.
7. that the ANSYS program provides numerous features that allow you to enhance the visual information presented to you. Some examples of the graphics capabilities of ANSYS are deformed shapes, result contours, sectional views, and animation.

## REFERENCES

*ANSYS Manual: Introduction to ANSYS*, Vol. I, Swanson Analysis Systems, Inc.

*ANSYS User's Manual: Procedures*, Vol. I, Swanson Analysis Systems, Inc.

*ANSYS User's Manual: Commands*, Vol. II, Swanson Analysis Systems, Inc.

*ANSYS User's Manual: Elements*, Volume III, Swanson Analysis Systems, Inc.

TABLE 6.1 Element types offered by ANSYS.

The ANSYS program offers nearly 100 different elements types. For detailed information on a specific element type, see the *Elements* volume (Vol. III) of the ANSYS User's Manual.

| Structural Point<br>Structural Mass<br>MASS21<br>1 node 3–D space<br>DOF: UX, UY, UZ, ROTX, ROTY, ROTZ | Structural 2–D Line<br>Spar <br>LINK1<br>2 nodes 2–D space<br>DOF: UX, UY | Structural 2–D Beam<br>Elastic Beam <br>BEAM3<br>2 nodes 2–D space<br>DOF: UX, UY, ROTZ | Plastic Beam <br>BEAM23<br>2 nodes 2–D space<br>DOF: UX, UY, ROTZ | Offset Tapered Unsymmetric Beam <br>BEAM54<br>2 nodes 2–D space<br>DOF: UX, UY, ROTZ |
|---|---|---|---|---|
| Structural 3–D Line<br>Spar <br>LINK8<br>2 nodes 3–D space<br>DOF: UX, UY, UZ | Tension–Only Spar <br>LINK10<br>2 nodes 3–D space<br>DOF: UX, UY, UZ | Linear Actuator 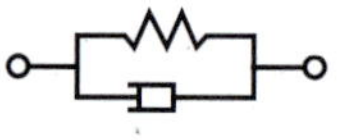<br>LINK11<br>2 nodes 3–D space<br>DOF: UX, UY, UZ | Structural 3–D Beam<br>Elastic Beam <br>BEAM4<br>2 nodes 3–D space<br>DOF: UX, UY, UZ, ROTX, ROTY, ROTZ | Thin Walled Plastic Beam <br>BEAM24<br>2 nodes 3–D space<br>DOF: UX, UY, UZ, ROTX, ROTY, ROTZ |
| Offset Tapered Unsymmetric Beam <br>BEAM44<br>2 nodes 3–D space<br>DOF: UX, UY, UZ, ROTX, ROTY, ROTZ | Structural Pipe<br>Elastic Straight Pipe <br>PIPE16<br>2 nodes 3–D space<br>DOF: UX, UY, UZ, ROTX, ROTY, ROTZ | Elastic Pipe Tee <br>PIPE17<br>4 nodes 3–D space<br>DOF: UX, UY, UZ, ROTX, ROTY, ROTZ | Curved Pipe (Elbow) <br>PIPE18<br>2 nodes 3–D space<br>DOF: UX, UY, UZ, ROTX, ROTY, ROTZ | Plastic Straight Pipe <br>PIPE20<br>2 nodes 3–D space<br>DOF: UX, UY, UZ, ROTX, ROTY, ROTZ |
| Immersed Pipe <br>PIPE59<br>2 nodes 3–D space<br>DOF: UX, UY, UZ, ROTX, ROTY, ROTZ | Plastic Curved Pipe <br>PIPE60<br>2 nodes 3–D space<br>DOF: UX, UY, UZ, ROTX, ROTY, ROTZ | Structural 2–D Solid<br>Triangular Solid 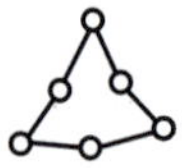<br>PLANE2<br>6 nodes 2–D space<br>DOF: UX, UY | Axisymmetric Harmonic Struct. Solid 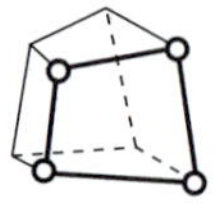<br>PLANE25<br>4 nodes 2–D space<br>DOF: UX, UY, UZ | Structural Solid 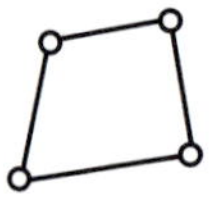<br>PLANE42<br>4 nodes 2–D space<br>DOF: UX, UY |

TABLE 6.1 (continued) Element types offered by ANSYS.

The ANSYS program offers nearly 100 different elements types. For detailed information on a specific element type, see the *Elements* volume (Vol. III) of the ANSYS User's Manual.

| | | | | |
|---|---|---|---|---|
| Structural Solid<br>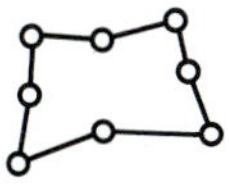<br>PLANE82<br>8 nodes 2–D space<br>DOF: UX, UY | Axisymmetric Harmonic Struct. Solid<br>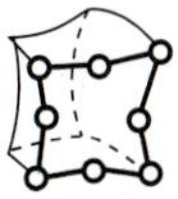<br>PLANE83<br>8 nodes 2–D space<br>DOF: UX, UY, UZ | Structural 3–D Solid<br>Structural Solid<br>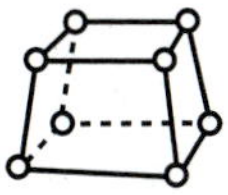<br>SOLID45<br>8 nodes 3–D space<br>DOF: UX, UY, UZ | Layered Solid<br>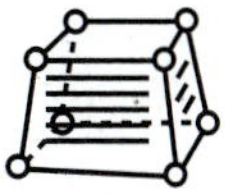<br>SOLID46<br>8 nodes 3–D space<br>DOF: UX, UY, UZ | Anisotropic Solid<br>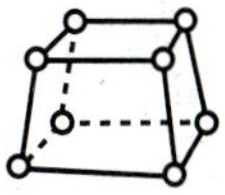<br>SOLID64<br>8 nodes 3–D space<br>DOF: UX, UY, UZ |
| Reinforced Solid<br>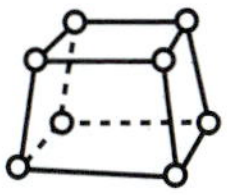<br>SOLID65<br>8 nodes 3–D space<br>DOF: UX, UY, UZ | Solid with Rotations<br>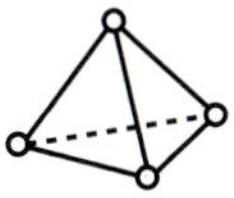<br>SOLID72<br>4 nodes 3–D space<br>DOF: UX, UY, UZ, ROTX, ROTY, ROTZ | Solid with Rotations<br>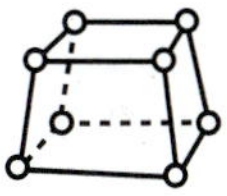<br>SOLID73<br>8 nodes 3–D space<br>DOF: UX, UY, UZ, ROTX, ROTY, ROTZ | Tetrahedral Solid<br>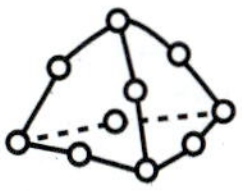<br>SOLID92<br>10 nodes 3–D space<br>DOF: UX, UY, UZ | Structural Solid<br>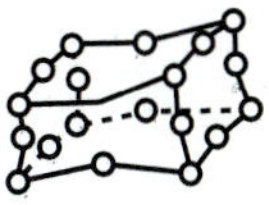<br>SOLID95<br>20 nodes 3–D space<br>DOF: UX, UY, UZ |
| Structural 2–D Shell<br>Plastic Axisymmetric Shell with Torsion<br><br>SHELL51<br>2 nodes 2–D space<br>DOF: UX, UY, UZ, ROTZ | Axisymmetric Harmonic Struct. Shell<br><br>SHELL61<br>2 nodes 2–D space<br>DOF: UX, UY, UZ, ROTZ | Structural 3–D Shell<br>Shear/Twist Panel<br>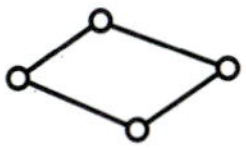<br>SHELL28<br>4 nodes 3–D space<br>DOF: UX, UY, UZ or ROTX, ROTY, ROTZ | Membrane Shell<br>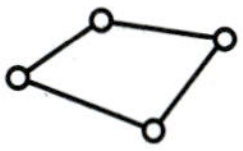<br>SHELL41<br>4 nodes 3–D space<br>DOF: UX, UY, UZ | Plastic Shell<br>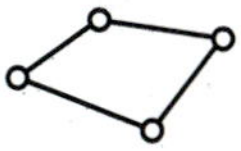<br>SHELL43<br>4 nodes 3–D space<br>DOF: UX, UY, UZ, ROTX, ROTY, ROTZ |
| Elastic Shell<br>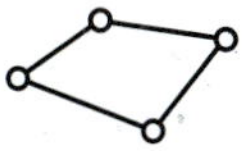<br>SHELL63<br>4 nodes 3–D space<br>DOF: UX, UY, UZ, ROTX, ROTY, ROTZ | 16–Layer Structural Shell<br>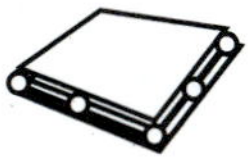<br>SHELL91<br>8 nodes 3–D space<br>DOF: UX, UY, UZ, ROTX, ROTY, ROTZ | Structural Shell<br>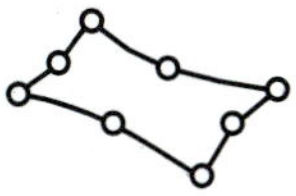<br>SHELL93<br>8 nodes 3–D space<br>DOF: UX, UY, UZ, ROTX, ROTY, ROTZ | 100–Layer Structural Shell<br>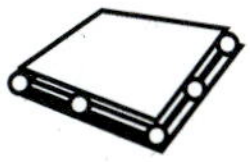<br>SHELL99<br>8 nodes 3–D space<br>DOF: UX, UY, UZ, ROTX, ROTY, ROTZ | |

TABLE 6.1 (continued) Element types offered by ANSYS.

The ANSYS program offers nearly 100 different elements types. For detailed information on a specific element type, see the *Elements* volume (Vol. III) of the ANSYS User's Manual.

| | | | | |
|---|---|---|---|---|
| Hyperelastic Solid<br>Hyperelastic Mixed U–P Solid<br>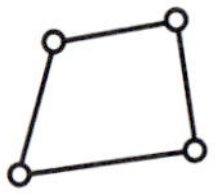<br>HYPER56<br>4 nodes 2–D space<br>DOF: UX, UY, UZ | Hyperelastic Mixed U–P Solid<br>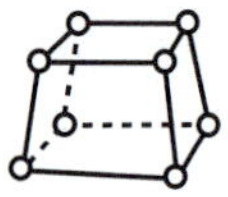<br>HYPER58<br>8 nodes 3–D space<br>DOF: UX, UY, UZ | Hyperelastic Mixed U–P Solid<br>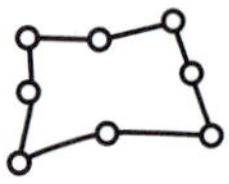<br>HYPER74<br>8 nodes 2–D space<br>DOF: UX, UY, UZ | Hyperelastic Solid<br>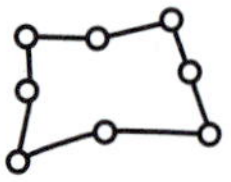<br>HYPER84<br>8 nodes 2–D space<br>DOF: UX, UY, UZ | Hyperelastic Solid<br>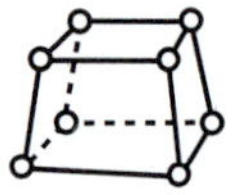<br>HYPER86<br>8 nodes 3–D space<br>DOF: UX, UY, UZ |
| Visco Solid<br>Viscoelastic Solid<br>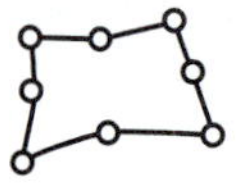<br>VISCO88<br>8 nodes 2–D space<br>DOF: UX, UY | Viscoelastic Solid<br>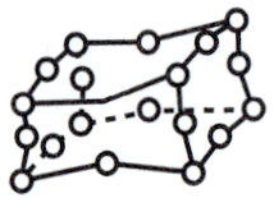<br>VISCO89<br>20 nodes 3–D space<br>DOF: UX, UY, UZ | Large Strain Solid<br>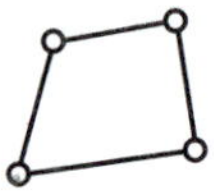<br>VISCO106<br>4 nodes 2–D space<br>DOF: UX, UY, UZ | Large Strain Solid<br>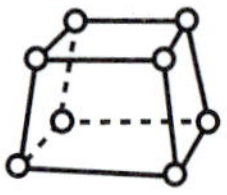<br>VISCO107<br>8 nodes 3–D space<br>DOF: UX, UY, UZ | Large Strain Solid<br>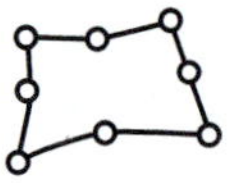<br>VISCO108<br>8 nodes 2–D space<br>DOF: UX, UY, UZ |
| Thermal Point<br>Thermal Mass<br><br>MASS71<br>1 node 3–D space<br>DOF: TEMP | Thermal Line<br>Radiation Link<br><br>LINK31<br>2 nodes 3–D space<br>DOF: TEMP | Conduction Bar<br><br>LINK32<br>2 nodes 2–D space<br>DOF: TEMP | Conduction Bar<br><br>LINK33<br>2 nodes 3–D space<br>DOF: TEMP | Convection Link<br><br>LINK34<br>2 nodes 3–D space<br>DOF: TEMP |
| Thermal 2–D Solid<br>Triangular Thermal Solid<br>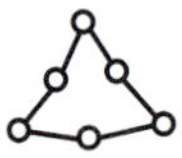<br>LANE35<br>6 nodes 2–D space<br>DOF: TEMP | Thermal Solid<br>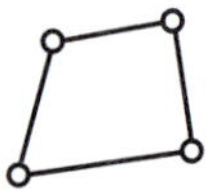<br>PLANE55<br>4 nodes 2–D space<br>DOF: TEMP | Axisymmetric Harmonic Thermal Solid<br>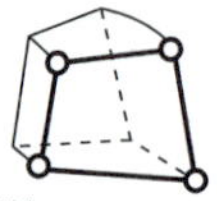<br>PLANE75<br>4 nodes 2–D space<br>DOF: TEMP | Thermal Solid<br>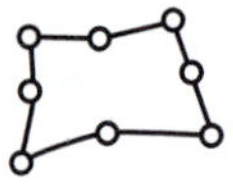<br>PLANE77<br>8 nodes 2–D space<br>DOF: TEMP | Axisymmetric Harmonic Thermal Solid<br>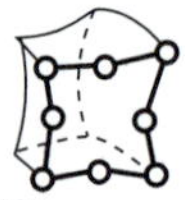<br>PLANE78<br>8 nodes 2–D space<br>DOF: TEMP |

TABLE 6.1 (continued) Element types offered by ANSYS.

The ANSYS program offers nearly 100 different elements types. For detailed information on a specific element type, see the *Elements* volume (Vol. III) of the ANSYS User's Manual.

| Thermal 3-D Solid | | | Thermal Shell | Fluid |
|---|---|---|---|---|
| Thermal Solid<br>SOLID70<br>8 nodes 3-D space<br>DOF: TEMP | Tetrahedral Thermal Solid<br>SOLID87<br>10 nodes 3-D space<br>DOF: TEMP | Thermal Solid<br>SOLID90<br>20 nodes 3-D space<br>DOF: TEMP | Thermal Shell<br>SHELL57<br>4 nodes 3-D space<br>DOF: TEMP | Acoustic Fluid<br>FLUID29<br>4 nodes 2-D space<br>DOF: UX, UY, PRES |
| Acoustic Fluid<br>FLUID30<br>8 nodes 3-D space<br>DOF: UX, UY, UZ, PRES | Dynamic Fluid Coupling<br>FLUID38<br>2 nodes 3-D space<br>DOF: UX, UY, UZ | Thermal-Fluid Pipe<br>FLUID66<br>2 nodes 3-D space<br>DOF: PRES, TEMP | Contained Fluid<br>FLUID79<br>4 nodes 2-D space<br>DOF: UX, UY | Contained Fluid<br>FLUID80<br>8 nodes 3-D space<br>DOF: UX, UY, UZ |
| | | | Thermal Electric | |
| Axisymmetric Harmonic Contained Fluid<br>FLUID81<br>4 nodes 2-D space<br>DOF: UX, UY, UZ | FLOTRAN CFD Fluid-Thermal<br>FLUID141<br>4 nodes 2-D space<br>DOF: VX, VY, VZ, PRES, TEMP, ENKE, ENDS | FLOTRAN CFD Fluid-Thermal<br>FLUID142<br>8 nodes 3-D space<br>DOF: VX, VY, VZ, PRES, TEMP, ENKE, ENDS | Thermal-Electric Solid<br>PLANE67<br>4 nodes 2-D space<br>DOF: TEMP, VOLT | Thermal-Electric Line<br>LINK68<br>2 nodes 3-D space<br>DOF: TEMP, VOLT |
| | Magnetic Electric | | | |
| Thermal-Electric Solid<br>SOLID69<br>8 nodes 3-D space<br>DOF: TEMP, VOLT | Current Source<br>SOURC36<br>3 nodes 3-D space<br>DOF: MAG | Magnetic Solid<br>PLANE53<br>8 nodes 2-D space<br>DOF: VOLT, AZ | Magnetic-Scalar Solid<br>SOLID96<br>8 nodes 3-D space<br>DOF: MAG | Magnetic Solid<br>SOLID97<br>8 nodes 3-D space<br>DOF: VOLT, AX, AY, AZ |

TABLE 6.1 (continued) Element types offered by ANSYS.

The ANSYS program offers nearly 100 different elements types. For detailed information on a specific element type, see the *Elements* volume (Vol. III) of the ANSYS User's Manual.

| | | | | |
|---|---|---|---|---|
| Magnetic Interface 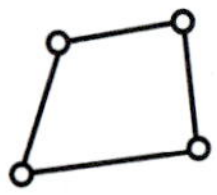<br>INTER115<br>4 nodes 3–D space<br>DOF: AX, AY, AZ, MAG | Electrostatic Solid 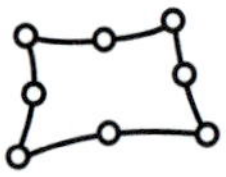<br>PLANE121<br>8 nodes 2–D space<br>DOF: VOLT | Electrostatic Solid 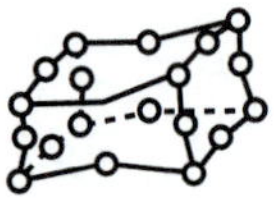<br>SOLID122<br>20 nodes 3–D space<br>DOF: VOLT | Tetrahedral Electrostatic Solid 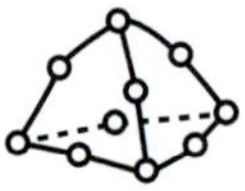<br>SOLID123<br>10 nodes 3–D space<br>DOF: VOLT | **Coupled–field**<br>Coupled–field Solid 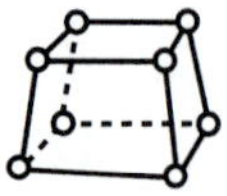<br>SOLID5<br>8 nodes 3–D space<br>DOF: UX, UY, UZ, TEMP, VOLT, MAG |
| Coupled–field Solid 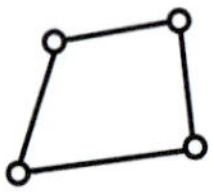<br>PLANE13<br>4 nodes 2–D space<br>DOF: UX, UY, TEMP, VOLT, AZ | Coupled–field Solid 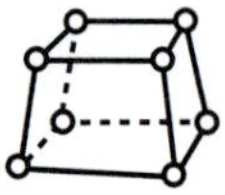<br>SOLID62<br>8 nodes 3–D space<br>DOF: UX, UY, UZ, AX, AY, AZ, VOLT | Tetrahedral Couple–field Solid <br>SOLID98<br>10 nodes 3–D space<br>DOF: UX, UY, UZ, TEMP, VOLT, MAG | **Contact**<br>Point–to–Point <br>CONTAC12<br>2 nodes 2–D space<br>DOF: UX, UY | Point–to–Ground <br>CONTAC26<br>3 nodes 2–D space<br>DOF: UX, UY |
| Point–to–Surface 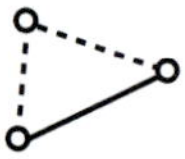<br>CONTAC48<br>3 nodes 2–D space<br>DOF: UX, UY, TEMP | Point–to–Surface 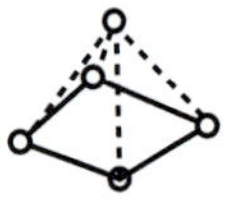<br>CONTAC49<br>5 nodes 3–D space<br>DOF: UX, UY, UZ, TEMP | Point–to–Point <br>CONTAC52<br>2 nodes 3–D space<br>DOF: UX, UY, UZ | **Combination**<br>Revolute Joint 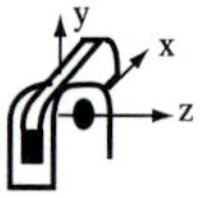<br>COMBIN7<br>5 nodes 3–D space<br>DOF: UX, UY, UZ, ROTX, ROTY, ROTZ | Spring–Damper <br>COMBIN14<br>2 nodes 3–D space<br>DOF: UX, UY, UZ, ROTX, ROTY, ROTZ PRES, TEMP |
| Control <br>COMBIN37<br>4 nodes 3–D space<br>DOF: UX, UY, UZ, ROTX, ROTY, ROTZ, PRES, TEMP | Nonlinear Spring <br>COMBIN39<br>2 nodes 3–D space<br>DOF: UX, UY, UZ, ROTX, ROTY, ROTZ, PRES, TEMP | Combination 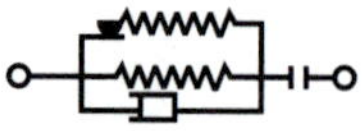<br>COMBIN40<br>2 nodes 3–D space<br>DOF: UX, UY, UZ, ROTX, ROTY, ROTZ, PRES, TEMP | **Matrix**<br>Stiffness, Mass or Damping Matrix 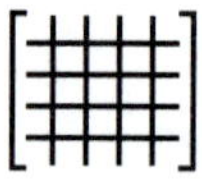<br>MATRIX27<br>2 nodes 3–D space<br>DOF: UX, UY, UZ, ROTX, ROTY, ROTZ | Superelement 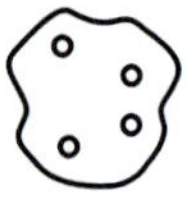<br>MATRIX50<br>2–D or 3–D space<br>DOF: Any |

TABLE 6.1 (continued) Element types offered by ANSYS.

The ANSYS program offers nearly 100 different elements types. For detailed information on a specific element type, see the *Elements* volume (Vol. III) of the ANSYS User's Manual.

| Infinite | | | | Surface |
|---|---|---|---|---|
| Infinite Boundary <br>INFIN9<br>2 nodes 2–D space<br>DOF: AZ, TEMP | Infinite Boundary 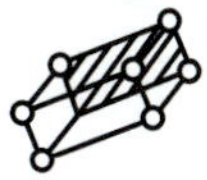<br>INFIN47<br>4 nodes 3–D space<br>DOF: MAG, TEMP | Infinite Boundary 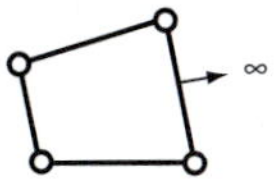<br>INFIN110<br>4 nodes 2–D space<br>DOF: AZ, VOLT, TEMP | Infinite Boundary 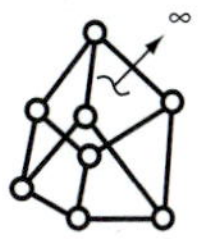<br>INFIN111<br>8 nodes 3–D space<br>DOF: MAG, AX, AY, AZ, VOLT, TEMP | Surface Effect <br>SURF19<br>3 nodes 2–D space<br>DOF: UX, UY, TEMP |
| Surface Effect 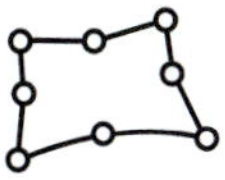<br>SURF22<br>8 nodes 3–D space<br>DOF: UX, UY, UZ, TEMP | | | | |
| | | | | |
| | | | | |

CHAPTER 7

# Analysis of Two-Dimensional Heat Transfer Problems

The main objective of this chapter is to introduce you to the analysis of two-dimensional heat transfer problems. General conduction problems and the treatment of various boundary conditions are discussed here. The main topics of Chapter 7 include the following:



## 7.1 GENERAL CONDUCTION PROBLEMS

In this chapter, we are concerned with determining how temperatures may vary with position in a medium as a result of either thermal conditions applied at the boundaries of the medium or heat generation within the medium. We are also interested in determining the heat flux at various points in a system, including its boundaries. Knowledge of temperature and heat flux fields is important in many engineering applications, including, for example, the cooling of electronic equipment, the design of thermal-fluid systems, and material and manufacturing processes. Knowledge of temperature distributions is also useful in determining thermal stresses and corresponding deflections in machine and structural elements. There are three modes of heat transfer: *conduction*, *convection*, and *radiation*. Conduction refers to that mode of heat transfer that occurs when there exists a temperature gradient in a medium. The energy is transported from the high-temperature region to the low-temperature region by molecular activities. Using a two-dimensional Cartesian frame of reference, we know that the rate of heat transfer by conduction is given by Fourier's Law:

$$q_X = -kA\frac{\partial T}{\partial X} \tag{7.1}$$

$$q_Y = -kA\frac{\partial T}{\partial Y} \tag{7.2}$$

$q_X$ and $q_Y$ are the $X$- and the $Y$-components of the heat transfer rate, $k$ is the thermal conductivity of the medium, $A$ is the cross-sectional area of the medium, and $\frac{\partial T}{\partial X}$ and $\frac{\partial T}{\partial Y}$ are the temperature gradients. Fourier's Law may also be expressed in terms of heat transfer rates per unit area as

$$q''_X = -k\frac{\partial T}{\partial X} \tag{7.3}$$

$$q''_Y = -k\frac{\partial T}{\partial Y} \tag{7.4}$$

where $q''_X = \frac{q_X}{A}$ and $q''_Y = \frac{q_Y}{A}$ are called heat fluxes in the $X$-direction and the $Y$-direction, respectively. It is important to realize that the direction of the total heat flow is always perpendicular to the *isotherms* (constant temperature lines or surfaces). This relationship is depicted in Figure 7.1.

Convective heat transfer occurs when a fluid in motion comes into contact with a surface whose temperature differs from the moving fluid. The overall heat transfer rate between the fluid and the surface is governed by Newton's Law of Cooling, which is

$$q = hA(T_s - T_f) \tag{7.5}$$

where $h$ is the heat transfer coefficient, $T_s$ is the surface temperature, and $T_f$ represents the temperature of the moving fluid. The value of the heat transfer coefficient for a particular situation is determined from experimental correlations that are available in many books about heat transfer.

All matters emit thermal radiation. This rule is true as long as the body in question is at a finite temperature. Simply stated, the amount of energy emitted by a surface is given by the equation

$$q'' = \varepsilon\sigma T_s^4 \tag{7.6}$$

where $q''$ represents the rate of thermal energy per unit area emitted by the surface, $\varepsilon$ is the emissivity of the surface $0 < \varepsilon < 1$, and $\sigma$ is the Stefan–Boltzman constant $(\sigma = 5.67 \times 10^{-8}\ \ \mathrm{W/m^2 \cdot K^4})$. It is important to note here that unlike conduction and

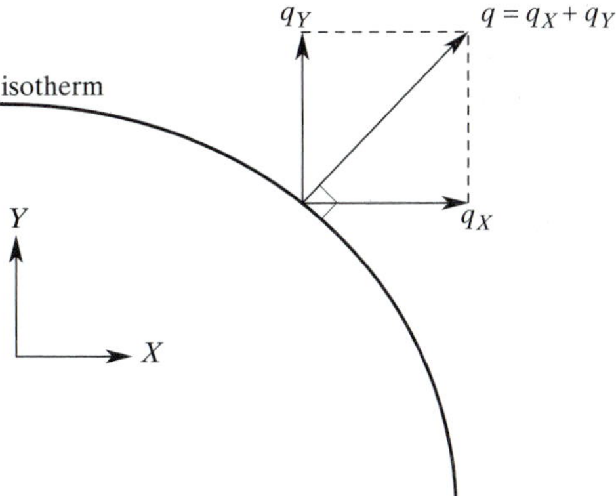


**FIGURE 7.1** The heat flux vector is always normal to the isotherms.

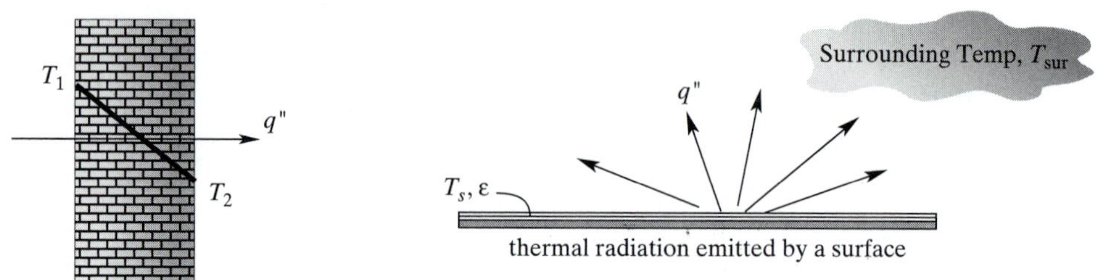


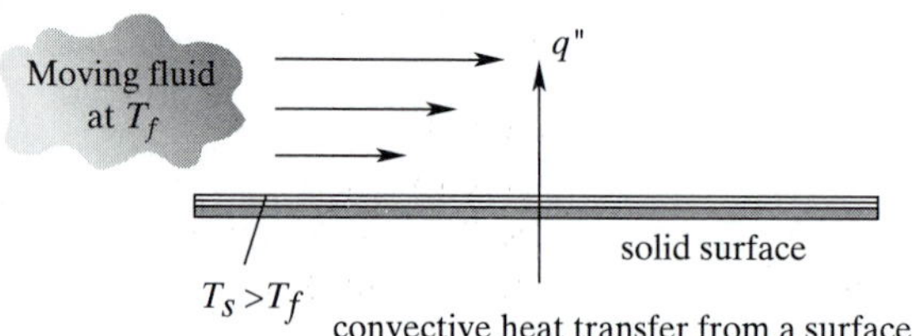


**FIGURE 7.2** Various modes of heat transfer.

convection modes, heat transfer by radiation can occur in a vacuum, and because all objects emit thermal radiation, it is the net energy exchange among the bodies that is of interest to us. The three modes of heat transfer are depicted in Figure 7.2.

In Chapter 1, it was explained that engineering problems are mathematical models of physical situations. Moreover, these mathematical models are differential equations that are derived by applying the fundamental laws and principles of nature to a system or a control volume. In heat transfer problems, these governing equations represent the balance of mass, momentum, and energy for a medium. Chapter 1 stated that when possible, the exact solutions of the governing differential equations should be sought because the exact solutions render the detailed behavior of a system. However, for many practical engineering problems, it is impossible to obtain exact solutions to the governing equations because either the geometry is too complex or the boundary conditions are too complicated.

The principle of the conservation of energy plays a significant role in the analysis of heat transfer problems. Consequently, you need to understand this principle fully in order to model a physical problem correctly. The principle of the conservation of energy states the following: The rate at which thermal and/or mechanical energy enters a system through its boundaries, minus the rate at which the energy leaves the system through its boundaries, plus the rate of energy generation within the volume of the system, must equal the rate at which energy is stored within the volume of the system. This statement is represented by Figure 7.3 and the equation

$$\dot{E}_{in} - \dot{E}_{out} + \dot{E}_{generation} = \dot{E}_{stored} \tag{7.7}$$

$\dot{E}_{in}$ and $\dot{E}_{out}$ represent the amount of energy crossing into and out of the surfaces of a system. The thermal energy generation rate $\dot{E}_{generation}$ represents the rate of the conversion of energy from electrical, chemical, nuclear, or electromagnetic forms to thermal en-

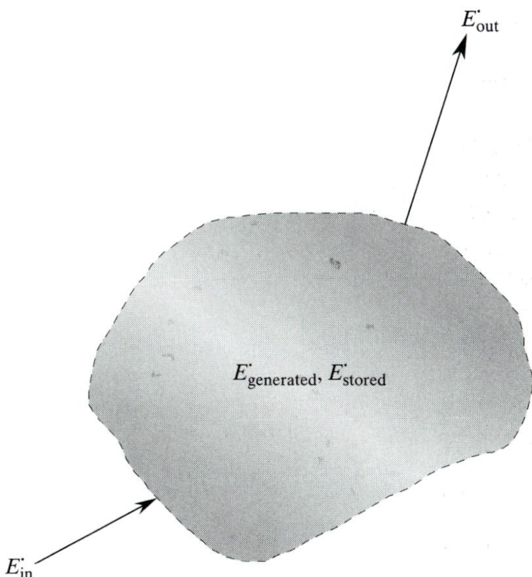


**FIGURE 7.3** The principle of the conservation of energy.

ergy within the volume of the system. An example of such conversion is the electric current running through a solid conductor. On the other hand, the energy storage term represents the increase or decrease in the amount of thermal internal energy within the volume of the system due to transient processes. It is important to understand the contribution of each term to the overall energy balance of a system in order to model an actual situation properly. A good understanding of the principle of the conservation of energy will also assist in the verification of the results of a model.

This chapter focuses on the conduction mode of heat transfer with possible convective or radiative boundary conditions. For now, we will focus on steady-state two-dimensional conduction problems. Applying the principle of the conservation of energy to a system represented in a Cartesian coordinate system results in the following heat diffusion equation:

$$k_X \frac{\partial^2 T}{\partial X^2} + k_Y \frac{\partial^2 T}{\partial Y^2} + \dot{q} = 0 \tag{7.8}$$

The derivation of Eq. (7.8) is shown in Figure 7.4. In Eq. (7.8), $\dot{q}$ represents the heat generation per unit volume, within a volume having a unit depth. There are several boundary conditions that occur in conduction problems:

1. A situation wherein heat loss or gain through a surface may be neglected. This situation, shown in Figure 7.5, is commonly referred to as an adiabatic surface or a perfectly insulated surface. In conduction problems, symmetrical lines also represent adiabatic lines. This type of boundary condition is represented by

$$\left.\frac{\partial T}{\partial X}\right|_{(X=0,\,Y)} = 0 \tag{7.9}$$

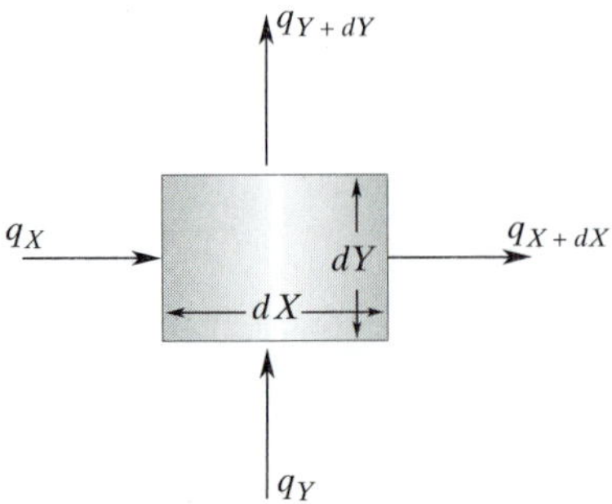


First, we begin by applying the principle of the conservation of energy to a small region (differential volume) in a medium:

$$\dot{E}_{in} - \dot{E}_{out} + \dot{E}_{generation} = \dot{E}_{stored}$$

$$q_X + q_Y - (q_{X+dX} + q_{Y+dY}) + \dot{q}\,dX dY(1) = 0$$

$$q_X + q_Y - \left( q_X + \frac{\partial q_X}{\partial X} dX + q_Y + \frac{\partial q_Y}{\partial Y} dY \right) + \dot{q}\,dX dY = 0$$

Simplifying, we get

$$-\frac{\partial q_X}{\partial X} dX - \frac{\partial q_Y}{\partial Y} dY + \dot{q}\,dX\,dY = 0$$

Making use of Fourier's Law, we have

$$q_X = -k_X A \frac{\partial T}{\partial X} = -k_X\, dY(1) \frac{\partial T}{\partial X}$$

$$q_Y = -k_Y A \frac{\partial T}{\partial X} = -k_Y\, dY(1) \frac{\partial T}{\partial X}$$

$$-\frac{\partial}{\partial X}\left(-k_X dY \frac{\partial T}{\partial X}\right) dX - \frac{\partial}{\partial Y}\left(-k_Y\, dX \frac{\partial T}{\partial X}\right) dY + \dot{q}\,dX\,dY = 0$$

and simplifying, we obtain

$$k_X \frac{\partial^2 T}{\partial X^2} + k_Y \frac{\partial^2 T}{\partial Y^2} + \dot{q} = 0$$

**FIGURE 7.4** The derivation of the equation of heat conduction under steady-state conditions.

2. A situation for which a constant heat flux is applied at a surface. This boundary condition, shown in Figure 7.6, is represented by the equation

$$-k \frac{\partial T}{\partial X}\bigg|_{X=0} = q_0'' \tag{7.10}$$

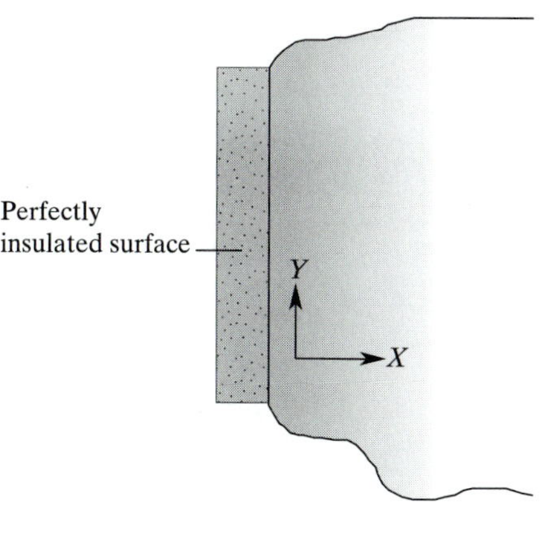


**FIGURE 7.5** An adiabatic, or perfectly insulated, surface.

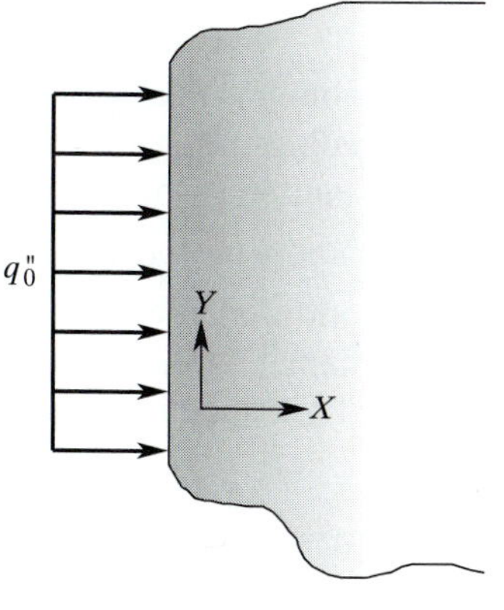


**FIGURE 7.6** A costant heat flux applied at a surface.

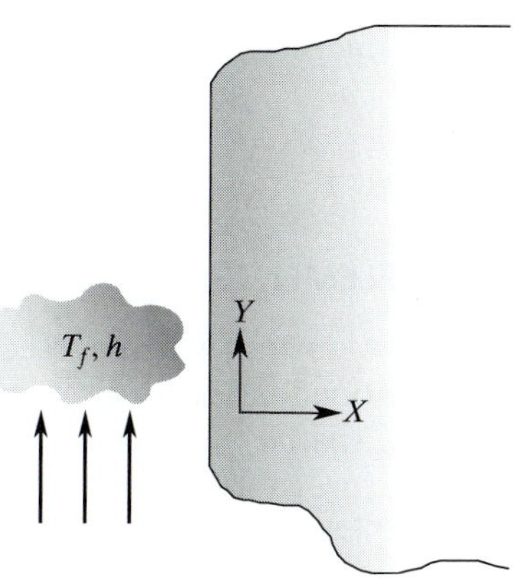


**FIGURE 7.7** Convection processes causing cooling or heating to take place at a surface.

**3.** A situation for which cooling or heating is taking place at a surface due to convection processes. This situation, shown in Figure 7.7, is represented by the equation

$$-k \left. \frac{\partial T}{\partial X} \right|_{(X=0,\,Y)} = h\left[T(0, y) - T_f\right] \tag{7.11}$$

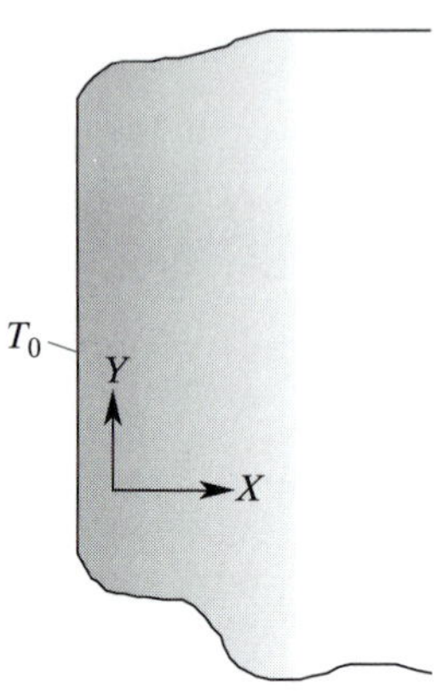


**FIGURE 7.8** Constant surface-temperature conditions occur due to phase change of a fluid in contact with a solid surface.

4. A situation wherein heating or cooling is taking place at a surface due to net radiation exchange with the surroundings. The expression for this condition will depend on the view factors and the emissivity of the surfaces involved.
5. A situation in which conditions 3 and 4 both exist simultaneously.
6. Constant surface-temperature conditions occur when a fluid in contact with a solid surface experiences phase change, as shown in Figure 7.8. Examples include condensation or evaporation of a fluid at constant pressure. This condition is represented by

$$T(0, Y) = T_0 \tag{7.12}$$

The modeling of actual situations with these boundary conditions will be discussed and illustrated with examples after we consider finite element formulations of two-dimensional heat conduction problems.

## 7.2 FORMULATION WITH RECTANGULAR ELEMENTS

Two-dimensional bilinear rectangular elements were covered in detail in Chapter 5. Recall that for problems with straight boundaries, linear rectangular shape functions offer simple means to approximate the spatial variation of a dependent variable, such as temperature. For convenience, the expression for a rectangular element in terms of its nodal temperatures and shape functions is repeated here (also see Figure 7.9). The expression is

$$T^{(e)} = \begin{bmatrix} S_i & S_j & S_m & S_n \end{bmatrix} \begin{Bmatrix} T_i \\ T_j \\ T_m \\ T_n \end{Bmatrix} \tag{7.13}$$

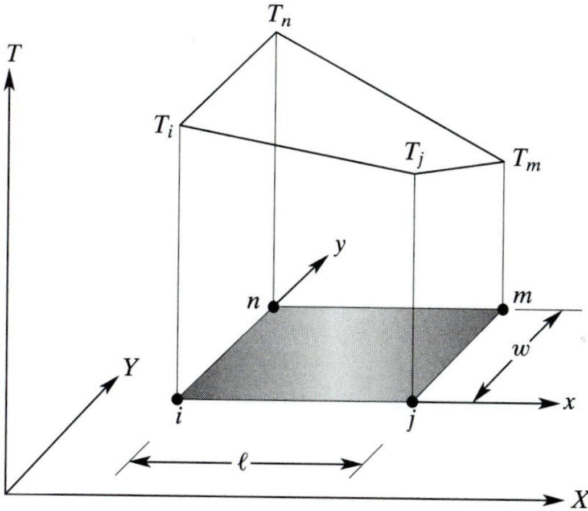


**FIGURE 7.9** A typical rectangular element.

where the shape functions $S_i, S_j, S_m$, and $S_n$ are given by:

$$S_i = \left(1 - \frac{x}{\ell}\right)\left(1 - \frac{y}{w}\right) \tag{7.14}$$

$$S_j = \frac{x}{\ell}\left(1 - \frac{y}{w}\right)$$

$$S_m = \frac{xy}{\ell w}$$

$$S_n = \frac{y}{w}\left(1 - \frac{x}{\ell}\right)$$

We now apply the Galerkin approach to the heat diffusion equation, Eq. (7.8) expressed in local coordinates $x, y$, yielding four residual equations:

$$\int_A S_i\left(k_x \frac{\partial^2 T}{\partial x^2} + k_y \frac{\partial^2 T}{\partial y^2} + q^{\cdot}\right) dA = 0 \tag{7.15}$$

$$\int_A S_j\left(k_x \frac{\partial^2 T}{\partial x^2} + k_y \frac{\partial^2 T}{\partial y^2} + q^{\cdot}\right) dA = 0$$

$$\int_A S_m\left(k_x \frac{\partial^2 T}{\partial x^2} + k_y \frac{\partial^2 T}{\partial y^2} + q^{\cdot}\right) dA = 0$$

$$\int_A S_n\left(k_x \frac{\partial^2 T}{\partial x^2} + k_y \frac{\partial^2 T}{\partial y^2} + q^{\cdot}\right) dA = 0$$

We can rewrite the four equations given by (7.15) in a compact matrix form as

$$\int_A [\mathbf{S}]^T\left(k_x \frac{\partial^2 T}{\partial x^2} + k_y \frac{\partial^2 T}{\partial y^2} + q^{\cdot}\right) dA = 0 \tag{7.16}$$

where the transpose of the shape functions is given by the following matrix:

$$[\mathbf{S}]^T = \begin{Bmatrix} S_i \\ S_j \\ S_m \\ S_n \end{Bmatrix} \tag{7.17}$$

Equation (7.16) consists of three main integrals:

$$\int_A [\mathbf{S}]^T\left(k_x\frac{\partial^2 T}{\partial x^2}\right)dA + \int_A [\mathbf{S}]^T\left(k_y\frac{\partial^2 T}{\partial y^2}\right)dA + \int_A [\mathbf{S}]^T q^{\cdot}\,dA = 0 \tag{7.18}$$

Let $C_1 = k_x$, $C_2 = k_y$, and $C_3 = q^{\cdot}$ so that we can later apply the results of the forthcoming derivation to other types of problems with similar forms of governing differential equations. As will be demonstrated later in Chapters 8 and 9, we will use the general findings of this chapter to analyze the torsion of solid members and ideal fluid flow problems. So making respective substitutions, we have

$$\int_A [\mathbf{S}]^T\left(C_1\frac{\partial^2 T}{\partial x^2}\right)dA + \int_A [\mathbf{S}]^T\left(C_2\frac{\partial^2 T}{\partial y^2}\right)dA + \int_A [\mathbf{S}]^T C_3\,dA = 0 \tag{7.19}$$

Evaluation of the integrals given by Eq. (7.19) will result in the elemental formulation. As was discussed in Chapter 4, the second derivative of a linear function assumed for temperature is equal to zero; therefore, we need to manipulate the second-order terms into first-order terms by using the chain rule in the following manner:

$$\frac{\partial}{\partial x}\left([\mathbf{S}]^T\frac{\partial T}{\partial x}\right) = [\mathbf{S}]^T\frac{\partial^2 T}{\partial x^2} + \frac{\partial[\mathbf{S}]^T}{\partial x}\frac{\partial T}{\partial x} \tag{7.20}$$

Rearranging Eq. (7.20), we have

$$[\mathbf{S}]^T\frac{\partial^2 T}{\partial x^2} = \frac{\partial}{\partial x}\left([\mathbf{S}]^T\frac{\partial T}{\partial x}\right) - \frac{\partial[\mathbf{S}]^T}{\partial x}\frac{\partial T}{\partial x} \tag{7.21}$$

Applying the results given by Eq. (7.21) to the first and the second terms in Eq. (7.19), we obtain

$$\int_A [\mathbf{S}]^T\left(C_1\frac{\partial^2 T}{\partial x^2}\right)dA = \int_A C_1\frac{\partial}{\partial x}\left([\mathbf{S}]^T\frac{\partial T}{\partial x}\right)dA - \int_A C_1\left(\frac{\partial[\mathbf{S}]^T}{\partial x}\frac{\partial T}{\partial x}\right)dA \tag{7.22}$$

$$\int_A [\mathbf{S}]^T\left(C_2\frac{\partial^2 T}{\partial y^2}\right)dA = \int_A C_2\frac{\partial}{\partial y}\left([\mathbf{S}]^T\frac{\partial T}{\partial y}\right)dA - \int_A C_2\left(\frac{\partial[\mathbf{S}]^T}{\partial y}\frac{\partial T}{\partial y}\right)dA \tag{7.23}$$

Using Green's Theorem, we can write the terms

$$\int_A C_1\frac{\partial}{\partial x}\left([\mathbf{S}]^T\frac{\partial T}{\partial x}\right)dA$$

and

$$\int_A C_2\frac{\partial}{\partial y}\left([\mathbf{S}]^T\frac{\partial T}{\partial y}\right)dA$$

in terms of integrals around the element boundary. We will come back to these terms later. For now, let us consider the

$$-\int_A C_1\left(\frac{\partial[\mathbf{S}]^T}{\partial x}\frac{\partial T}{\partial x}\right)dA$$

term in Eq. (7.22). This term can easily be evaluated. Evaluating the derivatives for a rectangular element, we obtain

$$\frac{\partial T}{\partial x}=\frac{\partial}{\partial x}[S_i \quad S_j \quad S_m \quad S_n]\begin{Bmatrix}T_i\\T_j\\T_m\\T_n\end{Bmatrix}=\frac{1}{\ell w}[(-w+y) \quad (w-y) \quad y \quad -y]\begin{Bmatrix}T_i\\T_j\\T_m\\T_n\end{Bmatrix} \tag{7.24}$$

Also evaluating $\frac{\partial[\mathbf{S}]^T}{\partial x}$ we have

$$\frac{\partial[\mathbf{S}]^T}{\partial x}=\frac{\partial}{\partial x}\begin{Bmatrix}S_i\\S_j\\S_m\\S_n\end{Bmatrix}=\frac{1}{\ell w}\begin{Bmatrix}-w+y\\w-y\\y\\-y\end{Bmatrix} \tag{7.25}$$

Substituting the results of Eqs. (7.24) and (7.25) into the term

$$-\int_A C_1\left(\frac{\partial[\mathbf{S}]^T}{\partial x}\frac{\partial T}{\partial x}\right)dA$$

we have

$$-\int_A C_1\left(\frac{\partial[\mathbf{S}]^T}{\partial x}\frac{\partial T}{\partial x}\right)dA=-C_1\int_A\frac{1}{(\ell w)^2}\begin{Bmatrix}-w+y\\w-y\\y\\-y\end{Bmatrix}$$

$$[(-w+y) \quad (w-y) \quad y \quad -y]\begin{Bmatrix}T_i\\T_j\\T_m\\T_n\end{Bmatrix}dA \tag{7.26}$$

Integrating yields

$$-C_1\int_A\frac{1}{(\ell w)^2}\begin{Bmatrix}-w+y\\w-y\\y\\-y\end{Bmatrix}[(-w+y) \quad (w-y) \quad y \quad -y]\begin{Bmatrix}T_i\\T_j\\T_m\\T_n\end{Bmatrix}dA$$

$$=-\frac{C_1 w}{6\ell}\begin{bmatrix}2&-2&-1&1\\-2&2&1&-1\\-1&1&2&-2\\1&-1&-2&2\end{bmatrix}\begin{Bmatrix}T_i\\T_j\\T_m\\T_n\end{Bmatrix} \tag{7.27}$$

In the same manner, we can evaluate the term

$$-\int_A C_2\left(\frac{\partial[\mathbf{S}]^T}{\partial y}\frac{\partial T}{\partial y}\right)dA$$

in Eq. (7.23) in the $y$-direction:

$$\frac{\partial T}{\partial y}=\frac{\partial}{\partial y}[S_i \quad S_j \quad S_m \quad S_n]\begin{Bmatrix}T_i\\T_j\\T_m\\T_n\end{Bmatrix}=\frac{1}{\ell w}[(-\ell+x) \quad -x \quad x \quad (\ell-x)]\begin{Bmatrix}T_i\\T_j\\T_m\\T_n\end{Bmatrix} \tag{7.28}$$

And evaluating $\frac{\partial[\mathbf{S}]^T}{\partial y}$ we have

$$\frac{\partial[\mathbf{S}]^T}{\partial x}=\frac{\partial}{\partial y}\begin{Bmatrix}S_i\\S_j\\S_m\\S_n\end{Bmatrix}=\frac{1}{\ell w}\begin{Bmatrix}-\ell+x\\-x\\x\\\ell-x\end{Bmatrix} \tag{7.29}$$

Substituting the results of Eqs. (7.28) and (7.29) into the term

$$-\int_A C_2\left(\frac{\partial[\mathbf{S}]^T}{\partial y}\frac{\partial T}{\partial y}\right)dA$$

we have:

$$-\int_A C_2\left(\frac{\partial[\mathbf{S}]^T}{\partial y}\frac{\partial T}{\partial y}\right)dA=-C_2\int_A\frac{1}{(\ell w)^2}\begin{Bmatrix}-\ell+x\\-x\\x\\\ell-x\end{Bmatrix}[(-\ell+x) \quad -x \quad x \quad (\ell-x)]\begin{Bmatrix}T_i\\T_j\\T_m\\T_n\end{Bmatrix}dA \tag{7.30}$$

Evaluation of the integral yields

$$-C_2\int_A\frac{1}{(\ell w)^2}\begin{Bmatrix}-\ell+x\\-x\\x\\\ell-x\end{Bmatrix}[(-\ell+x) \quad -x \quad x \quad (\ell-x)]\begin{Bmatrix}T_i\\T_j\\T_m\\T_n\end{Bmatrix}dA=$$

$$-\frac{C_2\ell}{6w}\begin{bmatrix}2&1&-1&-2\\1&2&-2&-1\\-1&-2&2&1\\-2&-1&1&2\end{bmatrix}\begin{Bmatrix}T_i\\T_j\\T_m\\T_n\end{Bmatrix} \tag{7.31}$$

Next, we will evaluate the thermal load term $\int_A [\mathbf{S}]^T C_3\, dA$:

$$\int_A [\mathbf{S}]^T C_3\, dA = C_3 \int_A \begin{Bmatrix} S_i \\ S_j \\ S_m \\ S_n \end{Bmatrix} dA = \frac{C_3 A}{4} \begin{Bmatrix} 1 \\ 1 \\ 1 \\ 1 \end{Bmatrix} \tag{7.32}$$

We now return to the terms

$$\int_A C_1 \frac{\partial}{\partial x}\left( [\mathbf{S}]^T \frac{\partial T}{\partial x} \right) dA$$

and

$$\int_A C_2 \frac{\partial}{\partial y}\left( [\mathbf{S}]^T \frac{\partial T}{\partial y} \right) dA$$

As mentioned earlier, we can use Green's Theorem to rewrite these area integrals in terms of line integrals around the element boundary:

$$\int_A C_1 \frac{\partial}{\partial x}\left( [\mathbf{S}]^T \frac{\partial T}{\partial x} \right) dA = \int_\tau C_1 [\mathbf{S}]^T \frac{\partial T}{\partial x} \cos\theta\, d\tau \tag{7.33}$$

$$\int_A C_2 \frac{\partial}{\partial y}\left( [\mathbf{S}]^T \frac{\partial T}{\partial y} \right) dA = \int_\tau C_2 [\mathbf{S}]^T \frac{\partial T}{\partial y} \sin\theta\, d\tau \tag{7.34}$$

$\tau$ represents the element boundary, and $\theta$ measures the angle to the unit normal. Equations (7.33) and (7.34) contribute to the derivative boundary conditions. To understand what is meant by derivative boundary conditions, consider an element with a convection boundary condition, as shown in Figure 7.10.

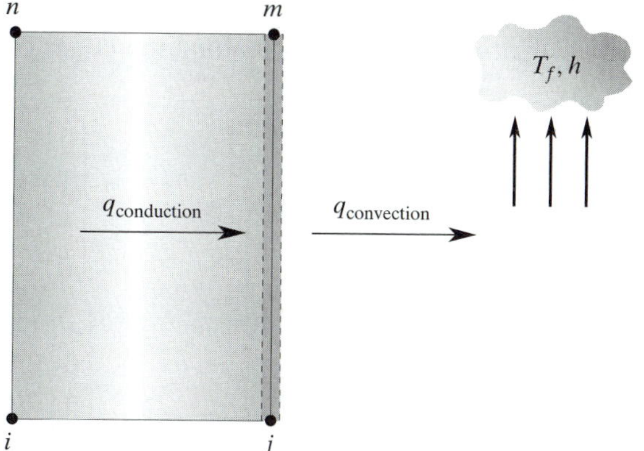


**FIGURE 7.10** A rectangular element with a convective boundary condition.

Neglecting radiation, the application of the conservation of energy in the $x$-direction to the $jm$ edge requires that the energy that reaches the $jm$ edge through conduction must be equal to the energy being convected away (by the fluid adjacent to the $jm$ edge). So,

$$-k\frac{\partial T}{\partial x} = h(T - T_f) \tag{7.35}$$

Substituting the right-hand side of Eq. (7.35) into Eq. (7.33), we get

$$\int_\tau C_1[\mathbf{S}]^T \frac{\partial T}{\partial x}\cos\theta d\tau = \int_\tau k[\mathbf{S}]^T \frac{\partial T}{\partial x}\cos\theta d\tau = -\int_\tau h[\mathbf{S}]^T(T - T_f)\cos\theta d\tau \tag{7.36}$$

The integral given by Eq. (7.36) has two terms:

$$-\int_\tau h[\mathbf{S}]^T(T - T_f)\cos\theta d\tau = -\int_\tau h[\mathbf{S}]^T T\cos\theta d\tau + \int_\tau h[\mathbf{S}]^T T_f\cos\theta d\tau \tag{7.37}$$

The terms $\int_\tau h[\mathbf{S}]^T T\cos\theta d\tau$ and $\int_\tau h[\mathbf{S}]^T T\sin\theta d\tau$, for convective boundary conditions along different edges of the rectangular element, contribute to conductance matrix:

$$[\mathbf{K}]^{(e)} = \frac{h\ell_{ij}}{6}\begin{bmatrix} 2 & 1 & 0 & 0 \\ 1 & 2 & 0 & 0 \\ 0 & 0 & 0 & 0 \\ 0 & 0 & 0 & 0 \end{bmatrix} \tag{7.38}$$

$$[\mathbf{K}]^{(e)} = \frac{h\ell_{jm}}{6}\begin{bmatrix} 0 & 0 & 0 & 0 \\ 0 & 2 & 1 & 0 \\ 0 & 1 & 2 & 0 \\ 0 & 0 & 0 & 0 \end{bmatrix} \tag{7.39}$$

$$[\mathbf{K}]^{(e)} = \frac{h\ell_{mn}}{6}\begin{bmatrix} 0 & 0 & 0 & 0 \\ 0 & 0 & 0 & 0 \\ 0 & 0 & 2 & 1 \\ 0 & 0 & 1 & 2 \end{bmatrix} \tag{7.40}$$

$$[\mathbf{K}]^{(e)} = \frac{h\ell_{ni}}{6}\begin{bmatrix} 2 & 0 & 0 & 1 \\ 0 & 0 & 0 & 0 \\ 0 & 0 & 0 & 0 \\ 1 & 0 & 0 & 2 \end{bmatrix} \tag{7.41}$$

Referring to Figure 7.9, note that in the above matrices, $\ell_{ij} = \ell_{mn} = \ell$ and $\ell_{jm} = \ell_{in} = w$. The terms $\int_\tau h[\mathbf{S}]^T T_f\cos\theta d\tau$ and $\int_\tau h[\mathbf{S}]^T T_f\sin\theta d\tau$ contribute to the elemental thermal load matrix. Evaluating these integrals along the edges of the rectangular element, we obtain

$$\{\mathbf{F}\}^{(e)} = \frac{hT_f\ell_{ij}}{2}\begin{Bmatrix} 1 \\ 1 \\ 0 \\ 0 \end{Bmatrix} \tag{7.42}$$

$$\{\mathbf{F}\}^{(e)} = \frac{hT_f\ell_{jm}}{2}\begin{Bmatrix} 0 \\ 1 \\ 1 \\ 0 \end{Bmatrix} \tag{7.43}$$

$$\{\mathbf{F}\}^{(e)} = \frac{hT_f\ell_{mn}}{2}\begin{Bmatrix} 0 \\ 0 \\ 1 \\ 1 \end{Bmatrix} \tag{7.44}$$

$$\{\mathbf{F}\}^{(e)} = \frac{hT_f\ell_{ni}}{2}\begin{Bmatrix} 1 \\ 0 \\ 0 \\ 1 \end{Bmatrix} \tag{7.45}$$

Let us summarize what we have done so far. The conductance matrix for a bilinear rectangular element is given by:

$$[\mathbf{K}]^{(e)} = \frac{k_x w}{6\ell}\begin{bmatrix} 2 & -2 & -1 & 1 \\ -2 & 2 & 1 & -1 \\ -1 & 1 & 2 & -2 \\ 1 & -1 & -2 & 2 \end{bmatrix} + \frac{k_y \ell}{6w}\begin{bmatrix} 2 & 1 & -1 & -2 \\ 1 & 2 & -2 & -1 \\ -1 & -2 & 2 & 1 \\ -2 & -1 & 1 & 2 \end{bmatrix}$$

Note that the elemental conductance matrix is composed of: (1) a conduction component in the $x$-direction; (2) a conduction component in the $y$-direction; and (3) a possible heat loss term by convection around the edge of a given element, as given by Eqs. (7.38)–(7.41). The load matrix for an element could have two components: (1) a component due to possible heat generation within a given element, and (2) a component due to possible convection heat loss along an element's edge(s), as given by Eqs. (7.42)–(7.45). The contribution of the heat generation to the element's thermal-load matrix is given by:

$$\{\mathbf{F}\}^{(e)} = \frac{q^{\cdot} A}{4}\begin{Bmatrix} 1 \\ 1 \\ 1 \\ 1 \end{Bmatrix}$$

It is worth noting here that in situations in which constant heat-flux boundary conditions occur along the edges of a rectangular element, the elemental load matrix is given by (see Problem 5):

$$\{\mathbf{F}\}^{(e)} = \frac{q''_o \ell_{ij}}{2}\begin{Bmatrix} 1 \\ 1 \\ 0 \\ 0 \end{Bmatrix} \qquad \{\mathbf{F}\}^{(e)} = \frac{q''_o \ell_{jm}}{2}\begin{Bmatrix} 0 \\ 1 \\ 1 \\ 0 \end{Bmatrix}$$

$$\{\mathbf{F}\}^{(e)} = \frac{q''_o \ell_{mn}}{2}\begin{Bmatrix} 0 \\ 0 \\ 1 \\ 1 \end{Bmatrix} \qquad \{\mathbf{F}\}^{(e)} = \frac{q''_o \ell_{ni}}{2}\begin{Bmatrix} 1 \\ 0 \\ 0 \\ 1 \end{Bmatrix}$$

The next step involves assembling elemental matrices to form the global matrices and solving the set of equations $[\mathbf{K}]\{\mathbf{T}\} = \{\mathbf{F}\}$ to obtain the nodal temperatures. We will demonstrate this step in Example 7.1. For now, let us turn our attention to the derivation of the elemental conductance and load matrices for a triangular element.

## 7.3 FORMULATION WITH TRIANGULAR ELEMENTS

As we discussed in Chapter 5, a major disadvantage associated with using rectangular elements is that they do not conform to curved boundaries. In contrast, triangular elements are better suited to approximate curved boundaries. For the sake of convenience, a triangular element is shown in Figure 7.11.

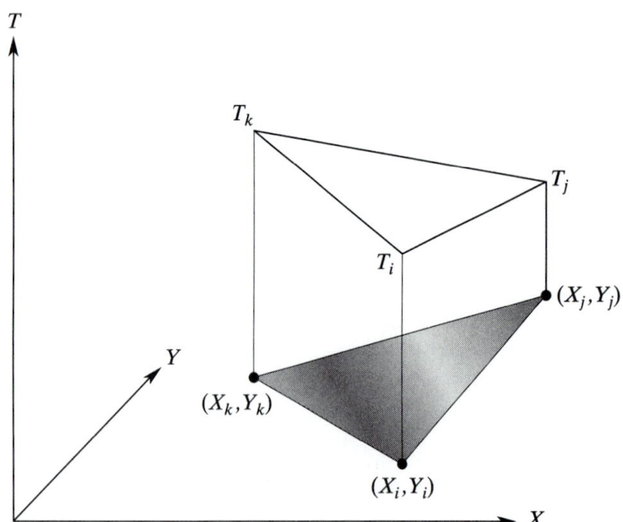


**FIGURE 7.11** A triangular element.

Recall that a triangular element is defined by three nodes and that we represent the variation of a dependent variable, such as temperature, over a triangular region using shape functions and the corresponding nodal temperatures by the equation

$$T^{(e)} = \begin{bmatrix} S_i & S_j & S_k \end{bmatrix} \begin{Bmatrix} T_i \\ T_j \\ T_k \end{Bmatrix} \tag{7.46}$$

where the shape functions $S_i$, $S_j$, and $S_k$ are

$$S_i = \frac{1}{2A}(\alpha_i + \beta_i X + \delta_i Y)$$

$$S_j = \frac{1}{2A}(\alpha_j + \beta_j X + \delta_j Y)$$

$$S_k = \frac{1}{2A}(\alpha_k + \beta_k X + \delta_k Y)$$

$A$ is the area of the element and is computed from the equation

$$2A = X_i(Y_j - Y_k) + X_j(Y_k - Y_i) + X_k(Y_i - Y_j)$$

Also,

$$\begin{aligned}
\alpha_i &= X_j Y_k - X_k Y_j & \beta_i &= Y_j - Y_k & \delta_i &= X_k - X_j \\
\alpha_j &= X_k Y_i - X_i Y_k & \beta_j &= Y_k - Y_i & \delta_j &= X_i - X_k \\
\alpha_k &= X_i Y_j - X_j Y_i & \beta_k &= Y_i - Y_j & \delta_k &= X_j - X_i
\end{aligned} \tag{7.47}$$

Employing the Galerkin approach, the three residual equations for a triangular element, in matrix form, are given by

$$\int_A [\mathbf{S}]^T\left(k_X\frac{\partial^2 T}{\partial X^2} + k_Y\frac{\partial^2 T}{\partial Y^2} + q^{\cdot}\right) dA = 0 \tag{7.48}$$

where

$$[\mathbf{S}]^T = \begin{Bmatrix} S_i \\ S_j \\ S_m \end{Bmatrix}$$

We now will proceed with steps similar to the ones we followed to formulate the conductance and thermal load matrices for rectangular elements. First, we rewrite the second-derivative expressions in terms of the first-derivative expressions using the chain rule. Evaluating the integral

$$-\int_A C_1\left(\frac{\partial[\mathbf{S}]^T}{\partial X}\frac{\partial T}{\partial X}\right) dA$$

for a triangular element, we obtain

$$\frac{\partial[\mathbf{S}]^T}{\partial X} = \frac{\partial}{\partial X}\begin{Bmatrix} S_i \\ S_j \\ S_k \end{Bmatrix} = \frac{1}{2A}\begin{Bmatrix} \beta_i \\ \beta_j \\ \beta_k \end{Bmatrix} \tag{7.49}$$

$$\frac{\partial T}{\partial X} = \frac{\partial}{\partial X}\begin{bmatrix} S_i & S_j & S_k \end{bmatrix}\begin{Bmatrix} T_i \\ T_j \\ T_k \end{Bmatrix} = \frac{1}{2A}\begin{bmatrix} \beta_i & \beta_j & \beta_k \end{bmatrix}\begin{Bmatrix} T_i \\ T_j \\ T_k \end{Bmatrix} \tag{7.50}$$

Substituting for the derivatives, we get

$$-\int_A C_1\left(\frac{\partial[\mathbf{S}]^T}{\partial X}\frac{\partial T}{\partial X}\right) dA = -C_1\int_A \frac{1}{4A^2}\begin{Bmatrix} \beta_i \\ \beta_j \\ \beta_k \end{Bmatrix}\begin{bmatrix} \beta_i & \beta_j & \beta_k \end{bmatrix}\begin{Bmatrix} T_i \\ T_j \\ T_k \end{Bmatrix} dA \tag{7.51}$$

and integrating, we are left with

$$-C_1\int_A \frac{1}{4A^2}\begin{Bmatrix} \beta_i \\ \beta_j \\ \beta_k \end{Bmatrix}\begin{bmatrix} \beta_i & \beta_j & \beta_k \end{bmatrix}\begin{Bmatrix} T_i \\ T_j \\ T_k \end{Bmatrix} dA = -\frac{C_1}{4A}\begin{bmatrix} \beta_i^2 & \beta_i\beta_j & \beta_i\beta_k \\ \beta_i\beta_j & \beta_j^2 & \beta_j\beta_k \\ \beta_i\beta_k & \beta_j\beta_k & \beta_k^2 \end{bmatrix}\begin{Bmatrix} T_i \\ T_j \\ T_k \end{Bmatrix} \tag{7.52}$$

In the same manner, we can evaluate the term

$$-\int_A C_2\left(\frac{\partial[\mathbf{S}]^T}{\partial Y}\frac{\partial T}{\partial Y}\right) dA$$

as

$$\frac{\partial[\mathbf{S}]^T}{\partial Y} = \frac{\partial}{\partial Y}\begin{Bmatrix} S_i \\ S_j \\ S_k \end{Bmatrix} = \frac{1}{2A}\begin{Bmatrix} \delta_i \\ \delta_j \\ \delta_k \end{Bmatrix} \tag{7.53}$$

$$\frac{\partial T}{\partial Y} = \frac{\partial}{\partial Y}\begin{bmatrix} S_i & S_j & S_k \end{bmatrix}\begin{Bmatrix} T_i \\ T_j \\ T_k \end{Bmatrix} = \frac{1}{2A}\begin{bmatrix} \delta_i & \delta_j & \delta_k \end{bmatrix}\begin{Bmatrix} T_i \\ T_j \\ T_k \end{Bmatrix} \tag{7.54}$$

Substituting for the derivatives and integrating, we have

$$-C_2\int_A \frac{1}{4A^2}\begin{Bmatrix} \delta_i \\ \delta_j \\ \delta_k \end{Bmatrix}\begin{bmatrix} \delta_i & \delta_j & \delta_k \end{bmatrix}\begin{Bmatrix} T_i \\ T_j \\ T_k \end{Bmatrix} dA = -\frac{C_2}{4A}\begin{bmatrix} \delta_i^2 & \delta_i\delta_j & \delta_i\delta_k \\ \delta_i\delta_j & \delta_j^2 & \delta_j\delta_k \\ \delta_i\delta_k & \delta_j\delta_k & \delta_k^2 \end{bmatrix}\begin{Bmatrix} T_i \\ T_j \\ T_k \end{Bmatrix} \tag{7.55}$$

For a triangular element, the thermal load matrix due to the heat generation term $C_3$ is

$$\int_A [\mathbf{S}]^T C_3\, dA = C_3\int_A \begin{Bmatrix} S_i \\ S_j \\ S_k \end{Bmatrix} dA = \frac{C_3 A}{3}\begin{Bmatrix} 1 \\ 1 \\ 1 \end{Bmatrix} \tag{7.56}$$

Evaluating the terms $\int_\tau h[\mathbf{S}]^T T\cos\theta\, d\tau$ and $\int_\tau h[\mathbf{S}]^T T\sin\theta\, d\tau$ for a convective boundary condition along the edges of the triangular element results in the equations

$$[\mathbf{K}]^{(e)} = \frac{h\ell_{ij}}{6}\begin{bmatrix} 2 & 1 & 0 \\ 1 & 2 & 0 \\ 0 & 0 & 0 \end{bmatrix} \tag{7.57}$$

$$[\mathbf{K}]^{(e)} = \frac{h\ell_{jk}}{6}\begin{bmatrix} 0 & 0 & 0 \\ 0 & 2 & 1 \\ 0 & 1 & 2 \end{bmatrix} \tag{7.58}$$

$$[\mathbf{K}]^{(e)} = \frac{h\ell_{ki}}{6}\begin{bmatrix} 2 & 0 & 1 \\ 0 & 0 & 0 \\ 1 & 0 & 2 \end{bmatrix} \tag{7.59}$$

Note that in the above matrices, $\ell_{ij}$, $\ell_{jk}$, and $\ell_{ki}$ represent the respective lengths of the three sides of the triangular element. The terms $\int_\tau h[\mathbf{S}]^T T_f\cos\theta\, d\tau$ and $\int_\tau h[\mathbf{S}]^T T_f\sin\theta\, d\tau$ contribute to the elemental thermal loads. Evaluating these integrals along the edges of the triangular element yields

$$\{\mathbf{F}\}^{(e)} = \frac{hT_f\ell_{ij}}{2}\begin{Bmatrix} 1 \\ 1 \\ 0 \end{Bmatrix} \tag{7.60}$$

$$\{\mathbf{F}\}^{(e)} = \frac{hT_f\ell_{jk}}{2}\begin{Bmatrix} 0 \\ 1 \\ 1 \end{Bmatrix} \tag{7.61}$$

$$\{\mathbf{F}\}^{(e)} = \frac{hT_f\ell_{ki}}{2}\begin{Bmatrix} 1 \\ 0 \\ 1 \end{Bmatrix} \tag{7.62}$$

Let us summarize the triangular formulation. The conductance matrix for a triangular element is

$$[\mathbf{K}]^{(e)} = \frac{k_X}{4A}\begin{bmatrix} \beta_i^2 & \beta_i\beta_j & \beta_i\beta_k \\ \beta_i\beta_j & \beta_j^2 & \beta_j\beta_k \\ \beta_i\beta_k & \beta_j\beta_k & \beta_k^2 \end{bmatrix} + \frac{k_Y}{4A}\begin{bmatrix} \delta_i^2 & \delta_i\delta_j & \delta_i\delta_k \\ \delta_i\delta_j & \delta_j^2 & \delta_j\delta_k \\ \delta_i\delta_k & \delta_j\delta_k & \delta_k^2 \end{bmatrix}$$

Note once again that the elemental conductance matrix for a triangular element is composed of: (1) a conduction component in the $X$-direction; (2) a conduction component in the $Y$-direction; and (3) a possible heat loss term by convection from the edge(s) of a given element, as given by Eq. (7.57)–(7.59). The thermal load matrix for a triangular element could have two components: (1) a component resulting from a possible heat-generation term within a given element, and (2) a component due to possible convection heat loss from the element's edge(s), as given by Eq. (7.60)–(7.62). The contribution of the heat generation to the element's load matrix is

$$\{\mathbf{F}\}^{(e)} = \frac{q^{\cdot}A}{3}\begin{Bmatrix} 1 \\ 1 \\ 1 \end{Bmatrix}$$

The development of constant heat flux boundary conditions for triangular elements is left as an exercise. (See Problem 6.)

Next, we use an example to demonstrate how to assemble the elemental information to obtain the global conductance matrix and the global load matrix.

## EXAMPLE 7.1

Consider a small industrial chimney constructed from concrete with a thermal conductivity value of $k = 1.4\ \mathrm{W/m \cdot K}$, as shown in Figure 7.12. The inside surface temperature of the chimney is assumed to be uniform at 100°C. The exterior surface is exposed to the surrounding air, which is at 30°C, with a corresponding natural convection heat transfer coefficient of $h = 20\ \mathrm{W/m^2 \cdot K}$. Determine the temperature distribution within the concrete under steady-state conditions.

We can make use of the symmetry of the problem, as shown in Figure 7.12, and only analyze a section of chimney containing 1/8 of the area. The selected section of the chimney is divided into nine nodes with five elements. Elements (1), (2), and (3) are squares, while elements (4) and (5) are triangular elements. Consult Table 7.1 while following the solution.

TABLE 7.1 The relationship between the elements and their corresponding nodes

| Element | $i$ | $j$ | $m$ or $k$ | $n$ |
|---|---|---|---|---|
| (1) | 1 | 2 | 4 | 3 |
| (2) | 3 | 4 | 7 | 6 |
| (3) | 4 | 5 | 8 | 7 |
| (4) | 2 | 5 | 4 | |
| (5) | 5 | 9 | 8 | |

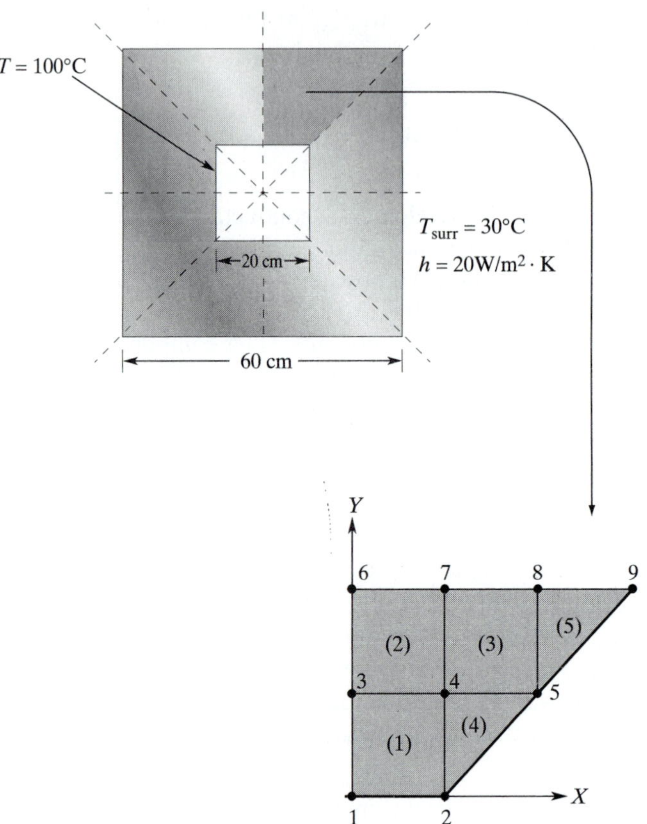


**FIGURE 7.12** A schematic of the chimney in Example 7.1.

The conductance matrix due to conduction in a rectangular element is given by:

$$[\mathbf{K}]^{(e)} = \frac{kw}{6\ell}\begin{bmatrix} 2 & -2 & -1 & 1 \\ -2 & 2 & 1 & -1 \\ -1 & 1 & 2 & -2 \\ 1 & -1 & -2 & 2 \end{bmatrix} + \frac{k\ell}{6w}\begin{bmatrix} 2 & 1 & -1 & -2 \\ 1 & 2 & -2 & -1 \\ -1 & -2 & 2 & 1 \\ -2 & -1 & 1 & 2 \end{bmatrix}$$

Elements (1), (2), and (3) all have the same dimensions; therefore,

$$[\mathbf{K}]^{(1)} = [\mathbf{K}]^{(2)} = [\mathbf{K}]^{(3)} = \frac{(1.4)(0.1)}{6(0.1)}\begin{bmatrix} 2 & -2 & -1 & 1 \\ -2 & 2 & 1 & -1 \\ -1 & 1 & 2 & -2 \\ 1 & -1 & -2 & 2 \end{bmatrix}$$

$$+ \frac{(1.4)(0.1)}{6(0.1)}\begin{bmatrix} 2 & 1 & -1 & -2 \\ 1 & 2 & -2 & -1 \\ -1 & -2 & 2 & 1 \\ -2 & -1 & 1 & 2 \end{bmatrix}$$

To help with assembly of the elements later, the corresponding node numbers for each element are shown on the top and the side of each matrix:

$$[\mathbf{K}]^{(1)} = \begin{array}{c} \begin{matrix} 1(i) & 2(j) & 4(m) & 3(n) \end{matrix} \\ \begin{bmatrix} 0.933 & -0.233 & -0.466 & -0.233 \\ -0.233 & 0.933 & -0.233 & -0.466 \\ -0.466 & -0.233 & 0.933 & -0.233 \\ -0.233 & -0.466 & -0.233 & 0.933 \end{bmatrix} \begin{matrix} 1 \\ 2 \\ 4 \\ 3 \end{matrix} \end{array}$$

$$[\mathbf{K}]^{(2)} = \begin{array}{c} \begin{matrix} 3(i) & 4(j) & 7(m) & 6(n) \end{matrix} \\ \begin{bmatrix} 0.933 & -0.233 & -0.466 & -0.233 \\ -0.233 & 0.933 & -0.233 & -0.466 \\ -0.466 & -0.233 & 0.933 & -0.233 \\ -0.233 & -0.466 & -0.233 & 0.933 \end{bmatrix} \begin{matrix} 3 \\ 4 \\ 7 \\ 6 \end{matrix} \end{array}$$

$$[\mathbf{K}]^{(3)} = \begin{array}{c} \begin{matrix} 4(i) & 5(j) & 8(m) & 7(n) \end{matrix} \\ \begin{bmatrix} 0.933 & -0.233 & -0.466 & -0.233 \\ -0.233 & 0.933 & -0.233 & -0.466 \\ -0.466 & -0.233 & 0.933 & -0.233 \\ -0.233 & -0.466 & -0.233 & 0.933 \end{bmatrix} \begin{matrix} 4 \\ 5 \\ 8 \\ 7 \end{matrix} \end{array}$$

For triangular elements (4) and (5), the conductance matrix is

$$[\mathbf{K}]^{(e)} = \frac{k}{4A}\begin{bmatrix} \beta_i^2 & \beta_i\beta_j & \beta_i\beta_k \\ \beta_i\beta_j & \beta_j^2 & \beta_j\beta_k \\ \beta_i\beta_k & \beta_j\beta_k & \beta_k^2 \end{bmatrix} + \frac{k}{4A}\begin{bmatrix} \delta_i^2 & \delta_i\delta_j & \delta_i\delta_k \\ \delta_i\delta_j & \delta_j^2 & \delta_j\delta_k \\ \delta_i\delta_k & \delta_j\delta_k & \delta_k^2 \end{bmatrix}$$

where the $\beta$- and $\delta$-terms are given by the relations of Eq. (7.47). Because the $\beta$- and $\delta$-terms are calculated from the difference of the coordinates of the involved nodes, it does not matter where we place the origin of the coordinate system $X, Y$. Evaluating the coefficients for element (4), we have

$$\beta_i = Y_j - Y_k = 0.1 - 0.1 = 0 \qquad \delta_i = X_k - X_j = 0 - 0.1 = -0.1$$

$$\beta_j = Y_k - Y_i = 0.1 - 0 = 0.1 \qquad \delta_j = X_i - X_k = 0 - 0 = 0$$

$$\beta_k = Y_i - Y_j = 0 - 0.1 = -0.1 \qquad \delta_k = X_j - X_i = 0.1 - 0 = 0.1$$

Evaluating the coefficients for element (5) renders the same results because the difference between the coordinates of its nodes is identical to that of element (4). Therefore, elements (4) and (5) will both have the following conductance matrix:

$$[\mathbf{K}]^{(4)} = [\mathbf{K}]^{(5)} = \frac{1.4}{4(0.005)}\begin{bmatrix} 0 & 0 & 0 \\ 0 & (0.1)^2 & (0.1)(-0.1) \\ 0 & (0.1)(-0.1) & (-0.1)^2 \end{bmatrix} + \frac{1.4}{4(0.005)}\begin{bmatrix} (-0.1)^2 & 0 & (-0.1)(0.1) \\ 0 & 0 & 0 \\ (-0.1)(0.1) & 0 & (0.1)^2 \end{bmatrix}$$

Showing the corresponding node numbers on the top and the side of each respective conductance matrix for elements (4) and (5), we obtain

$$[\mathbf{K}]^{(4)} = \begin{array}{c} \begin{array}{ccc} 2(i) & 5(j) & 4(k) \end{array} \\ \begin{bmatrix} 0.7 & 0 & -0.7 \\ 0 & 0.7 & -0.7 \\ -0.7 & -0.7 & 1.4 \end{bmatrix} \end{array} \begin{array}{c} 2 \\ 5 \\ 4 \end{array}$$

$$[\mathbf{K}]^{(5)} = \begin{array}{c} \begin{array}{ccc} 5(i) & 9(j) & 8(k) \end{array} \\ \begin{bmatrix} 0.7 & 0 & -0.7 \\ 0 & 0.7 & -0.7 \\ -0.7 & -0.7 & 1.4 \end{bmatrix} \end{array} \begin{array}{c} 5 \\ 9 \\ 8 \end{array}$$

As explained earlier, the convective boundary condition contributes to both the conductance matrix and the load matrix. The convective boundary condition contributes to the conductance matrices of elements (2) and (3) according to the relationship

$$[\mathbf{K}]^{(e)} = \frac{h\ell_{mn}}{6} \begin{bmatrix} 0 & 0 & 0 & 0 \\ 0 & 0 & 0 & 0 \\ 0 & 0 & 2 & 1 \\ 0 & 0 & 1 & 2 \end{bmatrix} \begin{matrix} i \\ j \\ m \\ n \end{matrix}$$

$$[\mathbf{K}]^{(2)} = [\mathbf{K}]^{(3)} = \frac{(20)(0.1)}{6} \begin{bmatrix} 0 & 0 & 0 & 0 \\ 0 & 0 & 0 & 0 \\ 0 & 0 & 2 & 1 \\ 0 & 0 & 1 & 2 \end{bmatrix} = \begin{bmatrix} 0 & 0 & 0 & 0 \\ 0 & 0 & 0 & 0 \\ 0 & 0 & 0.666 & 0.333 \\ 0 & 0 & 0.333 & 0.666 \end{bmatrix}$$

Including the nodal information, the conductance matrices for elements (2) and (3) are:

$$[\mathbf{K}]^{(2)} = \begin{array}{c} \begin{array}{cccc} 3 & 4 & 7 & 6 \end{array} \\ \begin{bmatrix} 0 & 0 & 0 & 0 \\ 0 & 0 & 0 & 0 \\ 0 & 0 & 0.666 & 0.333 \\ 0 & 0 & 0.333 & 0.666 \end{bmatrix} \end{array} \begin{array}{c} 3 \\ 4 \\ 7 \\ 6 \end{array}$$

$$[\mathbf{K}]^{(3)} = \begin{array}{c} \begin{array}{cccc} 4 & 5 & 8 & 7 \end{array} \\ \begin{bmatrix} 0 & 0 & 0 & 0 \\ 0 & 0 & 0 & 0 \\ 0 & 0 & 0.666 & 0.333 \\ 0 & 0 & 0.333 & 0.666 \end{bmatrix} \end{array} \begin{array}{c} 4 \\ 5 \\ 8 \\ 7 \end{array}$$

Heat loss by convection also occurs along *jk* edge of element (5); thus,

$$[\mathbf{K}]^{(e)} = \frac{h\ell_{jk}}{6}\begin{bmatrix} 0 & 0 & 0 \\ 0 & 2 & 1 \\ 0 & 1 & 2 \end{bmatrix}\begin{matrix} i \\ j \\ k \end{matrix}$$

$$[\mathbf{K}]^{(e)} = \frac{(20)(0.1)}{6}\begin{bmatrix} 0 & 0 & 0 \\ 0 & 2 & 1 \\ 0 & 1 & 2 \end{bmatrix} = \begin{bmatrix} 0 & 0 & 0 \\ 0 & 0.666 & 0.333 \\ 0 & 0.333 & 0.666 \end{bmatrix}$$

$$[\mathbf{K}]^{(5)} = \begin{matrix} 5 & 9 & 8 \end{matrix} \begin{bmatrix} 0 & 0 & 0 \\ 0 & 0.666 & 0.333 \\ 0 & 0.333 & 0.666 \end{bmatrix}\begin{matrix} 5 \\ 9 \\ 8 \end{matrix}$$

The convective boundary condition contributes to the thermal load matrices for elements (2) and (3) along their *mn* edge according to the relationship

$$\{\mathbf{F}\}^{(e)} = \frac{hT_f\ell_{mn}}{2}\begin{Bmatrix} 0 \\ 0 \\ 1 \\ 1 \end{Bmatrix} = \frac{(20)(30)(0.1)}{2}\begin{Bmatrix} 0 \\ 0 \\ 1 \\ 1 \end{Bmatrix} = \begin{Bmatrix} 0 \\ 0 \\ 30 \\ 30 \end{Bmatrix}$$

Including the nodal information, we have

$$\{\mathbf{F}\}^{(2)} = \begin{Bmatrix} 0 \\ 0 \\ 30 \\ 30 \end{Bmatrix}\begin{matrix} 3 \\ 4 \\ 7 \\ 6 \end{matrix}$$

$$\{\mathbf{F}\}^{(3)} = \begin{Bmatrix} 0 \\ 0 \\ 30 \\ 30 \end{Bmatrix}\begin{matrix} 4 \\ 5 \\ 8 \\ 7 \end{matrix}$$

The convective boundary condition contributes to the load matrix for element (5) along its *jk* edge according to the matrix

$$\{\mathbf{F}\}^{(e)} = \frac{hT_f\ell_{jk}}{2}\begin{Bmatrix} 0 \\ 1 \\ 1 \end{Bmatrix} = \frac{(20)(30)(0.1)}{2}\begin{Bmatrix} 0 \\ 1 \\ 1 \end{Bmatrix} = \begin{Bmatrix} 0 \\ 30 \\ 30 \end{Bmatrix}$$

Again, including the nodal information, we have

$$\{\mathbf{F}\}^{(5)} = \begin{Bmatrix} 0 \\ 30 \\ 30 \end{Bmatrix} \begin{matrix} 5 \\ 9 \\ 8 \end{matrix}$$

Next, we need to assemble all of the elemental matrices. Using the nodal information presented next to each element, the global conductance matrix becomes:

$$\begin{matrix} & \begin{matrix} 1 & \quad 2 & \quad 3 & \quad 4 & \quad 5 & \quad 6 & \quad 7 & \quad 8 & \quad 9 \end{matrix} \\ [\mathbf{K}]^{(G)} = & \begin{bmatrix} 0.933 & -0.233 & -0.233 & -0.466 & 0 & 0 & 0 & 0 & 0 \\ -0.233 & 1.633 & -0.466 & -0.933 & 0 & 0 & 0 & 0 & 0 \\ -0.233 & -0.466 & 1.866 & -0.466 & 0 & -0.233 & -0.466 & 0 & 0 \\ -0.466 & -0.933 & -0.466 & 4.199 & -0.933 & -0.466 & -0.466 & -0.466 & 0 \\ 0 & 0 & 0 & -0.933 & 2.333 & 0 & -0.466 & -0.933 & 0 \\ 0 & 0 & -0.233 & -0.466 & 0 & 1.599 & 0.1 & 0 & 0 \\ 0 & 0 & -0.466 & -0.466 & -0.466 & 0.1 & 3.198 & 0.1 & 0 \\ 0 & 0 & 0 & -0.466 & -0.933 & 0 & 0.1 & 3.665 & -0.367 \\ 0 & 0 & 0 & 0 & 0 & 0 & 0 & -0.367 & 1.366 \end{bmatrix} \begin{matrix} 1 \\ 2 \\ 3 \\ 4 \\ 5 \\ 6 \\ 7 \\ 8 \\ 9 \end{matrix} \end{matrix}$$

Applying the constant temperature boundary condition at nodes 1 and 2 results in the global matrix

$$[\mathbf{K}]^{(G)} = \begin{bmatrix} 1 & 0 & 0 & 0 & 0 & 0 & 0 & 0 & 0 \\ 0 & 1 & 0 & 0 & 0 & 0 & 0 & 0 & 0 \\ -0.233 & -0.466 & 1.866 & -0.466 & 0 & -0.233 & -0.466 & 0 & 0 \\ -0.466 & -0.933 & -0.466 & 4.199 & -0.933 & -0.466 & -0.466 & -0.466 & 0 \\ 0 & 0 & 0 & -0.933 & 2.333 & 0 & -0.466 & -0.933 & 0 \\ 0 & 0 & -0.233 & -0.466 & 0 & 1.599 & 0.1 & 0 & 0 \\ 0 & 0 & -0.466 & -0.466 & -0.466 & 0.1 & 3.198 & 0.1 & 0 \\ 0 & 0 & 0 & -0.466 & -0.933 & 0 & 0.1 & 3.665 & -0.367 \\ 0 & 0 & 0 & 0 & 0 & 0 & 0 & -0.367 & 1.366 \end{bmatrix}$$

Assembling the thermal load matrix, we have

$$\{\mathbf{F}\}^{(G)} = \begin{Bmatrix} 0 \\ 0 \\ 0 \\ 0 \\ 0 \\ 30 \\ 30 + 30 \\ 30 + 30 \\ 30 \end{Bmatrix}$$

and applying the constant temperature boundary condition at nodes 1 and 2 leads to the following final form of the thermal load matrix:

$$\{\mathbf{F}\}^{(G)} = \begin{Bmatrix} 100 \\ 100 \\ 0 \\ 0 \\ 0 \\ 30 \\ 60 \\ 60 \\ 30 \end{Bmatrix}$$

The final set of nodal equations is given by:

$$\begin{bmatrix} 1 & 0 & 0 & 0 & 0 & 0 & 0 & 0 & 0 \\ 0 & 1 & 0 & 0 & 0 & 0 & 0 & 0 & 0 \\ -0.233 & -0.466 & 1.866 & -0.466 & 0 & -0.233 & -0.466 & 0 & 0 \\ -0.466 & -0.933 & -0.466 & 4.199 & -0.933 & -0.466 & -0.466 & -0.466 & 0 \\ 0 & 0 & 0 & -0.933 & 2.333 & 0 & -0.466 & -0.933 & 0 \\ 0 & 0 & -0.233 & -0.466 & 0 & 1.599 & 0.1 & 0 & 0 \\ 0 & 0 & -0.466 & -0.466 & -0.466 & 0.1 & 3.198 & 0.1 & 0 \\ 0 & 0 & 0 & -0.466 & -0.933 & 0 & 0.1 & 3.665 & -0.367 \\ 0 & 0 & 0 & 0 & 0 & 0 & 0 & -0.367 & 1.366 \end{bmatrix}$$

$$\times \begin{Bmatrix} T_1 \\ T_2 \\ T_3 \\ T_4 \\ T_5 \\ T_6 \\ T_7 \\ T_8 \\ T_9 \end{Bmatrix} = \begin{Bmatrix} 100 \\ 100 \\ 0 \\ 0 \\ 0 \\ 30 \\ 60 \\ 60 \\ 30 \end{Bmatrix}$$

Solving the set of linear equations simultaneously leads to the following nodal solution:

$$[T]^T = [100 \quad 100 \quad 70.83 \quad 67.02 \quad 51.56 \quad 45.88 \quad 43.67 \quad 40.10 \quad 32.73]°\text{C}$$

To check for the accuracy of the results, first note that nodal temperatures are within the imposed boundary temperatures. Moreover, all temperatures at the outer edge are slightly above 30°C, with node 9 having the smallest value. This condition makes physical sense because node 9 is the outermost cornerpoint. As another check on the validity of the results, we can make sure that the conservation of energy, as applied to a control volume surrounding an arbitrary node, is

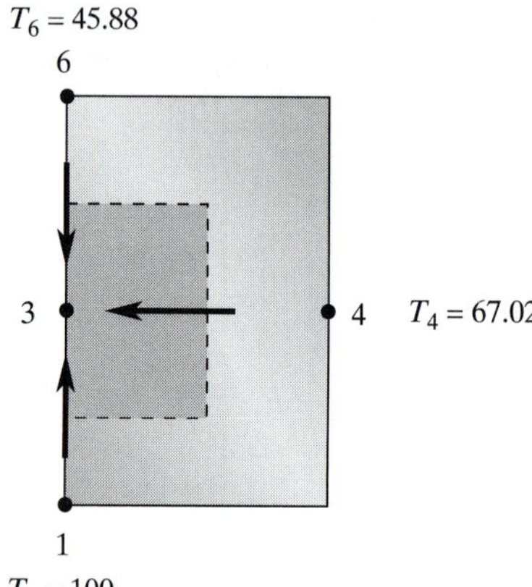


**FIGURE 7.13** Applying the principle of energy balance to node 3 to check the validity of our results.

satisfied. Are the energies flowing into and out of a node balanced out? As an example, let us consider node 3. Figure 7.13 shows the control volume surrounding node 3 used to apply the conservation of energy principle.

We start with the equation

$$\sum q = 0$$

and using Fourier's Law we have

$$k(0.1)\left(\frac{67.02 - T_3}{0.1}\right) + k(0.05)\left(\frac{45.88 - T_3}{0.1}\right) + k(0.05)\left(\frac{100 - T_3}{0.1}\right) = 0$$

Solving for $T_3$, we find that $T_3 = 69.98°C$. This value is reasonably close to the value of 70.83°C, particularly considering the coarseness of the element sizes. We will discuss the verification of results further with another example problem solved using ANSYS.

## 7.4 CONDUCTION ELEMENTS USED BY ANSYS

ANSYS offers many two-dimensional thermal-solid elements that are based on linear and quadratic quadrilateral and triangular shape functions:

**PLANE35** is a six-node triangular thermal-solid element. The element has one degree of freedom at each node—namely, temperature. Convection and heat fluxes may be input as surface loads at the element's faces. The output data for this element include nodal temperatures and other data, such as thermal gradients and thermal fluxes. This element is compatible with the eight-node PLANE77 element.

**PLANE55** is a four-node quadrilateral element used in modeling two-dimensional conduction heat transfer problems. The element has a single degree of freedom, which is temperature. Convection or heat fluxes may be input at the element's faces. Output data include nodal temperatures and element data, such as thermal gradient and thermal flux components.

**PLANE77** is an eight-node quadrilateral element used in modeling two-dimensional heat conduction problems. It is basically a higher order version of the two-dimensional four-node quadrilateral PLANE55 element. This element is better capable of modeling problems with curved boundaries. At each node, the element has a single degree of freedom, which is temperature. Output data include nodal temperatures and element data, such as thermal gradient and thermal flux components.

Keep in mind that although you generally achieve more accuracy of results with higher order elements, they require more computational time because numerical integration of elemental matrices is more involved.

## 7.5 AN EXAMPLE USING ANSYS

Consider a small chimney constructed from two different materials. The inner layer is constructed from concrete with a thermal conductivity $k = 0.07$ Btu/hr · in · °F. The outer layer of the chimney is constructed from bricks with a thermal conductivity value $k = 0.04$ Btu/hr · in · °F. The temperature of the hot gases on the inside surface of the chimney is assumed to be 140°F, with a convection heat transfer coefficient of 0.037 Btu/hr · in$^2$ · °F. The outside surface is exposed to the surrounding air, which is at 10°F, with a corresponding convection heat transfer coefficient $h = 0.012$ Btu/hr · in$^2$ · °F. The dimensions of the chimney are shown in Figure 7.14. Determine the temperature distribution within the concrete and within the brick layers under steady-state conditions. Also, plot the heat fluxes through each layer.

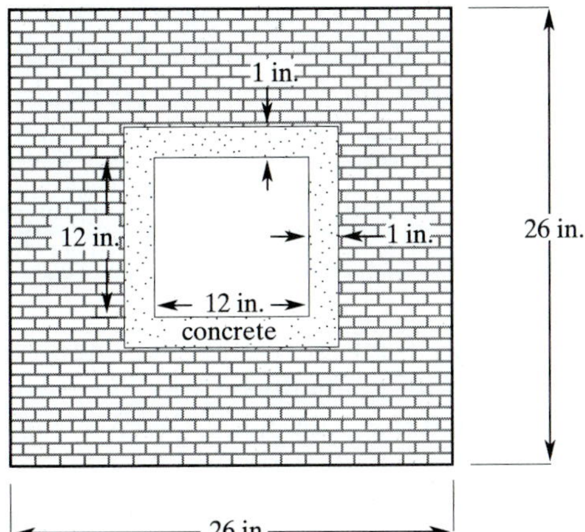


**FIGURE 7.14** A schematic of the chimney in the example problem of section 7.5.

The following steps demonstrate how to choose the appropriate element type, create the geometry of the problem, apply boundary conditions, and obtain nodal results for this problem using ANSYS.

Enter the **ANSYS** program by using the Launcher.

Type **xansys54** on the command line if you are running ANSYS on a UNIX platform, or consult your system administrator for information on how to run ANSYS on your computer system's platform.

Pick **Interactive** from the Launcher menu.

Type **Chimney** (or a file name of your choice) in the **Initial Jobname** entry field of the dialog box.

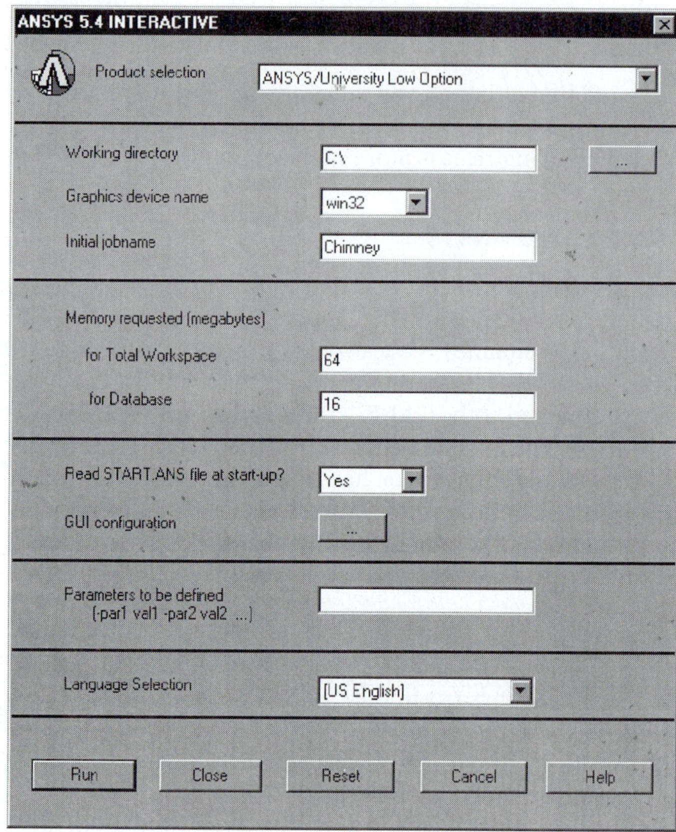


Pick **Run** to start the Graphic User Interface (GUI). A window will open with some disclaimer information. You will eventually be asked to press the <**Return**> key to start the graphics window and the main menu. Do so in order to proceed.

Create a title for the problem. This title will appear on ANSYS display windows to provide a simple way of identifying the displays. To create a title, issue the following command:

utility menu: **File** → **Change Title** ...

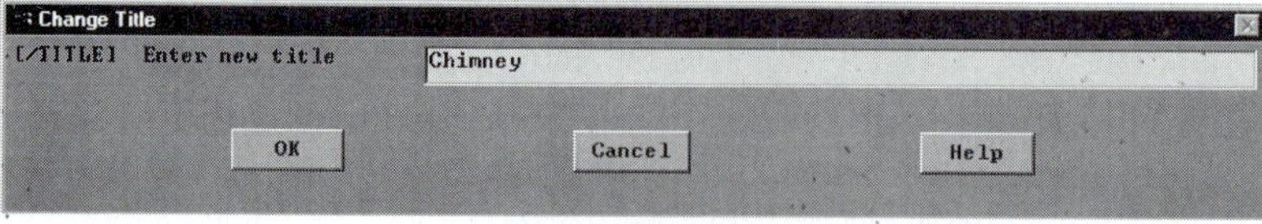

Define the element type and material properties with the commands

main menu: **Preprocessor** → **Element Type** → **Add/Edit/Delete** ...

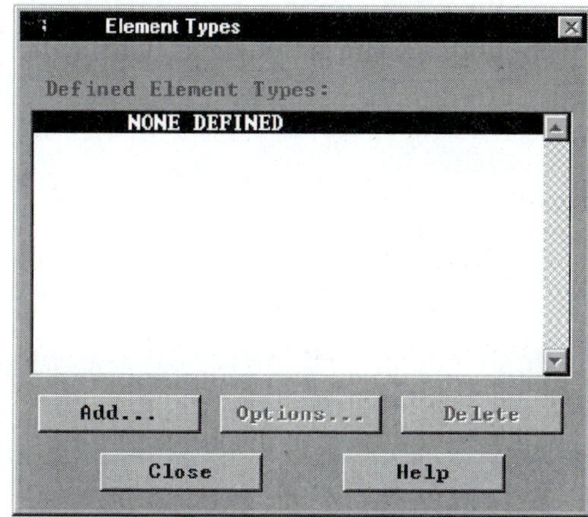


From the Library of Element Types, under Thermal Mass, choose **Solid**, then choose **Quad 4node 55**:

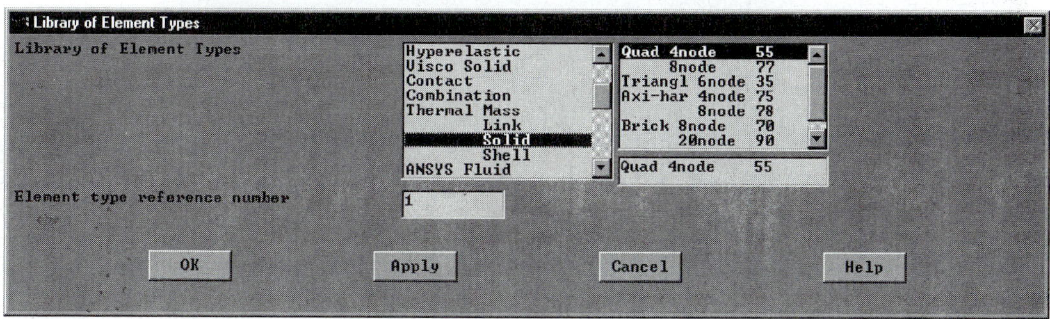


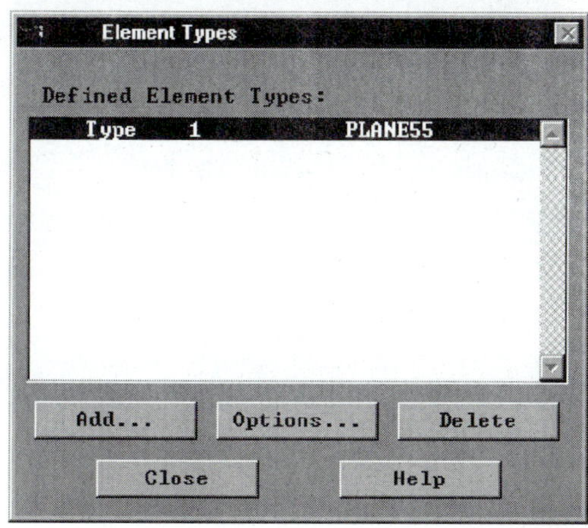

Assign the thermal conductivity values for concrete and brick. First, assign the value for concrete with the commands

main menu: **Preprocessor** → **Material Props** → **-Constant-Isotropic** …

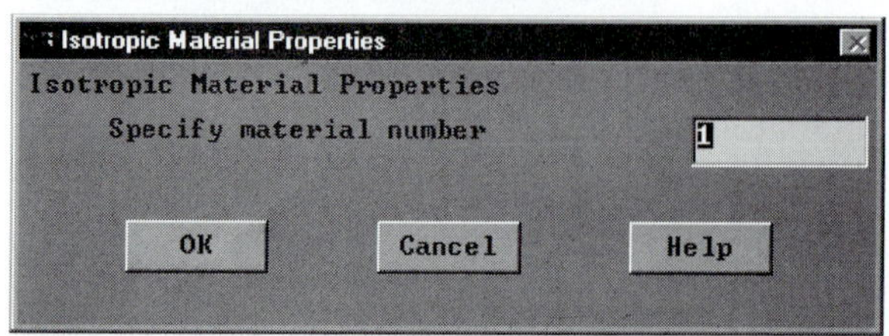


Isotropic Material Properties

Isotropic Material Properties

Properties for Material Number 1

| | | |
|---|---|---|
| Young's modulus | EX | |
| Density | DENS | |
| Thermal expansion coeff | ALPX | |
| Reference temperature | REFT | |
| Poisson's ratio (minor) | NUXY | |
| Poisson's ratio (major) | PRXY | |
| Shear modulus | GXY | |
| Friction coefficient | MU | |
| Damping multiplier | DAMP | |
| Thermal conductivity | KXX | 0.07 |
| Specific heat | C | |
| Enthalpy | ENTH | |
| Convection film coefficient | HF | |
| Emissivity | EMIS | |

OK Apply Cancel Help

Isotropic Material Properties

Isotropic Material Properties

Specify material number 2

OK Cancel Help

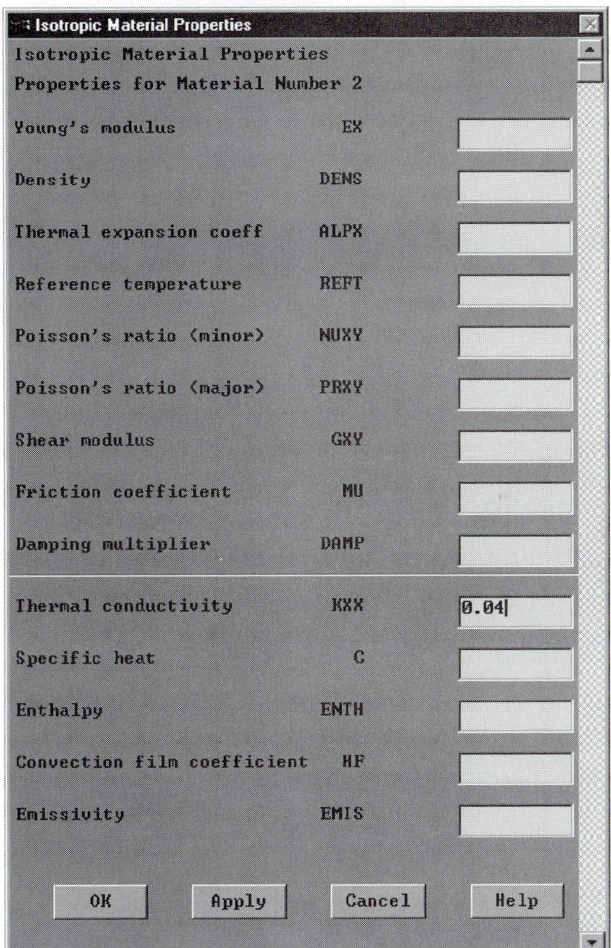


ANSYS Toolbar: **SAVE_DB**

Set up the graphics area (i.e., the workplane, zoom, etc.) with the commands:

utility menu: **Workplane** → **WP Settings** …

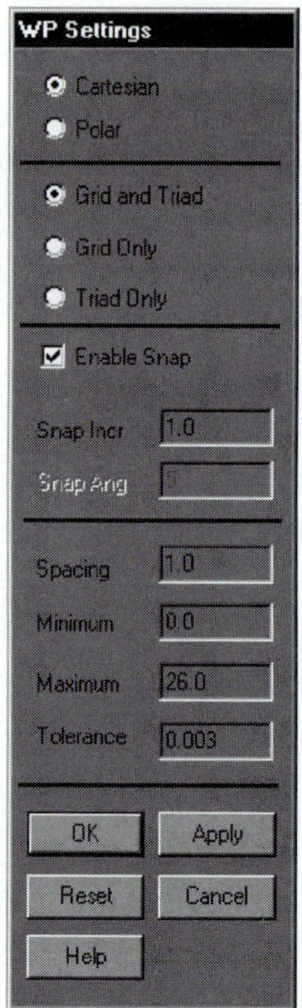


Toggle on the workplane by the issuing the command

utility menu: **Workplane** → **Display working plane**

Bring the workplane to view with the command

utility menu: **PlotCtrls** → **Pan, Zoom, Rotate** ...

Click on the **small circle** until you bring the workplane to view. Then create the brick section of the chimney by issuing the commands:

main menu: **Preprocessor** → **Modeling-Create** → **Areas-Rectangles** → **By 2 Corners** +

On the workplane, pick the respective locations of the corners of areas and apply:

**[WP = 0,0 lower left corner of the workplane]**

**[Expand the rubber band up 26.0 and right 26.0]**

**[WP = 6,6]**

**[Expand the rubber band up 14.0 and right 14.0]**

**OK**

To create the brick area of the chimney, subtract the two areas you have created with the commands:

main menu: **Preprocessor** → **-Modeling-Operate** → **-Booleans-Subtract** → **Areas** +

**[Pick area 1]**

**[Use the middle button anywhere in the ANSYS graphics window to apply]**

**[Pick area 2]**

**[anywhere in the ANSYS graphics window]**

**OK**

Next, create the area of concrete by issuing the following commands:

main menu: **Preprocessor** → **-Modeling-Create** → **-Areas-Rectangles** → **By 2 Corners** +

On the workplane, pick the respective locations of the corners of areas and apply:

**[WP = 6,6]**

**[Expand the rubber band up 14.0 and right 14.0]**

This is our area number 4.

**[WP = 7,7]**

**[Expand the rubber band up 12.0 and right 12.0]**

This is our area number 5.

**OK**

Next, subtract the two inside areas with commands:

main menu: **Preprocessor** → **-Modeling-Operate** → **-Booleans-Subtract** → **Areas** +

**[Pick the area number 4]**

**[Use the middle button anywhere in the ANSYS graphics window to apply]**

**[Pick the area number 5]**

**[anywhere in the ANSYS graphics window]**

**OK**

To check your work thus far, plot the areas. First, toggle off the workplane and turn on area numbering by the commands

utility menu: **Workplane → Display Working plane**

utility menu: **PlotCtrls → Numbering** …

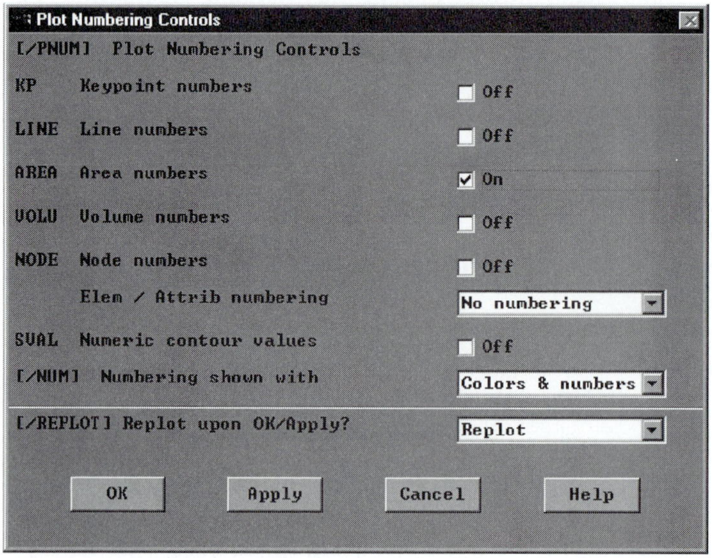

Plot Numbering Controls

[/PNUM] Plot Numbering Controls

| | | |
|---|---|---|
| KP | Keypoint numbers | Off |
| LINE | Line numbers | Off |
| AREA | Area numbers | ☑ On |
| VOLU | Volume numbers | Off |
| NODE | Node numbers | Off |
| | Elem / Attrib numbering | No numbering |
| SVAL | Numeric contour values | Off |
| [/NUM] | Numbering shown with | Colors & numbers |
| [/REPLOT] | Replot upon OK/Apply? | Replot |

OK Apply Cancel Help

utility menu: **Plot → Areas**

ANSYS Toolbar: **SAVE_DB**

We now want to mesh the areas to create elements and nodes. But first, we need to specify the element sizes. So issue the commands

main menu: **Preprocessor** → -**Meshing-SizeCntrls** → **-Global-Size** …

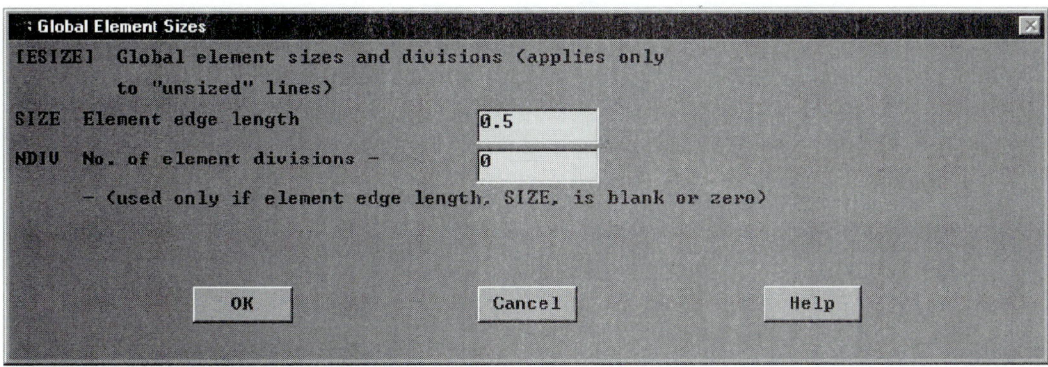


Next, glue areas to merge keypoints with the commands

main menu: **Preprocessor** → **-Modeling-Operate** → **-Boolean-Glue** → **Areas** +

Select **Pick All** to glue the areas. We also need to specify material attributes for the concrete and the brick areas before we proceed with meshing. So, issue the commands

main menu: **Preprocessor** → **-Attributes-Define** → **Picked Areas** +

**[Pick the concrete area]**

**[anywhere in the ANSYS graphics window to apply]**

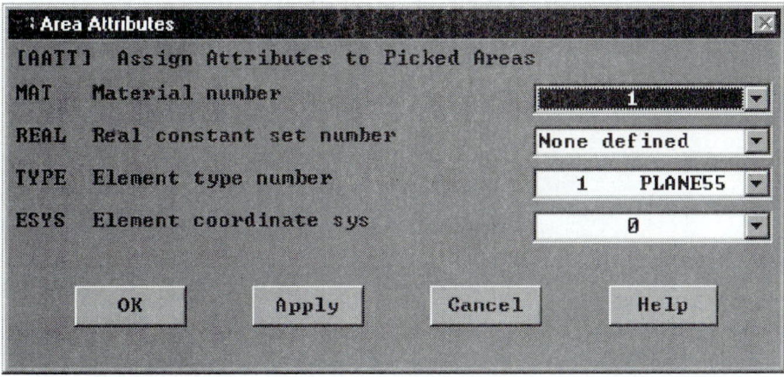


main menu: **Preprocessor** → **-Attributes-Define** → **Picked Areas** +

**[Pick the brick area]**

**[anywhere in the ANSYS graphics window to apply]**

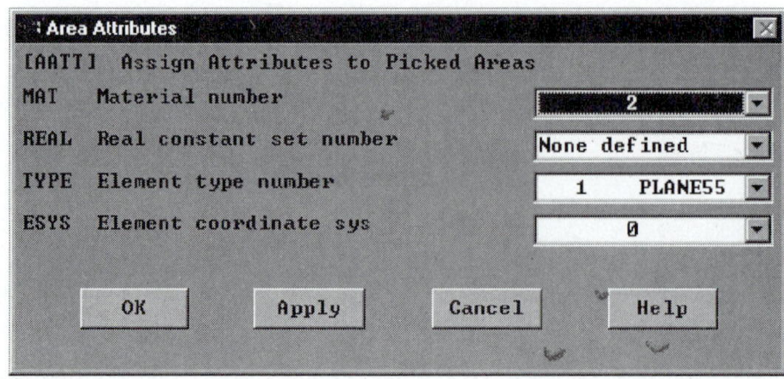


ANSYS Toolbar: **SAVE_DB**

We can proceed with meshing now. So, issue the following commands:

main menu: **Preprocessor** → **-Meshing-Mesh** → **-Areas-Free** +

Select **Pick All** and proceed. Then issue the command

utility menu: **PlotCtrls** → **Numbering** …

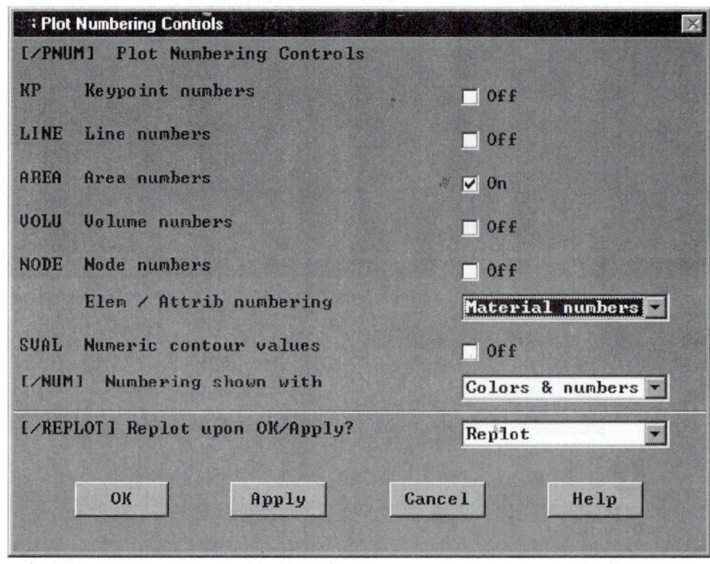


Apply boundary conditions using the command

main menu: **Solution** → **-Loads-Apply** → **-Thermal-Convection** → **On lines** +

Pick the convective lines of the concrete, and press the **OK** button to specify the convection coefficient and the temperature:

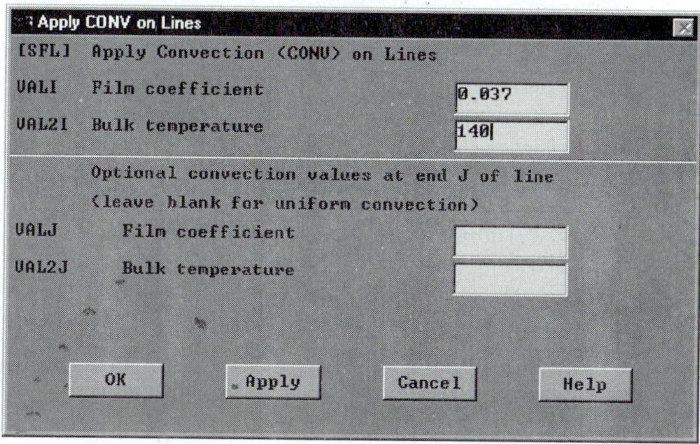


main menu: **Solution** → **-Loads-Apply** → **-Thermal-Convection** → **On lines** +

Pick the exterior lines of the brick layer, and press the **OK** button to specify the convection coefficient and the temperature:

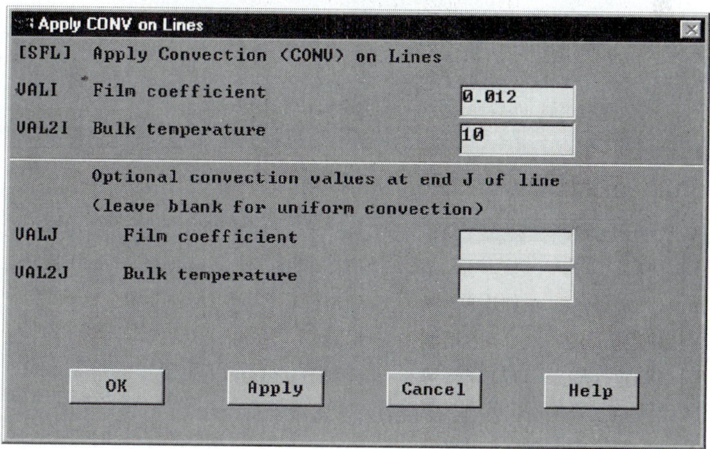


To see the applied convective boundary conditions, issue the command

utility menu: **PlotCtrls** → **Symbols** ...

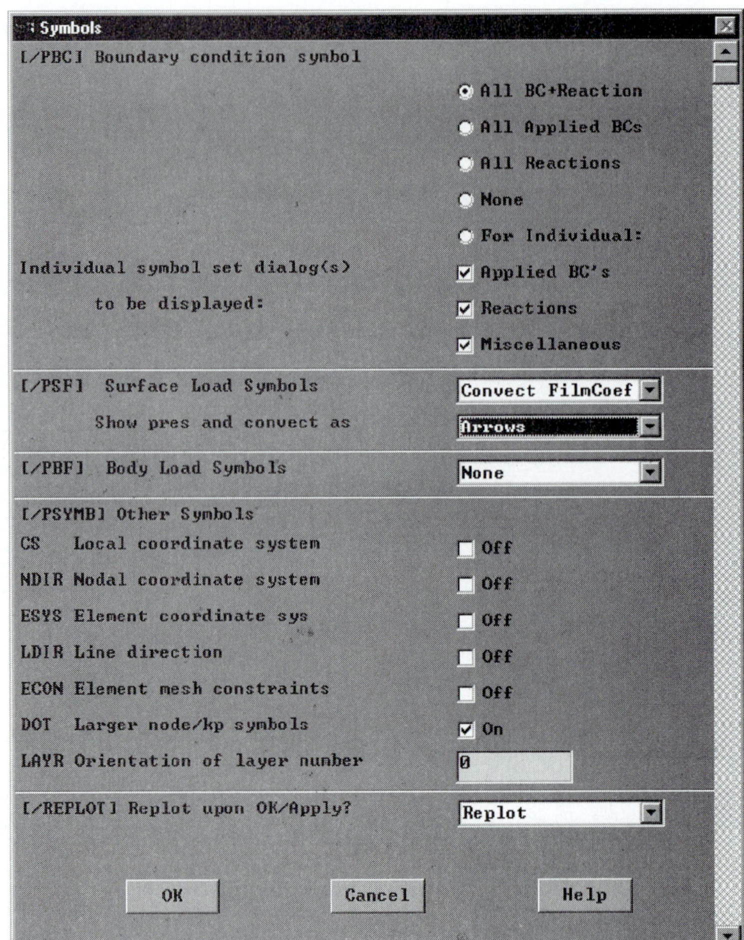


utility menu: **Plot** → **Lines**

ANSYS Toolbar: **SAVE_DB**

Now, solve the problem with the following commands:

main menu: **Solution** → **Solve-Current LS**

**OK**

**Close** (the solution is done!) window.

**Close** (the /STAT Command) window.

Begin the postprocessing phase. First obtain information, such as nodal temperatures and heat fluxes with the command

main menu: **General Postproc** → **Plot Results** → **-Contour Plot-Nodal Solu** ...

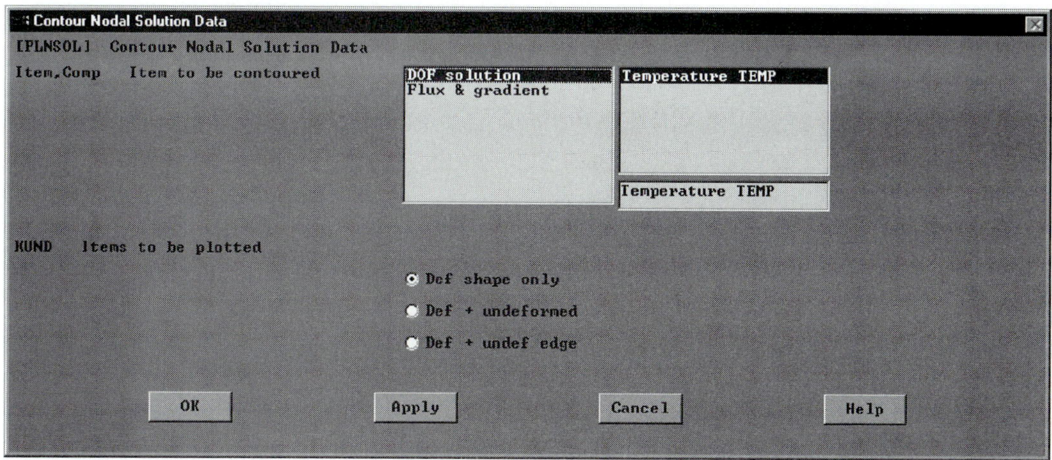


The contour plot of the temperature distribution is shown in Figure 7.15.

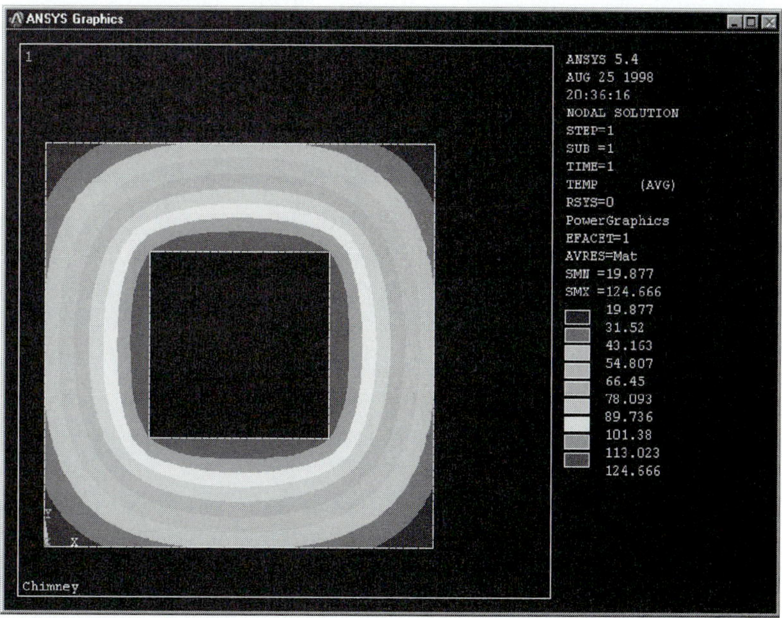


**FIGURE 7.15** Temperature contour plot.

Now use the following command to plot of the heat flow vectors (the plot is shown in Figure 7.16):

main menu: **General Postproc** → **Plot Results** → **-Vector Plot-Predefined** ...

Vector Plot of Predefined Vectors
[PLVECT] Vector Plot of Predefined Vectors
Item Vector item to be plotted
Flux & gradient
Thermal flux TF
Thermal grad TG
Thermal flux TF
Mode Vector or raster display
Vector Mode
Raster Mode
Loc Vector location for results
Elem Centroid
Elem Nodes
Edge Element edges
Hidden
[/VSCALE] Scaling of Vector Arrows
WN Window Number
Window 1
VRATIO Scale factor multiplier
1
KEY Vector scaling will be
Magnitude based
OPTION Vector plot based on
Undeformed Mesh
OK
Apply
Cancel
Help

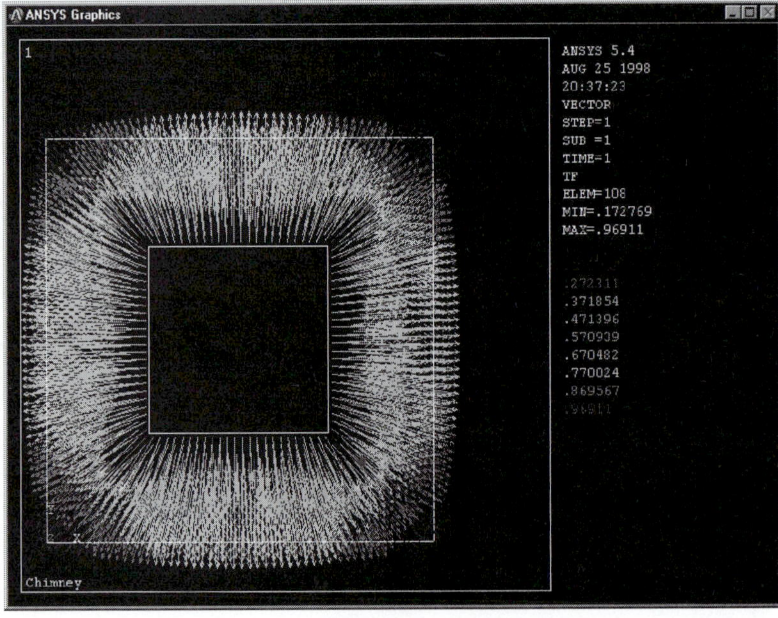


**FIGURE 7.16** Plot of the heat flow vectors.

Next, issue the following commands:

utility menu: **Plot** → **Areas**

main menu: **General Postproc** → **Path Operations** → **Define Path** → **On Working Plane** +

Pick the two points along the line marked as *A–A*, as shown in Figure 7.17, and press the **OK** button.

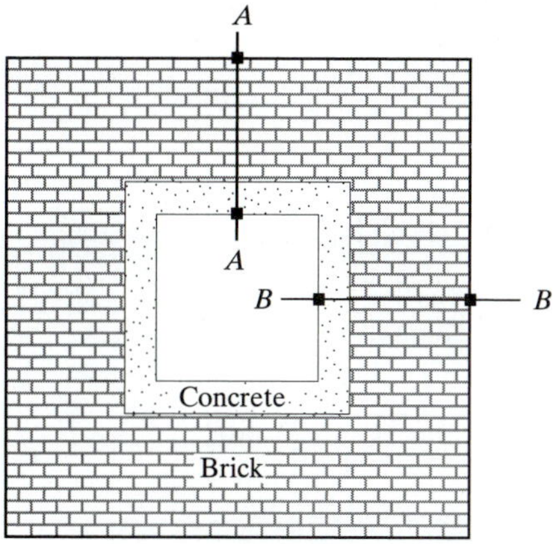


**FIGURE 7.17** Defining the path for path operation.

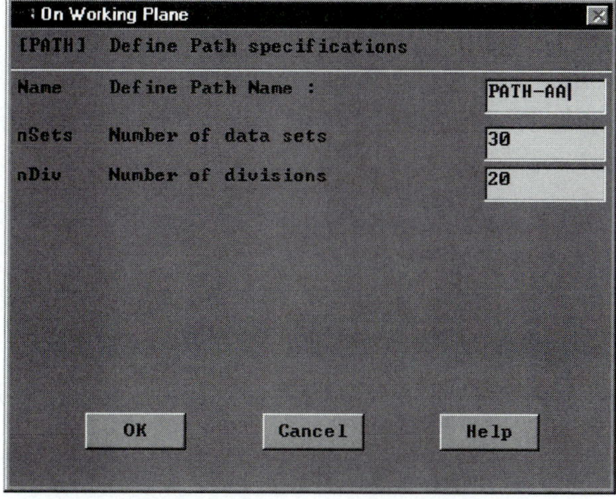


Then, issue the commands

main menu: **General Postproc** → **Path Operations** → **Map onto Path** …

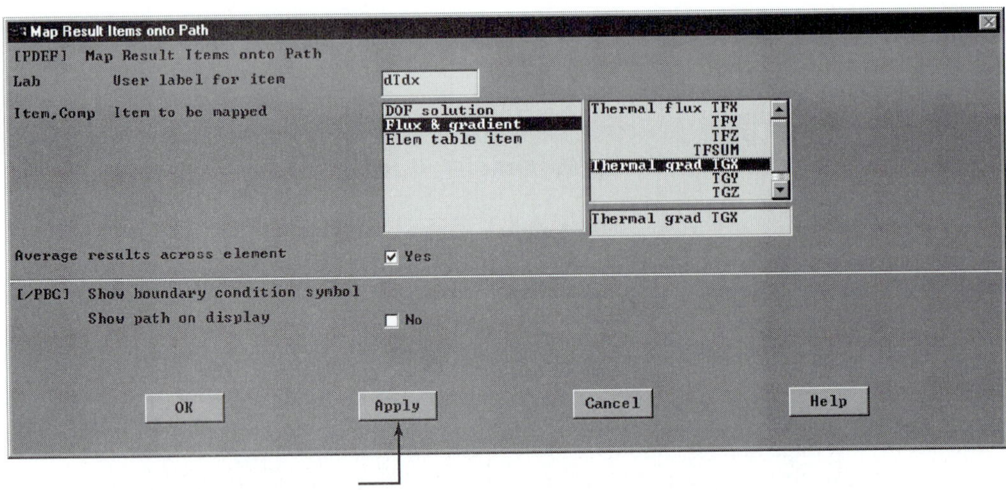
Map Result Items onto Path
[PDEF] Map Result Items onto Path
Lab User label for item
dTdx
Item,Comp Item to be mapped
DOF solution
Flux & gradient
Elem table item
Thermal flux TFX
TFY
TFZ
TFSUM
Thermal grad TGX
TGY
TGZ
Thermal grad TGX
Average results across element
Yes
[/PBC] Show boundary condition symbol
Show path on display
No
OK
Apply
Cancel
Help

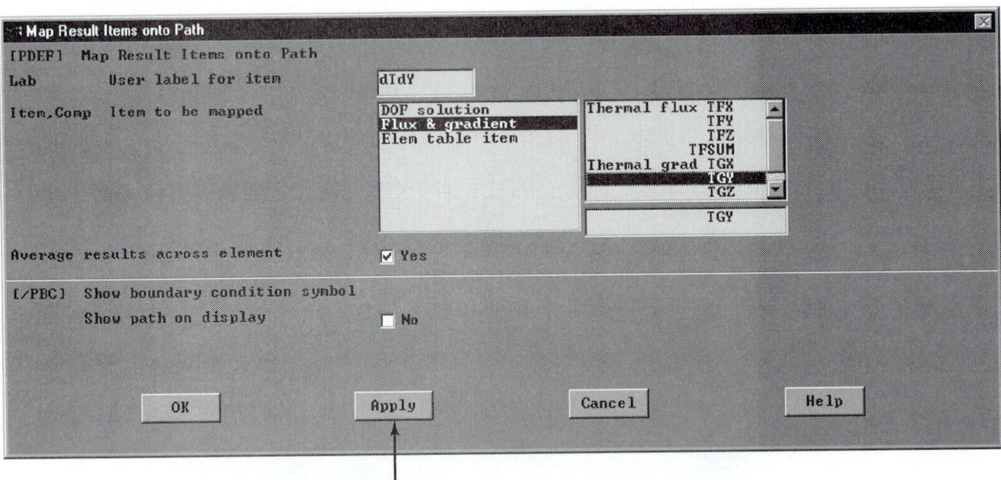
Map Result Items onto Path
[PDEF] Map Result Items onto Path
Lab User label for item
dTdY
Item,Comp Item to be mapped
DOF solution
Flux & gradient
Elem table item
Thermal flux TFX
TFY
TFZ
TFSUM
Thermal grad TGX
TGY
TGZ
TGY
Average results across element
Yes
[/PBC] Show boundary condition symbol
Show path on display
No
OK
Apply
Cancel
Help

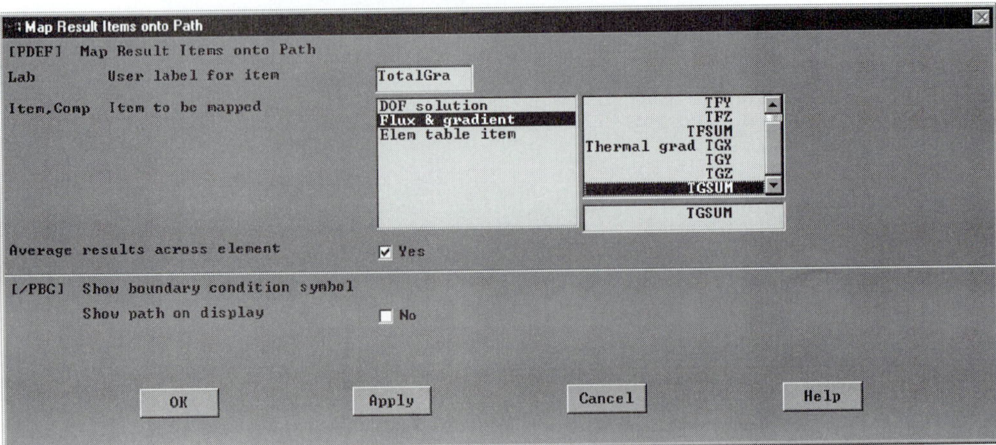
Map Result Items onto Path
[PDEF] Map Result Items onto Path
Lab User label for item
TotalGra
Item,Comp Item to be mapped
DOF solution
Flux & gradient
Elem table item
TFY
TFZ
TFSUM
Thermal grad TGX
TGY
TGZ
TGSUM
TGSUM
Average results across element
Yes
[/PBC] Show boundary condition symbol
Show path on display
No
OK
Apply
Cancel
Help

main menu: **General Postproc** → **Path Operations** → **-Plot Path Item-On Graph** ...

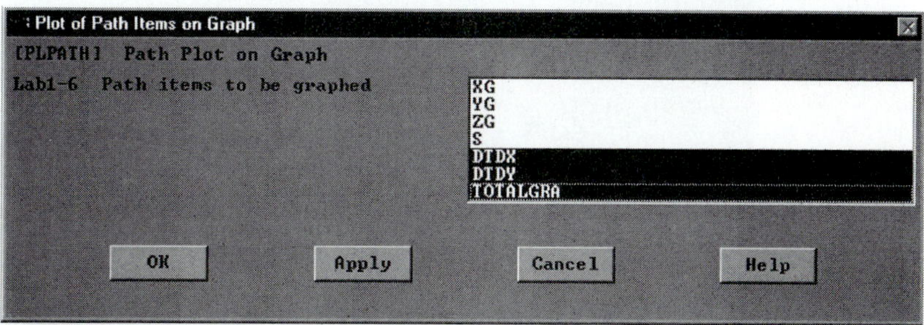


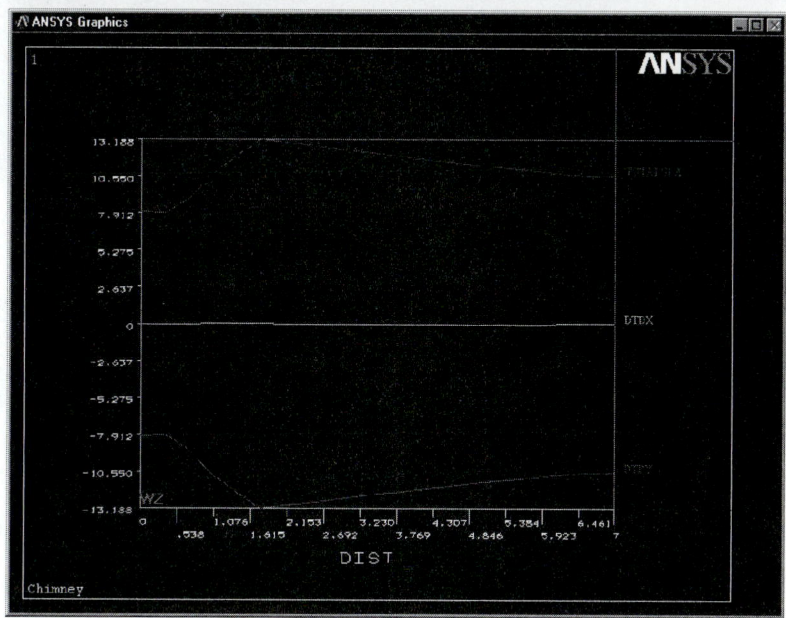


**FIGURE 7.18** The variation of temperature gradients along path *A–A*.

Finally, exit ANSYS and save everything:

ANSYS Toolbar: **Quit**

Exit from ANSYS
- Exit from ANSYS -
Save Geom+Loads
Save Geo+Ld+Solu
Save Everything
Quit - No Save!
OK Cancel Help

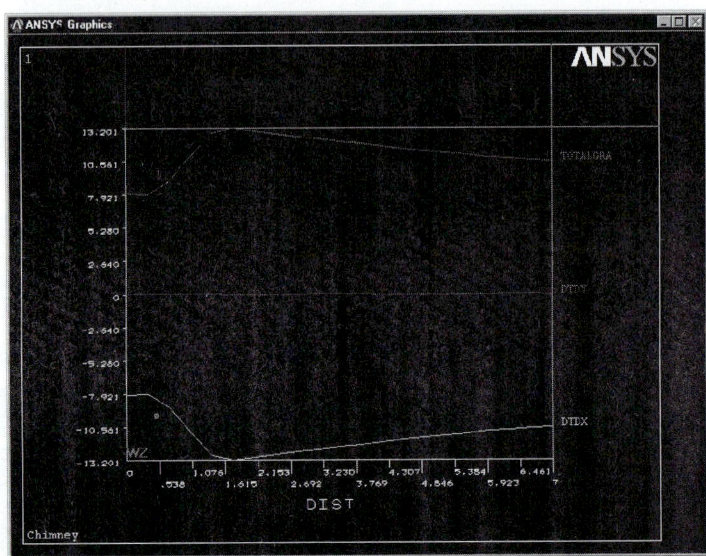


**FIGURE 7.19** The variation of temperature gradients along path *B–B*.

## 7.6 VERIFICATION OF RESULTS

First, let us discuss some simple, yet powerful, ways to verify your results visually. For symmetrical problems, you should always identify lines of symmetry created by geometrical and thermal conditions. Lines of symmetry are always adiabatic lines, meaning that no heat flows in the directions perpendicular to these lines. Because no heat flows in the directions perpendicular to lines of symmetry, they constitute heat flow lines. In other words, heat flows parallel to these lines. Consider the variation of the temperature gradients $\frac{\partial T}{\partial X}$ and $\frac{\partial T}{\partial Y}$ and their vector sum along path *A–A*, as shown in Figure 7.18. Note that path *A–A* (in Figure 7.17) is a line of symmetry and, therefore, constitutes an adiabatic line. Because of this fact, the magnitude of $\frac{\partial T}{\partial X}$ is zero along path *A–A* and $\frac{\partial T}{\partial Y}$ equals vector sum as shown in Figure 7.18. Comparing the variation of the temperature gradients $\frac{\partial T}{\partial X}$ and $\frac{\partial T}{\partial Y}$ and their vector sum along path *B–B* as shown in Figure 7.19 renders the conclusion that the magnitude of $\frac{\partial T}{\partial Y}$ is now zero and $\frac{\partial T}{\partial X}$ equals vector sum.

Another important visual inspection of the results requires that the isotherms (lines of constant temperatures) always be perpendicular to the adiabatic lines, or lines of symmetry. You can see this orthogonal relationship in the temperature contour plot of the chimney, as shown in Figure 7.15.

We can also perform a quantitative check on the validity of the results. For example, the conservation of energy applied to a control volume surrounding an arbitrary node must be satisfied. Are the energies flowing into and out of a node balanced out? This approach was demonstrated earlier with Example 7.1.

## SUMMARY

At this point you should:

1. understand the fundamental concepts of the three modes of heat transfer. You should also know the various types of boundary conditions that could occur in a conduction problem.

2. know how the conductance matrices and the load matrices for two-dimensional conduction problems were obtained. The conductance matrix for a bilinear rectangular element is:

$$[\mathbf{K}]^{(e)} = \frac{k_x w}{6\ell}\begin{bmatrix} 2 & -2 & -1 & 1 \\ -2 & 2 & 1 & -1 \\ -1 & 1 & 2 & -2 \\ 1 & -1 & -2 & 2 \end{bmatrix} + \frac{k_y \ell}{6w}\begin{bmatrix} 2 & 1 & -1 & -2 \\ 1 & 2 & -2 & -1 \\ -1 & -2 & 2 & 1 \\ -2 & -1 & 1 & 2 \end{bmatrix}$$

Heat loss by convection around the edge of a rectangular element can also contribute to the conductance matrix:

$$[\mathbf{K}]^{(e)} = \frac{h\ell_{ij}}{6}\begin{bmatrix} 2 & 1 & 0 & 0 \\ 1 & 2 & 0 & 0 \\ 0 & 0 & 0 & 0 \\ 0 & 0 & 0 & 0 \end{bmatrix} \qquad [\mathbf{K}]^{(e)} = \frac{h\ell_{jm}}{6}\begin{bmatrix} 0 & 0 & 0 & 0 \\ 0 & 2 & 1 & 0 \\ 0 & 1 & 2 & 0 \\ 0 & 0 & 0 & 0 \end{bmatrix}$$

$$[\mathbf{K}]^{(e)} = \frac{h\ell_{mn}}{6}\begin{bmatrix} 0 & 0 & 0 & 0 \\ 0 & 0 & 0 & 0 \\ 0 & 0 & 2 & 1 \\ 0 & 0 & 1 & 2 \end{bmatrix} \qquad [\mathbf{K}]^{(e)} = \frac{h\ell_{ni}}{6}\begin{bmatrix} 2 & 0 & 0 & 1 \\ 0 & 0 & 0 & 0 \\ 0 & 0 & 0 & 0 \\ 1 & 0 & 0 & 2 \end{bmatrix}$$

The load vector for a rectangular element could have many components. It could have a component due to a possible heat generation term within a given element:

$$\{\mathbf{F}\}^{(e)} = \frac{q^{\cdot}A}{4}\begin{Bmatrix} 1 \\ 1 \\ 1 \\ 1 \end{Bmatrix}$$

It could also have a possible convection heat loss term(s) along the edge(s):

$$\{\mathbf{F}\}^{(e)} = \frac{hT_f\ell_{ij}}{2}\begin{Bmatrix} 1 \\ 1 \\ 0 \\ 0 \end{Bmatrix} \qquad \{\mathbf{F}\}^{(e)} = \frac{hT_f\ell_{jm}}{2}\begin{Bmatrix} 0 \\ 1 \\ 1 \\ 0 \end{Bmatrix}$$

$$\{\mathbf{F}\}^{(e)} = \frac{hT_f\ell_{mn}}{2}\begin{Bmatrix} 0 \\ 0 \\ 1 \\ 1 \end{Bmatrix} \qquad \{\mathbf{F}\}^{(e)} = \frac{hT_f\ell_{ni}}{2}\begin{Bmatrix} 1 \\ 0 \\ 0 \\ 1 \end{Bmatrix}$$

The conductance matrix for a triangular element is

$$[\mathbf{K}]^{(e)} = \frac{k_x}{4A}\begin{bmatrix} \beta_i^2 & \beta_i\beta_j & \beta_i\beta_k \\ \beta_i\beta_j & \beta_j^2 & \beta_j\beta_k \\ \beta_i\beta_k & \beta_j\beta_k & \beta_k^2 \end{bmatrix} + \frac{k_y}{4A}\begin{bmatrix} \delta_i^2 & \delta_i\delta_j & \delta_i\delta_k \\ \delta_i\delta_j & \delta_j^2 & \delta_j\delta_k \\ \delta_i\delta_k & \delta_j\delta_k & \delta_k^2 \end{bmatrix}$$

Heat loss by convection around the edge of a triangular element can also contribute to the conductance matrix according to the equations

$$[\mathbf{K}]^{(e)} = \frac{h\ell_{ij}}{6}\begin{bmatrix} 2 & 1 & 0 \\ 1 & 2 & 0 \\ 0 & 0 & 0 \end{bmatrix} \qquad [\mathbf{K}]^{(e)} = \frac{h\ell_{jk}}{6}\begin{bmatrix} 0 & 0 & 0 \\ 0 & 2 & 1 \\ 0 & 1 & 2 \end{bmatrix}$$

$$[\mathbf{K}]^{(e)} = \frac{h\ell_{ki}}{6}\begin{bmatrix} 2 & 0 & 1 \\ 0 & 0 & 0 \\ 1 & 0 & 2 \end{bmatrix}$$

The load matrix for a triangular element could have many components. It could have a component due to a possible heat generation term within a given element:

$$\{\mathbf{F}\}^{(e)} = \frac{q^{\cdot}A}{3}\begin{Bmatrix} 1 \\ 1 \\ 1 \end{Bmatrix}$$

Also, it could have a possible convection heat loss term(s) along the edge(s):

$$\{\mathbf{F}\}^{(e)} = \frac{hT_f\ell_{ij}}{2}\begin{Bmatrix} 1 \\ 1 \\ 0 \end{Bmatrix} \qquad \{\mathbf{F}\}^{(e)} = \frac{hT_f\ell_{jk}}{2}\begin{Bmatrix} 0 \\ 1 \\ 1 \end{Bmatrix}$$

$$\{\mathbf{F}\}^{(e)} = \frac{hT_f\ell_{ki}}{2}\begin{Bmatrix} 1 \\ 0 \\ 1 \end{Bmatrix}$$

3. understand the contribution of convective boundary conditions to the conductance matrix and the forcing matrix.
4. always find ways to verify your results.

## REFERENCES

*ANSYS User's Manual: Procedures*, Vol. I, Swanson Analysis Systems, Inc.

*ANSYS User's Manual: Commands*, Vol. II, Swanson Analysis Systems, Inc.

*ANSYS User's Manual: Elements*, Vol. III, , Swanson Analysis Systems, Inc.

Incropera, F., and Dewitt, D., *Fundamentals of Heat and Mass Transfer*, 2d. ed., New York, John Wiley and Sons, 1985.

Segrlind, L., *Applied Finite Element Analysis*, 2d. ed., New York, John Wiley and Sons, 1984.

## PROBLEMS

1. Construct the conductance matrices for the elements shown in the accompanying figure. Also, assemble the elements to obtain the global conductance matrix. The properties and the boundary conditions for each element are shown in the figure.

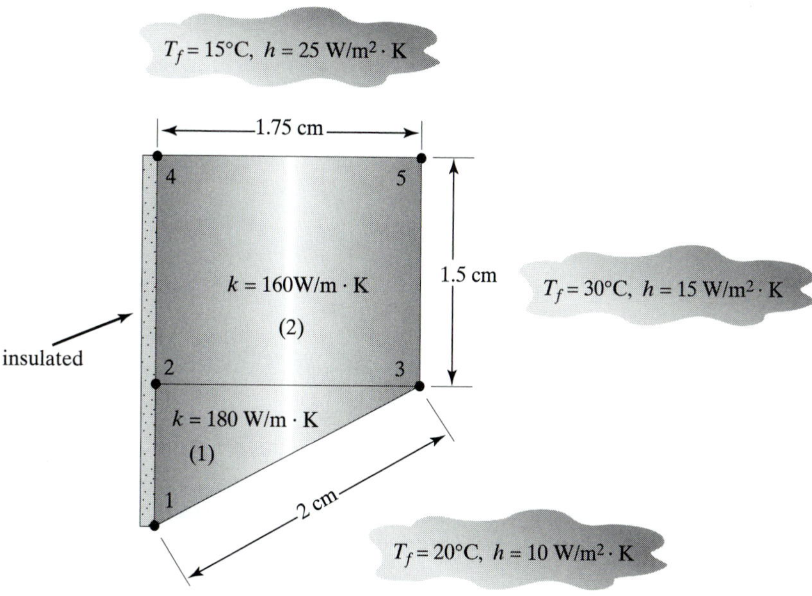


**2.** Construct the load matrix for each element in Problem 1. Also, assemble the elemental load matrices to construct the global load matrix.

**3.** Construct the conductance matrices shown in the accompanying figure. Also, assemble the elements to obtain the global conductance matrix. The properties and the boundary conditions for each element are shown in the figure.

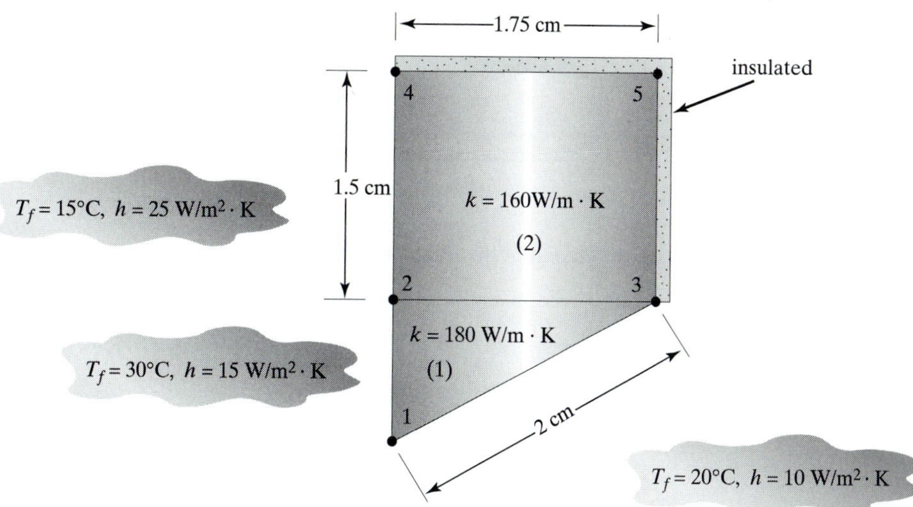


4. Construct the load matrix for each element in Problem 3. Also, assemble the elemental load matrices to construct the global load matrix.

**5.** Show that for a constant heat flux boundary condition $q''_o$, evaluation of the terms $\int_\tau [\mathbf{S}]^T q''_o \cos\theta\, d\tau$ and $\int_\tau [\mathbf{S}]^T q''_o \sin\theta\, d\tau$ along the edges of the rectangular element results in the elemental load matrices

$$\{\mathbf{F}\}^{(e)} = \frac{q''_o \ell_{ij}}{2}\begin{Bmatrix}1\\1\\0\\0\end{Bmatrix} \qquad \{\mathbf{F}\}^{(e)} = \frac{q''_o \ell_{jm}}{2}\begin{Bmatrix}0\\1\\1\\0\end{Bmatrix}$$

$$\{\mathbf{F}\}^{(e)} = \frac{q''_o \ell_{mn}}{2}\begin{Bmatrix}0\\0\\1\\1\end{Bmatrix} \qquad \{\mathbf{F}\}^{(e)} = \frac{q''_o \ell_{ni}}{2}\begin{Bmatrix}1\\0\\0\\1\end{Bmatrix}$$

**6.** Evaluate the constant heat flux boundary condition in Problem 5 for a triangular element.

**7.** Using the results of Problem 5, construct the load matrix for each element shown in the accompanying figure. Also, assemble the elemental matrices to construct the global load matrix. The boundary conditions are shown in the figure.

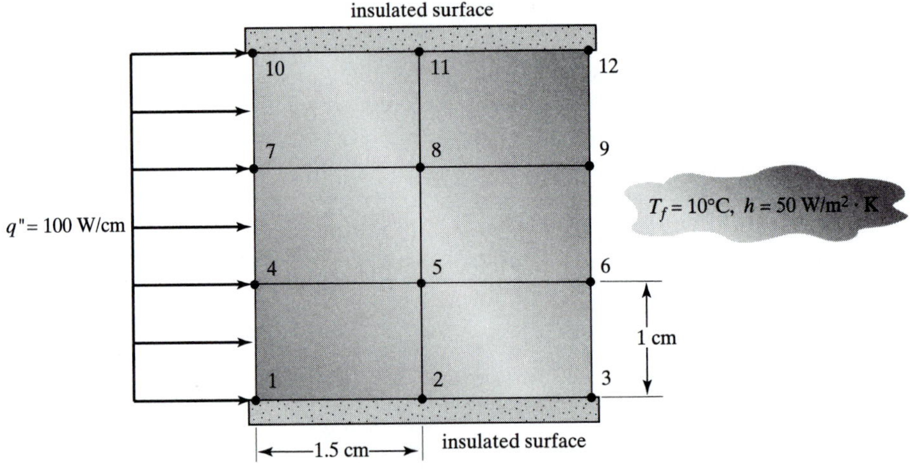


**8.** In the Galerkin formulation of two-dimensional fins, the convection heat loss from the periphery of the extended surface gives rise to the term $\int_A [S]^T hT\, dA$. The term contributes to the elemental conductance matrix. Show that for a bilinear rectangular element, the integral yields:

$$\int_A [\mathbf{S}]^T hT dA = \int_A [\mathbf{S}]^T h[S_i \quad S_j \quad S_m \quad S_n]\begin{Bmatrix}T_i\\T_j\\T_m\\T_n\end{Bmatrix} dA = \frac{hA}{36}\begin{bmatrix}4 & 2 & 1 & 2\\2 & 4 & 2 & 1\\1 & 2 & 4 & 2\\2 & 1 & 2 & 4\end{bmatrix}\begin{Bmatrix}T_i\\T_j\\T_m\\T_n\end{Bmatrix}$$

**9.** Evaluate the integral in Problem 8 for a triangular element. Show that

$$\int_A [\mathbf{S}]^T hT dA = \int_A [\mathbf{S}]^T h[S_i \quad S_j \quad S_k]\begin{Bmatrix}T_i\\T_j\\T_k\end{Bmatrix} dA = \frac{hA}{12}\begin{bmatrix}2 & 1 & 1\\1 & 2 & 1\\1 & 1 & 2\end{bmatrix}\begin{Bmatrix}T_i\\T_j\\T_k\end{Bmatrix}$$

**10.** Consider a small rectangular aluminum plate with dimensions of 20 cm × 10 cm and a thermal conductivity value of $k = 168 \text{ W/m} \cdot \text{K}$, as shown in the accompanying figure. The plate is exposed to the boundary conditions shown in the figure. Using manual calculations, determine the temperature distribution within the plate, under steady-state conditions. (*Hint*: Because of the existence of two axes of symmetry, you should model only a quarter of the plate.)

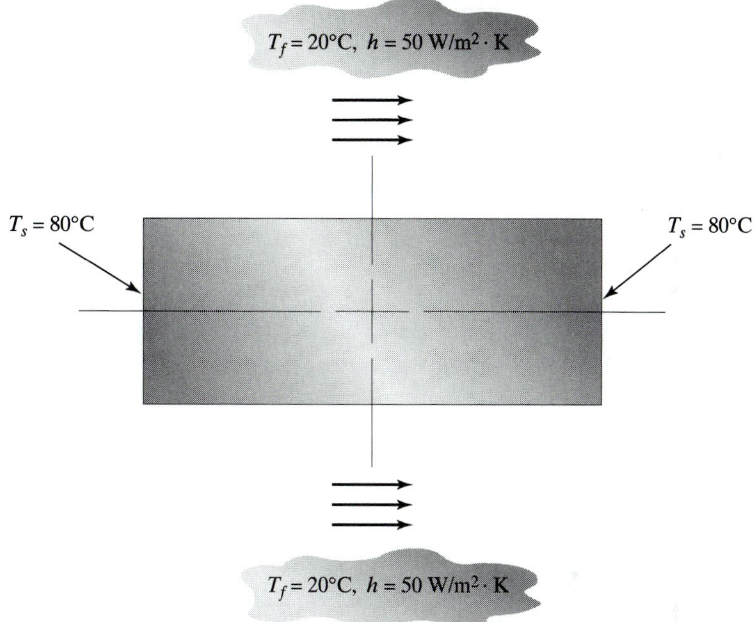


**11.** Aluminum fins with triangular profiles, shown in the accompanying figure, are used to remove heat from a surface whose temperature is 150°C. The temperature of the surrounding air is 20°C. The natural heat transfer coefficient associated with the surrounding air is $30 \text{ W/m} \cdot \text{K}$. The thermal conductivity of aluminum is $k = 168 \text{ W/m} \cdot \text{K}$. Using manual calculations, determine the temperature distribution along a fin. Approximate the heat loss for one such fin.

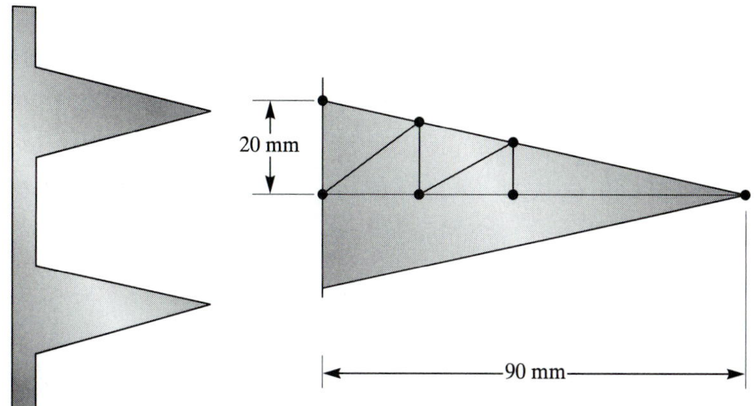


**12.** For the fin in Problem 11, use ANSYS to determine the temperature distribution within the fin. What is the overall heat loss through the fin? Compare these results to the results of your manual calculations.

**13.** Aluminum fins with parabolic profiles, shown in the accompanying figure, are used to remove heat from a surface whose temperature is 120°C. The temperature of the surrounding air is 20°C. The natural heat transfer coefficient associated with the surrounding air is 25 W/m · K. The thermal conductivity of aluminum is $k = 168$ W/m · K. Using ANSYS, determine the temperature distribution along a fin. Approximate the heat loss for one such fin.

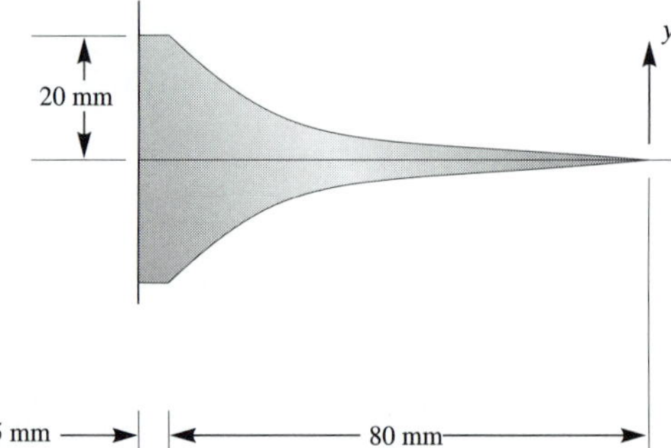


**14.** Using ANSYS, determine the temperature distribution in the window assembly shown in the accompanying figure. During the winter months, the inside air temperature is kept at 68°F, with a corresponding heat transfer coefficient of $h = 1.46$ Btu/hr · ft$^2$ · °F. Assume an outside air temperature of 10°F and a corresponding heat transfer coefficient of $h = 6$ Btu/hr · ft$^2$ · °F. What is the overall heat loss through the window assembly?

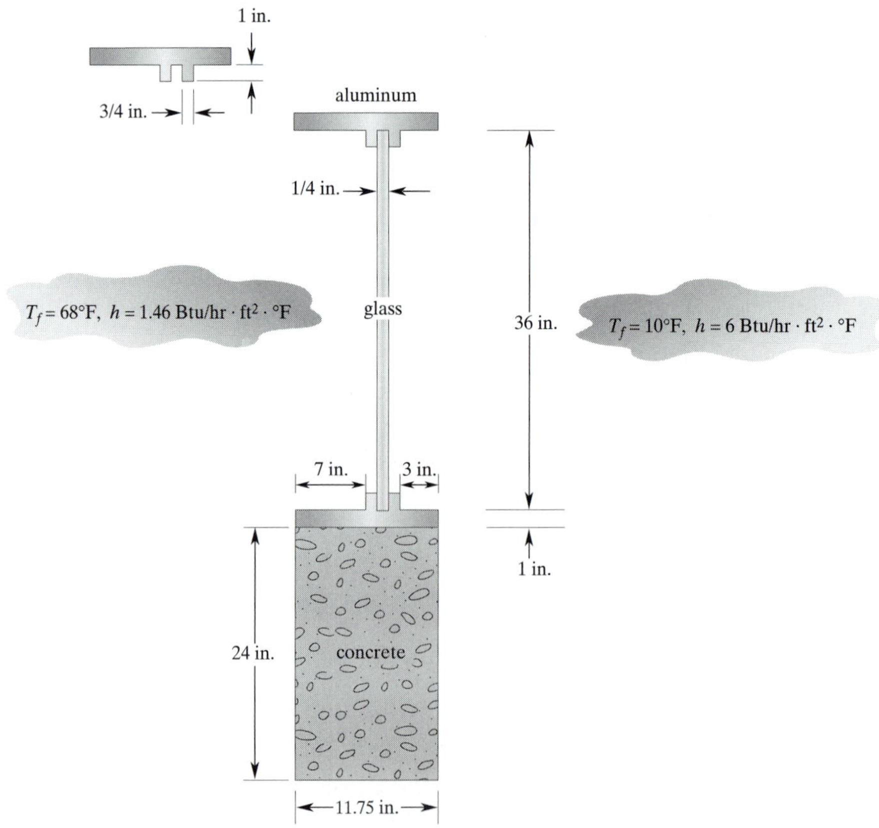

15. Aluminum fins ($k = 170\ \mathrm{W/m \cdot K}$) are commonly used to dissipate heat from electronic devices. An example of such a fin is shown in the accompanying figure. Using ANSYS, determine the temperature distribution within the fin. The base of the fin experiences a constant flux of $q' = 1000\ \mathrm{W/m}$. A fan forces air over the surfaces of the fin. The temperature of the surrounding air is 20°C with a corresponding heat transfer coefficient of $h = 40\ \mathrm{W/m^2 \cdot K}$.

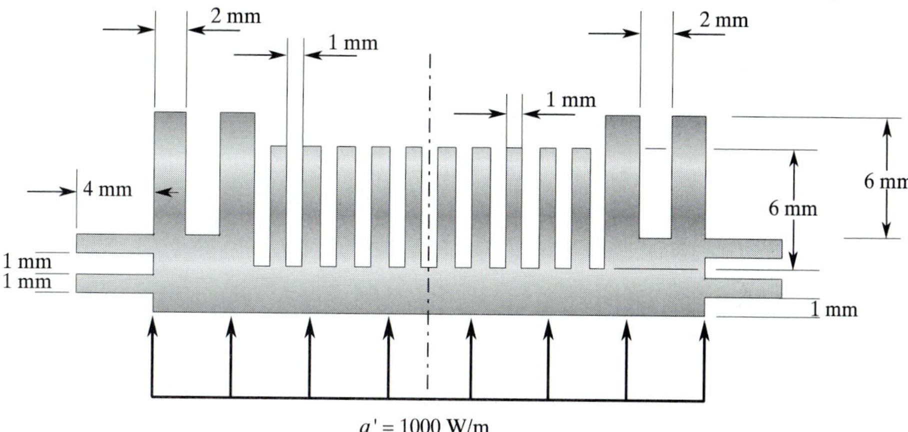


16. Hot water flows through pipes that are embedded in a concrete slab. A section of the slab is shown in the accompanying figure. The temperature of the water inside the pipe is 50°C, with a corresponding heat transfer coefficient of 200 W/m • K. With the conditions shown at the surface, use ANSYS to determine the temperature of the surface. Assuming that the heat transfer coefficient associated with the hot-water flow remains constant, find the water temperature at which the surface freezes. Neglect the thermal resistance through the pipe walls.

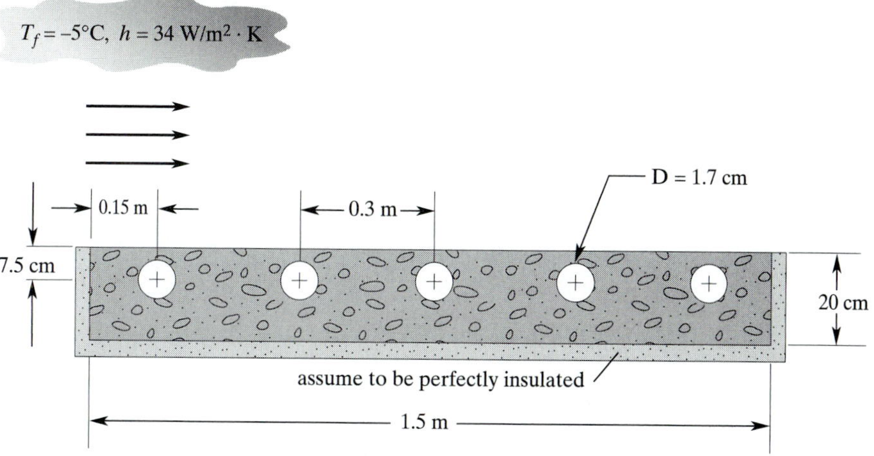


17. Consider the heat transfer through a basement wall with the dimensions given in the accompanying figure. The wall is constructed from concrete and has a thermal concuctivity of $k = 1.0$ Btu/hr • ft • °F. The nearby ground has an average thermal conductivity of $k = 0.85$ Btu/hr • ft • °F. Using ANSYS, determine the temperature distribution within the wall and the heat loss from the wall. The inside air is kept at 68°F with a corresponding heat transfer coefficient of $h = 1.46$ Btu/hr • ft$^2$ • °F. Assume an outside air temperature of 15°F, and a corresponding heat transfer coefficient of $h = 6$ Btu/hr • ft$^2$ • °F. Assume that at about four feet away from the wall, the horizontal component of the heat transfer in the soil becomes negligible.

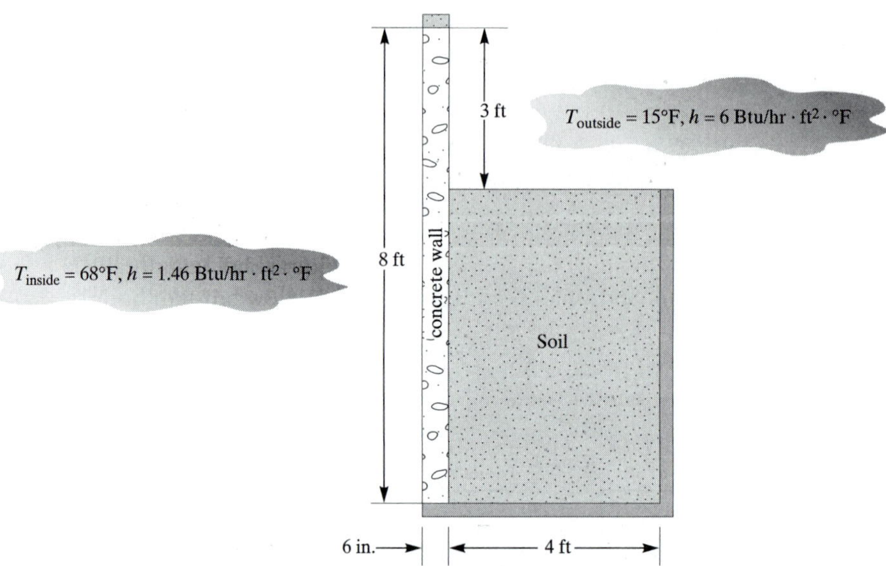


**18.** We would like to include the heat transfer rates through an uninsulated basement floor in our model in Problem 17. Considering the heat transfer model shown in the accompanying figure, determine the temperature distributions in the wall, the floor, and the soil and the heat loss from the floor and the wall. As shown in the figure, assume that at about four feet away from the wall and the floor, the horizontal and the vertical components of the heat transfer in the soil become negligible.

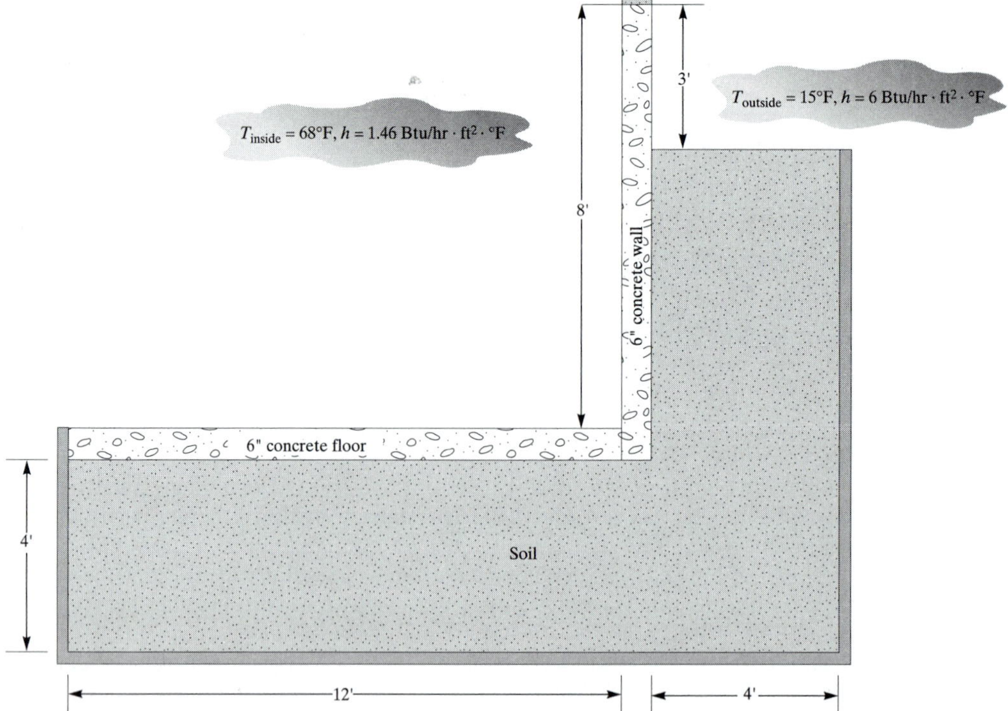

**19.** In order to enhance heat transfer rates, the inside surface of a tube is extended to form longitudinal fins, as shown in the accompanying figure. Determine the temperature distribution inside the tube wall, given the following data:

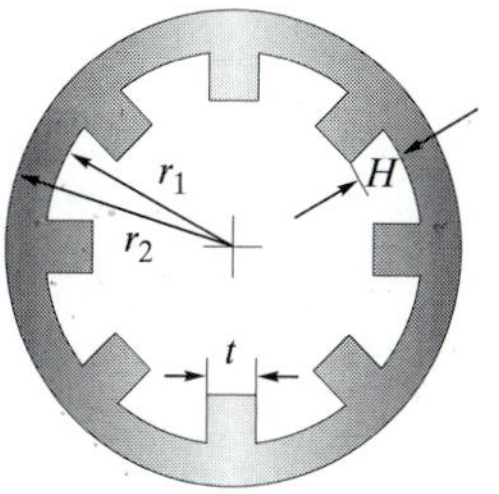


$r_1 = 2$ in $\qquad k = 400$ W/m·K

$r_2 = 2\frac{1}{4}$ in $\qquad T_{inside} = 80°C$

$t = \frac{3}{4}$ in $\qquad h_{inside} = 150$ W/m²·K

$H = \frac{3}{4}$ in $\qquad T_{outside} = 15°C$

$h_{outside} = 30$ W/m²·K

**20.** Consider the concentric-tube heat exchanger shown in the accompanying figure. A mixture of aqueous ethylene glycol solution arriving from a solar collector is passing through the inner tube. Water flows through the annulus as shown in the figure. The average temperature of the water at the section shown is 15°C, with a corresponding heat transfer coefficient of $h = 200\ \text{W/m}^2 \cdot \text{K}$. The average temperature of the ethylene glycol mixture is 48°C, with an associated heat transfer coefficient of $h = 150\ \text{W/m}^2 \cdot \text{K}$. In order to enhance the heat transfer rates between the fluids, the outside surface of the inner tube is extended to form longitudinal fins, as shown in the figure. Determine the temperature distribution inside the heat exchanger's walls, assuming that the outside of the heat exchanger is perfectly insulated. Also, determine the heat transfer rate between the fluids.

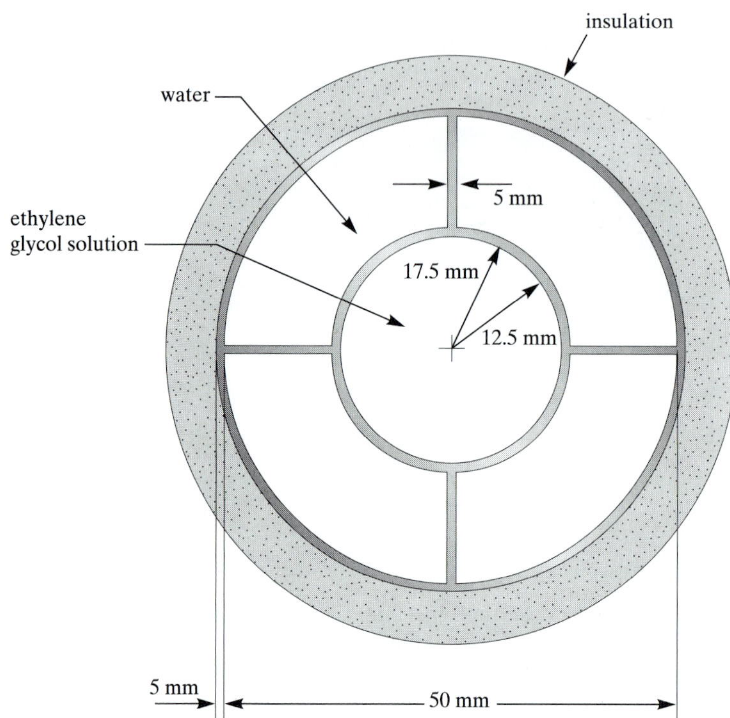


21. **Design Problem** At some ski resorts, in order to keep ice from forming on the surface of uphill roads leading to condominiums, hot water is pumped through pipes that are embedded beneath the surface of the road. You are to design a hydronic system to perform such a task. Choose your favorite ski resort and look up its design conditions, such as the ambient air temperature, soil temperature, etc. The system that you construct may consist of a series of tubes, pumps, a hot-water heater, valves, fittings, etc. Basic information sought includes: the type of pipes, their sizes, the spacing between the tubes, the configuration of the piping system, and the distance below the surface the pipes should be embedded. If time allows, you may also size the pump and the hot-water heater.

CHAPTER 8

# Analysis of Two-Dimensional Solid Mechanics Problems

The objective of this chapter is to introduce you to the analysis of two-dimensional solid mechanics problems. Structural members and machine components are generally subject to a push–pull, bending, or twisting type of loading. The components of common structures and machines normally include beams, columns, plates, and other members that can be modeled using two-dimensional approximations. The main topics discussed in Chapter 8 include the following:

**8.1** Torsion of Members With Arbitrary Cross-Section Shape
**8.2** Beams and Frames
**8.3** Plane-Stress Formulation
**8.4** Basic Failure Theory
**8.5** Examples Using ANSYS
**8.6** Verification of Results

## 8.1 TORSION OF MEMBERS WITH ARBITRARY CROSS-SECTION SHAPE

There are still many practicing engineers who generate finite element models for problems for which there exist simple analytical solutions. You should not be too quick to use the finite element method to solve simple torsional problems. This type of problem includes torsion of members with circular or rectangular cross sections. Let us briefly review the analytical solutions that are available for torsional problems. When studying the mechanics of materials, you were introduced to the torsion of long, straight members with circular cross sections. A problem is considered to be a torsional problem when the applied moment or torque twists the member about its longitudinal axis, as shown in Figure 8.1.

Over the elastic limit, the shear stress distribution within a member with a circular cross section, such as a shaft or a tube, is given by the equation

$$\tau = \frac{Tr}{J} \tag{8.1}$$

where $T$ is the applied torque, $r$ is the radial distance measured from the center of the shaft to a desired point in the cross section, and $J$ represents the polar moment of inertia of

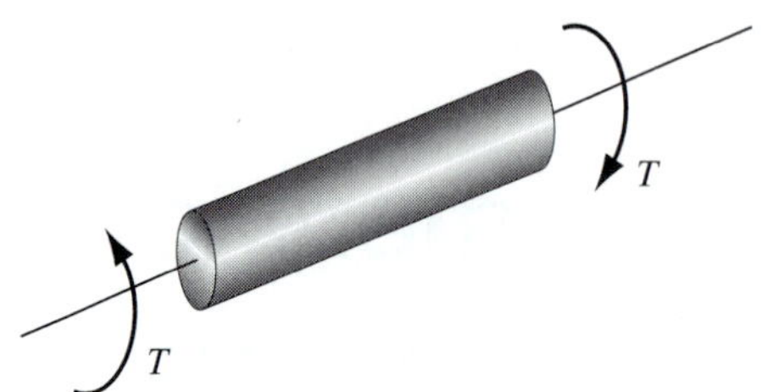


**FIGURE 8.1** Torsion of a shaft.

the cross-sectional area. It should be clear from examination of Eq. (8.1) that the maximum shear stress occurs at the outer surface of the shaft, where $r$ is equal to the radius of the shaft. Also, recall that the angle of twist caused by the applied torque can be determined from the equation

$$\theta = \frac{TL}{JG} \tag{8.2}$$

in which $L$ is the length of the member and $G$ is the shear modulus (modulus of rigidity) of the material. Furthermore, there are analytical solutions that can be applied to torsion of members with rectangular cross-sectional areas.* When a torque is applied to a straight bar with a rectangular cross-sectional area, within the elastic region of the material, the maximum shearing stress and an angle of twist caused by the torque are given by:

$$\tau_{max} = \frac{T}{c_1 wh^2} \tag{8.3}$$

$$\theta = \frac{TL}{c_2 Gwh^3} \tag{8.4}$$

$L$ is the length of the bar and $w$ and $h$ are the larger and smaller sides of the cross-section, respectively. (See Figure 8.2.) The values of coefficients $c_1$ and $c_2$ (given in Table 8.1) are dependent on the aspect ratio of the cross section. As the aspect ratio approaches

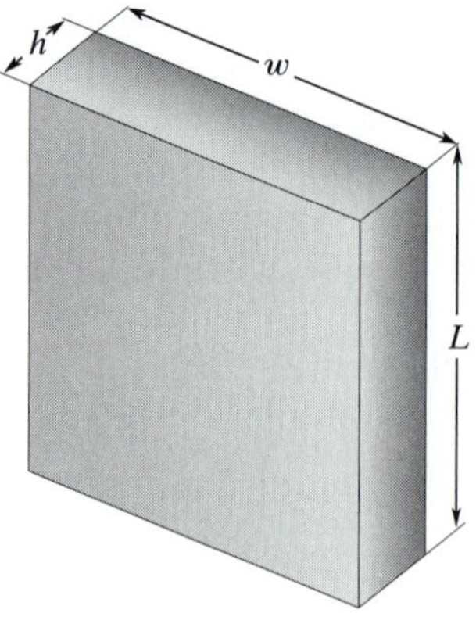


**FIGURE 8.2** A straight rectangular bar in torsion.

*See Timoshenko and Goadier (1970) for more detail.

large numbers ($W/h \rightarrow \infty$), $c_1 = c_2 = 0.3333$. This relationship is demonstrated in Table 8.1.

**TABLE 8.1** $c_1$ and $c_2$ values for a bar with a rectangular cross section

| $w/h$ | $c_1$ | $c_2$ |
|---|---|---|
| 1.0 | 0.208 | 0.141 |
| 1.2 | 0.219 | 0.166 |
| 1.5 | 0.231 | 0.196 |
| 2.0 | 0.246 | 0.229 |
| 2.5 | 0.258 | 0.249 |
| 3.0 | 0.267 | 0.263 |
| 4.0 | 0.282 | 0.281 |
| 5.0 | 0.291 | 0.291 |
| 10.0 | 0.312 | 0.312 |
| $\infty$ | 0.333 | 0.333 |

The maximum shear stress and the angle of twist for cross-sectional geometries with high aspect ratios ($w/h > 10$) are given by:

$$\tau_{max} = \frac{T}{0.333wh^2} \tag{8.5}$$

$$\theta = \frac{TL}{0.333Gwh^3} \tag{8.6}$$

These types of members are commonly referred to as thin-wall members. Examples of some thin-wall members are shown in Figure 8.3.

Therefore, if you come across a problem that fits these categories, solve it using the torsional formulae. Do not spend a great deal of time generating a finite element model.

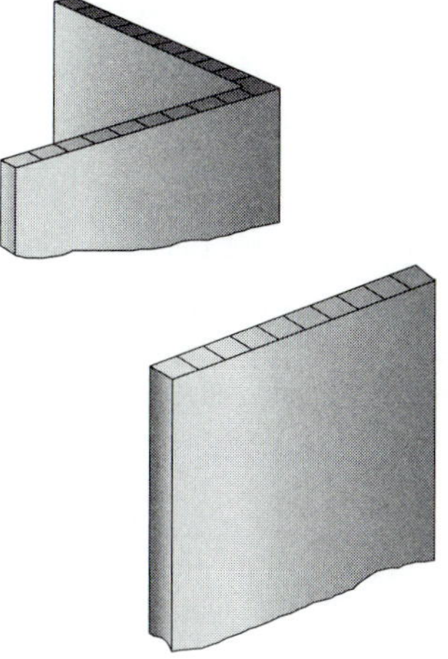

**FIGURE 8.3** Examples of thin-wall members.

### Finite Element Formulation of Torsional Problems

*Fung (1965)* discusses the elastic torsional behavior of noncircular shafts in detail. There are two basic theories: (1) St. Venant's formulation and (2) the Prandtl formulation. Here, we will use the Prandtl formulation. The governing differential equation for the elastic torsion of a shaft in terms of the stress function $\phi$ is

$$\frac{\partial^2 \phi}{\partial x^2} + \frac{\partial^2 \phi}{\partial y^2} + 2G\theta = 0 \tag{8.7}$$

where $G$ is the shear modulus of elasticity of the bar and $\theta$ represents the angle of twist per unit length. The shear stress components are related to the stress function $\phi$ according to the equations

$$\tau_{zx} = \frac{\partial \phi}{\partial y} \tag{8.8}$$

$$\tau_{zy} = -\frac{\partial \phi}{\partial x} \tag{8.9}$$

Note that with Prandtl's formulation, the applied torque does not directly appear in the governing equation. Instead, the applied torque is related to the stress function and is

$$T = 2\int_A \phi \, dA \tag{8.10}$$

In Eq. (8.10), $A$ represents the cross-sectional area of the shaft. Comparing the differential equation governing the torsional behavior of a member, Eq. (8.7), to the heat diffusion equation, Eq. (7.8), we note that both of these equations have the same form. Therefore, we can apply the results of Section 7.2 and Section 7.3 to torsional problems. However, when comparing the differential equations for torsional problems, we let $c_1 = 1$ and $c_2 = 1, c_3 = 2G\theta$. The stiffness matrix for a rectangular element then becomes

$$[\mathbf{K}]^{(e)} = \frac{w}{6\ell}\begin{bmatrix} 2 & -2 & -1 & 1 \\ -2 & 2 & 1 & -1 \\ -1 & 1 & 2 & -2 \\ 1 & -1 & -2 & 2 \end{bmatrix} + \frac{\ell}{6w}\begin{bmatrix} 2 & 1 & -1 & -2 \\ 1 & 2 & -2 & -1 \\ -1 & -2 & 2 & 1 \\ -2 & -1 & 1 & 2 \end{bmatrix} \tag{8.11}$$

where $w$ and $\ell$ are the length and the width, respectively, of the rectangular element, as shown in Figure 8.4. The load matrix for an element is

$$\{\mathbf{F}\}^{(e)} = \frac{2G\theta A}{4}\begin{Bmatrix} 1 \\ 1 \\ 1 \\ 1 \end{Bmatrix} \tag{8.12}$$

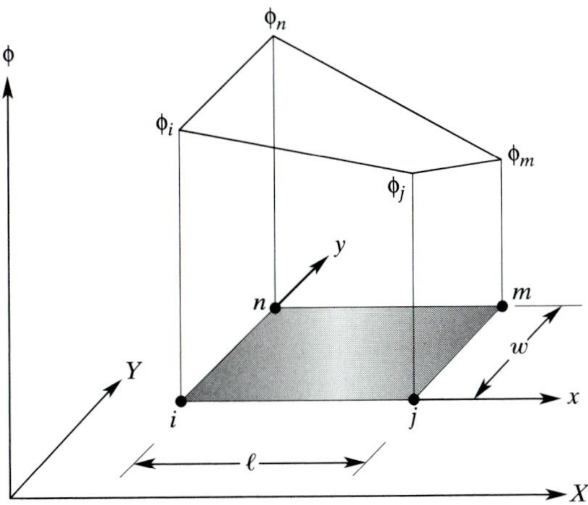


**FIGURE 8.4** Nodal values of the stress function for a rectangular element.

and for triangular elements, shown in Figure 8.5, the stiffness and load matrices are

$$[\mathbf{K}]^{(e)} = \frac{1}{4A}\begin{bmatrix} \beta_i^2 & \beta_i\beta_j & \beta_i\beta_k \\ \beta_i\beta_j & \beta_j^2 & \beta_j\beta_k \\ \beta_i\beta_k & \beta_j\beta_k & \beta_k^2 \end{bmatrix} + \frac{1}{4A}\begin{bmatrix} \delta_i^2 & \delta_i\delta_j & \delta_i\delta_k \\ \delta_i\delta_j & \delta_j^2 & \delta_j\delta_k \\ \delta_i\delta_k & \delta_j\delta_k & \delta_k^2 \end{bmatrix} \tag{8.13}$$

$$\{\mathbf{F}\}^{(e)} = \frac{2G\theta A}{3}\begin{Bmatrix} 1 \\ 1 \\ 1 \end{Bmatrix} \tag{8.14}$$

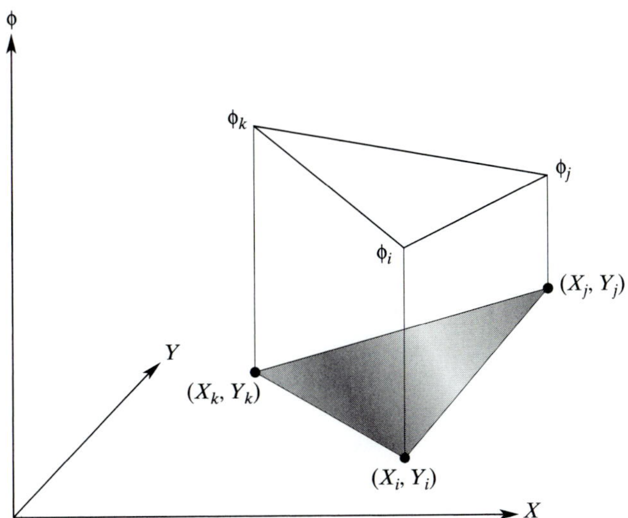


**FIGURE 8.5** Nodal values of the stress function for a triangular element.

where the area $A$ of the triangular element and the $\alpha$, $\beta$, and $\delta$-terms are given by:

$$2A = X_i(Y_j - Y_k) + X_j(Y_k - Y_i) + X_k(Y_i - Y_j)$$

$$\alpha_i = X_j Y_k - X_k Y_j \quad \beta_i = Y_j - Y_k \quad \delta_i = X_k - X_j$$

$$\alpha_j = X_k Y_i - X_i Y_k \quad \beta_j = Y_k - Y_i \quad \delta_j = X_i - X_k$$

$$\alpha_k = X_i Y_j - X_j Y_i \quad \beta_k = Y_i - Y_j \quad \delta_k = X_j - X_i$$

Next, we will consider the finite element formulation of beams and frames.

## 8.2 BEAMS AND FRAMES

Beams play significant roles in many engineering applications, including buildings, bridges, automobiles, and airplanes structures. A beam is defined as a structural member whose cross-sectional dimensions are relatively smaller than its length. Beams are commonly subjected to transverse loading, which is a type of loading that creates bending in the beam. A beam subjected to a distributed load is shown in Figure 8.6.

The deflection of the neutral axis at any location $x$ is represented by the variable $v$. For small deflections, the relationship between the normal stress $\sigma$ at a section, the bending moment at that section $M$, and the second moment of area $I$ is given by the flexure formula. The flexure formula is the equation

$$\sigma = -\frac{My}{I} \tag{8.15}$$

where $y$ locates a point in the cross section of the beam and represents the lateral distance from the neutral axis to that point. The deflection of the neutral axis $v$ is also related to the internal bending moment $M(x)$, the transverse shear $V(x)$, and the load $w(x)$ according to the equations

$$EI\frac{d^2 v}{dx^2} = M(x) \tag{8.16}$$

$$EI\frac{d^3 v}{dx^3} = \frac{dM(x)}{dx} = V(x) \tag{8.17}$$

$$EI\frac{d^4 v}{dx^4} = \frac{dV(x)}{dx} = w(x) \tag{8.18}$$

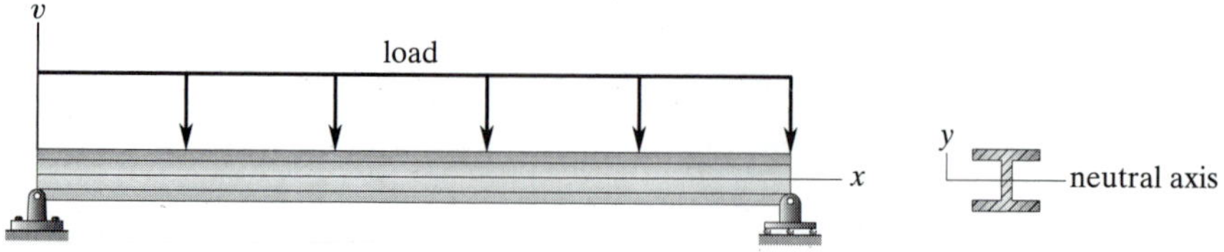


**FIGURE 8.6** A beam subjected to a distributed load.

Note that the standard beam sign convention is assumed in the previous equations; for example, the external load is assumed to be positive when it pushes upward on the beam. For your reference, the deflections and slopes of beams under some typical loads for simply supported and cantilevered supports are summarized in Table 8.2. Again, if you come across problems that can be analyzed using equations (8.16)-(8.18) and Table 8.2, solve them as such.

### Finite Element Formulation

In the following derivation, we will neglect the contribution of shear stresses to the strain energy. The strain energy for an arbitrary beam element $(e)$ then becomes:

$$\Lambda^{(e)} = \int_V \frac{\sigma\varepsilon}{2}\,dV = \int_V \frac{E\varepsilon^2}{2}\,dV = \frac{E}{2}\int_V \left(-y\frac{d^2v}{dx^2}\right)^2 dV \tag{8.19}$$

$$\Lambda^{(e)} = \frac{E}{2}\int_V \left(-y\frac{d^2v}{dx^2}\right)^2 dV = \frac{E}{2}\int_0^L \left(\frac{d^2v}{dx^2}\right)^2 dx \int_A y^2\,dA \tag{8.20}$$

Recognizing the integral $\int_A y^2\,dA$ as the second moment of the area $I$, we have:

$$\Lambda^{(e)} = \frac{EI}{2}\int_0^L \left(\frac{d^2v}{dx^2}\right)^2 dx \tag{8.21}$$

Before we proceed with integrating Eq. (8.21), we should define what we mean by a beam element. A simple beam element consists of two nodes. At each node, there are two degrees of freedom, a vertical displacement, and a rotation angle (slope), as shown in Figure 8.7.

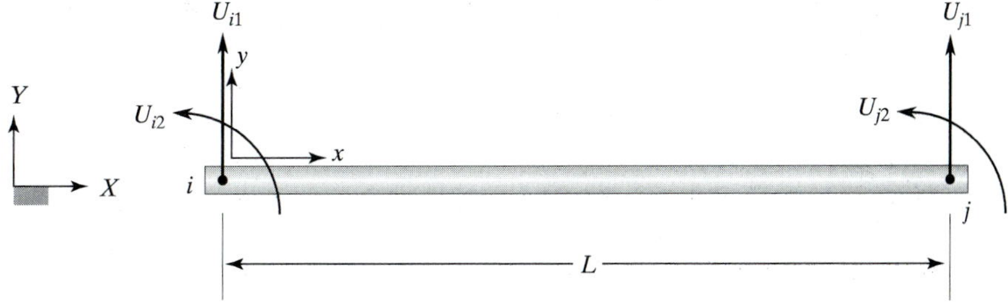


**FIGURE 8.7** A beam element.

There are four nodal values associated with a beam element. Therefore, we will use a third-order polynomial with four unknown coefficients to represent the displacement field. Moreover, we want the first derivatives of the shape functions to be continuous. The resulting shape functions are commonly referred to as Hermite shape functions. As you will see, they differ in some ways from the linear shape functions you have already studied in Chapters 3 and 5. We start with the third-order polynomial

$$v = c_1 + c_2x + c_3x^2 + c_4x^3 \tag{8.22}$$

TABLE 8.2 Deflections and slopes of beams under some typical loads and supports

| Beam Support and Load | Equation of Elastic Curve | Maximum Deflection | Slope |
|---|---|---|---|
|  | $v = \frac{-wx^2}{24EI}(x^2 - 4Lx + 6L^2)$ | $v_{max} = \frac{-wL^4}{8EI}$ | $\theta_{max} = \frac{-wL^3}{6EI}$ |
| 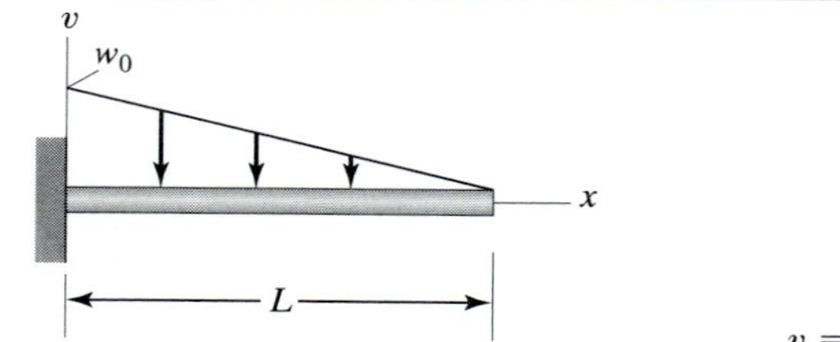 | $v = \frac{-w_0 x^2}{120LEI}(-x^3 + 5Lx^2 - 10L^2 x + 10L^3)$ | $v_{max} = \frac{-w_0 L^4}{30EI}$ | $\theta_{max} = \frac{-w_0 L^3}{24EI}$ |
|  | $v = \frac{-Px^2}{6EI}(3L - x)$ | $v_{max} = \frac{-PL^3}{3EI}$ | $\theta_{max} = \frac{-PL^2}{2EI}$ |

TABLE 8.2 (*continued*) Deflections and slopes of beams under some typical loads and supports

| Beam Support and Load | Equation of Elastic Curve | Maximum Deflection | Slope |
|---|---|---|---|
| 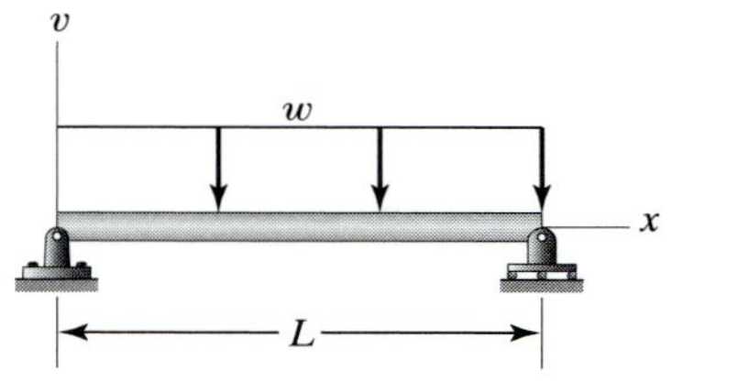  | $v = \frac{-wx}{24EI}(x^3 - 2Lx^2 + L^3)$ | $v_{max} = \frac{-5wL^4}{384EI}$ | $\theta_{max} = \frac{-wL^3}{24EI}$ |
| | $v = \frac{-Px}{48EI}(3L^2 - 4x^2)$ for $\left(x \le \frac{L}{2}\right)$ | $v_{max} = \frac{-PL^3}{48EI}$ | $\theta_{max} = \frac{-PL^2}{16EI}$ |

The element's end conditions are given by the following nodal values:

For node $i$: The vertical displacement at $x = 0$ $v = c_1 = U_{i1}$

For node $i$: The slope at $x = 0$ $\left.\dfrac{dv}{dx}\right|_{x=0} = c_2 = U_{i2}$

For node $j$: The vertical displacement at $x = L$ $v = c_1 + c_2 L + c_3 L^2 + c_4 L^3 = U_{j1}$

For node $j$: The slope at $x = L$ $\left.\dfrac{dv}{dx}\right|_{x=L} = c_2 + 2c_3 L + 3c_4 L^2 = U_{j2}$

We now have four equations with four unknowns. Solving for $c_1, c_2, c_3,$ and $c_4$; substituting into Eq. (8.22); and regrouping the $U_{i1}, U_{i2}, U_{j1}, U_{j2}$-terms results in the equation

$$v = S_{i1}U_{i1} + S_{i2}U_{i2} + S_{j1}U_{j1} + S_{j2}U_{j2} \tag{8.23}$$

where the shape functions are given by

$$S_{i1} = 1 - \frac{3x^2}{L^2} + \frac{2x^3}{L^3} \tag{8.24}$$

$$S_{i2} = x - \frac{2x^2}{L} + \frac{x^3}{L^2} \tag{8.25}$$

$$S_{j1} = \frac{3x^2}{L^2} - \frac{2x^3}{L^3} \tag{8.26}$$

$$S_{j2} = -\frac{x^2}{L} + \frac{x^3}{L^2} \tag{8.27}$$

It is clear that if we evaluate the shape functions, as given in Eqs. (8.24)–(8.27), at node $i$ at $x = 0$, we find that $S_{i1} = 1$ and $S_{i2} = S_{j1} = S_{j2} = 0$. Also, if we evaluate the slopes of the shape functions at $x = 0$, we find that $\dfrac{dS_{i2}}{dx} = 1$ and $\dfrac{dS_{i1}}{dx} = \dfrac{dS_{j1}}{dx} = \dfrac{dS_{j2}}{dx} = 0$. If we evaluate the shape functions at node $j$ at $x = L$ we find that $S_{j1} = 1$ and $S_{i1} = S_{i2} = S_{j2} = 0$, and if we evaluate the slopes of the shape functions at $x = L$, we determine that $\dfrac{dS_{j2}}{dx} = 1$ and $\dfrac{dS_{i1}}{dx} = \dfrac{dS_{i2}}{dx} = \dfrac{dS_{j1}}{dx} = 0$. These values are the properties of the Hermite third-order polynomials.

Now, we need to go back to Eq. (8.21) and substitute for the displacement field $v$ in terms of the shape functions and the nodal values. Let us begin by evaluating the equation

$$\frac{d^2 v}{dx^2} = \frac{d^2}{dx^2}\begin{bmatrix} S_{i1} & S_{i2} & S_{j1} & S_{j2} \end{bmatrix} \begin{Bmatrix} U_{i1} \\ U_{i2} \\ U_{j1} \\ U_{j2} \end{Bmatrix} \tag{8.28}$$

To simplify the next few steps of derivation and to avoid unnecessary mathematical operations, let us make use of matrix notations. First, let the second derivatives of the shape functions be defined in terms of the following relationships:

$$D_{i1} = \frac{d^2 S_{i1}}{dx^2}$$

$$D_{i2} = \frac{d^2 S_{i2}}{dx^2}$$

$$D_{j1} = \frac{d^2 S_{j1}}{dx^2}$$

$$D_{j2} = \frac{d^2 S_{j2}}{dx^2}$$

Then, Eq. (8.28) takes on the compact-matrix form of

$$\frac{d^2 v}{dx^2} = [\mathbf{D}]\{\mathbf{U}\} \tag{8.29}$$

The $\left(\frac{d^2 v}{dx^2}\right)^2$ term can be represented in terms of the $\{\mathbf{U}\}$ and $[\mathbf{D}]$ matrices as

$$\left(\frac{d^2 v}{dx^2}\right)^2 = \{\mathbf{U}\}^T[\mathbf{D}]^T[\mathbf{D}]\{\mathbf{U}\} \tag{8.30}$$

Thus, the strain energy for an arbitrary beam element is

$$\Lambda^{(e)} = \frac{EI}{2}\int_0^L \{\mathbf{U}\}^T[\mathbf{D}]^T[\mathbf{D}]\{\mathbf{U}\}\, dx \tag{8.31}$$

Recall that the total potential energy $\Pi$ for a body is the difference between the total strain energy and the work done by the external forces:

$$\Pi = \Sigma\Lambda^{(e)} - \Sigma FU \tag{8.32}$$

Also recall that the minimum total potential energy principle states that for a stable system, the displacement at the equilibrium position occurs such that the value of the system's total potential energy is a minimum. Thus, for a beam element, we have

$$\frac{\partial \Pi}{\partial U_k} = \frac{\partial}{\partial U_k}\Sigma\Lambda^{(e)} - \frac{\partial}{\partial U_k}\Sigma FU = 0 \quad \text{for } k = 1, 2, 3, 4 \tag{8.33}$$

where $U_k$ takes on the values of the nodal degrees of freedom $U_{i1}, U_{i2}, U_{j1}$, and $U_{j2}$. We begin minimizing the strain energy with respect to $U_{i1}, U_{i2}, U_{j1}$, and $U_{j2}$ to obtain the stiffness matrix. Starting with the strain energy part of the total potential energy, we get

$$\frac{\partial \Lambda^{(e)}}{\partial U_k} = EI\int_0^L [\mathbf{D}]^T[\mathbf{D}]\, dx\, \{\mathbf{U}\} \tag{8.34}$$

Evaluating Eq. (8.34) leads to the expression

$$\frac{\partial \Lambda^{(e)}}{\partial U_k} = EI \int_0^L [\mathbf{D}]^T[\mathbf{D}]\, dx \, \{\mathbf{U}\} = \frac{EI}{L^3}\begin{bmatrix} 12 & 6L & -12 & 6L \\ 6L & 4L^2 & -6L & 2L^2 \\ -12 & -6L & 12 & -6L \\ 6L & 2L^2 & -6L & 4L^2 \end{bmatrix} \begin{Bmatrix} U_{i1} \\ U_{i2} \\ U_{j1} \\ U_{j2} \end{Bmatrix}$$

The stiffness matrix for a beam element with two degrees of freedom at each node—the vertical displacement and rotation—is

$$[\mathbf{K}]^{(e)} = \frac{EI}{L^3}\begin{bmatrix} 12 & 6L & -12 & 6L \\ 6L & 4L^2 & -6L & 2L^2 \\ -12 & -6L & 12 & -6L \\ 6L & 2L^2 & -6L & 4L^2 \end{bmatrix} \tag{8.35}$$

## Load Matrix

There are two ways in which we can formulate the nodal load matrices: (1) by minimizing the work done by the load, and (2) by computing the beam's reaction forces. Consider a uniformly distributed load acting on a beam of length $L$, as shown in Figure 8.8. The reaction forces and moments at the endpoints are also shown in the figure.

Using the first approach, we can compute the work done by this type of loading from $\int_L wv\, dx$. The next step involves substituting for the displacement function in terms of the shape functions and nodal values, and then integrating and differentiating the work term with respect to the nodal displacements. This approach will be demonstrated in detail when we formulate the load matrix for a plane stress situation. Let us develop the load matrix using the alternate approach, starting with Eq. (8.18):

$$EI\frac{d^4 v}{dx^4} = \frac{dV(x)}{dx} = w(x)$$

For a uniformly distributed load, $w(x)$ is constant. Integrating this equation, we get

$$EI\frac{d^3 v}{dx^3} = -wx + c_1 \tag{8.36}$$

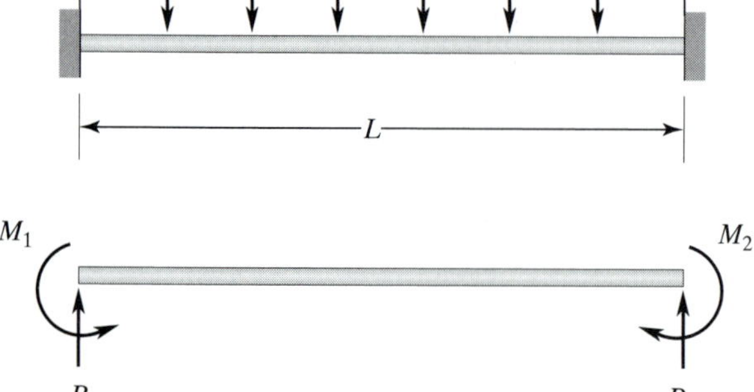


**FIGURE 8.8** A beam element subjected to a uniform distributed load.

Applying the boundary condition $EI \left.\dfrac{d^3 v}{dx^3}\right|_{x=0} = R_1$, we find that $c_1 = R_1$. Substituting for the value of $c_1$ and integrating again, we obtain

$$EI \frac{d^2 v}{dx^2} = -\frac{wx^2}{2} + R_1 x + c_2 \tag{8.37}$$

Applying the boundary condition $EI \left.\dfrac{d^2 v}{dx^2}\right|_{x=0} = -M_1$, we find that $c_2 = -M_1$. Substituting for the value of $c_2$ and integrating, we obtain

$$EI \frac{dv}{dx} = -\frac{wx^3}{6} + \frac{R_1 x^2}{2} - M_1 x + c_3 \tag{8.38}$$

Applying the boundary condition $\frac{dv}{dx}\big|_{x=0} = 0$, we find that $c_3 = 0$. Integrating one last time, we have

$$EIv = -\frac{wx^4}{24} + \frac{R_1 x^3}{6} - \frac{M_1 x^2}{2} + c_4 \tag{8.39}$$

Applying the boundary condition $v(0) = 0$, we determine that $c_4 = 0$. To obtain the values of $R_1$ and $M_1$, we can apply two additional boundary conditions to this problem: $\frac{dv}{dx}\big|_{x=L} = 0$ and $v(L) = 0$. Applying these conditions, we get

$$\left.\frac{dv}{dx}\right|_{x=L} = -\frac{wL^3}{6} + \frac{R_1 L^2}{2} - M_1 L = 0 \tag{8.40}$$

$$v(L) = -\frac{wL^4}{24} + \frac{R_1 L^3}{6} - \frac{M_1 L^2}{2} = 0 \tag{8.41}$$

Solving these equations simultaneously, we get $R_1 = \frac{wL}{2}$ and $M_1 = \dfrac{wL^2}{12}$. From the symmetry of the problem—that is, applying the statics equilibrium conditions—we find that the reactions at the other end of the beam are $R_2 = \frac{wL}{2}$ and $M_2 = \dfrac{wL^2}{12}$. All of the reactions are shown in Figure 8.9.

If we reverse the signs of the reactions at the endpoints, we can now represent the effect of a uniformly distributed load in terms of its equivalent nodal loads. Similarly, we

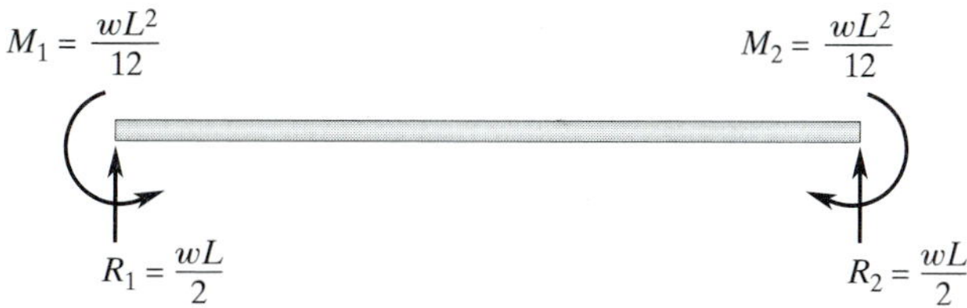


**FIGURE 8.9** Reaction results for a beam subjected to a uniformly distributed load.

TABLE 8.3 Equivalent nodal loading of beams

| Loading | Equivalent Nodal Loading |
|---|---|
| $w$; $L$ | $\frac{wL}{2}$, $\frac{wL}{2}$; $\frac{wL^2}{12}$, $\frac{wL^2}{12}$ |
| $w$; $L$ | $\frac{3wL}{20}$, $\frac{7wL}{20}$; $\frac{wL^2}{30}$, $\frac{wL^2}{20}$ |
| $P$; $\frac{L}{2}$, $\frac{L}{2}$ | $\frac{P}{2}$, $\frac{P}{2}$; $M = \frac{PL}{8}$, $M = \frac{PL}{8}$ |

can obtain the nodal load matrices for other loading situations. The relationships between the actual load and its equivalent nodal loads for some typical loading situations are summarized in Table 8.3.

## EXAMPLE 8.1

The beam shown in Figure 8.10 is a wide-flange W310 × 52 with a cross-sectional area of 6650 $\text{mm}^2$ and depth of 317 mm. The second moment of the area is $118.6 \times 10^6\ \text{mm}^4$. The beam is subjected to a uniformly distributed load of 25,000 N/m. The modulus of

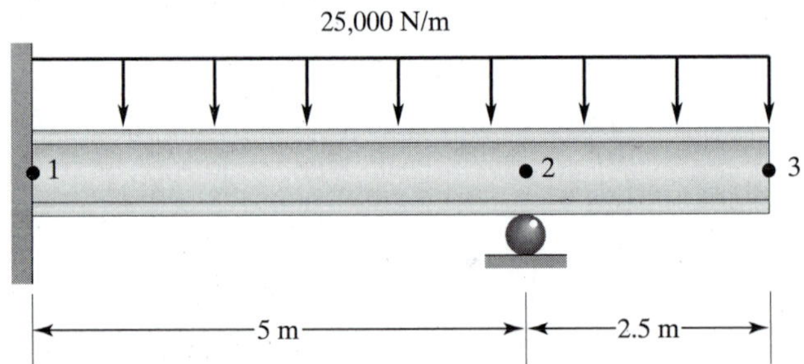


FIGURE 8.10 A schematic of the beam in Example 8.1.

elasticity of the beam is $E = 200$ GPa. Determine the vertical displacement at node 3 and the rotations at nodes 2 and 3. Also, compute the reaction forces and moment at nodes 1 and 2.

Note that this problem is statically indeterminate. We will use two elements to represent this problem. The stiffness matrices of the elements are computed from Eq. (8.35):

$$[\mathbf{K}]^{(e)} = \frac{EI}{L^3}\begin{bmatrix} 12 & 6L & -12 & 6L \\ 6L & 4L^2 & -6L & 2L^2 \\ -12 & -6L & 12 & -6L \\ 6L & 2L^2 & -6L & 4L^2 \end{bmatrix}$$

Substituting appropriate values for element (1), we have

$$[\mathbf{K}]^{(1)} = \frac{200 \times 10^9 \times 1.186 \times 10^{-4}}{5^3}\begin{bmatrix} 12 & 6(5) & -12 & 6(5) \\ 6(5) & 4(5)^2 & -6(5) & 2(5)^2 \\ -12 & -6(5) & 12 & -6(5) \\ 6(5) & 2(5)^2 & -6(5) & 4(5)^2 \end{bmatrix}$$

For convenience, the nodal degrees of freedom are shown alongside the stiffness matrices. For element (1), we have

$$[\mathbf{K}]^{(1)} = \begin{bmatrix} 2277120 & 5692800 & -2277120 & 5692800 \\ 5692800 & 18976000 & -5692800 & 9488000 \\ -2277120 & -5692800 & 2277120 & -5692800 \\ 5692800 & 9488000 & -5692800 & 18976000 \end{bmatrix}\begin{matrix} U_{11} \\ U_{12} \\ U_{21} \\ U_{22} \end{matrix}$$

Computing the stiffness matrix for element (2), we have

$$[\mathbf{K}]^{(2)} = \frac{200 \times 10^9 \times 1.186 \times 10^{-4}}{(2.5)^3}\begin{bmatrix} 12 & 6(2.5) & -12 & 6(2.5) \\ 6(2.5) & 4(2.5)^2 & -6(2.5) & 2(2.5)^2 \\ -12 & -6(2.5) & 12 & -6(2.5) \\ 6(2.5) & 2(2.5)^2 & -6(2.5) & 4(2.5)^2 \end{bmatrix}$$

Showing the nodal degrees of freedom alongside the stiffness matrix for element (2), we have

$$[\mathbf{K}]^{(2)} = \begin{bmatrix} 18216960 & 22771200 & -18216960 & 22771200 \\ 22771200 & 37952000 & -22771200 & 18976000 \\ -18216960 & -22771200 & 18216960 & -22771200 \\ 22771200 & 18976000 & -22771200 & 37952000 \end{bmatrix}\begin{matrix} U_{21} \\ U_{22} \\ U_{31} \\ U_{32} \end{matrix}$$

Assembling $[\mathbf{K}]^{(1)}$ and $[\mathbf{K}]^{(2)}$ to obtain the global stiffness matrix yields

$$[\mathbf{K}]^{(G)} = \begin{bmatrix} 2277120 & 5692800 & -2277120 & 5692800 & 0 & 0 \\ 5692800 & 18976000 & -5692800 & 9488000 & 0 & 0 \\ -2277120 & -5692800 & 20494080 & 17078400 & -18216960 & 22771200 \\ 5692800 & 9488000 & 17078400 & 56928000 & -22771200 & 18976000 \\ 0 & 0 & -18216960 & -22771200 & 18216960 & -22771200 \\ 0 & 0 & 22771200 & 18976000 & -22771200 & 37952000 \end{bmatrix}$$

Referring to Table 8.3, we can compute the load matrix for elements (1) and (2). The respective load matrices are:

$$\{\mathbf{F}\}^{(1)} = \begin{Bmatrix} -\dfrac{wL}{2} \\ -\dfrac{wL^2}{12} \\ -\dfrac{wL}{2} \\ \dfrac{wL^2}{12} \end{Bmatrix} = \begin{Bmatrix} -\dfrac{25 \times 10^3 \times 5}{2} \\ -\dfrac{25 \times 10^3 \times 5^2}{12} \\ -\dfrac{25 \times 10^3 \times 5}{2} \\ \dfrac{25 \times 10^3 \times 5^2}{12} \end{Bmatrix} = \begin{Bmatrix} -62500 \\ -52083 \\ -62500 \\ 52083 \end{Bmatrix}$$

$$\{\mathbf{F}\}^{(2)} = \begin{Bmatrix} -\dfrac{wL}{2} \\ -\dfrac{wL^2}{12} \\ -\dfrac{wL}{2} \\ \dfrac{wL^2}{12} \end{Bmatrix} = \begin{Bmatrix} -\dfrac{25 \times 10^3 \times 2.5}{2} \\ -\dfrac{25 \times 10^3 \times 2.5^2}{12} \\ -\dfrac{25 \times 10^3 \times 2.5}{2} \\ \dfrac{25 \times 10^3 \times 2.5^2}{12} \end{Bmatrix} = \begin{Bmatrix} -31250 \\ -13021 \\ -31250 \\ 13021 \end{Bmatrix}$$

Combining the two load matrices to obtain the global load matrix, we obtain

$$\{\mathbf{F}\}^{(G)} = \begin{Bmatrix} -62500 \\ -52083 \\ -62500 - 31250 \\ 52083 - 13021 \\ -31250 \\ 13021 \end{Bmatrix} = \begin{Bmatrix} -62500 \\ -52083 \\ -93750 \\ 39062 \\ -31250 \\ 13021 \end{Bmatrix}$$

Applying the boundary conditions $U_{11} = U_{12} = 0$ at node 1 and the boundary condition $U_{21} = 0$ at node 2, we have:

$$\begin{bmatrix} 1 & 0 & 0 & 0 & 0 & 0 \\ 0 & 1 & 0 & 0 & 0 & 0 \\ 0 & 0 & 1 & 0 & 0 & 0 \\ 5692800 & 9488000 & 17078400 & 56928000 & -22771200 & 18976000 \\ 0 & 0 & -18216960 & -22771200 & 18216960 & -22771200 \\ 0 & 0 & 22771200 & 18976000 & -22771200 & 37952000 \end{bmatrix} \begin{Bmatrix} U_{11} \\ U_{12} \\ U_{21} \\ U_{22} \\ U_{31} \\ U_{32} \end{Bmatrix} = \begin{Bmatrix} 0 \\ 0 \\ 0 \\ 39062 \\ -31250 \\ 13021 \end{Bmatrix}$$

Considering the applied boundary conditions, we reduce the global stiffness matrix and the load matrix to:

$$\begin{bmatrix} 56928000 & -22771200 & 18976000 \\ -22771200 & 18216960 & -22771200 \\ 18976000 & -22771200 & 37952000 \end{bmatrix} \begin{Bmatrix} U_{22} \\ U_{31} \\ U_{32} \end{Bmatrix} = \begin{Bmatrix} 39062 \\ -31250 \\ 13021 \end{Bmatrix}$$

Solving the three equations simultaneously results in the unknown nodal values. The displacement result is

$$[\mathbf{U}]^T = \begin{bmatrix} 0 & 0 & 0 & -0.0013723(\text{rad}) & -0.0085772(\text{m}) & -0.004117(\text{rad}) \end{bmatrix}$$

We can compute the nodal reaction forces and moments from the relationship

$$\{\mathbf{R}\} = [\mathbf{K}]\{\mathbf{U}\} - \{\mathbf{F}\} \tag{8.42}$$

where $\{\mathbf{R}\}$ is the reaction matrix. Substituting for the appropriate values in Eq. (8.42), we have

$$\begin{Bmatrix} R_1 \\ M_1 \\ R_2 \\ M_2 \\ R_3 \\ M_3 \end{Bmatrix} = \begin{bmatrix} 2277120 & 5692800 & -2277120 & 5692800 & 0 & 0 \\ 5692800 & 18976000 & -5692800 & 9488000 & 0 & 0 \\ -2277120 & -5692800 & 20494080 & 17078400 & -18216960 & 22771200 \\ 5692800 & 9488000 & 17078400 & 56928000 & -22771200 & 18976000 \\ 0 & 0 & -18216960 & -22771200 & 18216960 & -22771200 \\ 0 & 0 & 22771200 & 18976000 & -22771200 & 37952000 \end{bmatrix}$$

$$\begin{Bmatrix} 0 \\ 0 \\ 0 \\ -0.0013723 \\ -0.0085772 \\ -0.0041170 \end{Bmatrix} - \begin{Bmatrix} -62500 \\ -52083 \\ -93750 \\ 39062 \\ -31250 \\ 13021 \end{Bmatrix}$$

Performing the matrix operation results in the following reaction forces and moments at each node:

$$\begin{Bmatrix} R_1 \\ M_1 \\ R_2 \\ M_2 \\ R_3 \\ M_3 \end{Bmatrix} = \begin{Bmatrix} 54687(\text{N}) \\ 39062(\text{N}\cdot\text{m}) \\ 132814(\text{N}) \\ 0 \\ 0 \\ 0 \end{Bmatrix}$$

Note that by calculating the reaction matrix using the nodal displacement matrix, we can check the validity of our results. There is a reaction force and a reaction moment at node 1; there is a reaction force at node 2; there is no reaction moment at node 2, as expected; and there are no reaction forces or moments at node 3, as expected. The accuracy of the results is discussed further in Section 8.6.

---

### Frames

Frames represent structural members that may be rigidly connected with welded joints or bolted joints. For such structures, in addition to rotation and lateral displacement, we also need to be concerned about axial deformations. Here, we will focus on plane

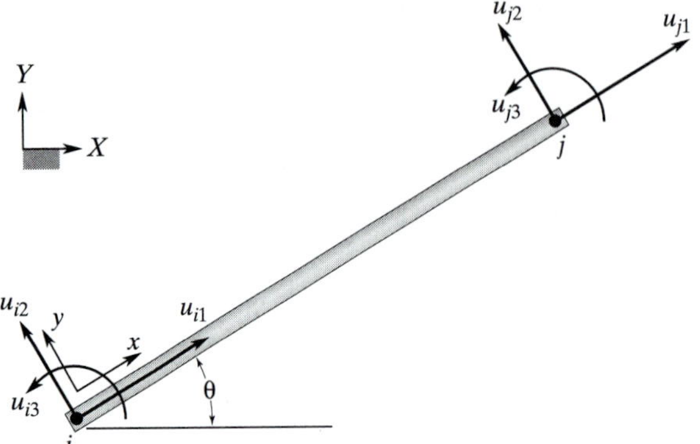


**FIGURE 8.11** A frame element.

frames. The frame element, shown in Figure 8.11, consists of two nodes. At each node, there are three degrees of freedom: a longitudinal displacement, a lateral displacement, and a rotation.

Referring to Figure 8.11, note that $u_{i1}$ represents the longitudinal displacement and $u_{i2}$ and $u_{i3}$ represent the lateral displacement and the rotation at node $i$, respectively. In the same manner, $u_{j1}$, $u_{j2}$, and $u_{j3}$ represent the longitudinal displacement, the lateral displacement, and the rotation at node $j$, respectively. In general, two frames of reference will be required to describe frame elements: a global coordinate system and a local frame of reference. We choose a fixed global coordinate system $(X, Y)$ for several uses: (1) to represent the location of each joint (node) and to keep track of the orientation of each element using angles such as $\theta$; (2) to apply the constraints and the applied loads in terms of their respective global components; and (3) to represent the solution. We will also need a local, or elemental, coordinate system to describe the axial-load behavior of an element. The relationship between the local coordinate system $(x, y)$ and the global coordinate system $(X, Y)$ is shown in Figure 8.11. Because there are three degrees of freedom associated with each node, the stiffness matrix for the frame element will be a $6 \times 6$ matrix. The local degrees of freedom are related to the global degrees of freedom through the transformation matrix, according to the relationship

$$[\mathbf{u}] = [\mathbf{T}][\mathbf{U}] \tag{8.43}$$

where the transformation matrix is

$$[\mathbf{T}] = \begin{bmatrix} \cos\theta & \sin\theta & 0 & 0 & 0 & 0 \\ -\sin\theta & \cos\theta & 0 & 0 & 0 & 0 \\ 0 & 0 & 1 & 0 & 0 & 0 \\ 0 & 0 & 0 & \cos\theta & \sin\theta & 0 \\ 0 & 0 & 0 & -\sin\theta & \cos\theta & 0 \\ 0 & 0 & 0 & 0 & 0 & 1 \end{bmatrix} \tag{8.44}$$

In the previous section, we developed the stiffness matrix attributed to bending for a beam element. This matrix accounts for lateral displacements and rotations at each node and is

$$[\mathbf{K}]^{(e)} = \frac{EI}{L^3}\begin{bmatrix} 0 & 0 & 0 & 0 & 0 & 0 \\ 0 & 12 & 6L & 0 & -12 & 6L \\ 0 & 6L & 4L^2 & 0 & -6L & 2L^2 \\ 0 & 0 & 0 & 0 & 0 & 0 \\ 0 & -12 & -6L & 0 & 12 & -6L \\ 0 & 6L & 2L^2 & 0 & -6L & 4L^2 \end{bmatrix} \begin{matrix} u_{i1} \\ u_{i2} \\ u_{i3} \\ u_{j1} \\ u_{j2} \\ u_{j3} \end{matrix} \qquad (8.45)$$

(columns: $u_{i1}$, $u_{i2}$, $u_{i3}$, $u_{j1}$, $u_{j2}$, $u_{j3}$)

To represent the contribution of each term to nodal degrees of freedom, the degrees of freedom are shown above and alongside the stiffness matrix in Eq. (8.45). In Chapter 4, we derived the stiffness matrix for members under axial loading as

$$[\mathbf{K}]^{(e)}_{\text{axial}} = \begin{bmatrix} \frac{AE}{L} & 0 & 0 & -\frac{AE}{L} & 0 & 0 \\ 0 & 0 & 0 & 0 & 0 & 0 \\ 0 & 0 & 0 & 0 & 0 & 0 \\ -\frac{AE}{L} & 0 & 0 & \frac{AE}{L} & 0 & 0 \\ 0 & 0 & 0 & 0 & 0 & 0 \\ 0 & 0 & 0 & 0 & 0 & 0 \end{bmatrix} \begin{matrix} u_{i1} \\ u_{i2} \\ u_{i3} \\ u_{j1} \\ u_{j2} \\ u_{j3} \end{matrix} \qquad (8.46)$$

(columns: $u_{i1}$, $u_{i2}$, $u_{i3}$, $u_{j1}$, $u_{j2}$, $u_{j3}$)

Adding Eqs. (8.45) and (8.46) results in the stiffness matrix for a frame element:

$$[\mathbf{K}]^{(e)}_{xy} = \begin{bmatrix} \frac{AE}{L} & 0 & 0 & -\frac{AE}{L} & 0 & 0 \\ 0 & \frac{12EI}{L^3} & \frac{6EI}{L^2} & 0 & -\frac{12EI}{L^3} & \frac{6EI}{L^2} \\ 0 & \frac{6EI}{L^2} & \frac{4EI}{L} & 0 & -\frac{6EI}{L^2} & \frac{2EI}{L} \\ -\frac{AE}{L} & 0 & 0 & \frac{AE}{L} & 0 & 0 \\ 0 & -\frac{12EI}{L^3} & -\frac{6EI}{L^2} & 0 & \frac{12EI}{L^3} & -\frac{6EI}{L^2} \\ 0 & \frac{6EI}{L^2} & \frac{2EI}{L} & 0 & -\frac{6EI}{L^2} & \frac{4EI}{L} \end{bmatrix} \qquad (8.47)$$

Note that we need to represent Eq. (8.47) with respect to the global coordinate system. To perform this task, we must substitute for the local displacements in terms of the global displacements in the strain energy equation, using the transformation matrix and performing the minimization. (See Problem 8.12.) These steps result in the relationship

$$[\mathbf{K}]^{(e)} = [\mathbf{T}]^T[\mathbf{K}]_{xy}[\mathbf{T}] \tag{8.48}$$

Next, we will demonstrate finite element modeling of frames with another example.

## EXAMPLE 8.2

Consider the overhang frame shown in Figure 8.12. The frame is made of steel, with $E = 30 \times 10^6$ lb/in$^2$. The cross-sectional areas and the second moment of areas for the two members are shown in Figure 8.12. The frame is fixed as shown in the figure, and we are interested in determining the deformation of the frame under the given distributed load.

We will model the problem using two elements. For element (1), the relationship between the local and the global coordinate systems is shown in Figure 8.13.

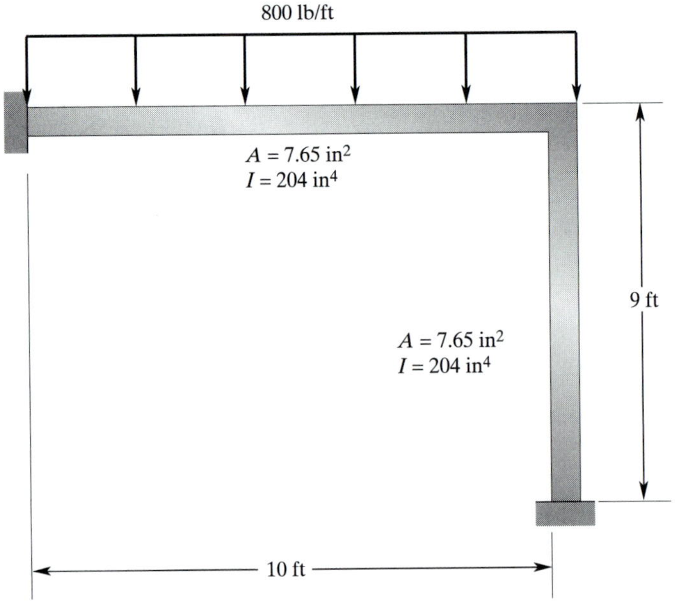


**FIGURE 8.12** An overhang frame supporting a distributed load.

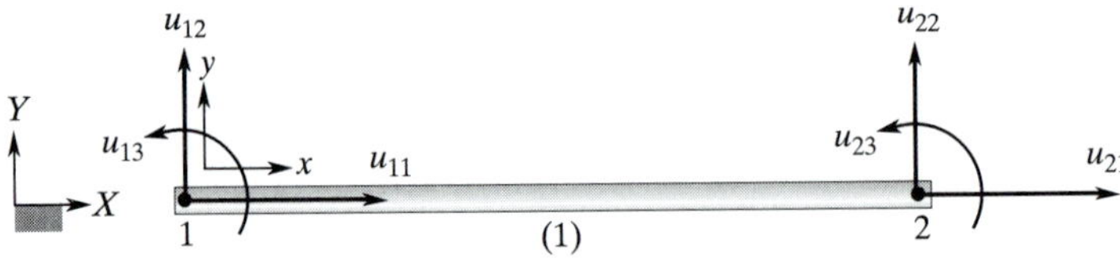


**FIGURE 8.13** The configuration of element (1).

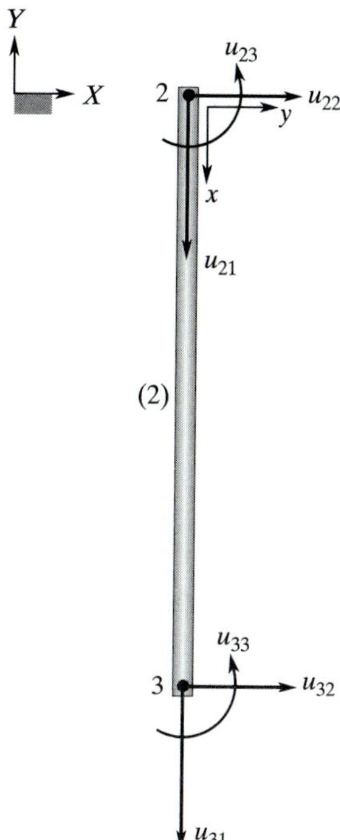


**FIGURE 8.14** The configuration of element (2).

Similarly, the relationship between the coordinate systems for element (2) is shown in Figure 8.14.

Note that for this problem, the boundary conditions are $U_{11} = U_{12} = U_{13} = U_{31} = U_{32} = U_{33} = 0$. For element (1), the local and the global frames of reference are aligned in the same direction; therefore, the stiffness matrix for element (1) can be computed from Eq. (8.47), resulting in:

$$[\mathbf{K}]^{(1)} = 10^3 \begin{bmatrix} 1912.5 & 0 & 0 & -1912.5 & 0 & 0 \\ 0 & 42.5 & 2550 & 0 & -42.5 & 2550 \\ 0 & 2550 & 204000 & 0 & -2550 & 102000 \\ -1912.5 & 0 & 0 & 1912.5 & 0 & 0 \\ 0 & -42.5 & -2550 & 0 & 42.5 & -2550 \\ 0 & 2550 & 102000 & 0 & -2550 & 204000 \end{bmatrix}$$

For element (2), the stiffness matrix represented with respect to the local coordinate system is:

$$[\mathbf{K}]^{(2)}_{xy} = 10^3 \begin{bmatrix} 2125 & 0 & 0 & -2125 & 0 & 0 \\ 0 & 58.299 & 3148.148 & 0 & -58.299 & 3148.148 \\ 0 & 3148.148 & 226666 & 0 & -3148.148 & 113333 \\ -2125 & 0 & 0 & 2125 & 0 & 0 \\ 0 & -58.299 & -3148.148 & 0 & 58.299 & -3148.148 \\ 0 & 3148.148 & 113333 & 0 & -3148.148 & 226666 \end{bmatrix}$$

For element (2), the transformation matrix is

$$[\mathbf{T}] = \begin{bmatrix} \cos(270) & \sin(270) & 0 & 0 & 0 & 0 \\ -\sin(270) & \cos(270) & 0 & 0 & 0 & 0 \\ 0 & 0 & 1 & 0 & 0 & 0 \\ 0 & 0 & 0 & \cos(270) & \sin(270) & 0 \\ 0 & 0 & 0 & -\sin(270) & \cos(270) & 0 \\ 0 & 0 & 0 & 0 & 0 & 1 \end{bmatrix}$$

$$[\mathbf{T}] = \begin{bmatrix} 0 & -1 & 0 & 0 & 0 & 0 \\ 1 & 0 & 0 & 0 & 0 & 0 \\ 0 & 0 & 1 & 0 & 0 & 0 \\ 0 & 0 & 0 & 0 & -1 & 0 \\ 0 & 0 & 0 & 1 & 0 & 0 \\ 0 & 0 & 0 & 0 & 0 & 1 \end{bmatrix}$$

The transpose of the transformation matrix is:

$$[\mathbf{T}]^T = \begin{bmatrix} 0 & 1 & 0 & 0 & 0 & 0 \\ -1 & 0 & 0 & 0 & 0 & 0 \\ 0 & 0 & 1 & 0 & 0 & 0 \\ 0 & 0 & 0 & 0 & 1 & 0 \\ 0 & 0 & 0 & -1 & 0 & 0 \\ 0 & 0 & 0 & 0 & 0 & 1 \end{bmatrix}$$

Substituting for $[\mathbf{T}]^T$, $[\mathbf{K}]^{(2)}_{xy}$, and $[\mathbf{T}]$ into Eq. (8.48), we have:

$$[\mathbf{K}]^{(2)} = 10^3 \begin{bmatrix} 0 & 1 & 0 & 0 & 0 & 0 \\ -1 & 0 & 0 & 0 & 0 & 0 \\ 0 & 0 & 1 & 0 & 0 & 0 \\ 0 & 0 & 0 & 0 & 1 & 0 \\ 0 & 0 & 0 & -1 & 0 & 0 \\ 0 & 0 & 0 & 0 & 0 & 1 \end{bmatrix} \begin{bmatrix} 2125 & 0 & 0 & -2125 & 0 & 0 \\ 0 & 58.299 & 3148.148 & 0 & -58.299 & 3148.148 \\ 0 & 3148.148 & 226666 & 0 & -3148.148 & 113333 \\ -2125 & 0 & 0 & 2125 & 0 & 0 \\ 0 & -58.299 & -3148.148 & 0 & 58.299 & -3148.148 \\ 0 & 3148.148 & 113333 & 0 & -3148.148 & 226666 \end{bmatrix}$$

$$\begin{bmatrix} 0 & -1 & 0 & 0 & 0 & 0 \\ 1 & 0 & 0 & 0 & 0 & 0 \\ 0 & 0 & 1 & 0 & 0 & 0 \\ 0 & 0 & 0 & 0 & -1 & 0 \\ 0 & 0 & 0 & 1 & 0 & 0 \\ 0 & 0 & 0 & 0 & 0 & 1 \end{bmatrix}$$

and performing the matrix operation, we obtain

$$[\mathbf{K}]^{(2)} = 10^3 \begin{bmatrix} 58.299 & 0 & 3148.148 & -58.299 & 0 & 3148.148 \\ 0 & 2125 & 0 & 0 & -2125 & 0 \\ 3148.148 & 0 & 226666 & -3148.148 & 0 & 113333 \\ -58.299 & 0 & -3148.148 & 58.299 & 0 & -3148.1480 \\ 0 & -2125 & 0 & 0 & 2125 & 0 \\ 3148.148 & 0 & 113333 & -3148.148 & 0 & 226666 \end{bmatrix}$$

Constructing the global stiffness matrix by assembling $[\mathbf{K}]^{(1)}$ and $[\mathbf{K}]^{(2)}$, we have

$$[\mathbf{K}]^{(G)} = 10^3 \begin{bmatrix} 1912.5 & 0 & 0 & -1912.5 & 0 & 0 & 0 & 0 & 0 \\ 0 & 42.5 & 2550 & 0 & -42.5 & 2550 & 0 & 0 & 0 \\ 0 & 2550 & 204000 & 0 & -2550 & 102000 & 0 & 0 & 0 \\ -1912.5 & 0 & 0 & 1912.5 + 58.299 & 0 & 0 + 3148.148 & -58.299 & 0 & 3148.148 \\ 0 & -42.5 & -2550 & 0 & 42.5 + 2125 & -2550 & 0 & -2125 & 0 \\ 0 & 2550 & 102000 & 0 + 3148.148 & -2550 & 204000 + 226666 & -3148.148 & 0 & 113333 \\ 0 & 0 & 0 & -58.299 & 0 & -3148.148 & 58.299 & 0 & -3148.1480 \\ 0 & 0 & 0 & 0 & -2125 & 0 & 0 & 2125 & 0 \\ 0 & 0 & 0 & 3148.148 & 0 & 113333 & -3148.148 & 0 & 226666 \end{bmatrix}$$

The load matrix is:

$$\{\mathbf{F}\}^{(1)} = \begin{Bmatrix} 0 \\ -\dfrac{wL}{2} \\ -\dfrac{wL^2}{12} \\ 0 \\ -\dfrac{wL}{2} \\ \dfrac{wL^2}{12} \end{Bmatrix} = \begin{Bmatrix} 0 \\ -\dfrac{800 \times 10}{2} \\ -\dfrac{800 \times 10^2 \times 12}{12} \\ 0 \\ -\dfrac{800 \times 10}{2} \\ \dfrac{800 \times 10^2 \times 12}{12} \end{Bmatrix} = \begin{Bmatrix} 0 \\ -4000 \\ -80000 \\ 0 \\ -4000 \\ 80000 \end{Bmatrix}$$

In the load matrix, the force terms have the units of lb, whereas the moment terms have the units of lb · in. Application of the boundary conditions ($U_{11} = U_{12} = U_{13} = U_{31} = U_{32} = U_{33} = 0$) reduces the 9×9 global stiffness matrix to the following 3×3 matrix:

$$10^3\begin{bmatrix} 1970.799 & 0 & 3148.148 \\ 0 & 2167.5 & -2550 \\ 3148.148 & -2550 & 430666 \end{bmatrix}\begin{Bmatrix} U_{21} \\ U_{22} \\ U_{23} \end{Bmatrix} = \begin{Bmatrix} 0 \\ -4000 \\ 80000 \end{Bmatrix}$$

Solving these equations simultaneously results in the following displacement matrix:

$$[\mathbf{U}]^T = [0 \quad 0 \quad 0 \quad -0.0002845(\text{in}) \quad -0.0016359(\text{in}) \quad 0.00017815(\text{rad}) \quad 0 \quad 0 \quad 0]$$

This problem will be revisited later in the chapter and solved with ANSYS.

## 8.3 PLANE STRESS FORMULATION

We begin by reviewing some of the fundamental concepts dealing with the elastic behavior of materials. Consider an infinitesimally small cube volume surrounding a point within a material. An enlarged version of this volume is shown in Figure 8.15. The faces of the cube are oriented in the directions of $(X, Y, Z)$ coordinate system.* The application of external forces creates internal forces and, subsequently, stresses within the material. The state of stress at a point can be defined in terms of the nine components on the positive faces and their counterparts on the negative surfaces, as shown in the figure. However, recall that because of equilibrium requirements, only six independent stress components are needed to characterize the general state of stress at a point. Thus the general state of stress at a point is defined by:

$$[\boldsymbol{\sigma}]^T = \begin{bmatrix}\sigma_{XX} & \sigma_{YY} & \sigma_{ZZ} & \tau_{XY} & \tau_{YZ} & \tau_{XZ}\end{bmatrix} \tag{8.49}$$

where $\sigma_{XX}$, $\sigma_{YY}$, and $\sigma_{ZZ}$ are the normal stresses and $\tau_{XY}$, $\tau_{YZ}$, and $\tau_{XZ}$ are the shear stress components, and they provide a measure of the intensity of the internal forces acting over areas of the cube faces. In many practical problems, we come across situa-

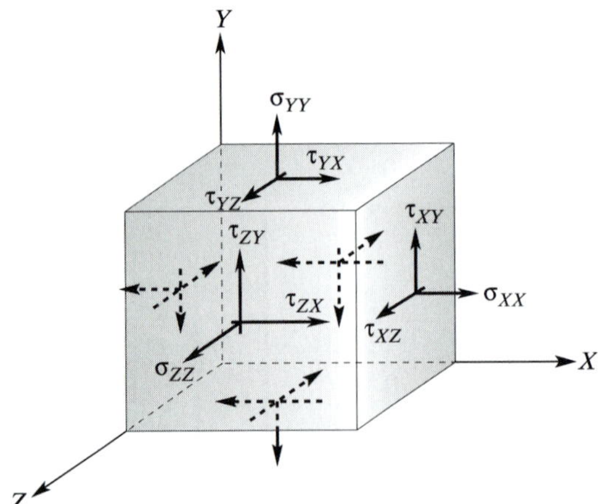


**FIGURE 8.15** The components of stress at a point.

*Note that throughout this section X, Y, Z and x, y, z coordinate systems are aligned.

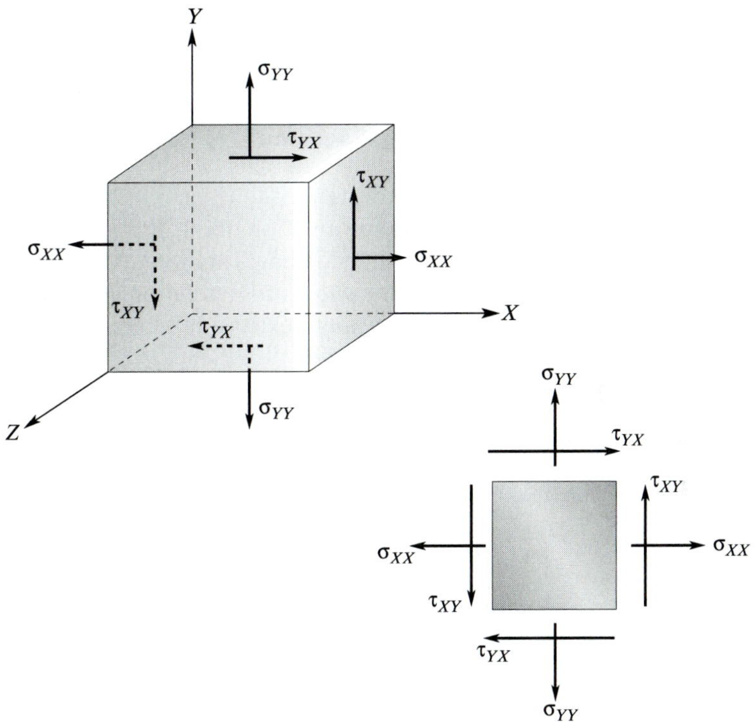


**FIGURE 8.16** Plane state of stress.

tions where there are no forces acting in the $Z$-direction and, consequently, no internal forces acting on the $Z$- faces. This situation is commonly referred to as a *plane stress* situation, as shown in Figure 8.16.

For a plane stress situation, the state of stress reduces to three components:

$$[\boldsymbol{\sigma}]^T = [\sigma_{XX} \quad \sigma_{YY} \quad \tau_{XY}] \tag{8.50}$$

We have just considered how an applied force can create stresses within a body. As you know, the applied force will also cause a body to undergo deformation, or change in its shape. We can use a displacement vector to measure the changes that occur in the position of a point within a body. The displacement vector $\vec{\delta}$ can be written in terms of its Cartesian components as

$$\vec{\delta} = u(x, y, z)\vec{i} + v(x, y, z)\vec{j} + w(x, y, z)\vec{k}$$

where the $i$, $j$, and $k$ components of the displacement vector represent the difference in the coordinates of the displacement of the point from its original position $(x, y, z)$ to a new position $(x', y', z')$ caused by loading, as given by the equations

$$u(x, y, z) = x' - x$$
$$v(x, y, z) = y' - y$$
$$w(x, y, z) = z' - z$$

To better measure the size and shape changes that occur locally within the material, we define normal and shear strains. The state of strain at a point is, therefore, characterized by six independent components:

$$[\boldsymbol{\varepsilon}]^T = \begin{bmatrix} \varepsilon_{xx} & \varepsilon_{yy} & \varepsilon_{zz} & \gamma_{xy} & \gamma_{yz} & \gamma_{xz} \end{bmatrix} \tag{8.51}$$

$\varepsilon_{xx}$, $\varepsilon_{yy}$, and $\varepsilon_{zz}$ are the normal strains, and $\gamma_{xy}$, $\gamma_{yz}$, and $\gamma_{xz}$ are the shear-strain components. These components provide information about the size and shape changes that occur locally in a given material due to loading. The situation in which no displacements occur in the $z$-direction is known as a *plane strain* situation. As you may recall from your study of the mechanics of materials, there exists a relationship between the strain and the displacement. These relationships are:

$$\varepsilon_{xx} = \frac{\partial u}{\partial x} \qquad \varepsilon_{yy} = \frac{\partial v}{\partial y} \qquad \varepsilon_{zz} = \frac{\partial w}{\partial z} \tag{8.52}$$

$$\gamma_{xy} = \frac{\partial u}{\partial y} + \frac{\partial v}{\partial x} \qquad \gamma_{yz} = \frac{\partial v}{\partial z} + \frac{\partial w}{\partial y} \qquad \gamma_{xz} = \frac{\partial u}{\partial z} + \frac{\partial w}{\partial x}$$

Over the elastic region of a material, there also exists a relationship between the state of stresses and strains, according to the generalized Hooke's Law. These relationships are:

$$\varepsilon_{xx} = \frac{1}{E}\left[\sigma_{xx} - \nu\left(\sigma_{yy} + \sigma_{zz}\right)\right] \tag{8.53}$$

$$\varepsilon_{yy} = \frac{1}{E}\left[\sigma_{yy} - \nu\left(\sigma_{xx} + \sigma_{zz}\right)\right]$$

$$\varepsilon_{zz} = \frac{1}{E}\left[\sigma_{zz} - \nu\left(\sigma_{xx} + \sigma_{yy}\right)\right]$$

$$\gamma_{xy} = \frac{1}{G}\tau_{xy} \qquad \gamma_{yz} = \frac{1}{G}\tau_{yz} \qquad \gamma_{zx} = \frac{1}{G}\tau_{zx}$$

where $E$ is the modulus of elasticity (Young's modulus), $\nu$ is Poisson's ratio, and $G$ is the shear modulus of elasticity (modulus of rigidity). For a plane stress situation, the generalized Hooke's Law reduces to

$$\begin{Bmatrix} \sigma_{xx} \\ \sigma_{yy} \\ \tau_{xy} \end{Bmatrix} = \frac{E}{1 - \nu^2}\begin{bmatrix} 1 & \nu & 0 \\ \nu & 1 & 0 \\ 0 & 0 & \dfrac{1 - \nu}{2} \end{bmatrix}\begin{Bmatrix} \varepsilon_{xx} \\ \varepsilon_{yy} \\ \gamma_{xy} \end{Bmatrix} \tag{8.54}$$

or, in a compact matrix form,

$$\{\boldsymbol{\sigma}\} = [\boldsymbol{\nu}]\{\boldsymbol{\varepsilon}\} \tag{8.55}$$

where

$$[\boldsymbol{\sigma}]^T = \begin{bmatrix}\sigma_{xx} & \sigma_{yy} & \tau_{xy}\end{bmatrix}$$

$$[\boldsymbol{\nu}] = \frac{E}{1-\nu^2}\begin{bmatrix} 1 & \nu & 0 \\ \nu & 1 & 0 \\ 0 & 0 & \dfrac{1-\nu}{2} \end{bmatrix}$$

$$[\boldsymbol{\varepsilon}] = \begin{Bmatrix} \varepsilon_{xx} \\ \varepsilon_{yy} \\ \gamma_{xy} \end{Bmatrix}$$

For a plane strain situation, the generalized Hooke's Law becomes:

$$\begin{Bmatrix} \sigma_{xx} \\ \sigma_{yy} \\ \tau_{xy} \end{Bmatrix} = \frac{E}{(1+\nu)(1-2\nu)}\begin{bmatrix} 1-\nu & \nu & 0 \\ \nu & 1-\nu & 0 \\ 0 & 0 & \dfrac{1}{2}-\nu \end{bmatrix}\begin{Bmatrix} \varepsilon_{xx} \\ \varepsilon_{yy} \\ \gamma_{xy} \end{Bmatrix} \tag{8.56}$$

Furthermore, for plane stress situations, the strain-displacement relationship becomes:

$$\varepsilon_{xx} = \frac{\partial u}{\partial x} \quad \varepsilon_{yy} = \frac{\partial v}{\partial y} \quad \gamma_{xy} = \frac{\partial u}{\partial y} + \frac{\partial v}{\partial x} \tag{8.57}$$

We have discussed throughout this text that the minimum total potential energy approach is very commonly used to generate finite element models in solid mechanics. External loads applied to a body will cause the body to deform. During the deformation, the work done by the external forces is stored in the material in the form of elastic energy, which is called strain energy. For a solid material under biaxial loading, the strain energy $\Lambda$ is

$$\Lambda^{(e)} = \frac{1}{2}\int_V \left(\sigma_{xx}\varepsilon_{xx} + \sigma_{yy}\varepsilon_{yy} + \tau_{xy}\gamma_{xy}\right) dV \tag{8.58}$$

Or, in a compact matrix form,

$$\Lambda^{(e)} = \frac{1}{2}\int_V [\boldsymbol{\sigma}]^T\{\boldsymbol{\varepsilon}\}\, dV \tag{8.59}$$

Substituting for stresses in terms of strains using Hooke's Law, Eq. (8.59) can be written as

$$\Lambda^{(e)} = \frac{1}{2}\int_V \{\boldsymbol{\varepsilon}\}^T[\boldsymbol{\nu}]\{\boldsymbol{\varepsilon}\}\, dV \tag{8.60}$$

We are now ready to look at finite element formulation of plane stress problems using triangular elements. We can represent the displacements $u$ and $v$ using a linear triangular element similar to the one shown in Figure 8.17.

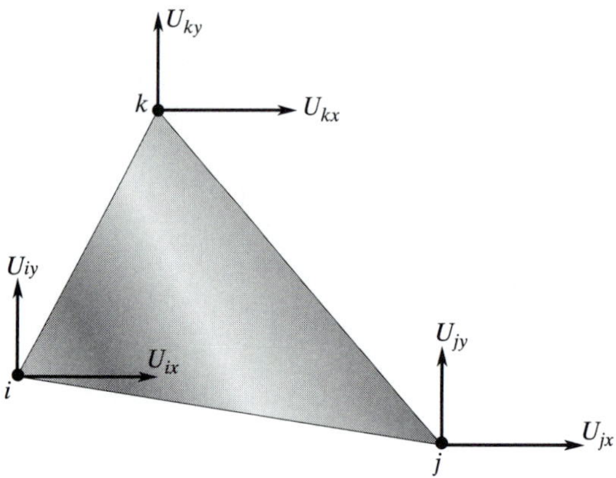


**FIGURE 8.17** A triangular element used in formulating plane stress problems.

The displacement variable, in terms of linear triangular shape functions and the nodal displacements, is

$$u = S_i U_{ix} + S_j U_{jx} + S_k U_{kx} \tag{8.61}$$

$$v = S_i U_{iy} + S_j U_{jy} + S_k U_{ky}$$

We can write the relations given by Eq. (8.61) in a matrix form:

$$\begin{Bmatrix} u \\ v \end{Bmatrix} = \begin{bmatrix} S_i & 0 & S_j & 0 & S_k & 0 \\ 0 & S_i & 0 & S_j & 0 & S_k \end{bmatrix} \begin{Bmatrix} U_{ix} \\ U_{iy} \\ U_{jx} \\ U_{jy} \\ U_{kx} \\ U_{ky} \end{Bmatrix} \tag{8.62}$$

The next step involves relating the strains to the displacement field and, subsequently, relating the strains to the nodal displacements using shape functions. Referring to the strain-displacement relations as given by Eq. (8.57), we need to take the derivatives of the components of the displacement field with respect to the $x$ and $y$ coordinates, which, in turn, means taking the derivatives of the appropriate shape functions with respect to $x$ and $y$. Performing these operations results in the following relations:

$$\varepsilon_{xx} = \frac{\partial u}{\partial x} = \frac{\partial}{\partial x}(S_i U_{ix} + S_j U_{jx} + S_k U_{kx}) = \frac{1}{2A}[\beta_i U_{ix} + \beta_j U_{jx} + \beta_k U_{kx}] \tag{8.63}$$

$$\varepsilon_{yy} = \frac{\partial v}{\partial y} = \frac{\partial}{\partial y}(S_i U_{iy} + S_j U_{jy} + S_k U_{ky}) = \frac{1}{2A}[\delta_i U_{iy} + \delta_j U_{jy} + \delta_k U_{kj}]$$

$$\gamma_{xy} = \frac{\partial u}{\partial y} + \frac{\partial v}{\partial x} = \frac{1}{2A}[\delta_i U_{ix} + \beta_i U_{iy} + \delta_j U_{jx} + \beta_j U_j + \delta_k U_{kx} + \beta_k U_{ky}]$$

Representing the relations of Eq. (8.63) in a matrix form, we have

$$\begin{Bmatrix} \varepsilon_{xx} \\ \varepsilon_{yy} \\ \gamma_{xy} \end{Bmatrix} = \frac{1}{2A}\begin{bmatrix} \beta_i & 0 & \beta_j & 0 & \beta_k & 0 \\ 0 & \delta_i & 0 & \delta_j & 0 & \delta_k \\ \delta_i & \beta_i & \delta_j & \beta_j & \delta_k & \beta_k \end{bmatrix}\begin{Bmatrix} U_{ix} \\ U_{iy} \\ U_{jx} \\ U_{jy} \\ U_{kx} \\ U_{ky} \end{Bmatrix} \tag{8.64}$$

and in a compact matrix form, Eq. (8.64) becomes

$$\{\boldsymbol{\varepsilon}\} = [\mathbf{B}]\{\mathbf{U}\} \tag{8.65}$$

where

$$\{\boldsymbol{\varepsilon}\} = \begin{Bmatrix} \varepsilon_{xx} \\ \varepsilon_{yy} \\ \gamma_{xy} \end{Bmatrix} \quad [\mathbf{B}] = \frac{1}{2A}\begin{bmatrix} \beta_i & 0 & \beta_j & 0 & \beta_k & 0 \\ 0 & \delta_i & 0 & \delta_j & 0 & \delta_k \\ \delta_i & \beta_i & \delta_j & \beta_j & \delta_k & \beta_k \end{bmatrix} \quad \{\mathbf{U}\} = \begin{Bmatrix} U_{ix} \\ U_{iy} \\ U_{jx} \\ U_{jy} \\ U_{kx} \\ U_{ky} \end{Bmatrix}$$

Substituting into the strain energy equation for the strain components in terms of the displacements, we obtain

$$\Lambda^{(e)} = \frac{1}{2}\int_V \{\boldsymbol{\varepsilon}\}^T[\boldsymbol{\nu}]\{\boldsymbol{\varepsilon}\}\, dV = \frac{1}{2}\int_V [\mathbf{U}]^T[\mathbf{B}]^T[\boldsymbol{\nu}][\mathbf{B}][\mathbf{U}]\, dV \tag{8.66}$$

Differentiating with respect to the nodal displacements, we obtain

$$\frac{\partial \Lambda^{(e)}}{\partial U_k} = \frac{\partial}{\partial U_k}\left(\frac{1}{2}\int_V [\mathbf{U}]^T[\mathbf{B}]^T[\boldsymbol{\nu}][\mathbf{B}][\mathbf{U}]\, dV\right) \quad \text{for } k = 1, 2, \ldots, 6 \tag{8.67}$$

Evaluation of Eq. (8.67) results in the expression $[\mathbf{K}]^{(e)}\{\mathbf{U}\}$. The expression for the stiffness matrix is thus

$$[\mathbf{K}]^{(e)} = \int_V [\mathbf{B}]^T[\boldsymbol{\nu}][\mathbf{B}]\, dV = V[\mathbf{B}]^T[\boldsymbol{\nu}][\mathbf{B}] \tag{8.68}$$

Here, $V$ is the volume of the element and is the product of the area of the element and its thickness.

### Load Matrix

To obtain the load matrix for a two-dimensional plane stress element, we must first compute the work done by the external forces, such as distributed loads or point loads. The work done by a concentrated load $Q$ is the product of the load component and the corresponding displacement component. We can represent the work done by concentrated loads in a compact matrix form as

$$W^{(e)} = \{\mathbf{U}\}^T\{\mathbf{Q}\} \tag{8.69}$$

A distributed load with $p_x$ and $p_y$ components does work according to the relationship

$$W^{(e)} = \int_A (up_x + vp_y)\, dA \tag{8.70}$$

where $u$ and $v$ are the displacements in the $x$ and $y$ directions, respectively, and $A$ represents the surface over which the distributed load components are acting. The magnitude of the surface $A$ is the product of the element thickness $t$ and the length of the edge over which the distributed load is applied. Using triangular elements to represent the displacements, we find that the work done by distributed loads becomes

$$W^{(e)} = \int_A \{\mathbf{U}\}^T[\mathbf{S}]^T\{\mathbf{p}\}\, dA \tag{8.71}$$

where

$$\{\mathbf{p}\} = \begin{Bmatrix} p_x \\ p_y \end{Bmatrix}$$

The next step in evaluating the load matrix involves the minimization process. In the case of the concentrated load, differentiation of Eq. (8.69) with respect to nodal displacements yields the components of the loads:

$$\{\mathbf{F}\}^{(e)} = \begin{Bmatrix} Q_{ix} \\ Q_{iy} \\ Q_{jx} \\ Q_{jy} \\ Q_{kx} \\ Q_{ky} \end{Bmatrix} \tag{8.72}$$

The differentiation of the work done by the distributed load with respect to the nodal displacements gives the load matrix

$$\{\mathbf{F}\}^{(e)} = \int_A [\mathbf{S}]^T\{\mathbf{p}\}\, dA \tag{8.73}$$

where

$$[\mathbf{S}]^T = \begin{bmatrix} S_i & 0 \\ 0 & S_i \\ S_j & 0 \\ 0 & S_j \\ S_k & 0 \\ 0 & S_k \end{bmatrix}$$

Consider an element subjected to a distributed load along its $ki$-edge, as shown in Figure 8.18.

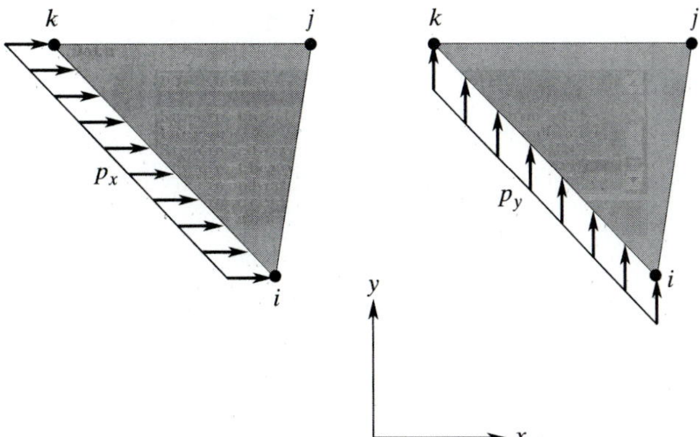


**FIGURE 8.18** A distributed load acting over the $ki$-edge of a triangular element.

Evaluating Eq. (8.73) along the $ki$-edge and realizing that along the $ki$-edge, $S_j = 0$, we have:

$$\{\mathbf{F}\}^{(e)} = \int_A \begin{bmatrix} S_i & 0 \\ 0 & S_i \\ S_j & 0 \\ 0 & S_j \\ S_k & 0 \\ 0 & S_k \end{bmatrix} \begin{Bmatrix} p_x \\ p_y \end{Bmatrix} dA = t \int_{\ell_{ki}} \begin{bmatrix} S_i & 0 \\ 0 & S_i \\ 0 & 0 \\ 0 & 0 \\ S_k & 0 \\ 0 & S_k \end{bmatrix} \begin{Bmatrix} p_x \\ p_y \end{Bmatrix} d\ell = \frac{tL_{ik}}{2} \begin{Bmatrix} p_x \\ p_y \\ 0 \\ 0 \\ p_x \\ p_y \end{Bmatrix} \quad (8.74)$$

Note that the effect of the distributed load in Figure 8.18 along the $ki$-edge is represented by two equal nodal forces at $i$ and $k$, with each force having $x$ and $y$ components. In a similar fashion, we can formulate the load matrix for a distributed load acting along other sides of the triangular element. Evaluation of the integral in Eq. (8.73) along the $ij$-edge and the $jk$-edge results in:

$$\{\boldsymbol{F}\}^{(e)} = \frac{tL_{ij}}{2} \begin{Bmatrix} p_x \\ p_y \\ p_x \\ p_y \\ 0 \\ 0 \end{Bmatrix} \qquad \{\boldsymbol{F}\}^{(e)} = \frac{tL_{jk}}{2} \begin{Bmatrix} 0 \\ 0 \\ p_x \\ p_y \\ p_x \\ p_y \end{Bmatrix} \quad (8.75)$$

It is worth noting that, generally speaking, linear triangular elements do not offer as accurate results as do the higher order elements. The purpose of the above derivation was to demonstrate the general steps involved in obtaining the elemental stiffness and load matrices. Next, we will derive the stiffness matrix for a quadrilateral element using isoparametric formulation.

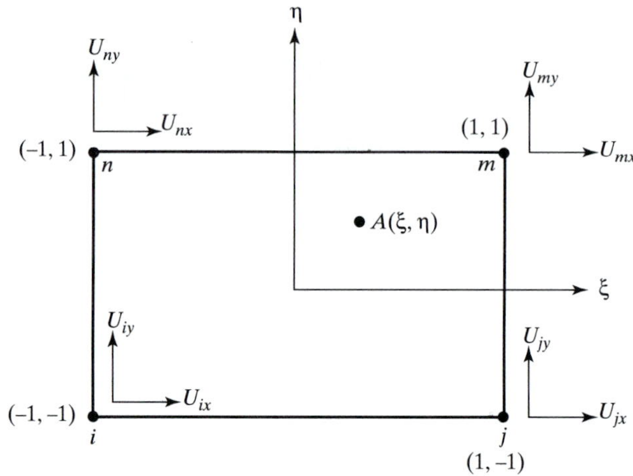


**FIGURE 5.12** A quadrilateral element used in formulating plane stress problems.

### Isoparametric Formulation: Using a Quadrilateral Element

As we discussed in Chapters 3 and 5, when we use a *single* set of parameters (a set of shape functions) to define the unknown variables $u$, $v$, $T$, and so on, as well as to express the position of any point within the element, we are using isoparametric formulation. An element expressed in such manner is called an isoparametric element. We will now turn our attention to the quadrilateral element previously shown as Figure 5.12 (repeated here for convenience). Using a quadrilateral element, we can express the displacement field within an element by Eq. (5.27):

$$u = S_i U_{ix} + S_j U_{jx} + S_m U_{mx} + S_n U_{nx} \quad (5.27)$$
$$v = S_i U_{iy} + S_j U_{jy} + S_m U_{my} + S_n U_{ny}$$

We can write the relations given by Eq. (5.27) in matrix form, given previously in Eq. (5.28):

$$\begin{Bmatrix} u \\ v \end{Bmatrix} = \begin{bmatrix} S_i & 0 & S_j & 0 & S_m & 0 & S_n & 0 \\ 0 & S_i & 0 & S_j & 0 & S_m & 0 & S_n \end{bmatrix} \begin{Bmatrix} U_{ix} \\ U_{iy} \\ U_{jx} \\ U_{jy} \\ U_{mx} \\ U_{my} \\ U_{nx} \\ U_{ny} \end{Bmatrix} \quad (5.28)$$

Note that using isoparametric formulation, we can use the same shape functions to describe the position of any point within the element by the relationships in Eq. (5.29):

$$x = S_i x_i + S_j x_j + S_m x_m + S_n x_n \quad (5.29)$$
$$y = S_i y_i + S_j y_j + S_m y_m + S_n y_n$$

The displacement field is related to the components of strains ($\varepsilon_{xx} = \frac{\partial u}{\partial x}$, $\varepsilon_{yy} = \frac{\partial v}{\partial y}$, and $\gamma_{xy} = \frac{\partial u}{\partial y} + \frac{\partial v}{\partial x}$) and, subsequently, to the nodal displacements through shape functions.

In Chapter 5, we also showed that using the Jacobian of the coordinate transformation, we can write the following, previously presented as Eq. (5.31):

$$\begin{Bmatrix} \dfrac{\partial f(x,y)}{\partial \xi} \\ \dfrac{\partial f(x,y)}{\partial \eta} \end{Bmatrix} = \overbrace{\begin{bmatrix} \dfrac{\partial x}{\partial \xi} & \dfrac{\partial y}{\partial \xi} \\ \dfrac{\partial x}{\partial \eta} & \dfrac{\partial y}{\partial \eta} \end{bmatrix}}^{[\mathbf{J}]} \begin{Bmatrix} \dfrac{\partial f(x,y)}{\partial x} \\ \dfrac{\partial f(x,y)}{\partial y} \end{Bmatrix} \tag{5.31}$$

The relationship of Eq. (5.31) was also presented as the following, previously shown as Eq. (5.32):

$$\begin{Bmatrix} \dfrac{\partial f(x,y)}{\partial x} \\ \dfrac{\partial f(x,y)}{\partial y} \end{Bmatrix} = [\mathbf{J}]^{-1} \begin{Bmatrix} \dfrac{\partial f(x,y)}{\partial \xi} \\ \dfrac{\partial f(x,y)}{\partial \eta} \end{Bmatrix} \tag{5.32}$$

For a quadrilateral element, the **J** matrix can be evaluated using Eqs. (5.29) and (5.7):

$$[\mathbf{J}] = \begin{bmatrix} \dfrac{\partial x}{\partial \xi} & \dfrac{\partial y}{\partial \xi} \\ \dfrac{\partial x}{\partial \eta} & \dfrac{\partial y}{\partial \eta} \end{bmatrix} = \begin{bmatrix} \dfrac{\partial}{\partial \xi}\left[S_i x_i + S_j x_j + S_m x_m + S_n x_n\right] & \dfrac{\partial}{\partial \xi}\left[S_i y_i + S_j y_j + S_m y_m + S_n y_n\right] \\ \dfrac{\partial}{\partial \eta}\left[S_i x_i + S_j x_j + S_m x_m + S_n x_n\right] & \dfrac{\partial}{\partial \eta}\left[S_i y_i + S_j y_j + S_m y_m + S_n y_n\right] \end{bmatrix} \tag{8.76}$$

$$[\mathbf{J}] = \frac{1}{4}\begin{bmatrix} \left[-(1-\eta)x_i + (1-\eta)x_j + (1+\eta)x_m - (1+\eta)x_n\right] & \left[-(1-\eta)y_i + (1-\eta)y_j + (1+\eta)y_m - (1+\eta)y_n\right] \\ \left[-(1-\xi)x_i - (1+\xi)x_j + (1+\xi)x_m + (1-\xi)x_n\right] & \left[-(1-\xi)y_i - (1+\xi)y_j + (1+\xi)y_m + (1-\xi)y_n\right] \end{bmatrix} = \begin{bmatrix} J_{11} & J_{12} \\ J_{21} & J_{22} \end{bmatrix} \tag{8.77}$$

Also recall that the inverse of a two-dimensional square matrix is given by:

$$[\mathbf{J}]^{-1} = \frac{1}{J_{11}J_{22} - J_{12}J_{21}}\begin{bmatrix} J_{22} & -J_{12} \\ -J_{21} & J_{11} \end{bmatrix} = \frac{1}{\det \mathbf{J}}\begin{bmatrix} J_{22} & -J_{12} \\ -J_{21} & J_{11} \end{bmatrix} \tag{8.78}$$

We can now proceed with the formulation of the stiffness matrix. The strain energy of an element is

$$\Lambda^{(e)} = \frac{1}{2}\int_V \{\boldsymbol{\varepsilon}\}^T[\boldsymbol{\nu}]\{\boldsymbol{\varepsilon}\}\, dV = \frac{1}{2}(t_e)\int_A \{\boldsymbol{\varepsilon}\}^T[\boldsymbol{\nu}]\{\boldsymbol{\varepsilon}\}\, dA \tag{8.79}$$

where $t_e$ is the thickness of the element. Recall the strain–displacement relationships in matrix form:

$$\{\boldsymbol{\varepsilon}\} = \begin{Bmatrix} \varepsilon_{xx} \\ \varepsilon_{yy} \\ \gamma_{xy} \end{Bmatrix} = \begin{Bmatrix} \dfrac{\partial u}{\partial x} \\ \dfrac{\partial v}{\partial y} \\ \dfrac{\partial u}{\partial y} + \dfrac{\partial v}{\partial x} \end{Bmatrix} \tag{8.80}$$

Evaluating the derivatives, we obtain

$$\begin{Bmatrix} \dfrac{\partial u}{\partial x} \\ \dfrac{\partial u}{\partial y} \end{Bmatrix} = \frac{1}{\det \mathbf{J}} \begin{bmatrix} J_{22} & -J_{12} \\ -J_{21} & J_{11} \end{bmatrix} \begin{Bmatrix} \dfrac{\partial u}{\partial \xi} \\ \dfrac{\partial u}{\partial \eta} \end{Bmatrix} \tag{8.81}$$

and

$$\begin{Bmatrix} \dfrac{\partial v}{\partial x} \\ \dfrac{\partial v}{\partial y} \end{Bmatrix} = \frac{1}{\det \mathbf{J}} \begin{bmatrix} J_{22} & -J_{12} \\ -J_{21} & J_{11} \end{bmatrix} \begin{Bmatrix} \dfrac{\partial v}{\partial \xi} \\ \dfrac{\partial v}{\partial \eta} \end{Bmatrix} \tag{8.82}$$

Combining Eqs. (8.80), (8.81), and (8.82) into a single relationship, we have

$$\{\boldsymbol{\varepsilon}\} = \begin{Bmatrix} \dfrac{\partial u}{\partial x} \\ \dfrac{\partial v}{\partial y} \\ \dfrac{\partial u}{\partial y} + \dfrac{\partial v}{\partial x} \end{Bmatrix} = \overbrace{\frac{1}{\det \mathbf{J}} \begin{bmatrix} J_{22} & -J_{12} & 0 & 0 \\ 0 & 0 & -J_{21} & J_{11} \\ -J_{21} & J_{11} & J_{22} & -J_{12} \end{bmatrix}}^{[\mathbf{A}]} \begin{Bmatrix} \dfrac{\partial u}{\partial \xi} \\ \dfrac{\partial u}{\partial \eta} \\ \dfrac{\partial v}{\partial \xi} \\ \dfrac{\partial v}{\partial \eta} \end{Bmatrix} \tag{8.83}$$

Note how we defined the $[\mathbf{A}]$ matrix, to be used later. Using Eq. (5.27), we can perform the following evaluation:

$$\begin{Bmatrix} \dfrac{\partial u}{\partial \xi} \\ \dfrac{\partial u}{\partial \eta} \\ \dfrac{\partial v}{\partial \xi} \\ \dfrac{\partial v}{\partial \eta} \end{Bmatrix} = \overbrace{\frac{1}{4} \begin{bmatrix} -(1-\eta) & 0 & (1-\eta) & 0 & (1+\eta) & 0 & -(1+\eta) & 0 \\ -(1-\xi) & 0 & -(1+\xi) & 0 & (1+\xi) & 0 & (1-\xi) & 0 \\ 0 & -(1-\eta) & 0 & (1-\eta) & 0 & (1+\eta) & 0 & -(1+\eta) \\ 0 & -(1-\xi) & 0 & -(1+\xi) & 0 & (1+\xi) & 0 & (1-\xi) \end{bmatrix}}^{[\mathbf{D}]} \overbrace{\begin{Bmatrix} U_{ix} \\ U_{iy} \\ U_{jx} \\ U_{jy} \\ U_{mx} \\ U_{my} \\ U_{nx} \\ U_{ny} \end{Bmatrix}}^{\{\mathbf{U}\}} \tag{8.84}$$

We can express the relationship in Eq. (8.84) in a compact matrix form as

$$\{\boldsymbol{\varepsilon}\} = [\mathbf{A}][\mathbf{D}]\{\mathbf{U}\} \tag{8.85}$$

Next, we need to transform the $dA$ term ($dA = dxdy$) in the strain energy integral into a product of natural coordinates. This transformation is achieved in the following manner:

$$\Lambda^{(e)} = \frac{1}{2}(t_e)\int_A \{\boldsymbol{\varepsilon}\}^T[\boldsymbol{\nu}]\{\boldsymbol{\varepsilon}\}\,dA = \frac{1}{2}(t_e)\int_{-1}^{1}\int_{-1}^{1}\{\boldsymbol{\varepsilon}\}^T[\boldsymbol{\nu}]\{\boldsymbol{\varepsilon}\}\overbrace{\det \mathbf{J}d\xi d\eta}^{dA} \tag{8.86}$$

Substituting for the strain matrix $[\boldsymbol{\varepsilon}]$ and the properties of the material matrix $[\boldsymbol{\nu}]$ into Eq. (8.86) and differentiating the strain energy of the element with respect to its nodal displacements, we find that the expression for the element stiffness matrix becomes:

$$[\mathbf{K}]^{(e)} = t_e\int_{-1}^{1}\int_{-1}^{1}[[\mathbf{A}][\mathbf{D}]]^T[\boldsymbol{\nu}][\mathbf{A}][\mathbf{D}]\det \mathbf{J}d\xi d\eta \tag{8.87}$$

Note that the resulting stiffness matrix is an $8 \times 8$ matrix. Furthermore, as discussed in Chapter 5, the integral of Eq. (8.87) is to be evaluated numerically, using the Gauss–Legendre formula.

## EXAMPLE 8.3

A two-dimensional triangular plane stress element made of steel, with modulus of elasticity $E = 200$ GPa and Poisson's ratio $\nu = 0.32$, is shown in Figure 8.19. The element is 3 mm thick, and the coordinates of nodes $i, j,$ and $k$ are given in centimeters in Figure 8.19. Determine the stiffness and load matrices under the given conditions.

The element stiffness matrix is

$$[\mathbf{K}]^{(e)} = V[\mathbf{B}]^T[\boldsymbol{\nu}][\mathbf{B}]$$

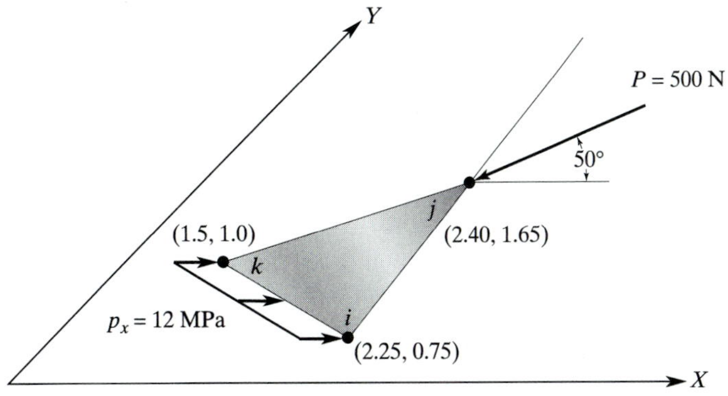


**FIGURE 8.19** The loading and nodal coordinates for the element in Example 8.3.

where

$$V = tA$$

$$[\mathbf{B}] = \frac{1}{2A}\begin{bmatrix} \beta_i & 0 & \beta_j & 0 & \beta_k & 0 \\ 0 & \delta_i & 0 & \delta_j & 0 & \delta_k \\ \delta_i & \beta_i & \delta_j & \beta_j & \delta_k & \beta_k \end{bmatrix}$$

$$[\boldsymbol{\nu}] = \frac{E}{1-\nu^2}\begin{bmatrix} 1 & \nu & 0 \\ \nu & 1 & 0 \\ 0 & 0 & \dfrac{1-\nu}{2} \end{bmatrix}$$

Thus,

$$\beta_i = Y_j - Y_k = 1.65 - 1.0 = 0.65 \qquad \delta_i = X_k - X_j = 1.50 - 2.40 = -0.9$$

$$\beta_j = Y_k - Y_i = 1.0 - 0.75 = 0.25 \qquad \delta_j = X_i - X_k = 2.25 - 1.5 = 0.75$$

$$\beta_k = Y_i - Y_j = 0.75 - 1.65 = -0.9 \quad \delta_k = X_j - X_i = 2.40 - 2.25 = 0.15$$

and

$$2A = X_i(Y_j - Y_k) + X_j(Y_k - Y_i) + X_k(Y_i - Y_j)$$

$$2A = 2.25(1.65 - 1.0) + 2.40(1.0 - 0.75) + 1.5(0.75 - 1.65) = 0.7125$$

Substituting appropriate values into the above matrices, we have

$$[\mathbf{B}] = \frac{1}{0.7125}\begin{bmatrix} 0.65 & 0 & 0.25 & 0 & -0.9 & 0 \\ 0 & -0.9 & 0 & 0.75 & 0 & 0.15 \\ -0.9 & 0.65 & 0.75 & 0.25 & 0.15 & -0.9 \end{bmatrix}$$

$$[\mathbf{B}]^T = \frac{1}{0.7125}\begin{bmatrix} 0.65 & 0 & -0.9 \\ 0 & -0.9 & 0.65 \\ 0.25 & 0 & 0.75 \\ 0 & 0.75 & 0.25 \\ -0.9 & 0 & 0.15 \\ 0 & 0.15 & -0.9 \end{bmatrix}$$

$$[\boldsymbol{\nu}] = \frac{200 \times 10^5 \dfrac{\text{N}}{\text{cm}^2}}{1-(0.32)^2}\begin{bmatrix} 1 & 0.32 & 0 \\ 0.32 & 1 & 0 \\ 0 & 0 & \dfrac{1-0.32}{2} \end{bmatrix} = \begin{bmatrix} 22281640 & 7130125 & 0 \\ 7130125 & 22281640 & 0 \\ 0 & 0 & 7575758 \end{bmatrix}$$

Carrying out the matrix operations results in the element stiffness matrix:

$$[\mathbf{K}]^{(e)} = \frac{(0.3)\left(\frac{0.7125}{2}\right)}{(0.7125)^2}\begin{bmatrix} 0.65 & 0 & -0.9 \\ 0 & -0.9 & 0.65 \\ 0.25 & 0 & 0.75 \\ 0 & 0.75 & 0.25 \\ -0.9 & 0 & 0.15 \\ 0 & 0.15 & -0.9 \end{bmatrix}\begin{bmatrix} 22281640 & 7130125 & 0 \\ 7130125 & 22281640 & 0 \\ 0 & 0 & 7575758 \end{bmatrix}$$

$$\begin{bmatrix} 0.65 & 0 & 0.25 & 0 & -0.9 & 0 \\ 0 & -0.9 & 0 & 0.75 & 0 & 0.15 \\ -0.9 & 0.65 & 0.75 & 0.25 & 0.15 & -0.9 \end{bmatrix}$$

Simplifying, we obtain

$$[\mathbf{K}]^{(e)} = \begin{bmatrix} 3273759 & -1811146 & -314288 & 372924 & -2959471 & 1438221 \\ -1811146 & 4473449 & 439769 & -2907167 & 1371376 & -1566282 \\ -314288 & 439769 & 1190309 & 580495 & -876020 & -1020265 \\ 372924 & -2907167 & 580495 & 2738296 & -953420 & 168871 \\ -2959471 & 1371376 & -876020 & -953420 & 3835491 & -417957 \\ 1438221 & -1566282 & -1020265 & 168871 & -417957 & 1397411 \end{bmatrix} \text{(N/cm)}$$

The load matrix due to the distributed load is

$$\{F\}^{(e)} = \frac{tL_{ik}}{2}\begin{Bmatrix} p_x \\ p_y \\ 0 \\ 0 \\ p_x \\ p_y \end{Bmatrix} = \frac{(0.3)\sqrt{(2.25 - 1.5)^2 + (0.75 - 1.0)^2}}{2}\begin{Bmatrix} 1200 \\ 0 \\ 0 \\ 0 \\ 1200 \\ 0 \end{Bmatrix} = \begin{Bmatrix} 142 \\ 0 \\ 0 \\ 0 \\ 142 \\ 0 \end{Bmatrix}$$

The load matrix due to the concentrated load is

$$\{F\}^{(e)} = \begin{Bmatrix} 0 \\ 0 \\ Q_{jx} \\ Q_{jy} \\ 0 \\ 0 \end{Bmatrix} = \begin{Bmatrix} 0 \\ 0 \\ -500\cos(50) \\ -500\sin(50) \\ 0 \\ 0 \end{Bmatrix} = \begin{Bmatrix} 0 \\ 0 \\ -321 \\ -383 \\ 0 \\ 0 \end{Bmatrix}$$

The complete load matrix for the element is

$$\{F\}^{(e)} = \begin{Bmatrix} 142 \\ 0 \\ -321 \\ -383 \\ 142 \\ 0 \end{Bmatrix} \text{(N)}$$

## 8.4 BASIC FAILURE THEORIES

One of the goals of most structural solid analyses is to check for failure. The prediction of failure is quite complex in nature; consequently, many investigators have been studying this topic. This section presents a brief overview of some failure theories. For an in-depth review of failure theories, you are encouraged to study a good text on the mechanics of materials or on machine design. (For a good example of such a text, see *Shigley and Mischke (1989)*).

Using ANSYS, you can calculate the distribution of the stress components $\sigma_x$, $\sigma_y$, and $\tau_{xy}$, as well as the principal stresses $\sigma_1$, and $\sigma_2$ within the material. But how would you decide whether or not the solid part you are analyzing will permanently deform or fail under the applied loading? You may recall from your previous study of the mechanics of materials that to compensate for what we do not know about the exact behavior of a material and/or to account for future loading for which we may have not accounted, but to which someone may subject the part, we introduce a *Factor of Safety* (F.S.), which is defined as

$$\text{F.S.} = \frac{P_{\text{max}}}{P_{\text{allowable}}} \tag{8.88}$$

where $P_{\text{max}}$ is the load that can cause failure. For certain situations, it is also customary to define the factor of safety in terms of the ratio of maximum stress that causes failure to the allowable stresses if the applied loads are linearly related to the stresses. But how do we apply the knowledge of stress distributions in a material to predict failure? Let us begin by reviewing how the *principal stresses* and *maximum shear stresses* are computed. The in-plane principal stresses at a point are determined from the values of $\sigma_{xx}$, $\sigma_{yy}$, and $\tau_{xy}$ at that point using the equation

$$\sigma_{1,2} = \frac{\sigma_x + \sigma_y}{2} \pm \sqrt{\left(\frac{\sigma_x - \sigma_y}{2}\right)^2 + \tau_{xy}^2} \tag{8.89}$$

The maximum in-plane shear stress at the point is determined from the relationship

$$\tau_{\text{max}} = \sqrt{\left(\frac{\sigma_x - \sigma_y}{2}\right)^2 + \tau_{xy}^2} \tag{8.90}$$

There are a number of failure criteria, including the maximum-normal-stress theory, the maximum-shear-stress theory, and the distortion-energy theory. The distortion-energy theory, often called the von Mises–Hencky theory, is one of the most commonly used criteria to predict failure of ductile materials. This theory is used to define the start of yielding. For design purposes, the von Mises stress $\sigma_v$ is calculated according to the equation

$$\sigma_v = \sqrt{\sigma_1^2 - \sigma_1\sigma_2 + \sigma_2^2} \tag{8.91}$$

A safe design is one that keeps the von Mises stresses in the material below the yield strength of the material. The relationship among the von Mises stress, the yield strength, and the factor of safety is

$$\sigma_v = \frac{S_Y}{\text{F.S.}} \tag{8.92}$$

where $S_Y$ is the yield strength of the material, obtained from a tension test. Most brittle materials have a tendency to fail abruptly without any yielding. For a brittle material under plane stress conditions, the maximum-normal-stress theory states that the material will fail if any point within the material experiences principal stresses exceeding the ultimate normal strength of the material. This idea is represented by the equations

$$|\sigma_1| = S_{\text{ultimate}} \qquad |\sigma_2| = S_{\text{ultimate}} \tag{8.93}$$

where $S_{\text{ultimate}}$ is the ultimate strength of the material, obtained from a tension test. The maximum-normal-stress theory may not produce reasonable predictions for materials with different tension and compression properties; in such structures, consider using the *Mohr failure criteria* instead.

## 8.5 EXAMPLES USING ANSYS

ANSYS offers a number of elements that can be used to model two-dimensional solid-structural problems. Some of these elements were introduced in Chapter 5. The two-dimensional solid-structural elements in ANSYS include: BEAM3, PLANE2, PLANE42, and PLANE82.

**BEAM3** is a uniaxial element with tension, compression, and bending capabilities. The element has three degrees of freedom at each node: translation in the $x$ and $y$-directions and rotation about the $z$-axis. The element input data include node locations, the cross-sectional area, the second moment of area, the height, and the material properties. Output data include nodal displacements and additional elemental output. Examples of elemental output include axial stress, bending stress at the top or bottom of the beam's cross section, maximum (axial + bending), and minimum(axial − bending). **BEAM4** is a three-dimensional version of BEAM3.

**PLANE2** is a six-node triangular structural-solid element. The element has quadratic displacement behavior, with two degrees of freedom at each node: translation in the nodal $x$ and $y$-directions. The element input data can include thickness if the KEYOPTION 3 (plane stress with thickness input) is selected. Surface-pressure loads may be applied to element faces. Output data include nodal displacements and elemental data, such as directional stresses and principal stresses.

**PLANE42** is a four-node quadrilateral element used to model solid problems. The element is defined by four nodes, with two degrees of freedom at each node: translation in the $x$ and $y$-directions. The element input data can include thickness if the KEYOPTION 3 (plane stress with thickness input) is selected. Surface-pressure loads may be applied to element faces. Output data include nodal displacements and elemental data, such as directional stresses and principal stresses.

**PLANE82** is an eight-node quadrilateral element used to model two-dimensional structural-solid problems. It is a higher order version of the two-dimensional, four-node quadrilateral PLANE42 element. This element offers more accuracy when modeling problems with curved boundaries. At each node, there are two degrees of freedom: translation in the *x* and *y*-directions. The element input data can include thickness if the KEYOPTION 3 (plane stress with thickness input) is selected. Surface-pressure loads may be applied to element faces. Output data include nodal displacements and elemental data, such as directional stresses and principal stresses.

## EXAMPLE 8.2 (REVISITED)

Let us consider the overhang frame again, in order to solve this problem using ANSYS. Recall that the frame is made of steel with $E = 30 \times 10^6$ lb/in$^2$. The respective cross-sectional areas and the second moments of areas for the two members are shown in Figure 8.12 (repeated here for your convenience). The members have a depth of 12.22 in. The frame is fixed as shown in the figure. We are interested in determining the deflections and the rotation of the frame under the given distributed load.

Enter the **ANSYS** program by using the Launcher. Type **xansys54** on the command line, or consult your system administrator for the appropriate command name to launch ANSYS from your computer system.

Pick **Interactive** from the Launcher menu.

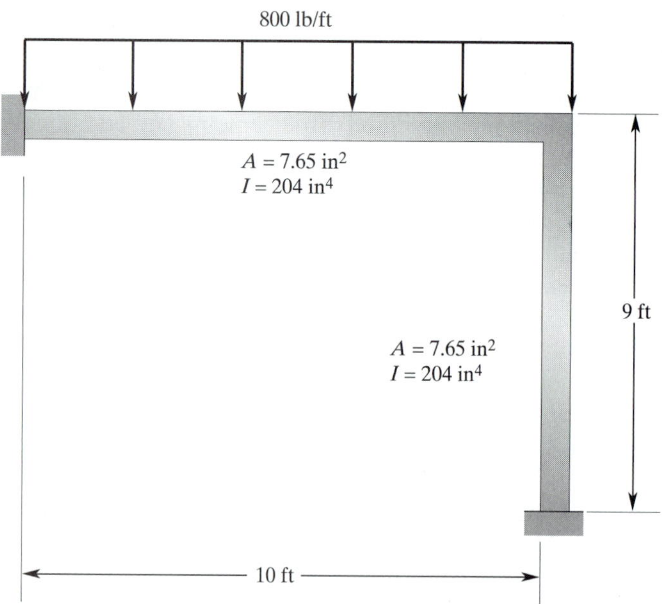


**FIGURE 8.12** An overhang frame supporting a distributed load.

Type **Frame2D** (or a file name of your choice) in the **Initial Jobname** entry field of the dialog box.

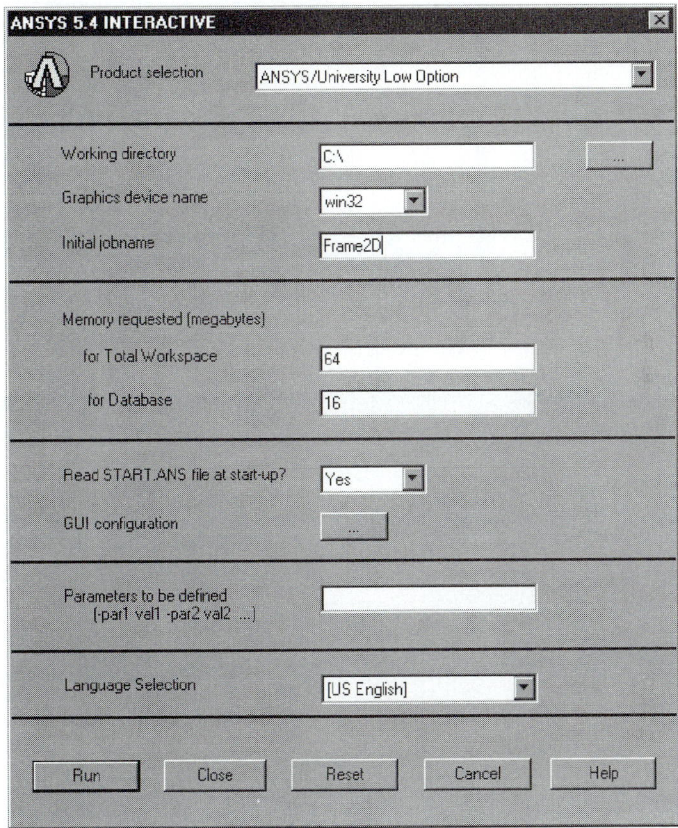


Pick **Run** to start the Graphic User Interface (GUI). A window will open with some disclaimer information. You will eventually be asked to press the **Return** key to start the graphics window and the main menu. Do so in order to proceed.

Create a title for the problem. This title will appear on ANSYS display windows to provide a simple way to identify the displays. So, use the following command sequences:

utility menu: **File** → **Change Title** ...

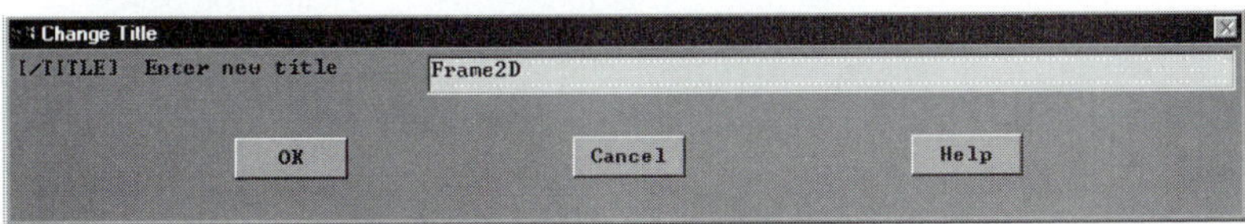


main menu: **Preprocessor** → **Element Type** → **Add/Edit/Delete** ...

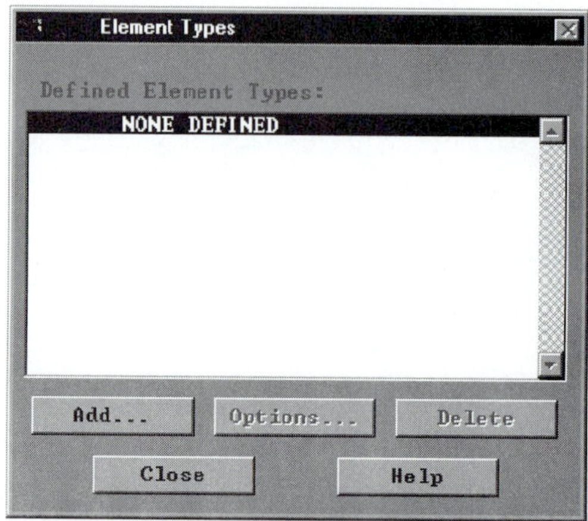


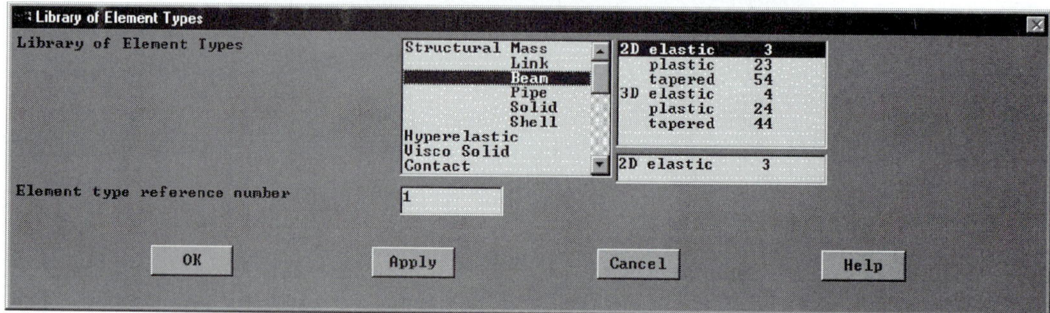


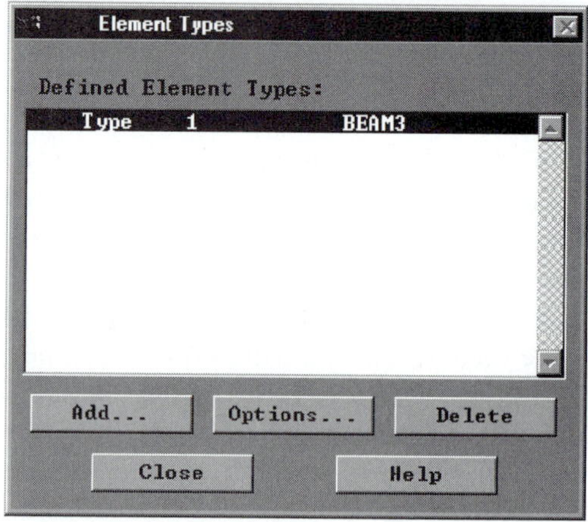


Assign the modulus of elasticity by using the following commands:

main menu: **Preprocessor** → **Material Props** → **-Constant-Isotropic** …

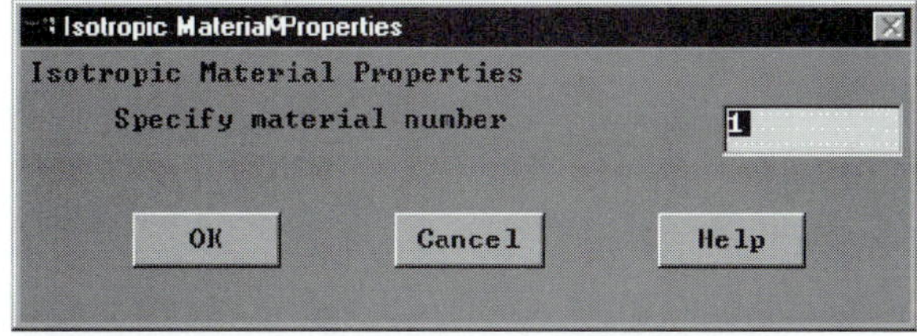


Isotropic Material Properties

Isotropic Material Properties
Properties for Material Number 1

| | | |
|---|---|---|
| Young's modulus | EX | 30e6 |
| Density | DENS | |
| Thermal expansion coeff | ALPX | |
| Reference temperature | REFT | |
| Poisson's ratio (minor) | NUXY | |
| Poisson's ratio (major) | PRXY | |
| Shear modulus | GXY | |
| Friction coefficient | MU | |
| Damping multiplier | DAMP | |
| Thermal conductivity | KXX | |
| Specific heat | C | |
| Enthalpy | ENTH | |
| Convection film coefficient | HF | |
| Emissivity | EMIS | |

OK Apply Cancel Help

main menu: **Preprocessor** → **Real Constants** …

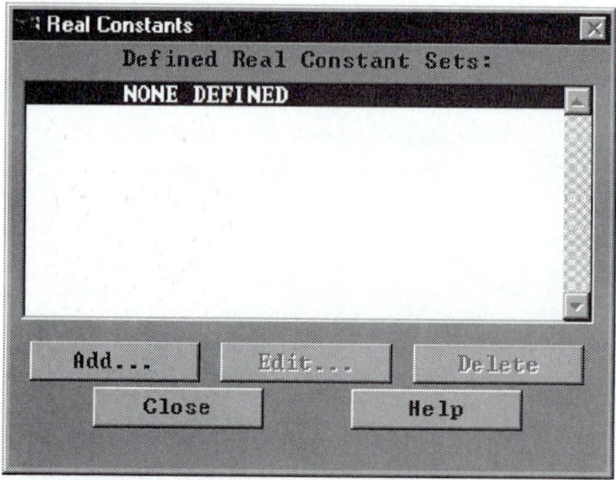
Real Constants
Defined Real Constant Sets:
NONE DEFINED
Add...
Edit...
Delete
Close
Help

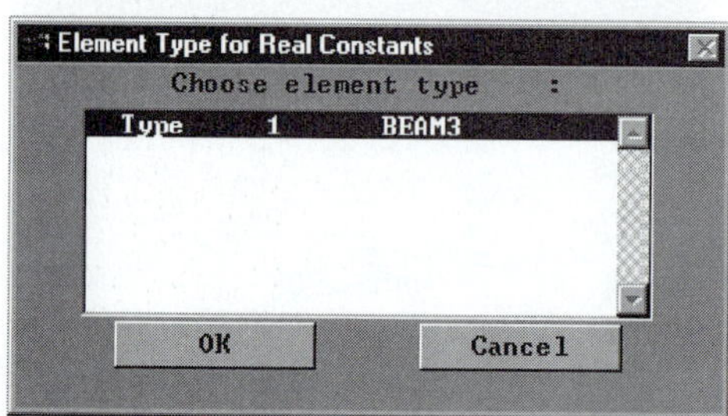
Element Type for Real Constants
Choose element type :
Type 1 BEAM3
OK
Cancel

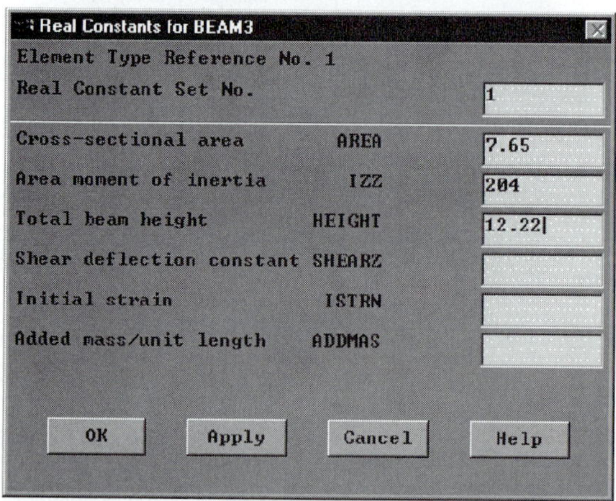
Real Constants for BEAM3
Element Type Reference No. 1
Real Constant Set No.
1
Cross-sectional area AREA
7.65
Area moment of inertia IZZ
204
Total beam height HEIGHT
12.22
Shear deflection constant SHEARZ
Initial strain ISTRN
Added mass/unit length ADDMAS
OK
Apply
Cancel
Help

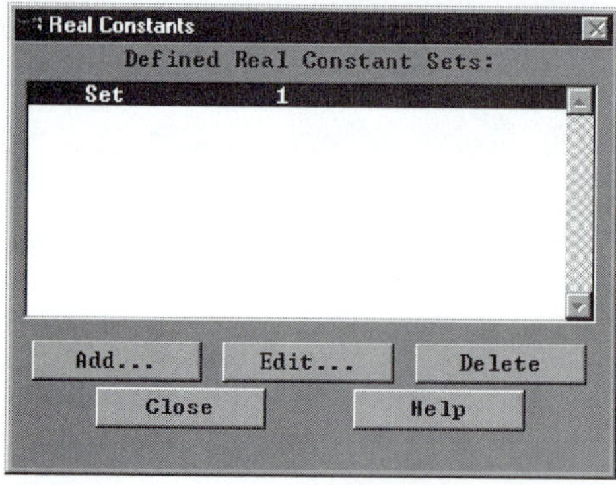
Real Constants
Defined Real Constant Sets:
Set 1
Add...
Edit...
Delete
Close
Help

ANSYS Toolbar: **SAVE_DB**

Set up the graphics area (i.e., work plane, zoom, etc.) with the following commands:

utility menu: **Workplane → Wp Settings** …

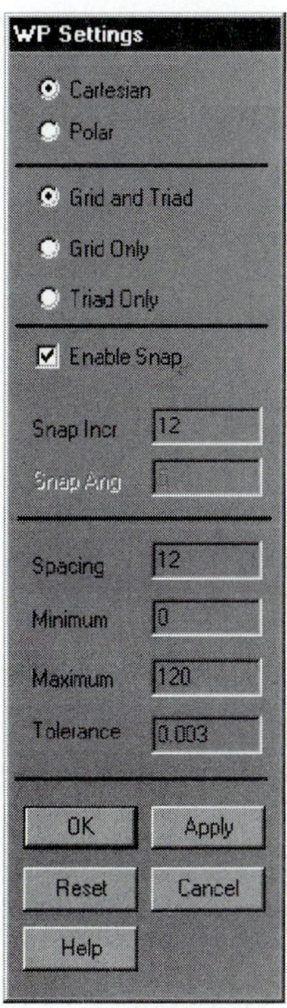


utility menu: **Workplane → Display Working Plane**

Bring the workplane to view by the command

utility menu: **PlotCtrls → Pan, Zoom, Rotate** …

Click on the small circle until you bring the workplane to view. Then create the nodes and elements:

main menu: **Preprocessor → -Modeling-Create → Nodes → On Working Plane+**

**[WP = 0,108]**

**[WP = 120,108]**

**[WP = 120,0]**

**OK**

main menu: **Preprocessor → -Modeling-Create → Elements → -Auto Numbered-Thru Nodes +**

**[pick node 1]**

**[pick node 2]**

**[apply anywhere in the ANSYS graphics window]**

**[pick node 2]**

**[pick node 3]**

**[anywhere in the ANSYS graphics window]**

**OK**

utility menu: **Plot → Elements**

Toolbar: **SAVE_DB**

Apply boundary conditions with the following commands:

main menu: **Solution → -Loads-Apply → -Structural-Displacements → On Nodes +**

**[pick node 1]**

**[pick node 3]**

**[anywhere in the ANSYS graphics window]**

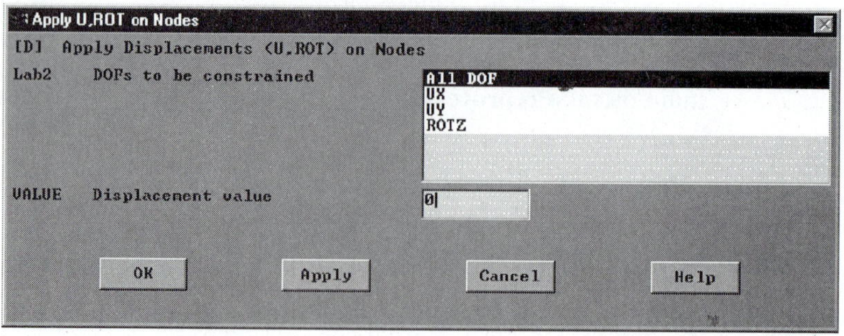


main menu: **Solution → -Loads-Apply → -Structural-Pressure → On Beams +**

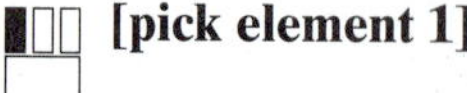

**[pick element 1]**

**[anywhere in the ANSYS graphics window]**

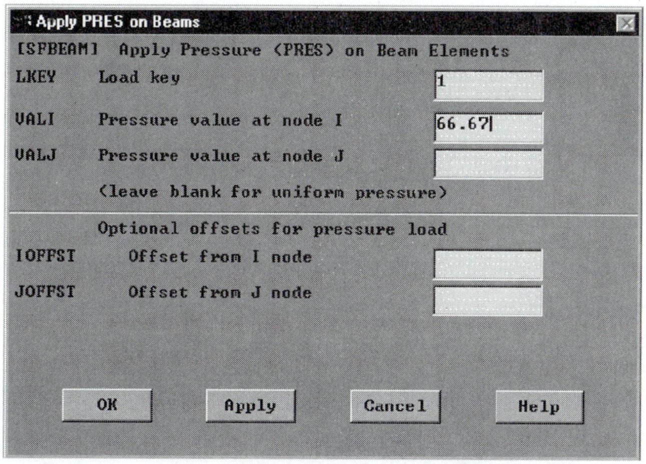

Apply PRES on Beams

[SFBEAM] Apply Pressure (PRES) on Beam Elements

| | | |
|---|---|---|
| LKEY | Load key | 1 |
| VALI | Pressure value at node I | 66.67 |
| VALJ | Pressure value at node J | |
| | (leave blank for uniform pressure) | |
| | Optional offsets for pressure load | |
| IOFFST | Offset from I node | |
| JOFFST | Offset from J node | |

OK Apply Cancel Help

To see the applied distributed load and boundary conditions, use the following commands:

utility menu: **PlotCtrls → Symbols** …

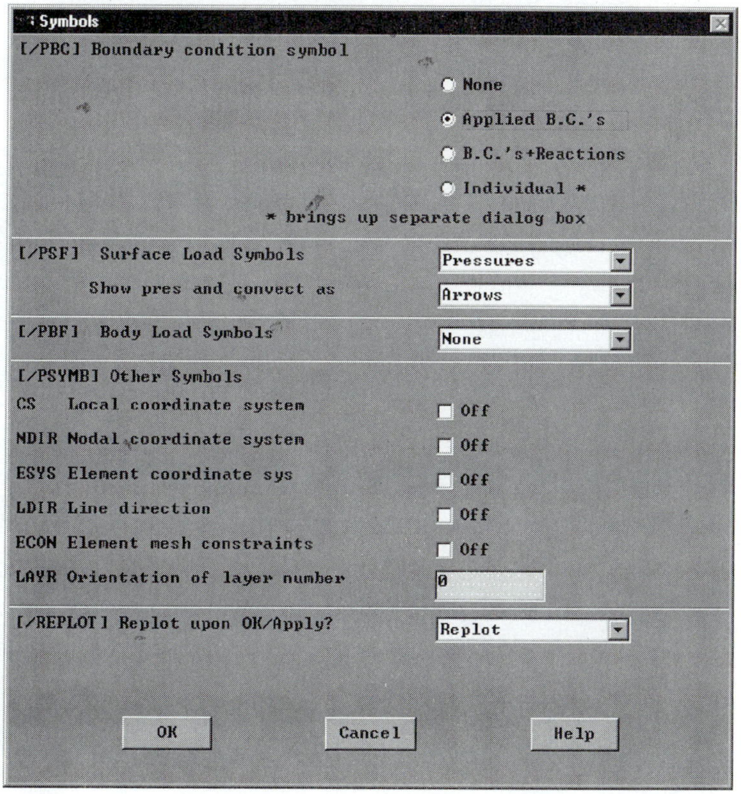

Symbols

[/PBC] Boundary condition symbol

- None
- Applied B.C.'s
- B.C.'s+Reactions
- Individual *

* brings up separate dialog box

| | | |
|---|---|---|
| [/PSF] | Surface Load Symbols | Pressures |
| | Show pres and convect as | Arrows |
| [/PBF] | Body Load Symbols | None |
| [/PSYMB] | Other Symbols | |
| CS | Local coordinate system | Off |
| NDIR | Nodal coordinate system | Off |
| ESYS | Element coordinate sys | Off |
| LDIR | Line direction | Off |
| ECON | Element mesh constraints | Off |
| LAYR | Orientation of layer number | 0 |
| [/REPLOT] | Replot upon OK/Apply? | Replot |

OK Cancel Help

utility menu: **Plot** → **Elements**

ANSYS Toolbar: **SAVE_DB**

Solve the problem:

main menu: **Solution** → **-Solve-Current LS**

**OK**

**Close** (the solution is done!) window.

**Close** (the /STAT Command) window.

Begin the postprocessing phase and plot the deformed shape with the following commands:

main menu: **General Postproc** → **Plot Results** → **Deformed Shape** ...

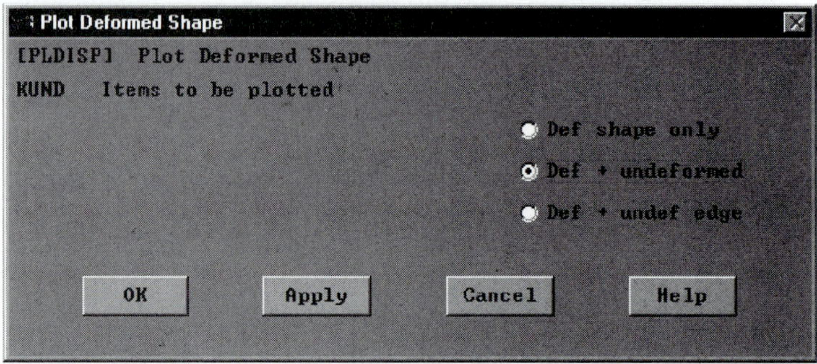


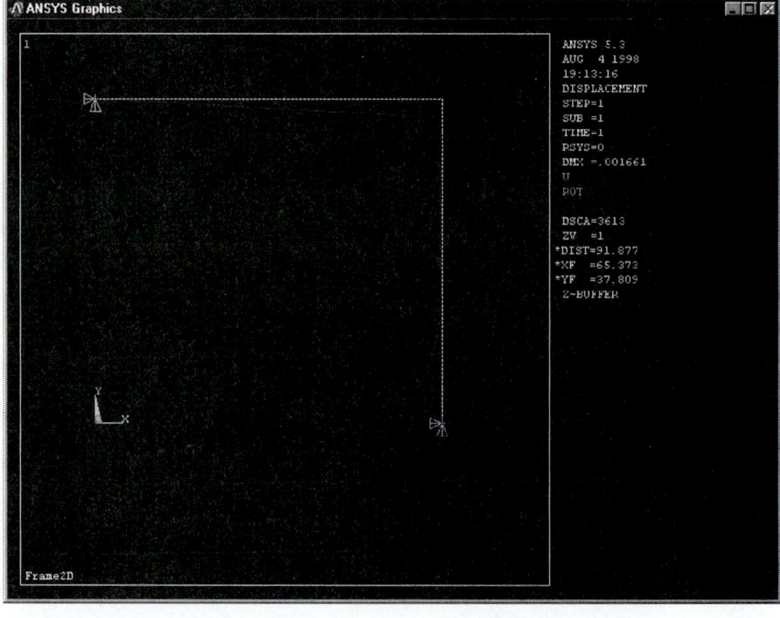

List the nodal displacements with the following commands:

main menu: **General Postproc** → **List Results** → **Nodal Solution** ...

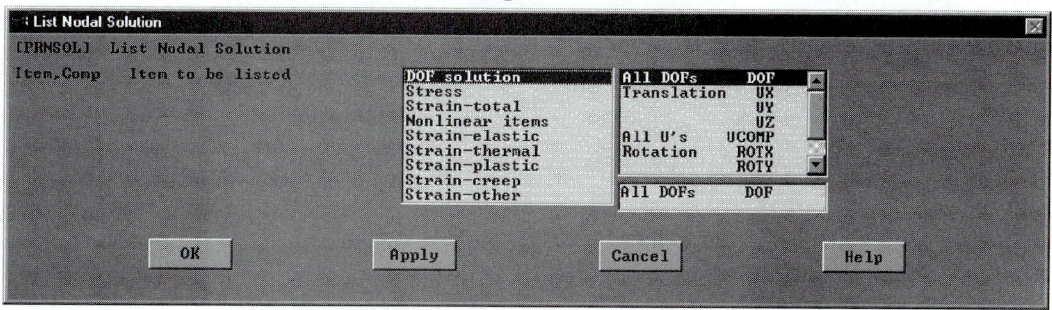


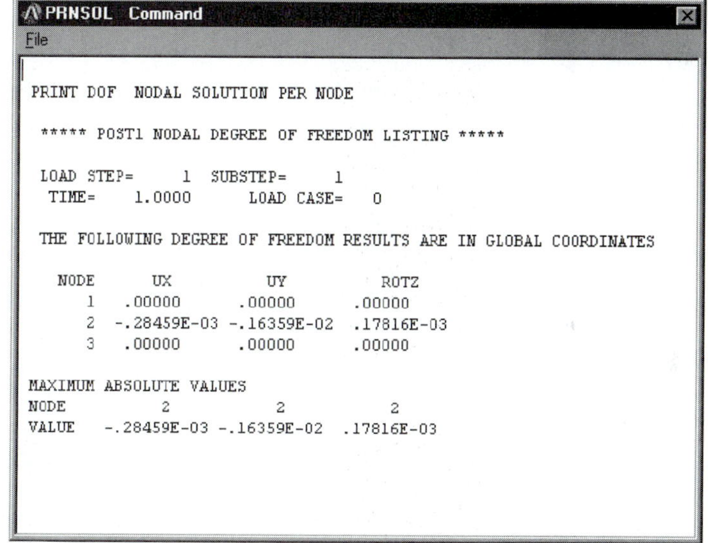


List the reactions with the following commands:

main menu: **General Postproc** → **List Results** → **Reaction Solution** ...

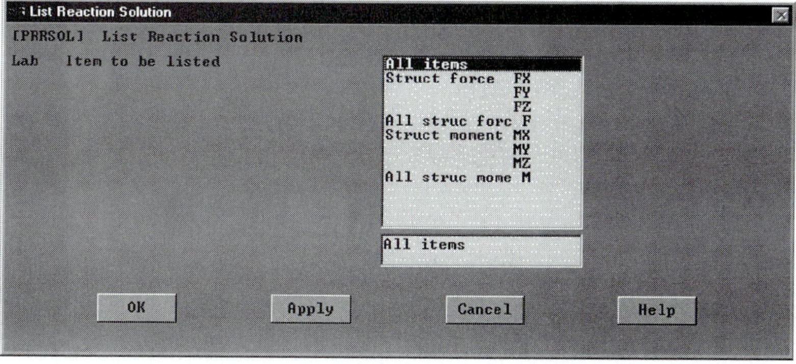

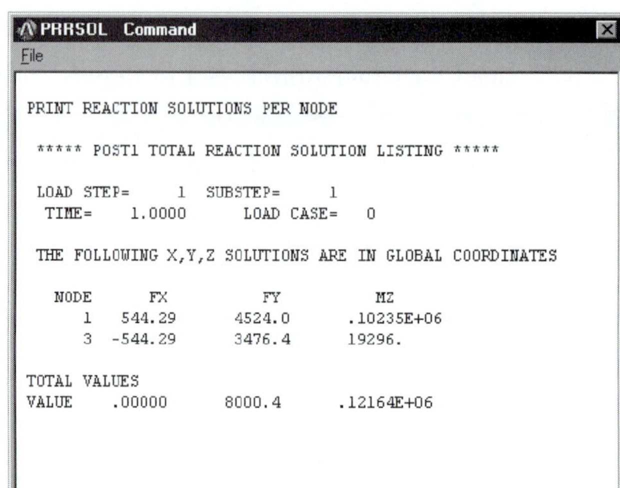


Exit ANSYS and save everything:

ANSYS Toolbar: **QUIT**

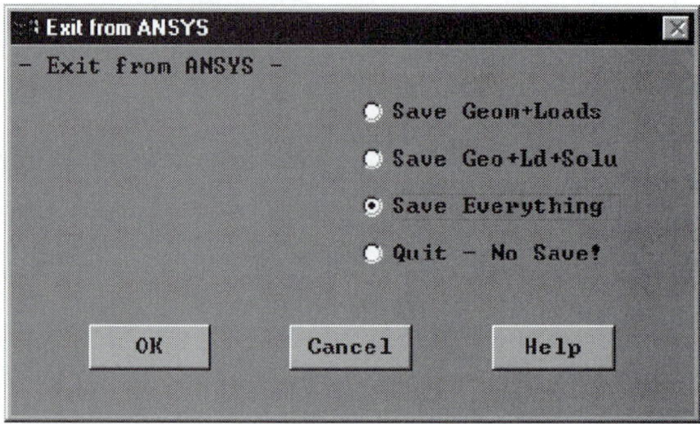


---

## EXAMPLE 8.4

The bicycle wrench shown in Figure 8.20 is made of steel with a modulus of elasticity $E = 200$ GPa and a Poisson's ratio $\nu = 0.32$. The wrench is 3 mm thick. Determine the von Mises stresses under the given distributed load and boundary conditions.

The following steps demonstrate how to (1) create the geometry of the problem, (2) choose the appropriate element type, (3) apply boundary conditions, and (4) obtain nodal results:

Enter the **ANSYS** program by using the Launcher. Type **xansys54** on the command line, or consult your system administrator for the appropriate command name to launch ANSYS from your computer system.

Pick **Interactive** from the Launcher menu.

Type **Bikewh** (or a file name of your choice) in the **Initial Jobname** entry field of the dialog box.

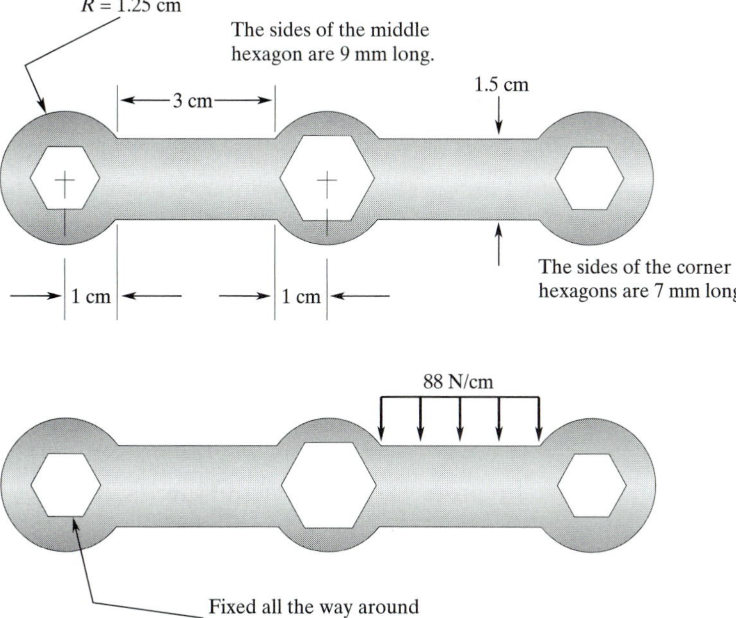


**FIGURE 8.20** A schematic for the bicycle wrench in Example 8.4.

ANSYS 5.4 INTERACTIVE

Product selection: ANSYS/University Low Option

Working directory: C:\

Graphics device name: win32

Initial jobname: Bikewh

Memory requested (megabytes)

for Total Workspace: 32

for Database: 16

Read START.ANS file at start-up? Yes

GUI configuration: ...

Parameters to be defined (-par1 val1 -par2 val2 ...)

Language Selection: [US English]

Run | Close | Reset | Cancel | Help

Pick **Run** to start the Graphic User Interface (GUI). A window will open with some disclaimer information. You will eventually be asked to press the **Return** key to start the graphics window and the main menu. Do so in order to proceed.

Create a title for the problem. This title will appear on ANSYS display windows to provide a simple way to identify the displays. So, issue the command

utility menu: **File** → **Change Title** ...

Define the element type and material properties with the following commands:

main menu: **Preprocessor** → **Element Type** → **Add/Edit/Delete** ...

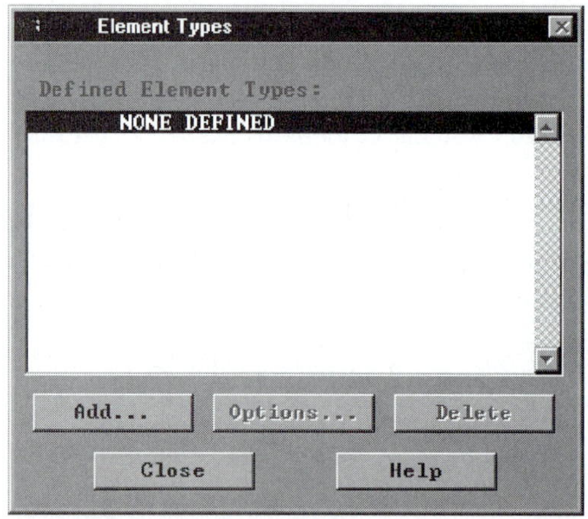


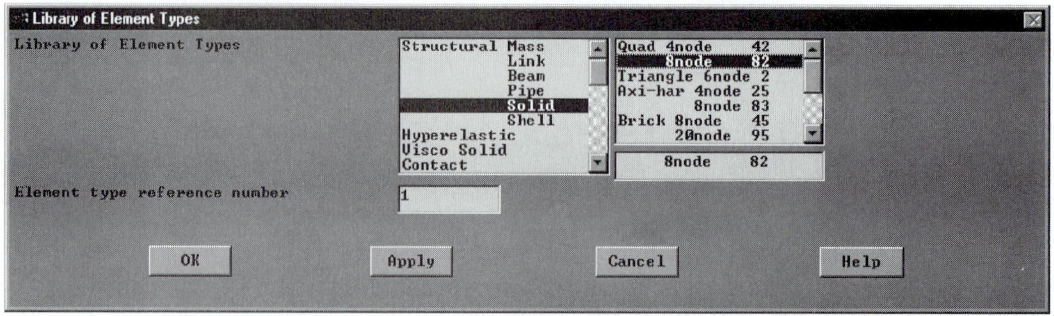

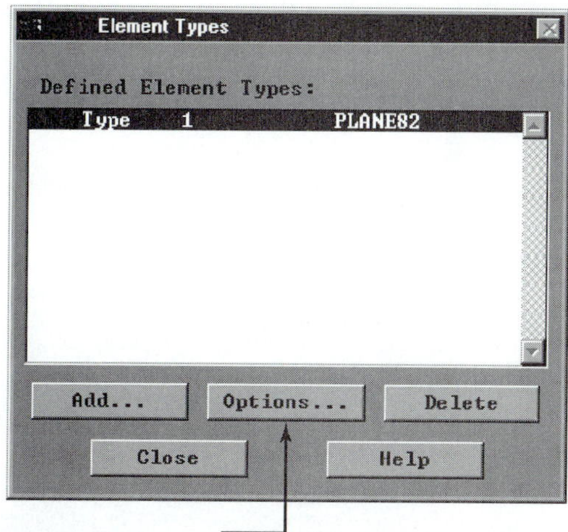


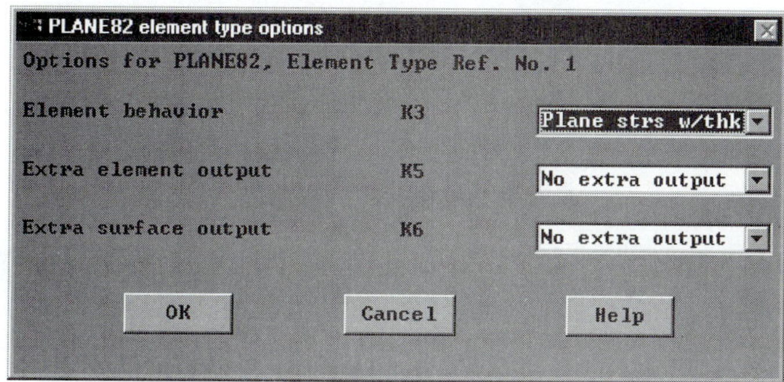


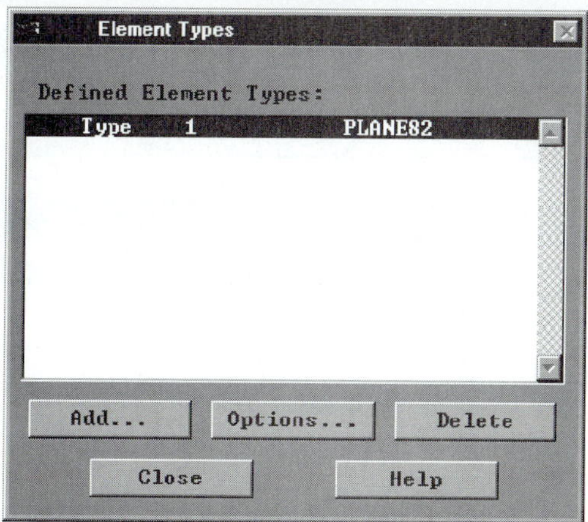


Assign the thickness of the wrench with the following commands:

main menu: **Preprocessor** → **Real Constants** ...

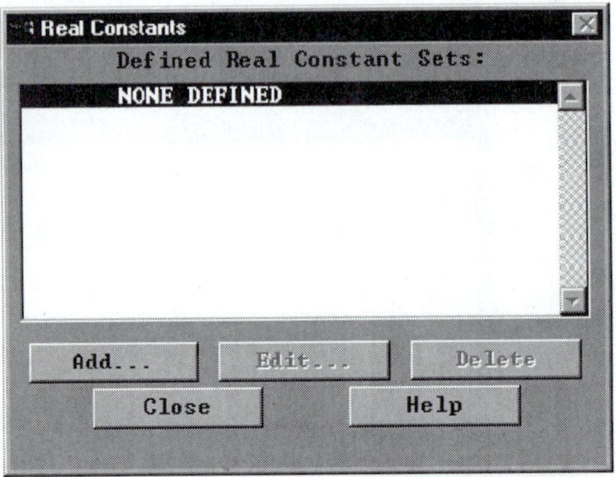


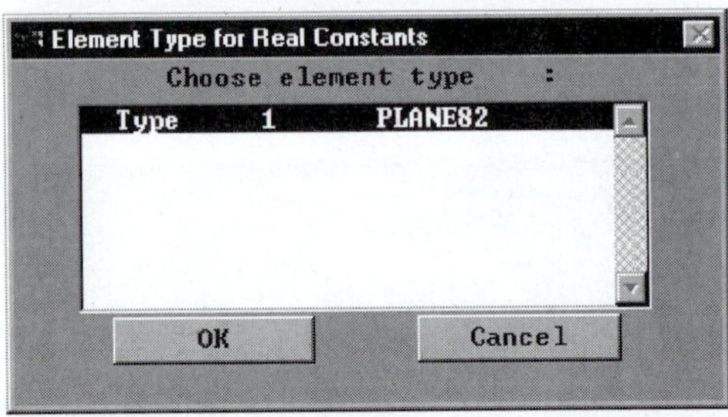


Real Constants for PLANE82

Element Type Reference No. 1

Real Constant Set No. 1

Real Constant for Plane Stress with Thickness (KEYOPT(3)=3)

Thickness THK 0.3

OK Apply Cancel Help

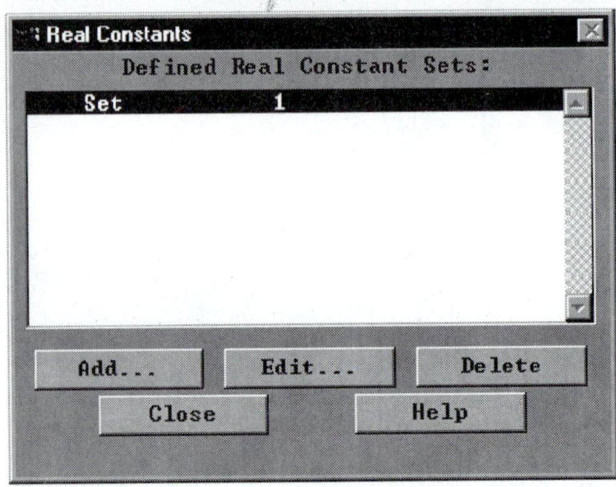


Assign the modulus of elasticity and Poisson's ratio values by using the following commands:

main menu: **Preprocessor** → **Material Props** → **-Constant-Isotropic** ...

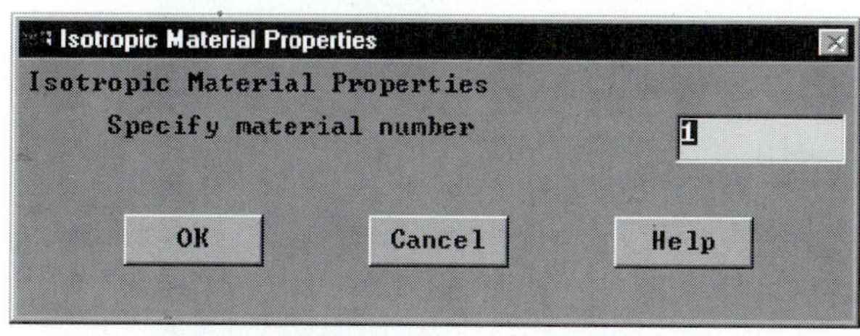


Isotropic Material Properties

Isotropic Material Properties

Properties for Material Number 1

| | | |
|---|---|---|
| Young's modulus | EX | 200e5 |
| Density | DENS | |
| Thermal expansion coeff | ALPX | |
| Reference temperature | REFT | |
| Poisson's ratio (minor) | NUXY | 0.3 |
| Poisson's ratio (major) | PRXY | |
| Shear modulus | GXY | |
| Friction coefficient | MU | |
| Damping multiplier | DAMP | |
| Thermal conductivity | KXX | |
| Specific heat | C | |
| Enthalpy | ENTH | |
| Convection film coefficient | HF | |
| Emissivity | EMIS | |

OK Apply Cancel Help

ANSYS Toolbar: **SAVE_DB**

Set up the graphics area (i.e., workplane, zoom, etc.) with the following commands:

utility menu: **Workplane** → **Wp Settings** ...

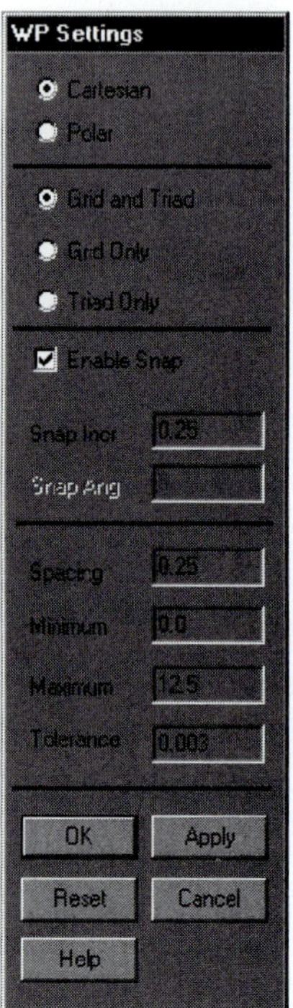


Toggle on the workplane by using the command

utility menu: **Workplane** → **Display Working Plane**

Bring the workplane to view by using the command

utility menu: **PlotCtrls** → **Pan, Zoom, Rotate** ...

Click on the small circle until you bring the work plane to view. Then, create the geometry with the following commands:

main menu: **Preprocessor** → **-Modeling-Create** → **-Areas-Rectangles** → **By 2 Corners** +

On the workplane, create the two rectangles:

Use the mouse buttons as shown below, or type the values in the appropriate fields.

**[WP = 2.25, 0.5]**

**[Expand the rubber band up 1.5 and right 3.0]**

**[WP = 7.25, 0.5]**

**[Expand the rubber band up 1.5 and right 3.0]**

**OK**

Create the circles with the following commands:

main menu: **Preprocessor** → **-Modeling-Create** → **-Areas-Circle** → **Solid Circle** +

**[WP = 1.25,1.25]**

**[Expand the rubber band to a radius of 1.25]**

**[WP = 6.25,1.25]**

**[Expand the rubber band to a radius of 1.25]**

**[WP = 11.25, 1.25]**

**[Expand the rubber band to a radius of 1.25]**

**OK**

Add the areas together with the commands:

main menu: **Preprocessor** → **-Modeling-Operate** → **-Booleans-Add** → **Areas** +

Click on the **Pick All** button, and then create the hexagons. First, change the **Snap Incr** in the WP Settings dialog box to 0.1 with the command

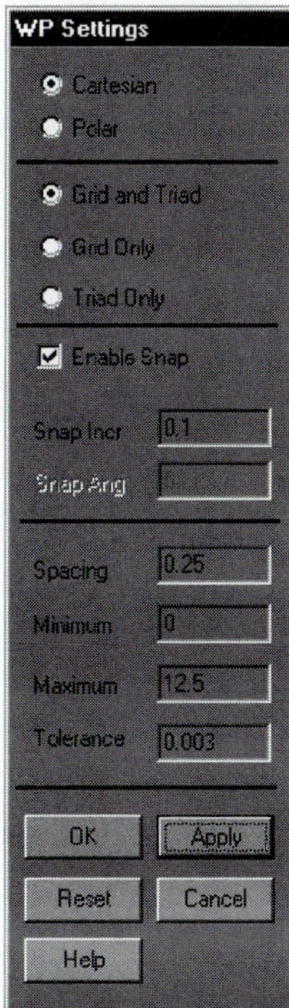

utility menu: **PlotCtrls** → **Pan, Zoom, Rotate** ...

Click on the **Box Zoom**, and put a box zoom around the left circle, then using the following commands create the hexagon:

main menu: **Preprocessor** → **-Modeling-Create** → **-Areas-Polygon** → **Hexagon** +

Use the mouse buttons as shown below, or type the values in the appropriate fields:

**[1.25, 1.25]**

**[Expand the hexagon to WP Rad = 0.7, Ang = 120]**

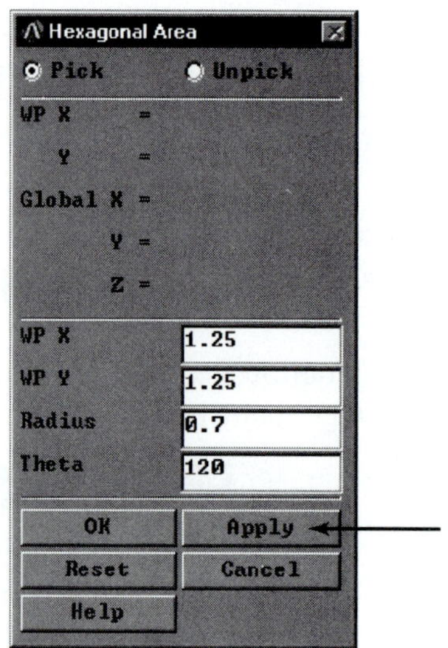


Then, issue the command

utility menu: **PlotCtrls** → **Pan, Zoom, Rotate** ...

Click on the **Fit** button. Then, click on the **Box Zoom**, and put a box zoom around the center circle. Use the mouse buttons as shown below, or type the values in the appropriate fields:

**[6.25, 1.25]**

**[Expand the hexagon to WP Rad = 0.9, Ang = 120]**

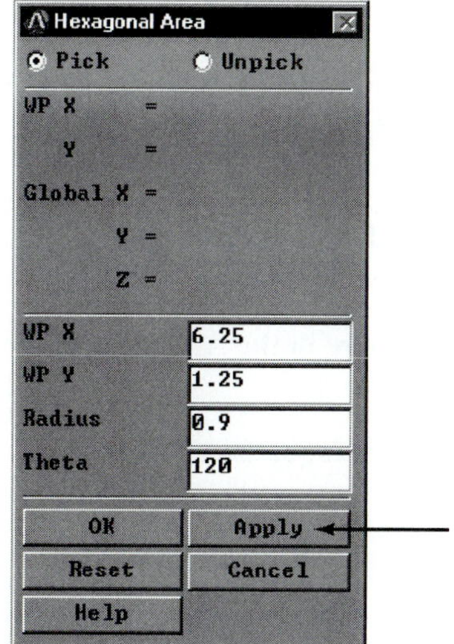


utility menu: **PlotCtrls → Pan, Zoom, Rotate** ...

Click on the **Fit** button. Then, Click on the **Box Zoom**, and put a box zoom around the right-end circle. Use the mouse buttons as shown below, or type the values in appropriate fields:

**[11.25, 1.25]**

**[Expand the hexagon to WP Rad = 0.7, Ang = 120]**

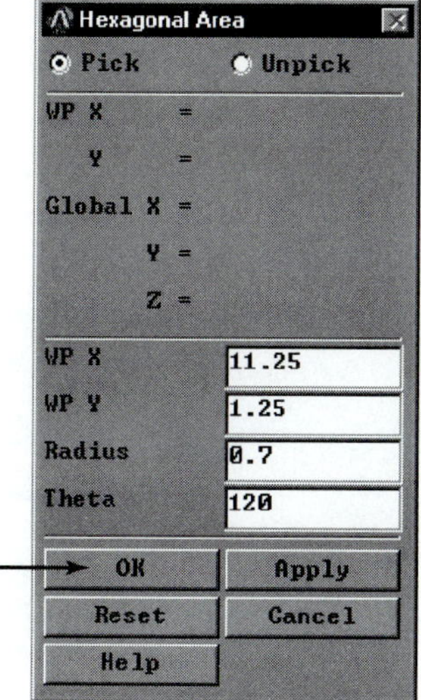

ANSYS Toolbar: **SAVE_DB**

Subtract the areas of the hexagons to create the driver holes:

main menu: **Preprocessor** → **-Modeling-Operate** → **-Booleans-Subtract** → **Areas** +

**[Pick the solid area of the wrench]**

**[Apply anywhere in the ANSYS graphics area]**

**[pick the left hexagon area]**

**[pick the center hexagon area]**

**[pick the right hexagon area]**

**[Apply anywhere in the ANSYS graphics area]**

**OK**

Now you can toggle off the workplane grids with the following command:

utility menu: **Workplane** → **Display Working Plane**

ANSYS Toolbar: **SAVE_DB**

You are now ready to mesh the area of the bracket to create elements and nodes. So, issue the following commands:

main menu: **Preprocessor** → **-Meshing-Size Cntrls** → **-Manual Size** **-Global-Size** ...

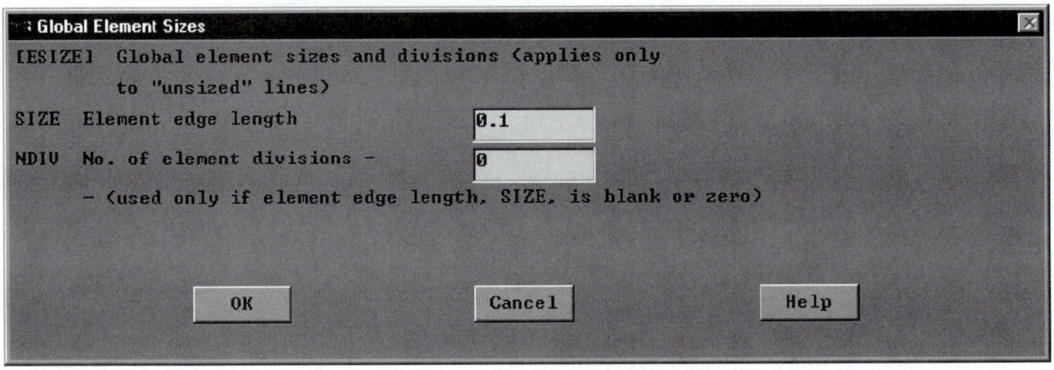

Global Element Sizes

[ESIZE] Global element sizes and divisions (applies only to "unsized" lines)

SIZE Element edge length 0.1

NDIV No. of element divisions - 0

- (used only if element edge length, SIZE, is blank or zero)

OK Cancel Help

ANSYS Toolbar: **SAVE_DB**

main menu: **Preprocessor** → **-Meshing-Mesh** → **-Areas-Free** +

Click on the **Pick All** button.

**OK**

Apply the boundary conditions and the load:

main menu: **Solution** → **-Loads-Apply** → **-Structural-Displacements** → **On Keypoints** +

Pick the six corner keypoints of the left hexagon:

**OK**

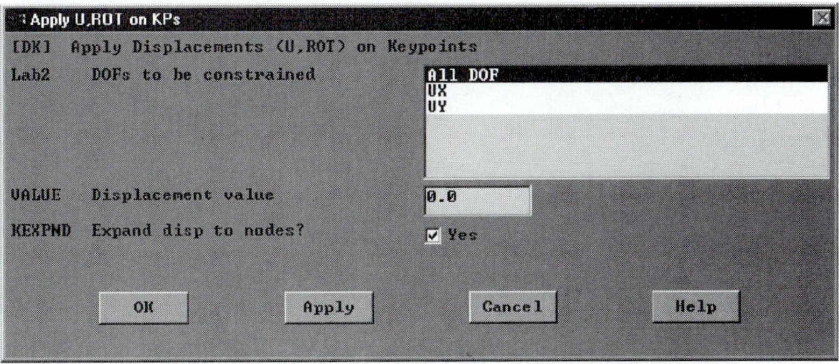


main menu: **Solution** → **-Loads-Apply** → **-Structural-Pressure** → **On Lines** +

Pick the appropriate horizontal line, as shown in the problem statement:

**OK**

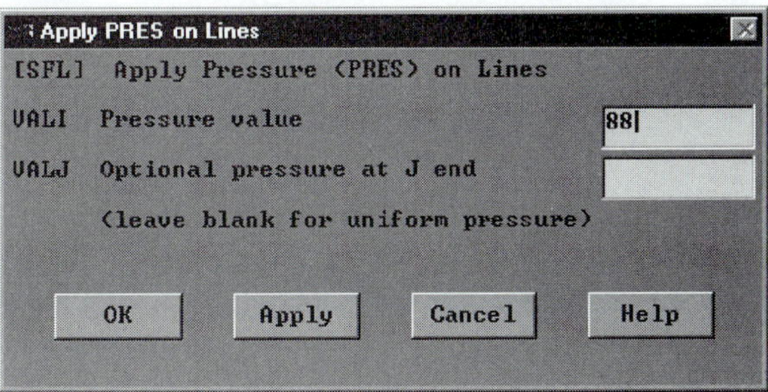


Solve the problem:

main menu: **Solution** → **-Solve-Current LS**

**OK**

**Close** (the solution is done!) window.

**Close** (the /STAT Command) window.

Begin the postprocessing phase and plot the deformed shape with the following commands:

main menu: **General Postproc** → **Plot Results** → **Deformed Shape** ...

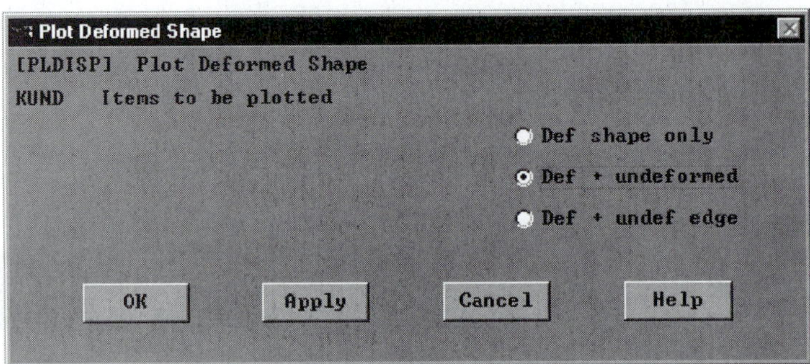


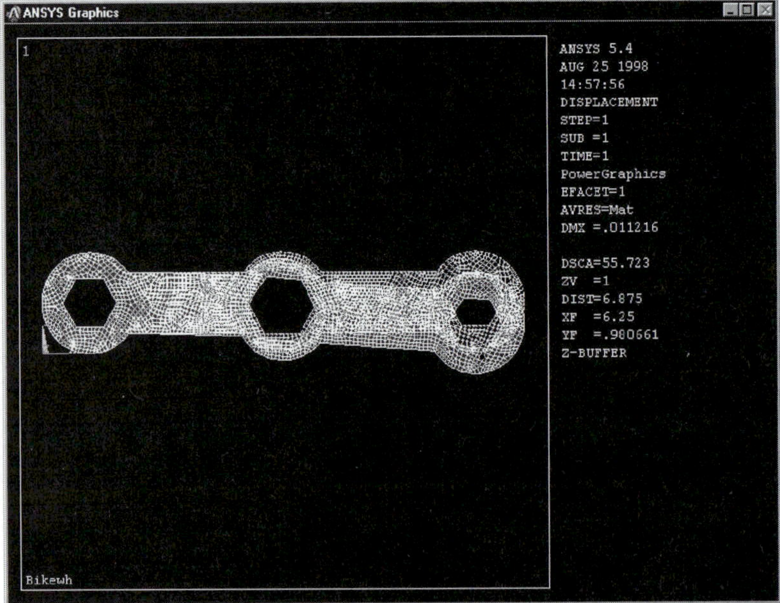


Plot the von Mises stresses with the following commands:

main menu: **General Postproc** → **Plot Results** → **-Contour Plot-Nodal Solu** ...

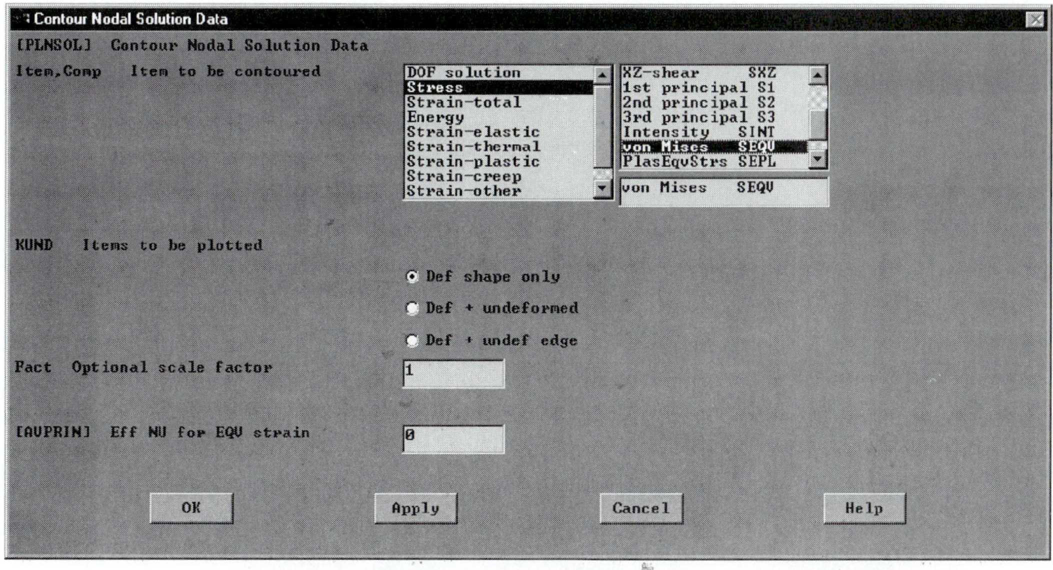


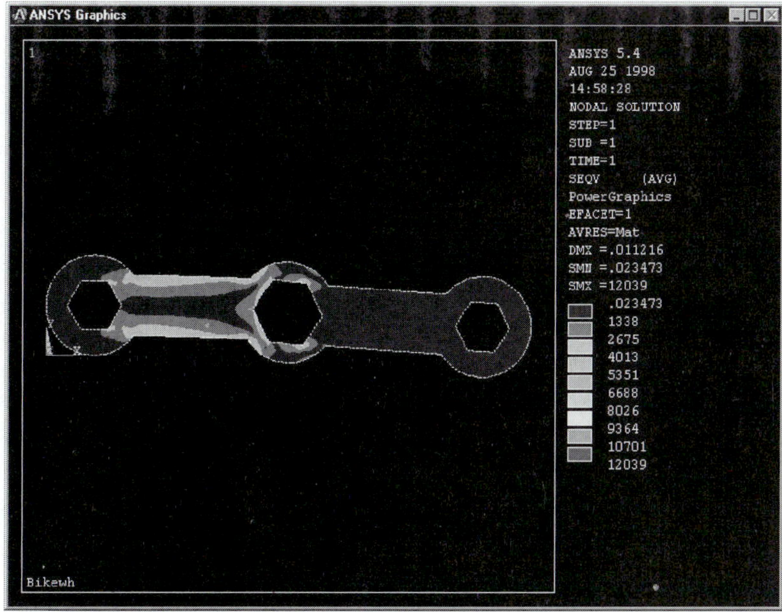


Exit ANSYS and save everything:

ANSYS Toolbar: **QUIT**

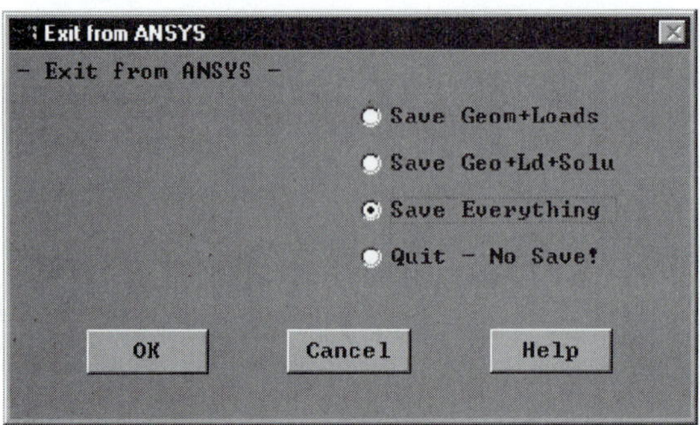


## 8.6 VERIFICATION OF RESULTS

It is always necessary to compute the reaction forces and moments for beam and frame problems. The nodal reaction forces and moments can be computed from the relationship

$$\{\mathbf{R}\} = [\mathbf{K}]\{\mathbf{U}\} - \{\mathbf{F}\}$$

We computed the reaction matrix for Example 8.1, repeated here:

$$\begin{Bmatrix} R_1 \\ M_1 \\ R_2 \\ M_2 \\ R_3 \\ M_3 \end{Bmatrix} = \begin{Bmatrix} 54687(\text{N}) \\ 39062(\text{N}\cdot\text{m}) \\ 132814(\text{N}) \\ 0 \\ 0 \\ 0 \end{Bmatrix}$$

Earlier, we discussed how to check the validity of results qualitatively. It was mentioned that the results indicated that there is a reaction force and a reaction moment at node 1; there is a reaction force at node 2; there is no reaction moment at node 2, as expected; and there are no reaction forces or moments at node 3, as expected for the given problem. Let us also perform a quantitative check on the accuracy of the results. We can use the computed reaction forces and moments against the external loading to check for static equilibrium (see Figure 8.21):

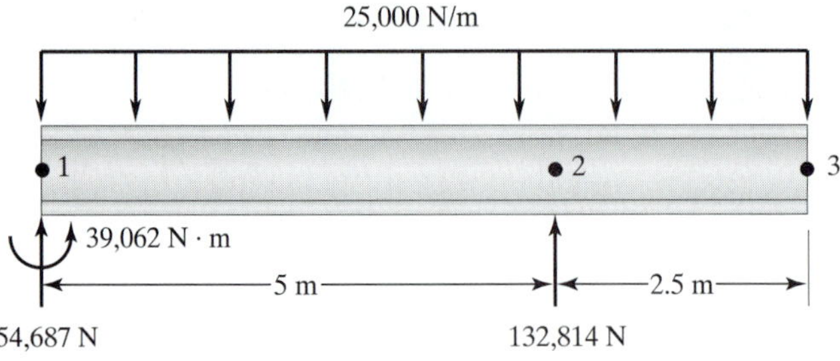


**FIGURE 8.21** The free body diagram for Example 8.1.

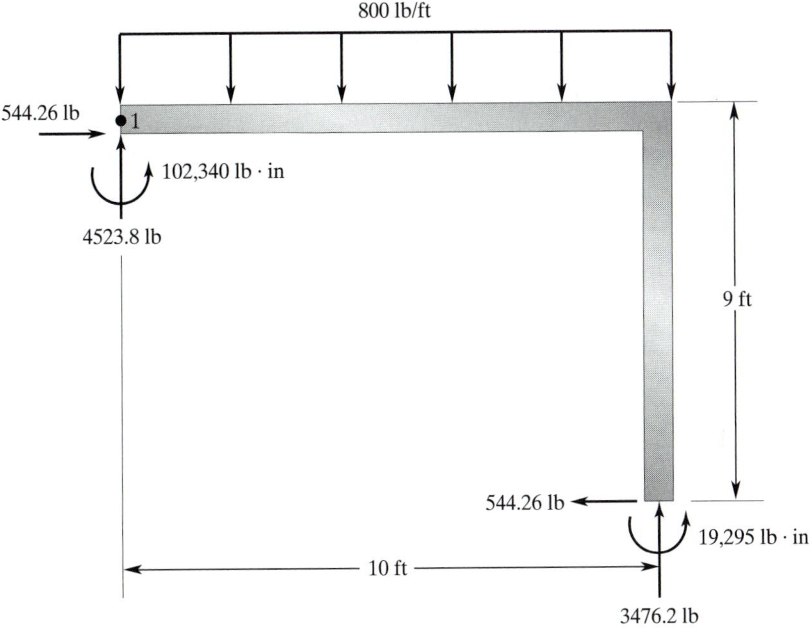


**FIGURE 8.22** The free body diagram for Example 8.2.

$$\Sigma F_Y = 0 \qquad 13{,}2814 + 54{,}687 - (25{,}000)(7.5) = -1 \approx 0$$

and

$$\Sigma M_{\text{node 2}} = 0 \quad 39{,}062 - 54{,}687(5) + (25{,}000)(7.5)(1.25) = 2 \approx 0$$

Similarly, in reference to Example 8.2, we find that the reaction results generated using ANSYS are shown in Figure 8.22. Checking for static equilibrium, we find that

$$\Sigma F_X = 0 \qquad 544.26 - 544.26 = 0$$

$$\Sigma F_Y = 0 \qquad 4523.8 + 3476.2 - (800)(10) = 0$$

$$\Sigma M_{\text{node 1}} = 0 \qquad 102{,}340 + 19{,}295 + 3476.2(10)(12) - (544.26)(9)(12) - (800)(10)(5)(12) = -1.08 \approx 0$$

Now we turn our attention to Example 8.4. There are a number of ways you can check the validity of the results of this problem. You can print the reaction forces and check the value of their sum against the applied force. Are statics equilibrium conditions satisfied? Using the path operations of ANSYS, you can also cut an arbitrary section through the wrench and visually assess the *x* and *y*-components of the local stresses and shear stresses along the section. You can integrate the stress information along the path to obtain the internal forces and compare their values to the applied force. Are statics equilibrium conditions satisfied? These questions are left to you to confirm.

## SUMMARY

At this point you should:

1. know that it is wise to use simple analytical solutions, rather than finite element modeling, for a simple problem whenever appropriate. Use finite element modeling only when it is necessary to do so. Simple analytical solutions are particularly appropriate when you are solving basic torsional or beam problems.
2. know that the stiffness matrix for torsional problems is similar to the conductance matrix obtained for two-dimensional conduction problems. The stiffness matrix and the load matrix using a rectangular element are:

$$[\mathbf{K}]^{(e)} = \frac{w}{6\ell}\begin{bmatrix} 2 & -2 & -1 & 1 \\ -2 & 2 & 1 & -1 \\ -1 & 1 & 2 & -2 \\ 1 & -1 & -2 & 2 \end{bmatrix} + \frac{\ell}{6w}\begin{bmatrix} 2 & 1 & -1 & -2 \\ 1 & 2 & -2 & -1 \\ -1 & -2 & 2 & 1 \\ -2 & -1 & 1 & 2 \end{bmatrix}$$

$$\{\mathbf{F}\}^{(e)} = \frac{2G\theta A}{4}\begin{Bmatrix} 1 \\ 1 \\ 1 \\ 1 \end{Bmatrix}$$

For triangular elements, the stiffness and load matrices are, respectively:

$$[\mathbf{K}]^{(e)} = \frac{1}{4A}\begin{bmatrix} \beta_i^2 & \beta_i\beta_j & \beta_i\beta_k \\ \beta_i\beta_j & \beta_j^2 & \beta_j\beta_k \\ \beta_i\beta_k & \beta_j\beta_k & \beta_k^2 \end{bmatrix} + \frac{1}{4A}\begin{bmatrix} \delta_i^2 & \delta_i\delta_j & \delta_i\delta_k \\ \delta_i\delta_j & \delta_j^2 & \delta_j\delta_k \\ \delta_i\delta_k & \delta_j\delta_k & \delta_k^2 \end{bmatrix}$$

$$\{\mathbf{F}\}^{(e)} = \frac{2G\theta A}{3}\begin{Bmatrix} 1 \\ 1 \\ 1 \end{Bmatrix}$$

3. know that the stiffness matrix for a beam element with two degrees of freedom at each node (the vertical displacement and rotation) is

$$[\mathbf{K}]^{(e)} = \frac{EI}{L^3}\begin{bmatrix} 12 & 6L & -12 & 6L \\ 6L & 4L^2 & -6L & 2L^2 \\ -12 & -6L & 12 & -6L \\ 6L & 2L^2 & -6L & 4L^2 \end{bmatrix}$$

4. know how to compute the load matrix for a beam element by consulting Table 8.3 for equivalent nodal forces.

5. know that the stiffness matrix for a frame element (with local and global coordinate systems aligned) consisting of two nodes with three degrees of freedom at each node (axial displacement, lateral displacement and rotation) is

$$[\mathbf{K}]^{(e)} = \begin{bmatrix} \frac{AE}{L} & 0 & 0 & -\frac{AE}{L} & 0 & 0 \\ 0 & \frac{12EI}{L^3} & \frac{6EI}{L^2} & 0 & -\frac{12EI}{L^3} & \frac{6EI}{L^2} \\ 0 & \frac{6EI}{L^2} & \frac{4EI}{L} & 0 & -\frac{6EI}{L^2} & \frac{2EI}{L} \\ -\frac{AE}{L} & 0 & 0 & \frac{AE}{L} & 0 & 0 \\ 0 & -\frac{12EI}{L^3} & -\frac{6EI}{L^2} & 0 & \frac{12EI}{L^3} & -\frac{6EI}{L^2} \\ 0 & \frac{6EI}{L^2} & \frac{2EI}{L} & 0 & -\frac{6EI}{L^2} & \frac{4EI}{L} \end{bmatrix} \quad (8.47)$$

Note that for members that are not horizontal, the local degrees of freedom are related to the global degrees of freedom through the transformation matrix, according to the relationship

$$\{\mathbf{u}\} = [\mathbf{T}]\{\mathbf{U}\}$$

where the transformation matrix is:

$$[\mathbf{T}] = \begin{bmatrix} \cos\theta & \sin\theta & 0 & 0 & 0 & 0 \\ -\sin\theta & \cos\theta & 0 & 0 & 0 & 0 \\ 0 & 0 & 1 & 0 & 0 & 0 \\ 0 & 0 & 0 & \cos\theta & \sin\theta & 0 \\ 0 & 0 & 0 & -\sin\theta & \cos\theta & 0 \\ 0 & 0 & 0 & 0 & 0 & 1 \end{bmatrix}$$

6. know how to compute the stiffness matrix for a frame element with an arbitrary orientation with respect to the global coordinate system using the relationship

$$[\mathbf{K}]^{(e)} = [\mathbf{T}]^T[\mathbf{K}]_{xy}[\mathbf{T}]$$

7. know how to compute the load matrix for a frame element by consulting Table 8.3 for equivalent nodal forces.

8. know that the stiffness matrix for a plane stress triangular element is

$$[\mathbf{K}]^{(e)} = V[\mathbf{B}]^T[\nu][\mathbf{B}]$$

where

$$V = tA$$

$$[\mathbf{B}] = \frac{1}{2A}\begin{bmatrix} \beta_i & 0 & \beta_j & 0 & \beta_k & 0 \\ 0 & \delta_i & 0 & \delta_j & 0 & \delta_k \\ \delta_i & \beta_i & \delta_j & \beta_j & \delta_k & \beta_k \end{bmatrix} \quad [\boldsymbol{\nu}] = \frac{E}{1-\nu^2}\begin{bmatrix} 1 & \nu & 0 \\ \nu & 1 & 0 \\ 0 & 0 & \dfrac{1-\nu}{2} \end{bmatrix}$$

and

$$\beta_i = Y_j - Y_k \quad \delta_i = X_k - X_j$$

$$\beta_j = Y_k - Y_i \quad \delta_j = X_i - X_k$$

$$\beta_k = Y_i - Y_j \quad \delta_k = X_j - X_i$$

$$2A = X_i(Y_j - Y_k) + X_j(Y_k - Y_i) + X_k(Y_i - Y_j)$$

9. know that the load matrix due to a distributed load along the element's edges is

$$\{\mathbf{F}\}^{(e)} = \frac{tL_{ij}}{2}\begin{Bmatrix} p_x \\ p_y \\ p_x \\ p_y \\ 0 \\ 0 \end{Bmatrix} \qquad \{\mathbf{F}\}^{(e)} = \frac{tL_{jk}}{2}\begin{Bmatrix} 0 \\ 0 \\ p_x \\ p_y \\ p_x \\ p_y \end{Bmatrix} \qquad \{\mathbf{F}\}^{(e)} = \frac{tL_{ik}}{2}\begin{Bmatrix} p_x \\ p_y \\ 0 \\ 0 \\ p_x \\ p_y \end{Bmatrix}$$

10. understand how an element's stiffness matrix is obtained through the isoparametric formulation.

## REFERENCES

*ANSYS User's Manual: Procedures*, Vol. I, Swanson Analysis Systems, Inc.

*ANSYS User's Manual: Commands*, Vol. II, Swanson Analysis Systems, Inc.

*ANSYS User's Manual: Elements*, Vol. III, Swanson Analysis Systems, Inc.

Beer P., and Johnston, E. R., *Mechanics of Materials*, 2d ed., New York, McGraw-Hill, 1992.

Fung, Y. C., *Foundations of Solid Mechanics*, Englewood Cliffs, NJ, Prentice-Hall, 1965.

Hibbleer, R. C., *Mechanics of Materials*, 2d. ed., New York, Macmillan, 1994.

Segrlind, L., *Applied Finite Element Analysis*, 2d ed., New York, John Wiley and Sons, 1984.

Shigley, J. E., and Mischke, C. R., *Mechanical Engineering Design*, 5th ed., New York, McGraw-Hill, 1989.

Timoshenko, S. P., and Goodier J. N., *Theory of Elasticity*, 3d ed., New York, McGraw-Hill, 1970.

## PROBLEMS

**1.** The beam shown in the accompanying figure is a wide-flange W 18 $\times$ 35, with a cross-sectional area of 10.3 in$^2$ and a depth of 17.7 in. The second moment of area is 510 in$^4$. The beam is subjected to a uniformly distributed load of 2000 lb/ft. The modulus of elasticity of the beam is $E = 29 \times 10^6$ lb/in$^2$. Using manual calculations, determine the vertical displacement at node 3 and the rotations at nodes 2 and 3. Also, compute the reaction forces at nodes 1 and 2 and reaction moment at node 1.

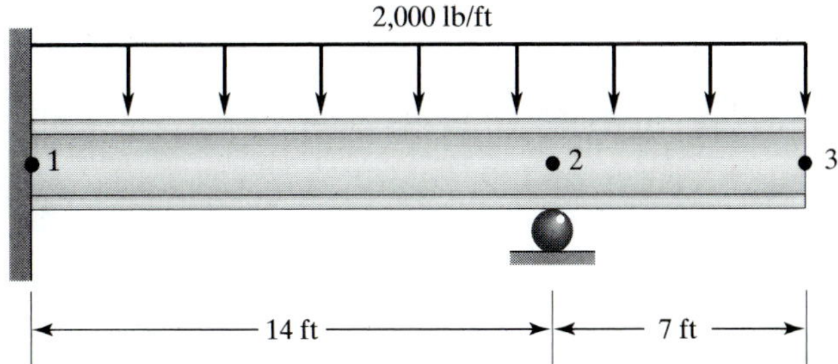


**2.** The beam shown in the accompanying figure is a wide-flange W 16 $\times$ 31 with a cross-sectional area of 9.12 in$^2$ and a depth of 15.88 in. The second moment of area is 375 in$^4$. The beam is subjected to a uniformly distributed load of 1000 lb/ft and a point load of 500 lb. The modulus of elasticity of the beam is $E = 29 \times 10^6$ lb/in$^2$. Using manual calculations, determine the vertical displacement at node 3 and the rotations at nodes 2 and 3. Also, compute the reaction forces at nodes 1 and 2 and reaction moment at node 1.

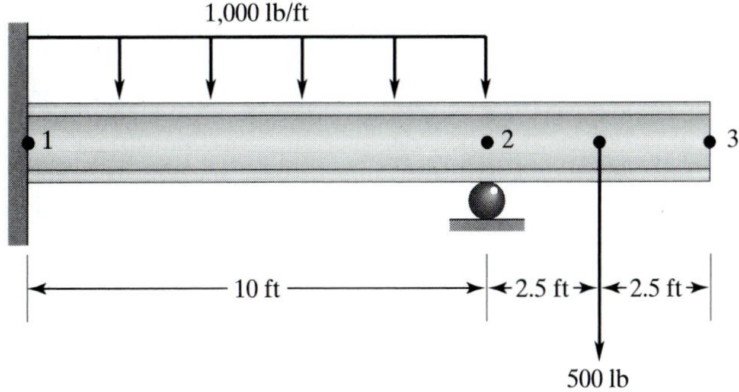


**3.** The lamp frame shown in the accompanying figure has hollow square cross sections and is made of steel, with $E = 29 \times 10^6$ lb/in$^2$. Using hand calculations, determine the endpoint deflection of the cross member where the lamp is attached.

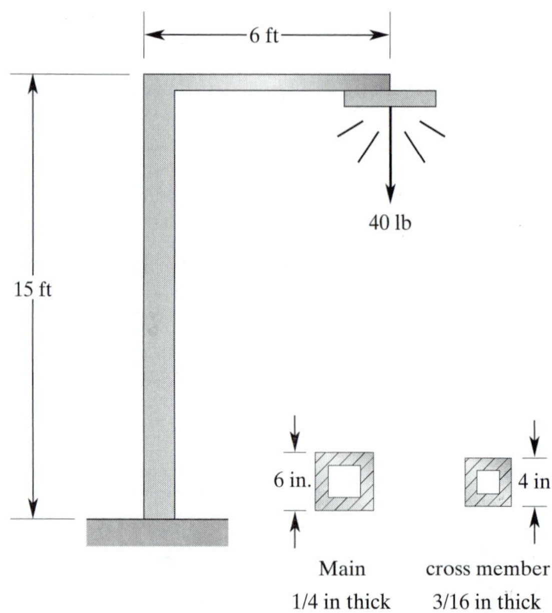


4. A park picnic-table top is supported by two identical metal frames; one such frame is shown in the accompanying figure. The frames are embedded in the ground and have hollow circular cross-sectional areas. The table top is designed to support a distributed load of 250 $lb/ft^2$. Using ANSYS, size the cross section of the frame to support the load safely.

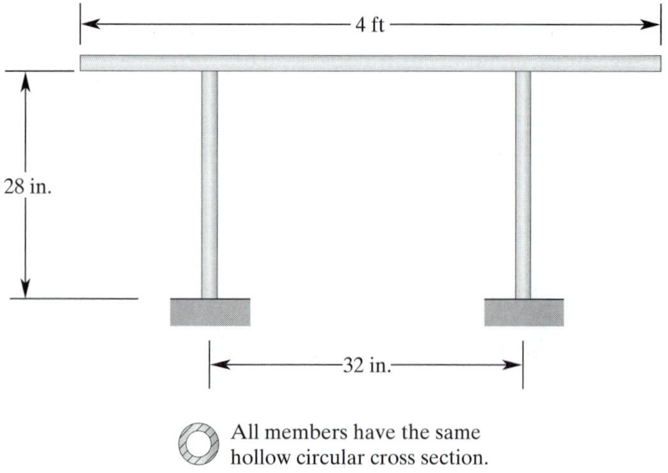


5. The frame shown in the accompanying figure is used to support a load of 2000 lb. The main vertical section of the frame has an annular cross section with an area of 8.63 $in^2$ and a polar radius of gyration of 2.75 in. The outer diameter of the main tubular section is 6 in. All other members also have annular cross sections with respective areas of 2.24 $in^2$ and polar radii of

gyration of 1.91 in. The outer diameter of these members is 4 in. Using ANSYS, determine the deflections at the points where the load is applied. The frame is made of steel, with a modulus elasticity of $E = 29 \times 10^6$ lb/in$^2$.

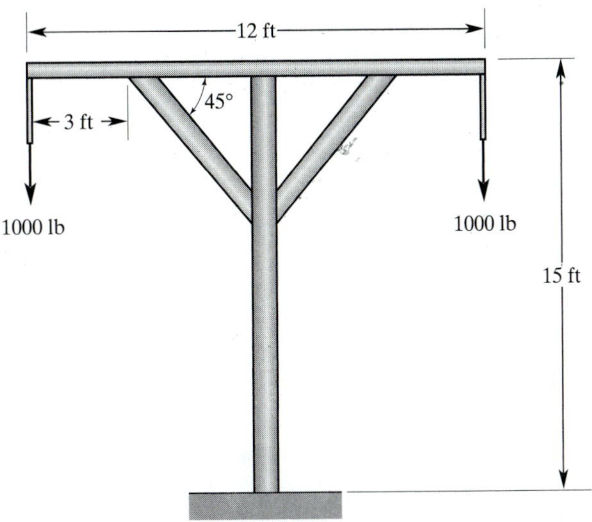


6. Using ANSYS, verify the stress-concentration chart for a flat bar with a circular hole under axial loading. Refer to a textbook on the mechanics of materials or textbook on machine design for the appropriate chart. Recall that the stress-concentration factor $k$ is defined as

$$k = \frac{\sigma_{\text{max}}}{\sigma_{\text{avg}}}$$

and for this case, its value changes from approximately 3.0 to 2.0, depending on the size of the hole.

7. Consider one of the many steel brackets $(E = 29 \times 10^6 \text{ lb/in}^2, \nu = 0.3)$ used to support bookshelves. The thickness of the bracket is 1/8 in. The dimensions of the bracket are shown in the accompanying figure. The bracket is loaded uniformly along its top surface, and it is fixed along its left edge. Under the given loading and the constraints, plot the deformed shape; also, determine the von Mises stresses in the bracket.

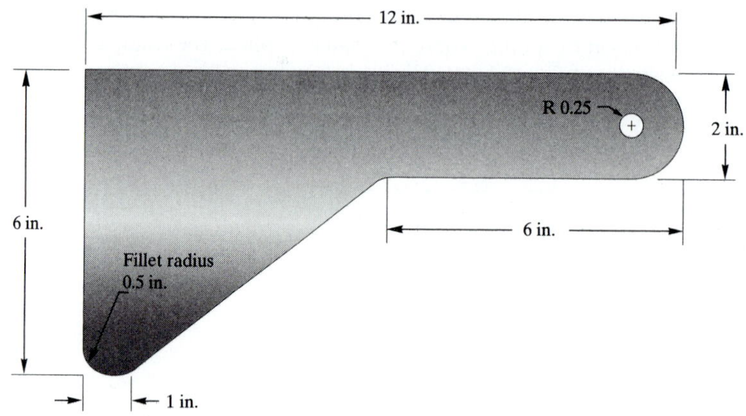


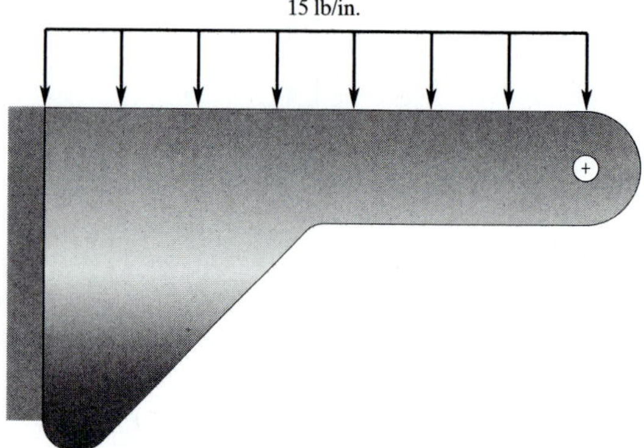


8. A $\frac{1}{8}$-in-thick plate supports a load of 100 lb, as shown in the accompanying figure. The plate is made of steel, with $E = 29 \times 10^6$ lb/in$^2$ and $\nu = 0.3$. Using ANSYS, determine the principal stresses in the plate. When modeling, distribute the load over part of the bottom portion of the hole.

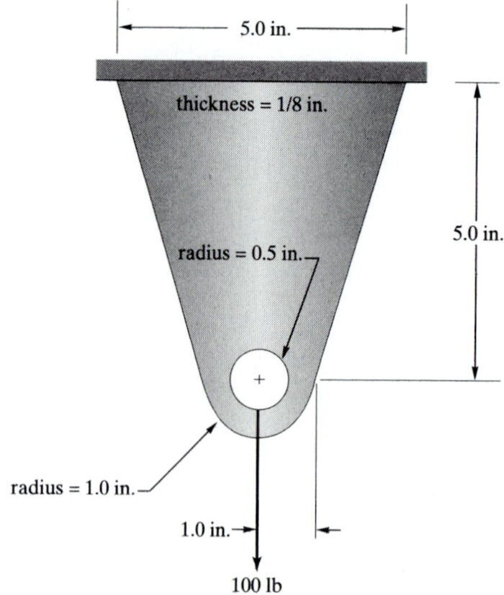

9. Elements (1) and (2) are subjected to the distributed loads shown in the accompanying figure. Replace the distributed loads by equivalent loads at nodes 3, 4, and 5.

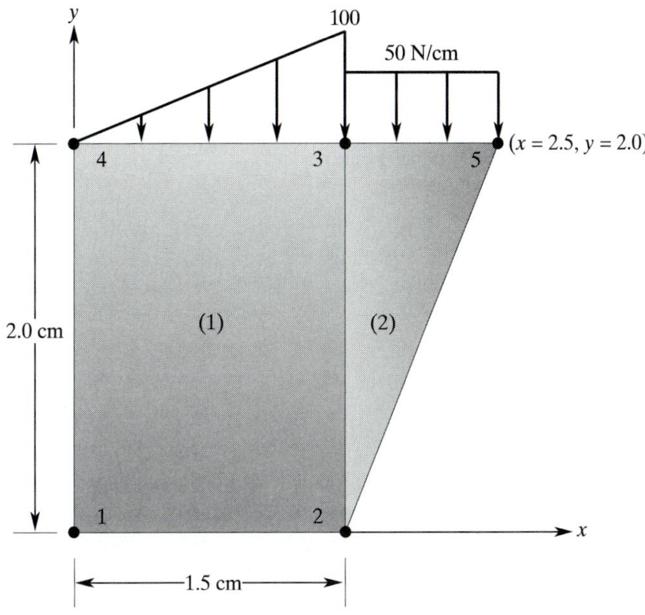


10. Using a steel sample similar to the one shown in the accompanying figure, perform a numerical tension test over the elastic region of the material. Plot the stress–strain diagram over the elastic region.

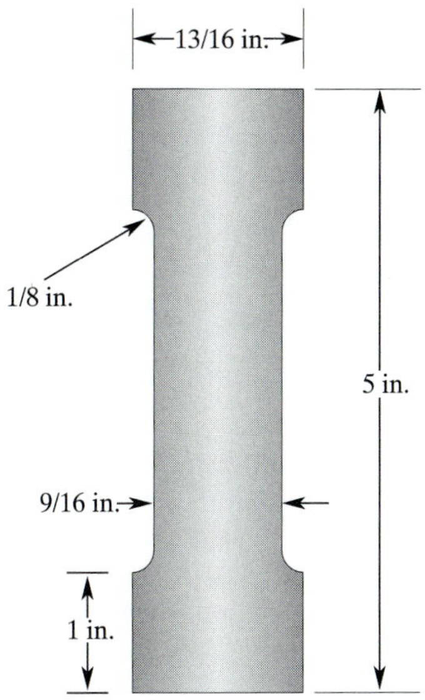

**11.** Verify the equivalent nodal loading for a beam element subjected to a triangular load, as shown in the accompanying figure.

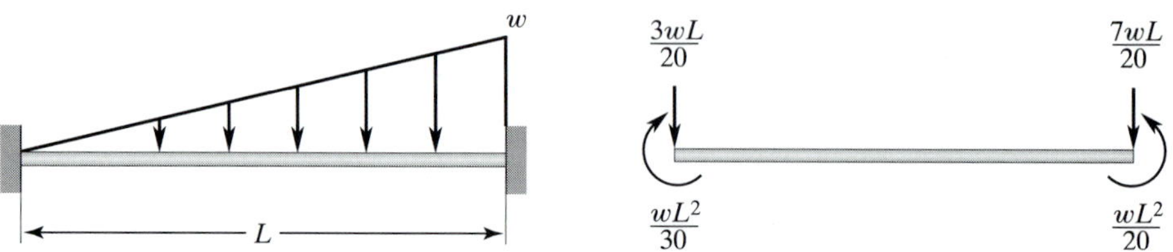


**12.** Referring to the section in this chapter discussing the frame elements, show that the stiffness matrix represented with respect to the global coordinate system is related to the stiffness matrix described with respect to the frame's local coordinate system, according to the relationship

$$[\mathbf{K}]^{(e)} = [\mathbf{T}]^T[\mathbf{K}]_{xy}[\mathbf{T}]$$

**13. Example 1.4 (revisited).** A steel plate is subjected to an axial load, as shown in the accompanying figure. The plate is 1/16 in thick, and it has a modulus of elasticity $E = 29 \times 10^6$ lb/in$^2$. Recall that we approximated the deflections and average stresses along the plate using the concept of one-dimensional direct formulation. Using ANSYS, determine the deflection and the $x$ and $y$-components of the stress distributions in the plate. Also, determine the location of the maximum-stress-concentration regions. Plot the variation of the $x$-component of the stress at sections $A$–$A$, $B$–$B$, and $C$–$C$. Compare the results of the direct-formulation model to the results obtained from ANSYS. Furthermore, recall that for the given problem, it was mentioned that the way in which you apply the external load to your finite element model will influence the stress-distribution results. Experiment with applying the load over an increasingly large load-contact surface area. Discuss your results.

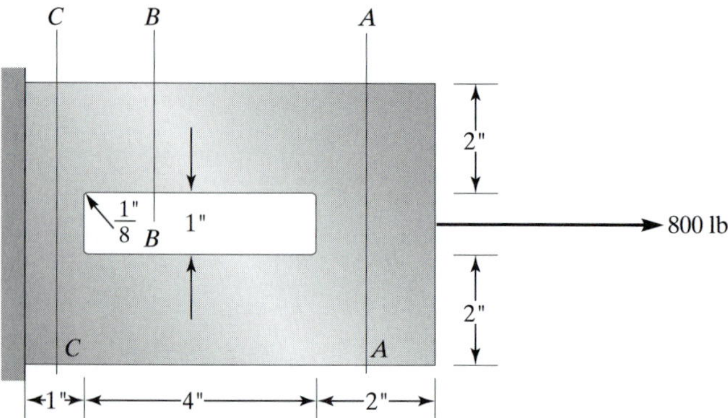


**14.** Consider a plate with a variable cross section supporting a load of 1500 lb, as shown in the accompanying figure. Using ANSYS, determine the deflection and the $x$- and $y$-components of the stress distribution in the plate. The plate is made of a material with a modulus of elasticity $E = 10.6 \times 10^3$ ksi. In Problem 24 of Chapter 1, you were asked to analyze this problem using simple direct formulation. Compare the results of your direct-formulation model to the results obtained from ANSYS. Experiment with applying the load over an increasingly large load-contact surface area. Discuss your results.

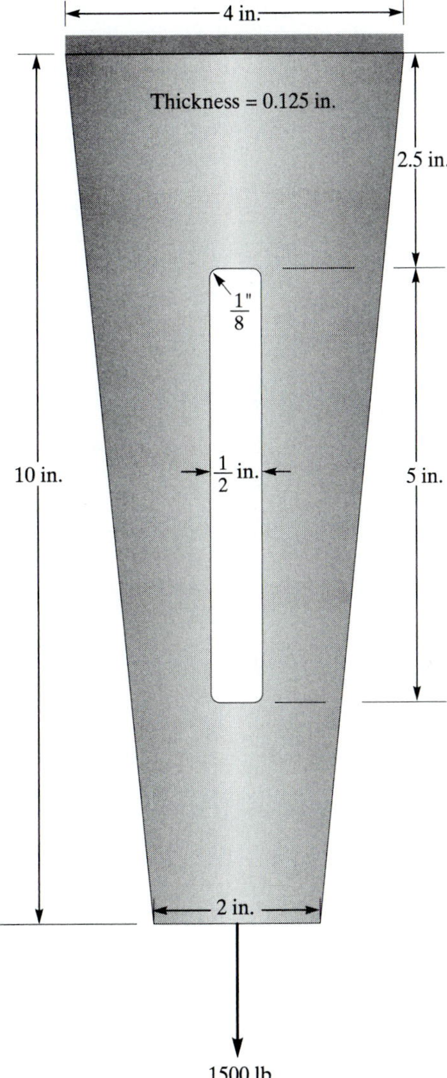


**15.** A thin steel plate with the profile given in the accompanying figure is subjected to an axial load. Using ANSYS, determine the deflection and the $x$ and $y$-components of the stress distributions in the plate. The plate has a thickness of 0.125 in and a modulus of elasticity of $E = 28 \times 10^3$ ksi. In Problem 4 of Chapter 1, you were asked to analyze this problem using simple direct formulation. Compare the results of your direct-formulation model to the results obtained from ANSYS. Experiment with applying the load over an increasingly larger load-contact surface area. Discuss your results.

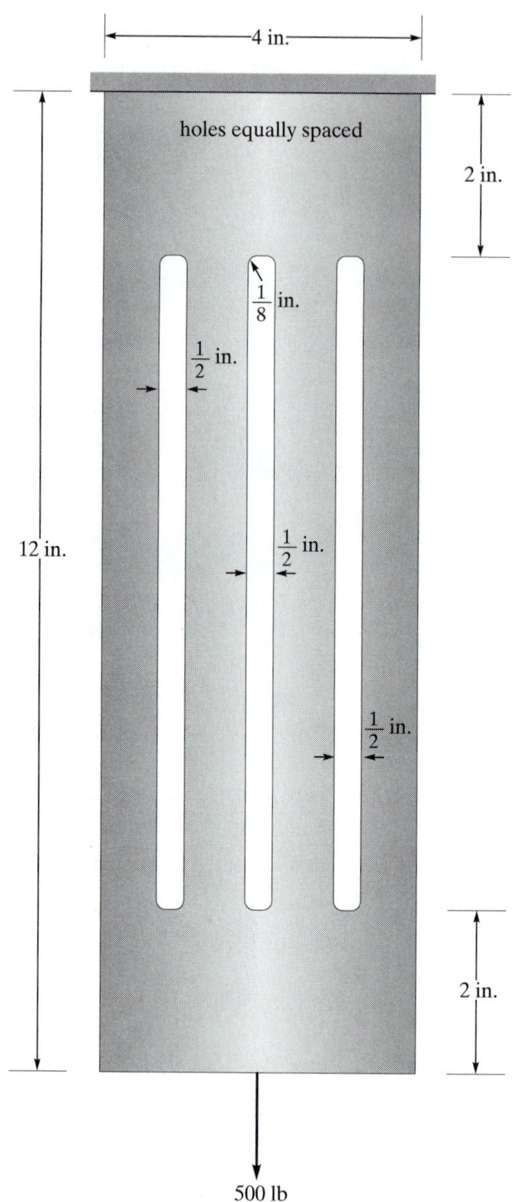


**16.** Use ANSYS to solve Problem 3. Determine the magnitude and the location of the maximum tensile and compressive stresses.

**17.** The frame shown in the accompanying figure is used to support a load of 500 lb/ft. Using ANSYS, size the cross sections of each member if standard-size steel square tubing is to be used. Use three different sizes. The deflection of the centerpoint is to be kept under 0.05 in.

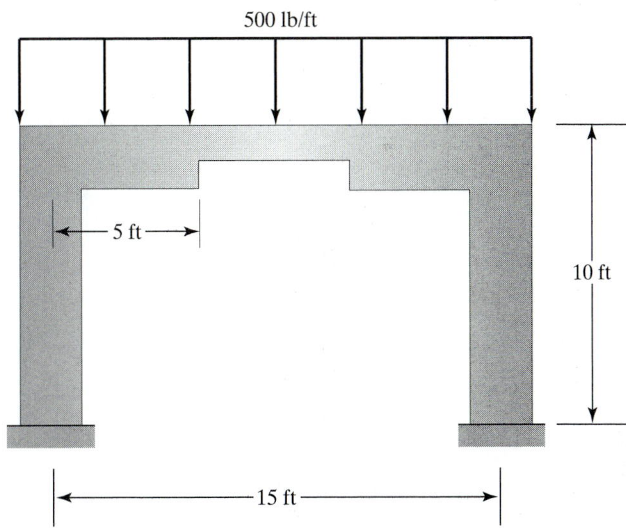


**18.** The frame shown in the accompanying figure is used to support the load given in the figure. Using ANSYS, size the members if standard sizes of steel I-beams are to be used.

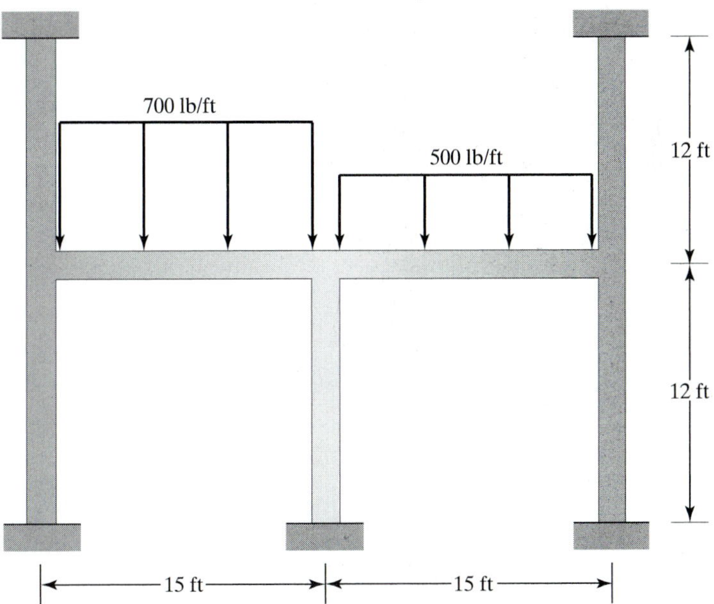


**19.** Consider the torsion of a steel bar $\left(G = 11 \times 10^3 \text{ ksi}\right)$ having an equilateral-triangular cross section, as shown in the accompanying figure. Assuming that $\theta = 0.0005$ rad/in and using

ANSYS, determine the location(s) and magnitude of the maximum shear stress. Compare the solution generated with ANSYS to the exact solution obtained from the equation

$$\tau_{max} = \frac{GL\theta}{2}$$

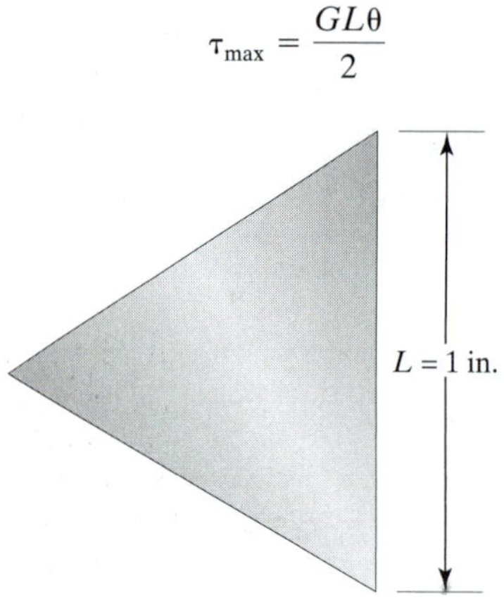


**20.** Consider the torsion of a steel wide-flange member (W 4 × 13 and $G = 11 \times 10^3$ ksi) with dimensions shown in the accompanying figure. Assuming $\theta = 0.00035$ rad/in and using ANSYS, plot the shear stress distributions. Could you have solved this problem using the thin-wall member assumption and, thus, avoid resorting to a finite element model?

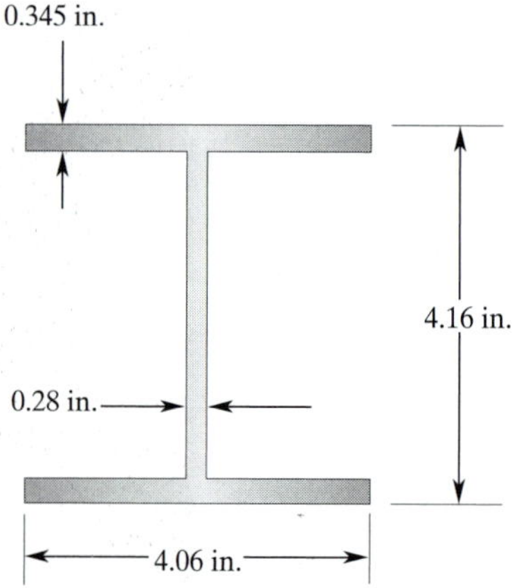


**21.** Verify the equivalent nodal loading for a beam element subjected to the load shown in the accompanying figure.

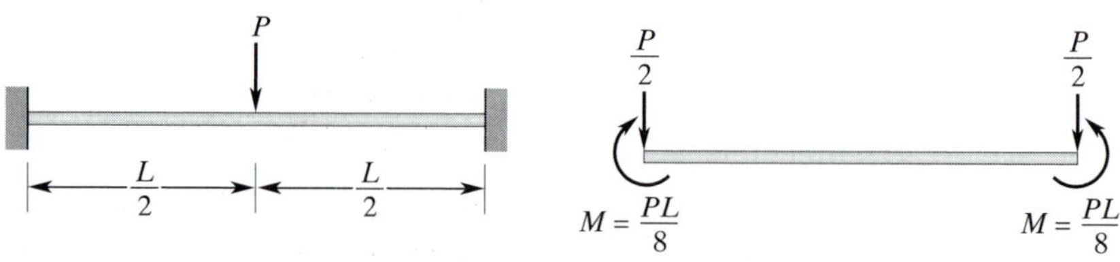

**22. Design Project** The purpose of this project is twofold: (1) to provide a basis for the application of solid-design principles using finite element methods and (2) to foster competitiveness among students. Each student is to design and construct a structural model from a $\frac{3}{8} \times 6 \times 6$ in sheet of plexiglas material that adheres to the specifications and rules given later in this problem and that is capable of competing in three areas: (1) maximum failure load per model weight, (2) predication of failure load using ANSYS, and (3) workmanship. A sketch of a *possible* model is shown in the accompanying figure. Each end of the model will have a diameter hole (eye) of $d > 1/2''$ drilled through it perpendicular to the axis of loading, for which pins can be inserted and the model loaded in tension. The dimension $a$ must also be set such that $a > 1''$. The distance between the eyes will be $\ell > 2''$. The maximum thickness of the member in the region of the eye will be $t < 3/8''$. This requirement will ensure that the model fits into the loading attachment. A dimension of $b < 1''$ from the center of the eyes to the outer edge in the direction of loading must be maintained so that the loading attachment can be utilized. The maximum width is limited to $w < 6''$, and the maximum height is limited to $h < 6''$. Any configuration may be used. Two sheets of $\frac{3}{8} \times 6 \times 6$ in plexiglas will be provided. You can use one sheet to experiment and one sheet for your final design. Write a brief report discussing the evolution of your final design.

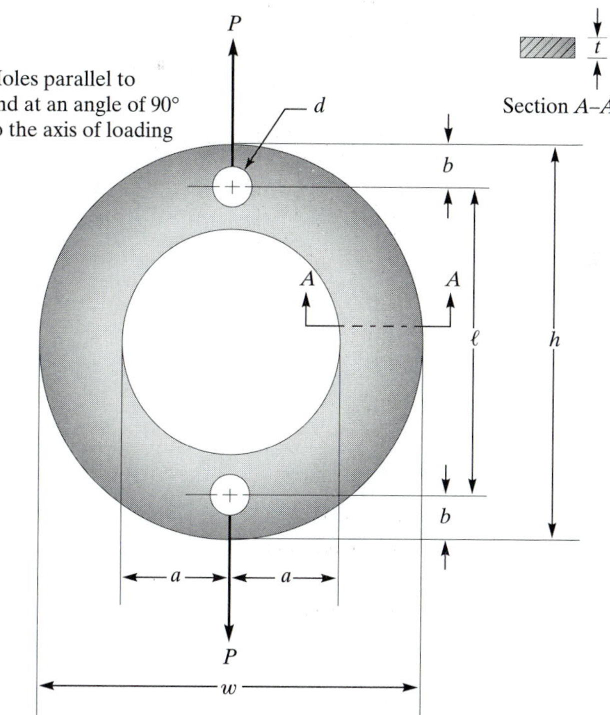


**23. Design Project** Size the members of the bridge shown in the accompanying figure for a case in which traffic is backed up with a total of four trucks equally spaced on the bridge. A typical truck has a payload weight of 64,000 lb and a cab weight of 8000 lb. As a starting point, you may use one cross section for all beam elements. You may also assume one cross section for all truss members. The roadbed weighs 1500 lb/ft and is supported by I-beams. Use standard steel I-beam sizes. Design your own truss configuration. In your analysis, you may assume that the concrete column does not deflect significantly. Write a brief report discussing how you came up with the final design.

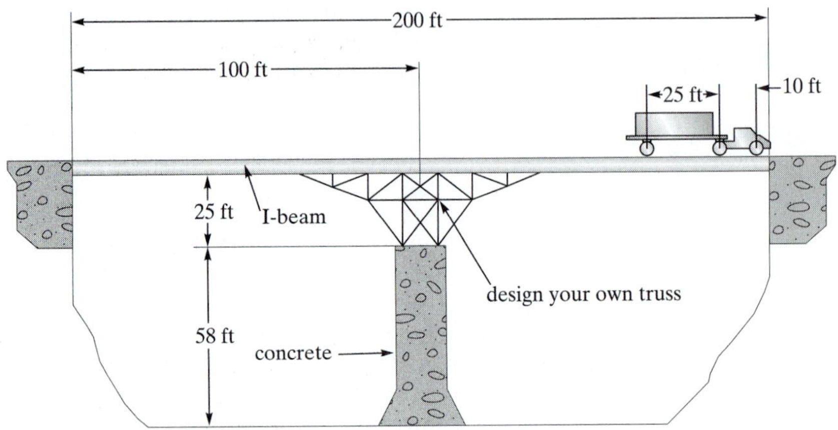

CHAPTER 9

# Analysis of Fluid Mechanics Problems

The main objective of this chapter is to introduce you to the analysis of fluid mechanics problems. First, we will discuss the direct formulation of pipe-network problems. Then, we will consider finite element formulation of ideal fluid behavior (inviscid flow). Finally, we will briefly look at the flow of fluid through porous media and finite element formulation of underground seepage flows. The main topics discussed in Chapter 9 include the following:

**9.1** Direct Formulation of Flow Through Pipes
**9.2** Ideal Fluid Flow
**9.3** Groundwater Flow
**9.4** Examples Using ANSYS
**9.5** Verification of Results

## 9.1 DIRECT FORMULATION OF FLOW THROUGH PIPES

We begin by reviewing fundamental concepts of fluid flow through pipes. The internal flow through a conduit may be classified as laminar or turbulent flow. In laminar flow situations, a thin layer of dye injected into a pipe will show as a straight line. No mixing of fluid layers will be visible. This situation does not hold for turbulent flow, in which the bulk mixing of adjacent fluid layers will occur. Laminar and turbulent flow are depicted in Figure 9.1. Laminar flow typically occurs when the Reynolds number of the flowing fluid is less than 2100. The Reynolds number is defined as

$$\mathrm{Re} = \frac{\rho V D}{\mu} \tag{9.1}$$

where $\rho$ and $\mu$ are the density and the dynamic viscosity of the fluid, respectively. $V$ represents the average fluid velocity, and $D$ represents the diameter of the pipe. The flow is said to be in a transition region when the Reynolds number is typically between 2100 and 4000. The behavior of the fluid flow is unpredictable in the transition region. The flow is generally considered to be turbulent when the Reynolds number is greater than 4000. The conservation of mass for a steady flow requires that the mass flow rate at any section of the pipe remains constant according to the equation

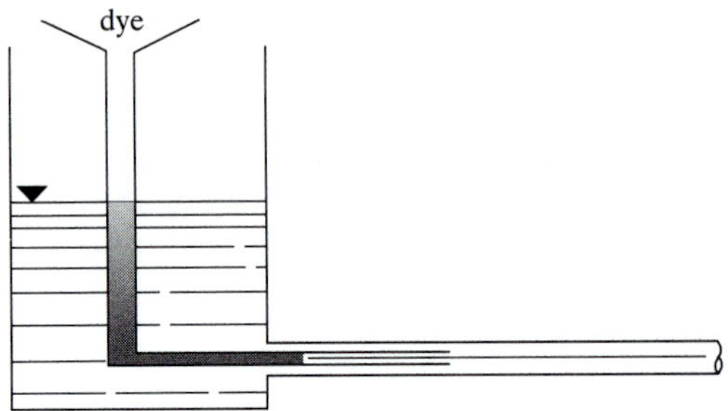


(a) Laminar flow in a pipe.

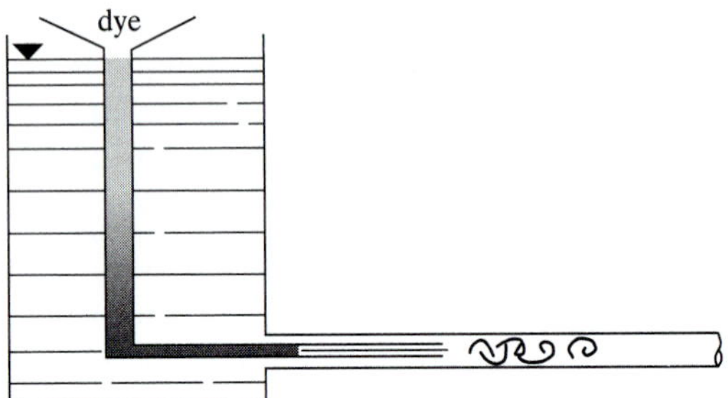


(b) Turbulent flow in a pipe.

**FIGURE 9.1** Laminar and turbulent flows.

$$\dot{m_1} = \dot{m_2} = \rho_1 V_1 A_1 = \rho_2 V_2 A_2 = \text{constant} \tag{9.2}$$

Again, ρ is the density of the fluid, $V$ is the average fluid velocity at a section, and $A$ represents the cross-sectional area of the flow as shown in Figure 9.2.

For an incompressible flow—a flow situation where the density of the fluid remains constant—the volumetric flow rate $Q$ through a conduit at any section of the conduit is also constant:

$$Q_1 = Q_2 = V_1 A_1 = V_2 A_2 \tag{9.3}$$

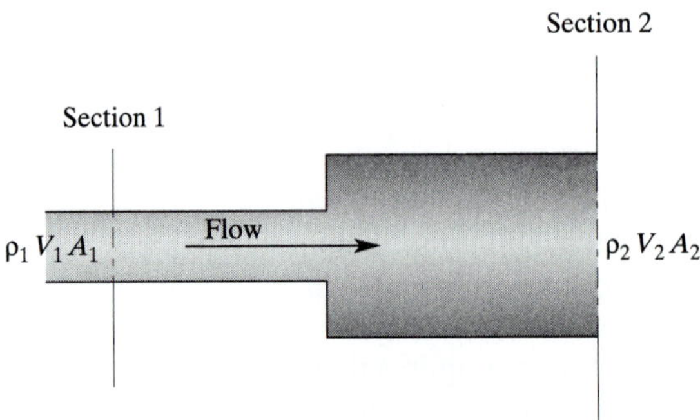


**FIGURE 9.2** Flow of fluid through a conduit with variable cross section.

For a fully developed laminar flow, there exists a relationship between the volumetric flow rate and the pressure drop $P_1 - P_2$ along a pipe of length $L$. This relationship is given by:

$$Q = \frac{\pi D^4}{128\mu}\left(\frac{P_1 - P_2}{L}\right) \tag{9.4}$$

The pressure drop for a turbulent flow is commonly expressed in terms of head loss, which is defined as

$$H_{\text{loss}} = \frac{P_1 - P_2}{\rho g} = f\frac{L}{D}\frac{V^2}{2g} \tag{9.5}$$

where $f$ is the friction factor, which depends on the surface roughness of the pipe and the Reynolds number. For turbulent flows, we can also obtain a relationship between the volumetric flow rate and the pressure drop by substituting for $V$ in terms of the flow rate in Eq. (9.5) and rearranging terms:

$$Q^2 = \frac{1}{f}\frac{\pi^2 D^5}{8\rho}\left(\frac{P_1 - P_2}{L}\right) \tag{9.6}$$

When we compare turbulent flow to laminar flow, we note that for turbulent flow, the relationship between the flow rate and pressure drop is nonlinear.

## Pipes in Series

For flow of a fluid through a piping network consisting of a series of pipes with respective diameters $D_1, D_2, D_3, \ldots$, as shown in Figure 9.3, the conservation of mass (continuity equation) requires that under steady-state conditions, the mass flow rate through each pipe be the same:

$$\dot{m}_1 = \dot{m}_2 = \dot{m}_3 = \ldots = \text{constant} \tag{9.7}$$

Moreover, for an incompressible flow, the volumetric flow rate through each pipe that is part of a piping network in series is constant. That is,

$$Q_1 = Q_2 = Q_3 = \ldots = \text{constant} \tag{9.8}$$

Expressing the flow rates in terms of the average fluid velocity in each pipe, we obtain

$$V_1 D_1^2 = V_2 D_2^2 = V_3 D_3^2 = \ldots = \text{constant} \tag{9.9}$$

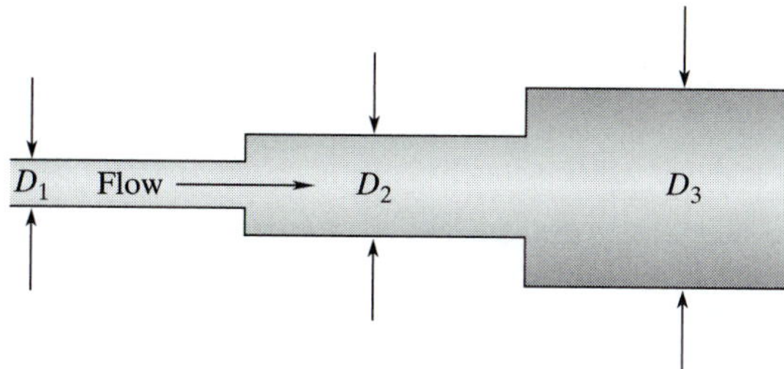


**FIGURE 9.3** Pipes in series.

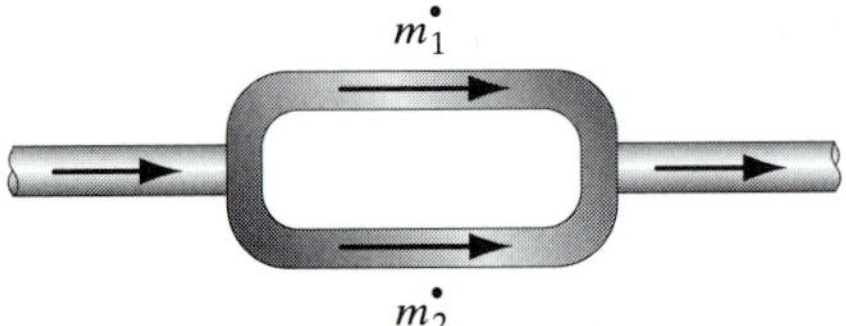


**FIGURE 9.4** Pipes in parallel.

For pipes in series, the total pressure drop through a network is determined from the sum of the pressure drops in each pipe:

$$\Delta P_{\text{total}} = \Delta P_1 + \Delta P_2 + \Delta P_3 + \ldots \tag{9.10}$$

For flow of a fluid through a piping network consisting of pipes in parallel arrangement, as shown in Figure 9.4, the conservation of mass (continuity equation) requires that

$$\dot{m}_{\text{total}} = \dot{m}_1 + \dot{m}_2 \tag{9.11}$$

Moreover, for an incompressible flow,

$$Q_{\text{total}} = Q_1 + Q_2 \tag{9.12}$$

For pipes in parallel configuration, the pressure drop in each parallel branch is the same, and is related according to:

$$\Delta P_{\text{total}} = \Delta P_1 = \Delta P_2 \tag{9.13}$$

### Finite Element Formulation

Consider an incompressible laminar flow of a viscous fluid through a network of piping systems, as shown in Figure 9.5. We start by subdividing the problem into nodes and elements. This example may be represented by a model that has four nodes and four elements.

The behavior of the fluid flow inside a pipe section is modeled by an element with two nodes. The elemental description is given by the relationship between the flow rate and the pressure drop as given by Eq. (9.4), such that

$$Q = \frac{\pi D^4}{128\mu}\left(\frac{P_i - P_{i+1}}{L}\right) = C(P_i - P_{i+1}) \tag{9.14}$$

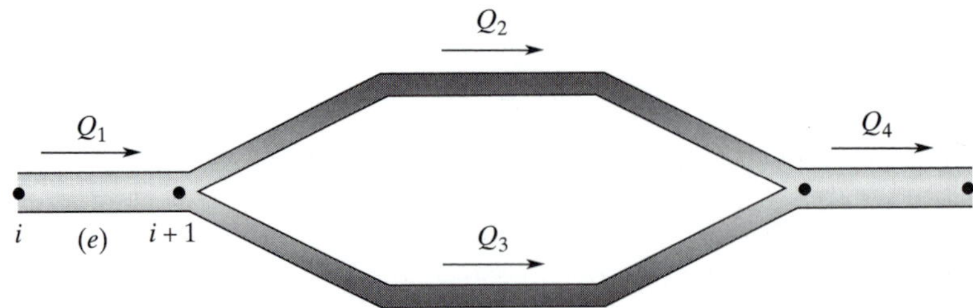


**FIGURE 9.5** A network problem: an incompressible laminar flow of a viscous fluid through a network of piping systems.

where the flow-resistance coefficient $C$ is given by

$$C = \frac{\pi D^4}{128L\mu} \tag{9.15}$$

Because there are two nodes associated with each element, we need to create two equations for each element. These equations must involve nodal pressure and the element's flow resistance. Consider the flow rates $Q_i$ and $Q_{i+1}$ and the nodal pressures $P_i$ and $P_{i+1}$ of an element, which are related according to the equations

$$Q_i = C(P_i - P_{i+1})$$

$$Q_{i+1} = C(P_{i+1} - P_i) \tag{9.16}$$

The equations given by (9.16) were formulated such that the conservation of mass is satisfied as well. The sum of $Q_i$ and $Q_{i+1}$ is zero, which implies that under steady-state conditions, what flows into a given node also flows out. Equations (9.16) can be expressed in matrix form by

$$\begin{bmatrix} Q_i \\ Q_{i+1} \end{bmatrix} = \begin{bmatrix} C & -C \\ -C & C \end{bmatrix} \begin{bmatrix} P_i \\ P_{i+1} \end{bmatrix} = \begin{bmatrix} \dfrac{\pi D^4}{128L\mu} & -\dfrac{\pi D^4}{128L\mu} \\ -\dfrac{\pi D^4}{128L\mu} & \dfrac{\pi D^4}{128L\mu} \end{bmatrix} \begin{bmatrix} P_i \\ P_{i+1} \end{bmatrix} \tag{9.17}$$

The element's flow-resistance matrix is then given by

$$[\mathbf{R}]^{(e)} = \begin{bmatrix} \dfrac{\pi D^4}{128L\mu} & -\dfrac{\pi D^4}{128L\mu} \\ -\dfrac{\pi D^4}{128L\mu} & \dfrac{\pi D^4}{128L\mu} \end{bmatrix} \tag{9.18}$$

Applying the elemental description given by Eq. (9.17) to all elements and assembling them will lead to the formation of the global flow matrix, the flow-resistance matrix, and the pressure matrix.

## EXAMPLE 9.1

Oil with dynamic viscosity of $\mu = 0.3\ \text{N} \cdot \text{s/m}^2$ and density of $\rho = 900\ \text{kg/m}^3$ flows through the piping network shown in Figure 9.6. The 2–4–5 branch was added in parallel to the 2–3–5 branch to allow for the flexibility of performing maintenance on one branch while the oil flows through the other branch. The dimensions of the piping system are shown in Figure 9.6. Determine the pressure distribution in the system if both branches are on line. The flow rate at node 1 is $5 \times 10^{-4}\ \text{m}^3/\text{s}$. The pressure at node 1 is 39182 Pa (g) and the pressure at node 6 is $-3665$ Pa (g). For the given conditions, the flow is laminar throughout the system. How does the flow divide in each branch?

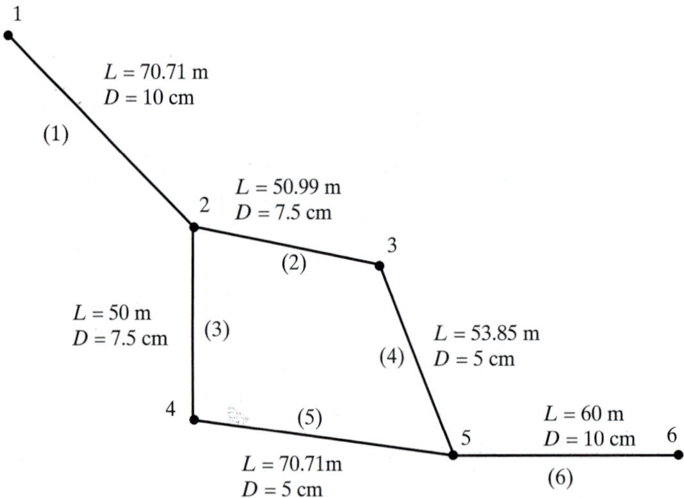


**FIGURE 9.6** The piping network of Example 9.1.

The elemental flow resistance is given by Eq. (9.18) as

$$[\mathbf{R}]^{(e)} = \begin{bmatrix} \dfrac{\pi D^4}{128L\mu} & -\dfrac{\pi D^4}{128L\mu} \\ -\dfrac{\pi D^4}{128L\mu} & \dfrac{\pi D^4}{128L\mu} \end{bmatrix}$$

We model the given network using six elements and six nodes. Evaluating the respective resistance matrices for elements (1)–(6), we obtain

$$[\mathbf{R}]^{(1)} = 10^{-9}\begin{bmatrix} 115.70 & -115.70 \\ -115.70 & 115.70 \end{bmatrix}\begin{matrix} 1 \\ 2 \end{matrix} \qquad [\mathbf{R}]^{(2)} = 10^{-9}\begin{bmatrix} 50.76 & -50.76 \\ -50.76 & 50.76 \end{bmatrix}\begin{matrix} 2 \\ 3 \end{matrix}$$

$$[\mathbf{R}]^{(3)} = 10^{-9}\begin{bmatrix} 51.77 & -51.77 \\ -51.77 & 51.77 \end{bmatrix}\begin{matrix} 2 \\ 4 \end{matrix} \qquad [\mathbf{R}]^{(4)} = 10^{-9}\begin{bmatrix} 9.50 & -9.50 \\ -9.50 & 9.50 \end{bmatrix}\begin{matrix} 3 \\ 5 \end{matrix}$$

$$[\mathbf{R}]^{(5)} = 10^{-9}\begin{bmatrix} 7.23 & -7.23 \\ -7.23 & 7.23 \end{bmatrix}\begin{matrix} 4 \\ 5 \end{matrix} \qquad [\mathbf{R}]^{(6)} = 10^{-9}\begin{bmatrix} 136.35 & -136.35 \\ -136.35 & 136.35 \end{bmatrix}\begin{matrix} 5 \\ 6 \end{matrix}$$

Note that in order to aid us in assembling the elemental resistance matrices into the global resistance matrix, the corresponding nodes are shown alongside of each element's resistance matrix. So, we have

$$10^{-9}\begin{bmatrix} 115.7 & -115.7 & 0 & 0 & 0 & 0 \\ -115.7 & 115.7+50.76+51.77 & -50.76 & -51.77 & 0 & 0 \\ 0 & -50.76 & 50.76+9.50 & 0 & -9.50 & 0 \\ 0 & -51.77 & 0 & 51.77+7.23 & -7.23 & 0 \\ 0 & 0 & -9.50 & -7.23 & 9.50+7.23+136.35 & -136.35 \\ 0 & 0 & 0 & 0 & -136.35 & 136.35 \end{bmatrix}\begin{matrix} 1 \\ 2 \\ 3 \\ 4 \\ 5 \\ 6 \end{matrix}$$

Applying the boundary conditions $P_1 = 39182$ and $P_2 = -3665$, we obtain

$$\begin{bmatrix} 1 & 0 & 0 & 0 & 0 & 0 \\ -115.7 & 218.23 & -50.76 & -51.77 & 0 & 0 \\ 0 & -50.76 & 60.26 & 0 & -9.50 & 0 \\ 0 & -51.77 & 0 & 59.0 & -7.23 & 0 \\ 0 & 0 & -9.50 & -7.23 & 153.08 & -136.35 \\ 0 & 0 & 0 & 0 & 0 & 1 \end{bmatrix} \begin{bmatrix} P_1 \\ P_2 \\ P_3 \\ P_4 \\ P_5 \\ P_6 \end{bmatrix} = \begin{bmatrix} 39182 \\ 0 \\ 0 \\ 0 \\ 0 \\ -3665 \end{bmatrix}$$

Solving the systems of equations simultaneously results in the nodal pressure values:

$$[\mathbf{P}]^T = [39182 \quad 34860 \quad 29366 \quad 30588 \quad 2 \quad -3665]\text{Pa}$$

The flow rate in each branch is determined from Eq. (9.14):

$$Q = \frac{\pi D^4}{128\mu}\left(\frac{P_i - P_{i+1}}{L}\right) = C(P_i - P_{i+1})$$

$$Q^{(2)} = 50.76 \times 10^{-9}(34860 - 29366) = 2.79 \times 10^{-4} \text{ m}^3/\text{s}$$

$$Q^{(3)} = 51.77 \times 10^{-9}(34860 - 30588) = 2.21 \times 10^{-4} \text{ m}^3/\text{s}$$

$$Q^{(4)} = 9.50 \times 10^{-9}(29366 - 2) = 2.79 \times 10^{-4} \text{ m}^3/\text{s}$$

$$Q^{(5)} = 7.23 \times 10^{-9}(30588 - 2) = 2.21 \times 10^{-4} \text{ m}^3/\text{s}$$

The verification of these results is discussed in Section 9.5.

---

## 9.2 IDEAL FLUID FLOW

All fluids have some viscosity; however, in certain flow situations it may be reasonable to neglect the effects of viscosity and the corresponding shear stresses. The assumption may be made as a first approximation to simplify the behavior of real fluids with relatively small viscosity. Also, in many external viscous flow situations, we can divide the flow into two regions: (1) a thin layer close to a solid boundary—called the boundary layer region—where the effects of viscosity are important and (2) a region outside the boundary layer where the viscous effects are negligible, in which the fluid is considered to be inviscid. This concept is demonstrated for the case of the flow of air over an airfoil in Figure 9.7. For inviscid flow situations, the only forces considered are those resulting from pressure and the inertial forces acting on a fluid element.

Before discussing finite element formulation of ideal fluid problems, let us review some fundamental information. For a two-dimensional flow field, the fluid velocity is

$$\vec{V} = v_x\vec{i} + v_y\vec{j} \tag{9.19}$$

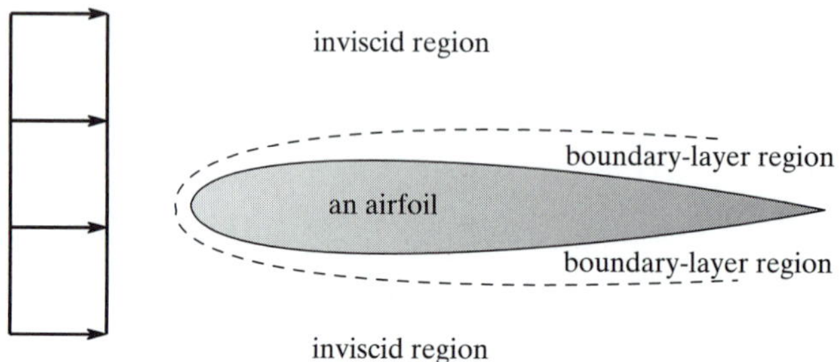


**FIGURE 9.7** The flow of air over an airfoil.

where $v_x$ and $v_y$ are the $x$- and $y$-components of the fluid's velocity vector, respectively. The conservation of mass (continuity equation) for a two-dimensional incompressible fluid can be expressed in the differential form in terms of fluid's velocity components as

$$\frac{\partial v_x}{\partial x} + \frac{\partial v_y}{\partial y} = 0 \tag{9.20}$$

### The Stream Function and Stream Lines

For a steady flow, a streamline represents the trajectory of a fluid particle. The streamline is a line that is tangent to the velocity of a fluid particle. Streamlines provide a means for visualizing the flow patterns. The stream function $\psi(x, y)$ is defined such that it will satisfy the continuity equation Eq. (9.20) according to the following relationships:

$$v_x = \frac{\partial \psi}{\partial y} \quad \text{and} \quad v_y = -\frac{\partial \psi}{\partial x} \tag{9.21}$$

Note that upon substitution of Eq. (9.21) into Eq. (9.20), the conservation of mass is satisfied. Along a line of constant $\psi(x, y)$, we have:

$$d\psi = 0 = \frac{\partial \psi}{\partial x} dx + \frac{\partial \psi}{\partial y} dy = -v_y dx + v_x dy \tag{9.22}$$

or

$$\frac{dy}{dx} = \frac{v_y}{v_x} \tag{9.23}$$

Consider Figure 9.8, in which the flow of a fluid around a sharp corner is shown. For this flow situation, the velocity field is represented by

$$\vec{V} = cx\vec{i} - cy\vec{j}$$

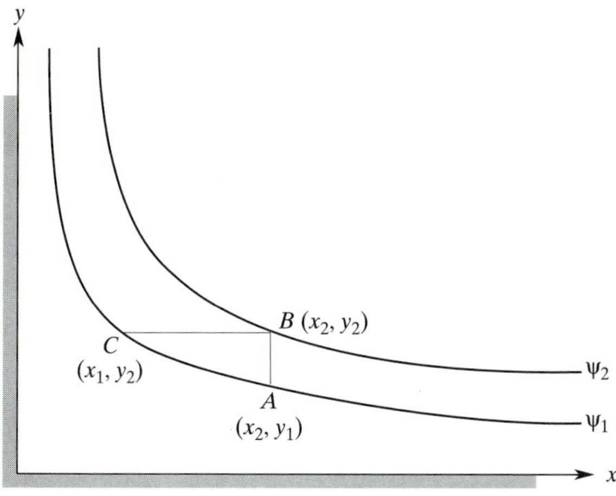


**FIGURE 9.8** The flow of fluid around a sharp corner.

To obtain the expression for the stream function, we make use of Eq. (9.23):

$$\frac{v_y}{v_x} = \frac{dy}{dx} = \frac{-cy}{cx}$$

Integrating, we have

$$\int \frac{dy}{y} = -\int \frac{dx}{x}$$

Evaluating the integral results in the stream function, which is given by

$$xy = \text{constant}$$

or

$$\psi = xy$$

We can plot the streamlines by assigning various values to $\psi$. Note that the individual values of streamlines are not important; it is the difference between their values that is important. The difference between the values of two streamlines provides a measure of volumetric flow rate between the streamlines. To demonstrate this idea, let us refer back to Figure 9.8. Along, the $A$–$B$ section, we can write:

$$\frac{Q}{w} = \int_{y_1}^{y_2} v_x\, dy = \int_{y_1}^{y_2} \frac{\partial \psi}{\partial y}\, dy = \int_{\psi_1}^{\psi_2} d\psi = \psi_2 - \psi_1 \tag{9.24}$$

Similarly, along the $B$–$C$ section, we have:

$$\frac{Q}{w} = \int_{x_1}^{x_2} -v_y\,dx = \int_{x_1}^{x_2} \frac{\partial \Psi}{\partial x}\,dx = \int_{\Psi_1}^{\Psi_2} d\Psi = \Psi_2 - \Psi_1 \tag{9.25}$$

Therefore, the difference between the values of the streamlines represents the volumetric flow rate per unit width $w$.

### The Irrotational Flow, Potential Function, and Potential Lines

As mentioned earlier, there are many flow situations for which the effects of viscosity may be neglected. Moreover at low speeds, the fluid elements within inviscid flow situations may have an angular velocity of zero (no rotation). These types of flow situations are referred to as irrotational flows. A two-dimensional flow is considered to be irrotational when

$$\frac{\partial v_y}{\partial x} - \frac{\partial v_x}{\partial y} = 0 \tag{9.26}$$

We can also define a potential function $\phi$ such that the spatial gradients of the potential function are equal to the components of the velocity field:

$$v_x = \frac{\partial \phi}{\partial x} \qquad v_y = \frac{\partial \phi}{\partial y} \tag{9.27}$$

Along a line of constant potential function, we have:

$$d\phi = 0 = \frac{\partial \phi}{\partial x}dx + \frac{\partial \phi}{\partial y}dy = v_x dx + v_y dy = 0 \tag{9.28}$$

$$\frac{dy}{dx} = -\frac{v_x}{v_y} \tag{9.29}$$

By comparing Eqs. (9.29) and (9.23), we can see that the streamlines and the velocity potential lines are orthogonal to each other. It is clear that the potential function complements the stream function. Using the relationship in Eq. (9.27) to substitute for $v_x$ and $v_y$ in the continuity equation Eq. (9.20), we have

$$\frac{\partial^2 \phi}{\partial x^2} + \frac{\partial^2 \phi}{\partial y^2} = 0 \tag{9.30}$$

Using the definitions of stream functions as given by Eq. (9.21) and substituting for $v_x$ and $v_y$ in Eq. (9.26), we have

$$\frac{\partial^2 \psi}{\partial x^2} + \frac{\partial^2 \psi}{\partial y^2} = 0 \tag{9.31}$$

Equations (9.30) and (9.31), which are forms of Laplace's equation, govern the motion of an ideal irrotational flow. Typically, for potential flow situations, the boundary conditions are the known free-stream velocities, and at the solid surface boundary, the fluid cannot have velocity normal to the surface. The latter condition is given by the equation

$$\frac{\partial \phi}{\partial n} = 0 \tag{9.32}$$

Here, $n$ represents a direction normal to the solid surface. Comparing the differential equation governing the irrotational flow behavior of an inviscid fluid, Eq. (9.30), to the heat diffusion equation, Eq. (7.8), we note that both of these equations have the same form; therefore, we can apply the results of Sections 7.2 and 7.3 to the potential flow problems. However, when comparing the differential equations for irrotational flow problems, we let $C_1 = 1, C_2 = 1$, and $C_3 = 0$. Later in this chapter, we will use ANSYS to analyze the flow of an ideal fluid around a cylinder.

Now, let us briefly discuss the analysis of viscous flows. As mentioned earlier, all real fluids have viscosity. The analysis of a complex viscous flow is generally performed by solving the governing equations of motion for a specific boundary condition using the finite differencing approach. However, in recent years, we have made some advances in the finite element formulation of viscous fluid flow problems. *Bathe* (*1996*) discusses a Galerkin procedure for the analysis of the two-dimensional laminar flow of an incompressible fluid. For more details on the formulation of viscous laminar flows, also see Section 7.1 of the theory volume of ANSYS documents.

## 9.3 GROUNDWATER FLOW

The study of fluid flow and heat transfer in porous media is important in many engineering applications, including problems related to oil-recovery methods, groundwater hydrology, solar energy storage, and geothermal energy. The flow of fluid through an unbounded porous medium is governed by Darcy's Law. Darcy's Law relates the pressure drop to the mean fluid velocity according to the relationship

$$U_D = -\frac{k}{\mu}\frac{dP}{dx} \tag{9.33}$$

where $U_D$ is the mean fluid velocity, $k$ is the permeability of the porous medium, and $\mu$ is the viscosity of the fluid. For two-dimensional flows, it is customary to use the hydraulic head $\phi$ to define the components of the fluid velocities. Consider the seepage flow of water under a dam, as shown in Figure 9.9.

The two-dimensional flow of fluid through the soil is governed by Darcy's Law, which is given by:

$$k_x \frac{\partial^2 \phi}{\partial x^2} + k_y \frac{\partial^2 \phi}{\partial y^2} = 0 \tag{9.34}$$

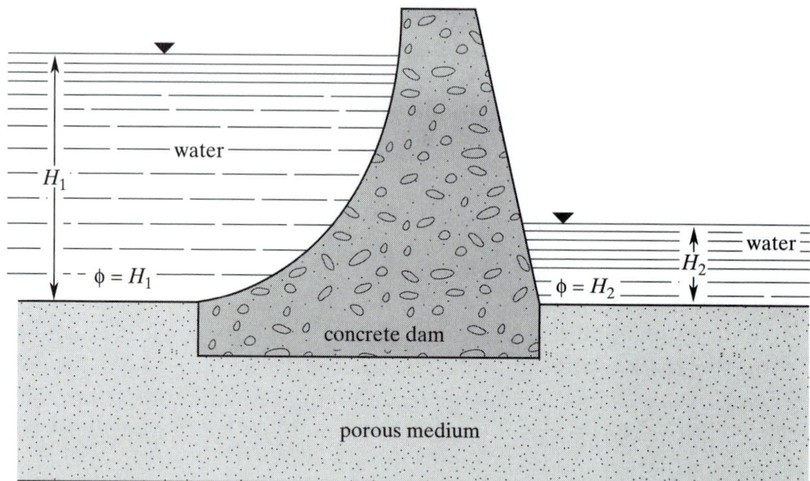


**FIGURE 9.9** The seepage flow of water through a porous medium under a dam.

The components of the seepage velocity are

$$v_x = -k_x \frac{\partial \phi}{\partial x} \quad \text{and} \quad v_y = -k_y \frac{\partial \phi}{\partial y} \tag{9.35}$$

where $k_x$ and $k_y$ are the permeability coefficients and $\phi$ represents the hydraulic head. Comparing the differential equation governing the groundwater seepage flow, Eq. (9.34), to the heat diffusion equation, Eq. (7.8), we note that both of these equations have the same form; therefore, we can apply the results of Sections 7.2 and 7.3 to the groundwater flow problems. However, when comparing the differential equations for the groundwater seepage flow problems, we let $C_1 = k_x, C_2 = k_y$, and $C_3 = 0$.

The permeability matrix for a rectangular element is

$$[\mathbf{K}]^{(e)} = \frac{k_x w}{6\ell}\begin{bmatrix} 2 & -2 & -1 & 1 \\ -2 & 2 & 1 & -1 \\ -1 & 1 & 2 & -2 \\ 1 & -1 & -2 & 2 \end{bmatrix} + \frac{k_y \ell}{6w}\begin{bmatrix} 2 & 1 & -1 & -2 \\ 1 & 2 & -2 & -1 \\ -1 & -2 & 2 & 1 \\ -2 & -1 & 1 & 2 \end{bmatrix} \tag{9.36}$$

where $w$ and $\ell$ are the length and the width, respectively, of the rectangular element, as shown in Figure 9.10. In addition, for a typical see page flow problem, the magnitude of the hydraulic head is generally known at certain surfaces, as shown in Figure 9.9. The known hydraulic head will then serve as a given boundary condition.

The nodal values of a hydraulic head for a triangular element are depicted in Figure 9.11. For triangular elements, the permeability matrix is

$$[\mathbf{K}]^{(e)} = \frac{k_x}{4A}\begin{bmatrix} \beta_i^2 & \beta_i\beta_j & \beta_i\beta_k \\ \beta_i\beta_j & \beta_j^2 & \beta_j\beta_k \\ \beta_i\beta_k & \beta_j\beta_k & \beta_k^2 \end{bmatrix} + \frac{k_y}{4A}\begin{bmatrix} \delta_i^2 & \delta_i\delta_j & \delta_i\delta_k \\ \delta_i\delta_j & \delta_j^2 & \delta_j\delta_k \\ \delta_i\delta_k & \delta_j\delta_k & \delta_k^2 \end{bmatrix} \tag{9.37}$$

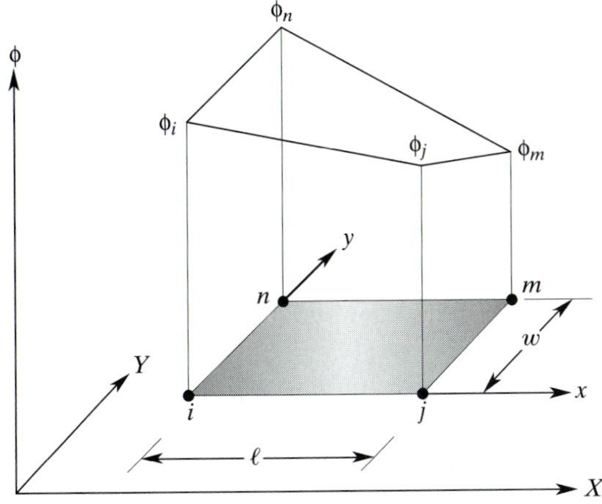


**FIGURE 9.10** Nodal values of a hydraulic head for a rectangular element.

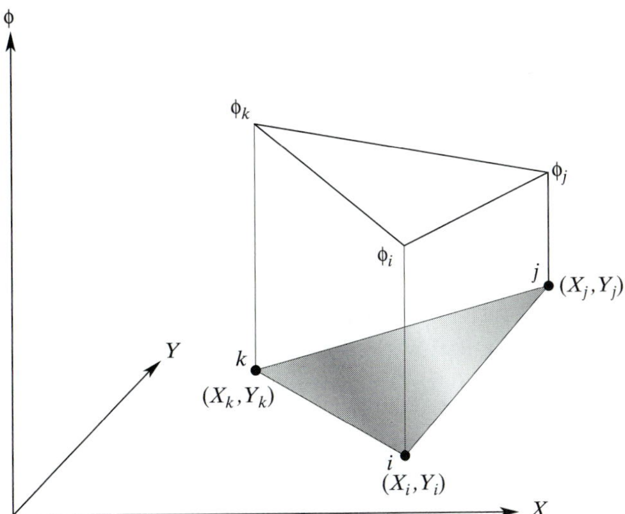


**FIGURE 9.11** Nodal values of a hydraulic head for a triangular element.

where the area $A$ of the triangular element and the α-, β-, and δ-terms are given by:

$$2A = X_i(Y_j - Y_k) + X_j(Y_k - Y_i) + X_k(Y_i - Y_j)$$

$$\alpha_i = X_j Y_k - X_k Y_j \quad \beta_i = Y_j - Y_k \quad \delta_i = X_k - X_j$$

$$\alpha_j = X_k Y_i - X_i Y_k \quad \beta_j = Y_k - Y_i \quad \delta_j = X_i - X_k$$

$$\alpha_k = X_i Y_j - X_j Y_i \quad \beta_k = Y_i - Y_j \quad \delta_k = X_j - X_i$$

Next we discuss ANSYS elements.

## 9.4 EXAMPLES USING ANSYS

ANSYS offers a number of elements for modeling fluid mechanics problems. Examples of those elements include: FLUID15, FLUID66, and FLUID79.

**FLUID15** is a two-dimensional plane fluid flow element with heat transfer capability. The Navier–Stokes equations, the continuity equation, and the energy equation for incompressible laminar flow are discretized. The element has four corner nodes with three degrees of freedom at each node. The degrees of freedom are the respective velocities in the nodal $x$- and $y$-directions and the temperature. Pressure is also computed at the centroid of each element. The elemental input data include the node locations, the fluid density, thermal conductivity, and viscosity.

**FLUID66** is a thermal-flow element with the ability to conduct heat and transport fluid between its two primary nodes. FLUID66 has two degrees of freedom at each node: temperature and pressure. It can also account for convections taking place with two additional optional nodes. The element is defined by its two primary nodes. The elemental input data include the node locations, the fluid density, the convective heat transfer coefficient, thermal conductivity, specific heat, and viscosity.

**FLUID 79** is a modification of the two-dimensional structural solid element PLANE42. This element is used to model fluids contained within vessels having no net flow rate. This element is defined by four nodes, with two degrees of freedom at each node: translation in the nodal $x$- and $y$-directions. The elemental input data include the node locations, the fluid's elastic (bulk) modulus, and viscosity. The viscosity is used to compute a damping matrix for dynamic analysis. Pressure may be input as surface loads on the element faces.

As the theory in the previous sections suggested, because of the similarities among the governing differential equations, in addition to the elements listed above, you can use thermal solid elements (e.g., **PLANE35**, a six-node triangular element; **PLANE55**, a four-node quadrilateral element; or **PLANE77**, an eight-node quadrilateral element) to model irrotational fluid flow or groundwater flow problems. However, when using the solid thermal elements, make sure that the appropriate values are supplied to the property fields. Examples 9.2 and 9.3 demonstrate this point.

### EXAMPLE 9.2

Consider an ideal flow of air around a cylinder, as shown in Figure 9.12. The radius of the cylinder is 5 cm, and the velocity of the approach is $U = 10$ cm/s. Using ANSYS,

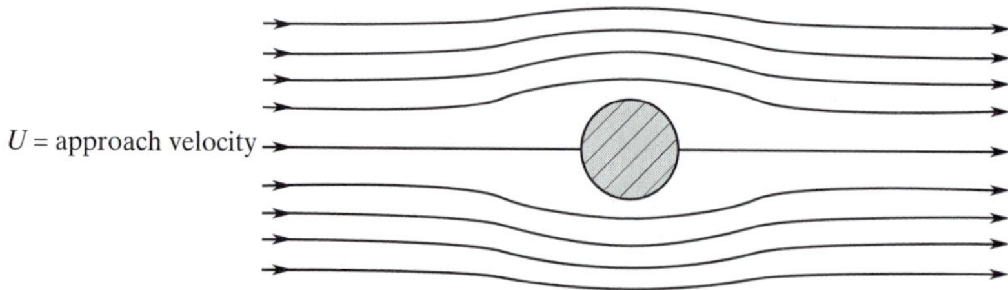


**FIGURE 9.12** An ideal flow of air around a cylinder.

determine the velocity distribution around the cylinder. Assume that the free-stream velocity remains constant at a distance of five diameters downstream and upstream of the cylinder.

Enter the **ANSYS** program by using the Launcher. Type **xansys54** on the command line, or consult your system administrator for the appropriate command name to launch ANSYS from your computer system.

Pick **Interactive** from the Launcher menu.

Type **FlowCYL** (or a file name of your choice) in the **Initial Jobname** entry field of the dialog box.

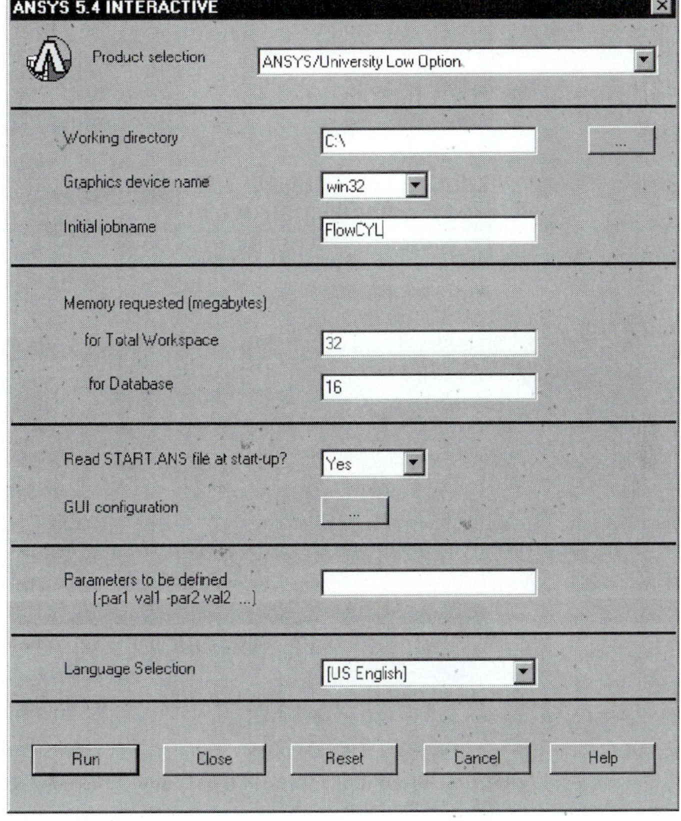


Pick **Run** to start the Graphic User Interface (GUI). A window will open with some disclaimer information. You will eventually be asked to press the **Return** key to start the graphics window and the main menu. Do so in order to proceed.

Create a title for the problem. This title will appear on ANSYS display windows to provide a simple way of identifying the displays. So, issue the following command sequence:

utility menu: **File** → **Change Title** ...

main menu: **Preprocessor** → **Element Types** → **Add/Edit/Delete** ...

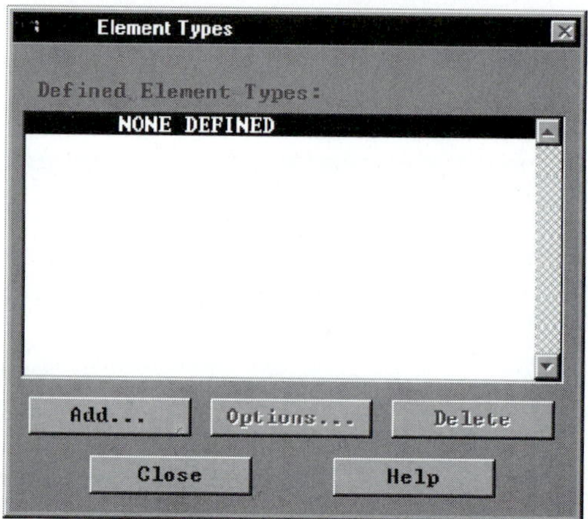


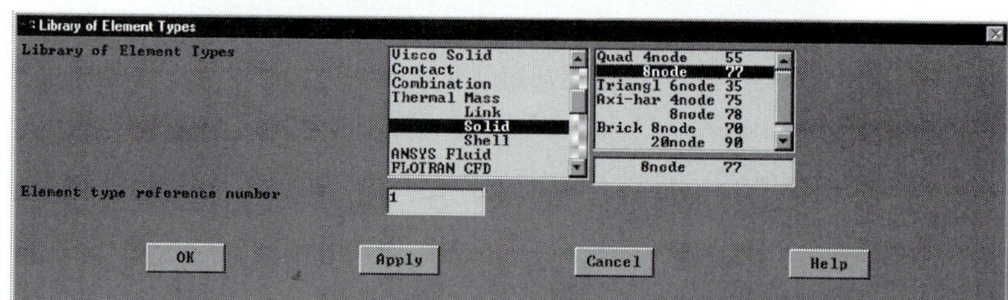


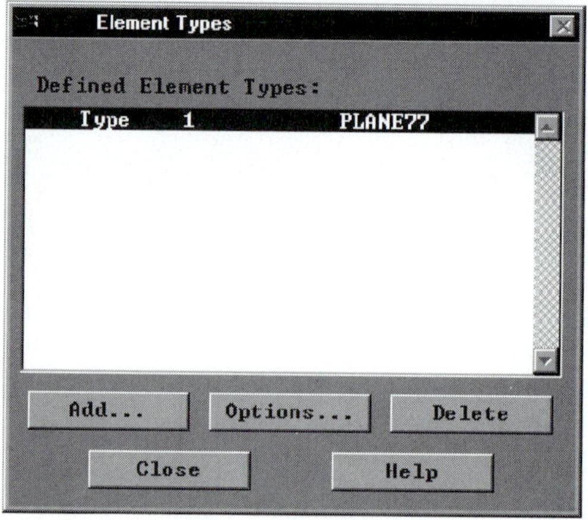


main menu: **Preprocessor** → **Material Props** → **-Constant-Isotropic** ...

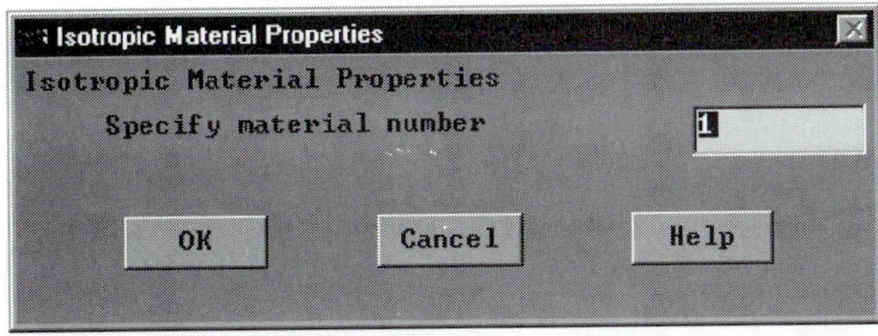


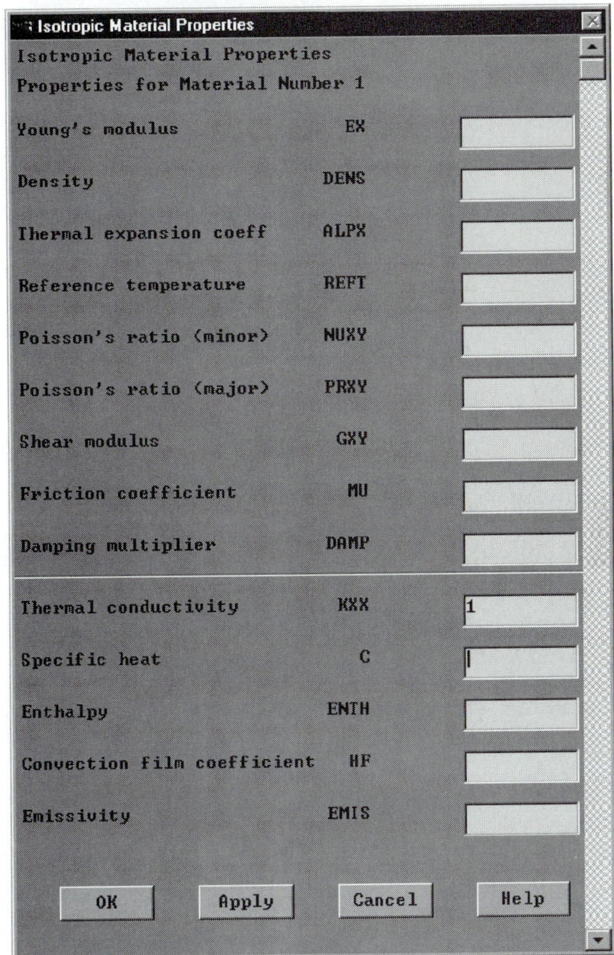


ANSYS Toolbar: **SAVE_DB**

Set up the graphics area (i.e. workplane, zoom, etc.) with the following commands:

utility menu: **Workplane** → **Wp Settings** ...

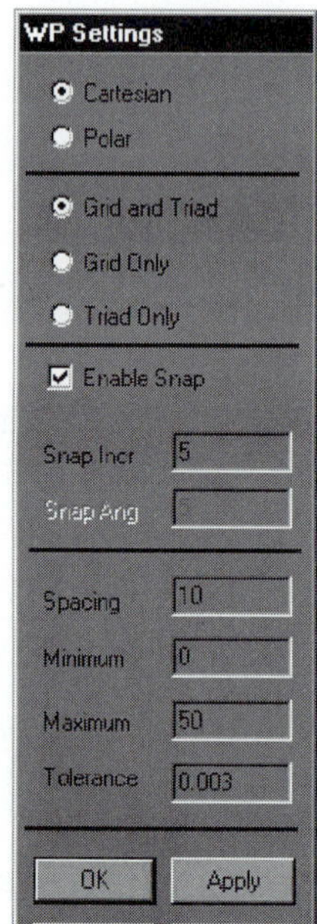


Toggle on the workplane by using the command:

utility menu: **Workplane** → **Display Working Plane**

Bring the workplane to view by using the command:

utility menu: **PlotCtrls** → **Pan, Zoom, Rotate** ...

Click on the **small circle** until you bring the workplane to view. Then, create the geometry with the following commands:

main menu: **Preprocessor** → **-Modeling-Create** → **Areas-Rectangle** → **By 2 Corners+**

**[WP = 0,0]**

**[Expand the rubber up 50 and right 50]**

**OK**

Create the cross section of the cylinder to be removed later:

main menu: **Preprocessor** → **-Modeling-Create** → **Areas-Circle** → **Solid Circle+**

**[WP = 25, 25]**

**[Expand the rubber to $r = 5.0$]**

**OK**

main menu: **Preprocessor** → **-Modeling-Operate** → **-Booleans-Subtract** → **Areas+**

Pick Area1 (the rectangle) and apply; then, pick Area2 (the circle) and apply.

**OK**

We now want to mesh the areas to create elements and nodes, but first, we need to specify the element sizes. So, issue the following commands:

main menu: **Preprocessor** → **-Meshing-Size Cntrls** → **-Global-Size** ...

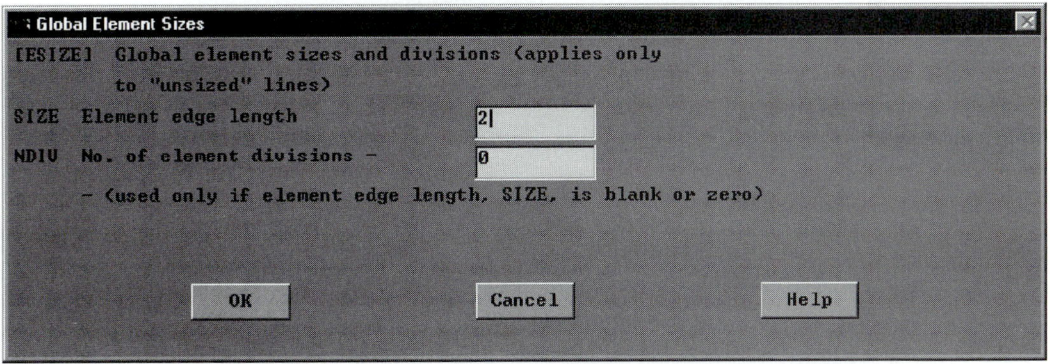


main menu: **Preprocessor** → **-Meshing-Mesh** → **-Areas-Free** +

**Pick All**

Apply boundary conditions with the following commands:

main menu: **Solution** → **-Loads-Apply** → **-Thermal-Heat Flux** → **On Lines** +.

Pick the left vertical edge of the rectangle.

**OK**

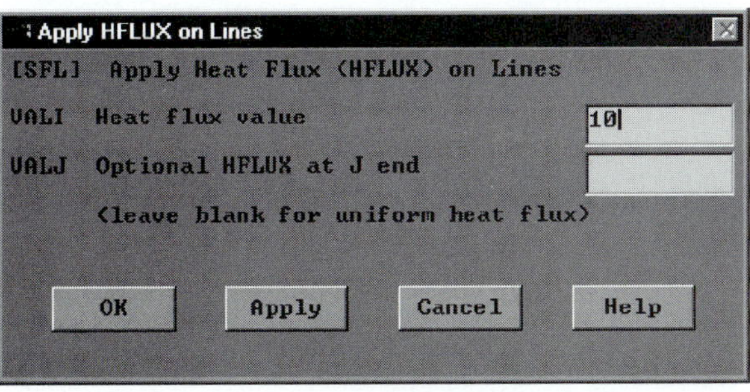

main menu: **Solution** → **-Loads-Apply** → **-Thermal-Heat Flux** → **On Lines** +

Pick the right vertical edge of the rectangle.

**OK**

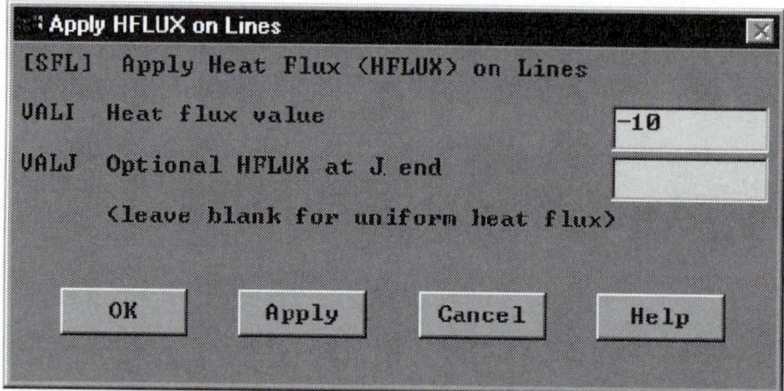


utility menu: **PlotCtrls** → **Symbols** ...

Symbols
[/PBC] Boundary condition symbol
All BC+Reaction
All Applied BCs
All Reactions
None
For Individual:
Individual symbol set dialog(s) to be displayed:
Applied BC's
Reactions
Miscellaneous
[/PSF] Surface Load Symbols — Heat Fluxes
Show pres and convect as — Arrows
[/PBF] Body Load Symbols — None
[/PSYMB] Other Symbols
CS Local coordinate system — Off
NDIR Nodal coordinate system — Off
ESYS Element coordinate sys — Off
LDIR Line direction — Off
ECON Element mesh constraints — Off
DOT Larger node/kp symbols — On
LAYR Orientation of layer number — 0
[/REPLOT] Replot upon OK/Apply? — Replot
OK
Cancel
Help

utility menu: **Plot** → **Lines**

ANSYS Toolbar: **SAVE_DB**

Solve the problem:

main menu: **Solution** → **-Solve-Current LS**

**OK**

**Close** (the solution is done!) window.

**Close** (the /STAT Command) window.

postprocessing phase, obtain information such as velocities (see Figure 9.13):

main menu: **General Postproc** → **Plot Results** → **-Vector Plot-Predefined** ...

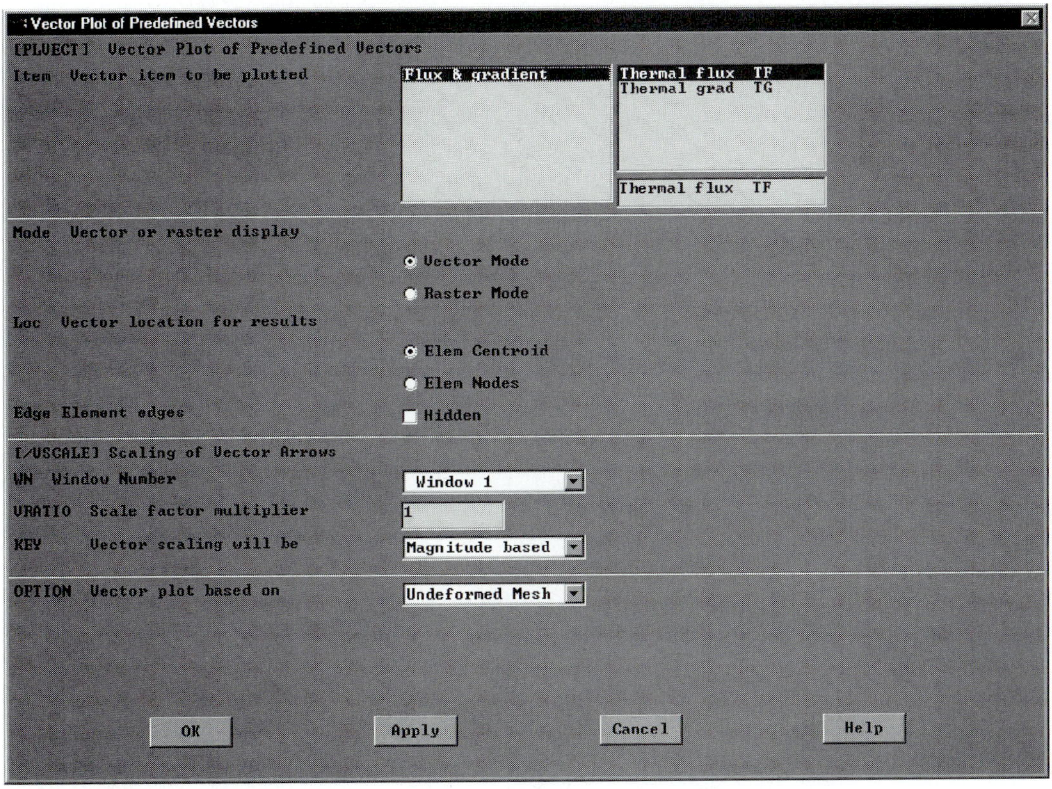

Vector Plot of Predefined Vectors

[PLVECT] Vector Plot of Predefined Vectors

Item Vector item to be plotted — Flux & gradient — Thermal flux TF / Thermal grad TG — Thermal flux TF

Mode Vector or raster display — Vector Mode / Raster Mode

Loc Vector location for results — Elem Centroid / Elem Nodes

Edge Element edges — Hidden

[/VSCALE] Scaling of Vector Arrows

WN Window Number — Window 1

VRATIO Scale factor multiplier — 1

KEY Vector scaling will be — Magnitude based

OPTION Vector plot based on — Undeformed Mesh

OK Apply Cancel Help

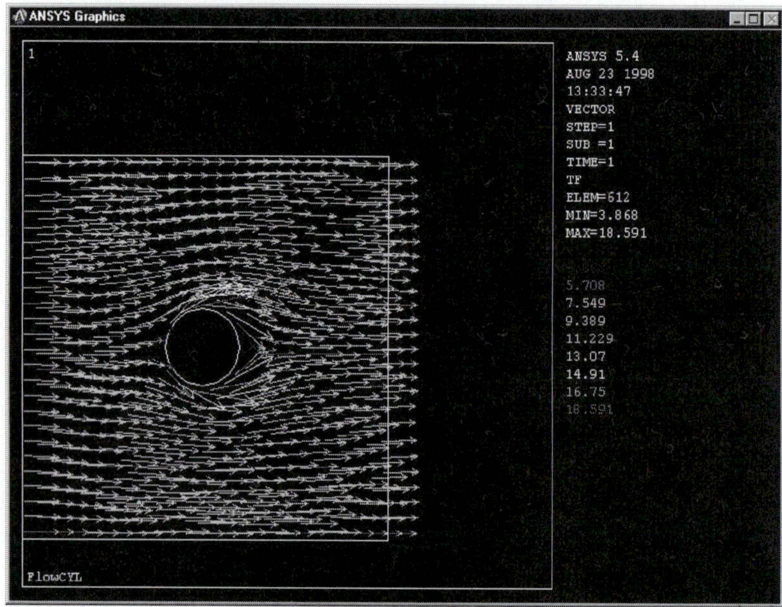


**FIGURE 9.13** The velocity vectors.

utility menu: **Plot** → **Areas**

main menu: **General Postproc** → **Path Operations** → **Define Path** → **On Working Plane** +

Pick the two points along the line marked *A–B*, as shown in Figure 9.14.

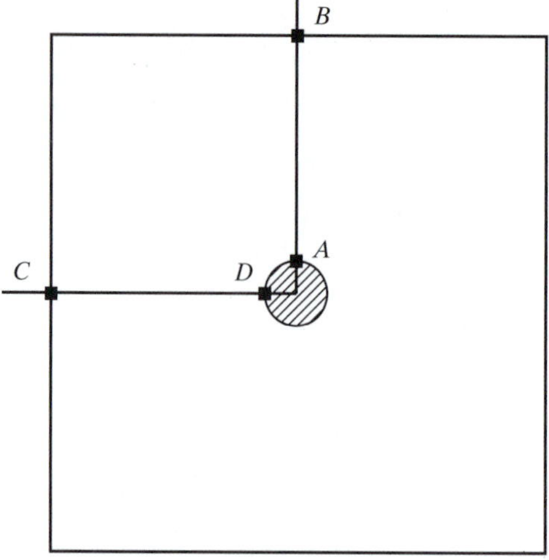


**FIGURE 9.14** Defining the path for path operation.

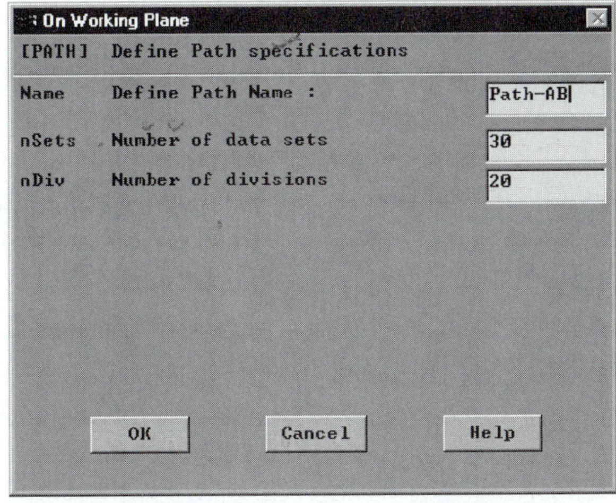


main menu: **General Postproc** → **Path Operations** → **Map onto Path** ...

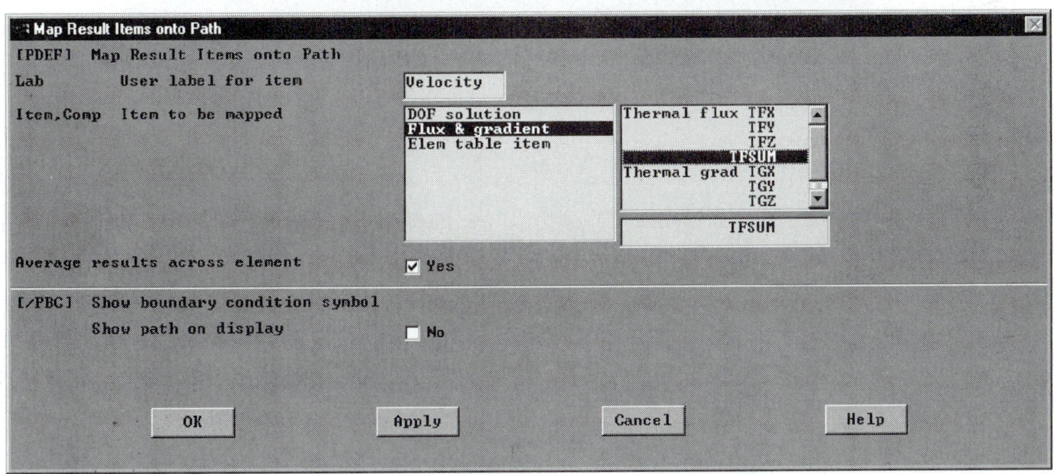


Now, plot the results (see Figure 9.15):

main menu: **General Postproc** → **Path Operations** → **-Plot Path Items-On Graph** ...

Plot of Path Items on Graph
[PLPATH] Path Plot on Graph
Lab1-6 Path items to be graphed
XG
YG
ZG
S
VELOCITY
OK
Apply
Cancel
Help

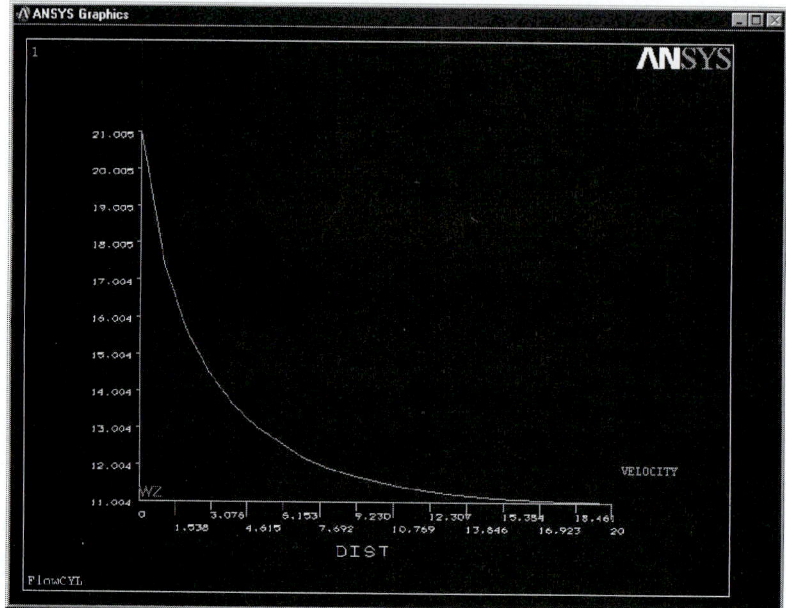


**FIGURE 9.15** The variation of fluid velocity along path *A–B*.

utility menu: **Plot → Areas**

main menu: **General Postproc → Path Operations → Define Path → On Working Plane** +

Pick the two points along the line marked as *C–D*, as shown in Figure 9.14.

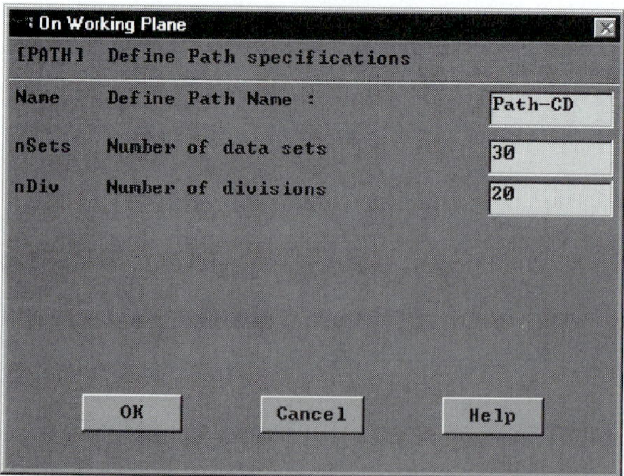


**OK**

main menu: **General Postproc** → **Path Operations** → **Map onto Path** ...

Map Result Items onto Path
[PDEF] Map Result Items onto Path
Lab User label for item Velocity
Item,Comp Item to be mapped
DOF solution
Flux & gradient
Elem table item
Thermal flux TFX
TFY
TFZ
TFSUM
Thermal grad TGX
TGY
TGZ
TFSUM
Average results across element Yes
[/PBC] Show boundary condition symbol
Show path on display No
OK Apply Cancel Help

Now, plot the results (see Figure 9.16):

main menu: **General Postproc** → **Path Operations** → **-Plot Path Items-On Graph** ...

Plot of Path Items on Graph
[PLPATH] Path Plot on Graph
Lab1-6 Path items to be graphed
XG
YG
ZG
S
VELOCITY
OK Apply Cancel Help

Exit and save the results:

ANSYS Toolbar: **QUIT**

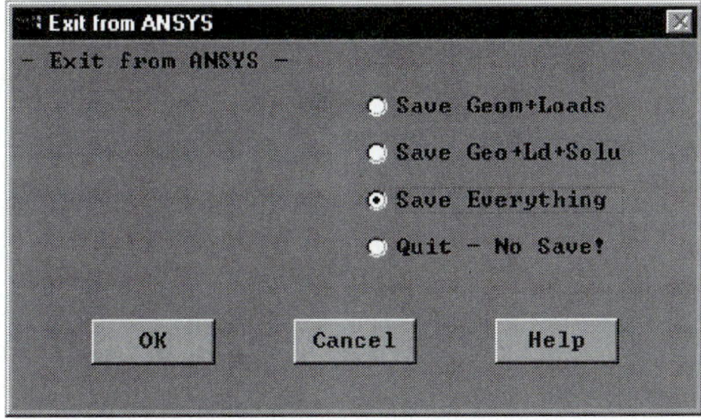

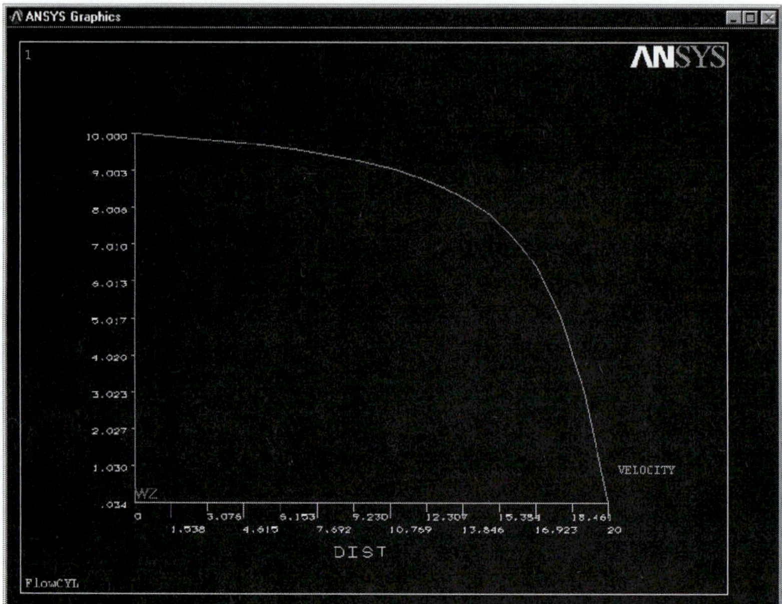


**FIGURE 9.16** The variation of fluid velocity along the $C$–$D$ section.

## EXAMPLE 9.3

Consider the seepage flow of water under the concrete dam shown in Figure 9.17. The permeability of the porous soil under the dam is approximated as $K = 15$ m/day. Determine the seepage velocity distribution in the porous soil.

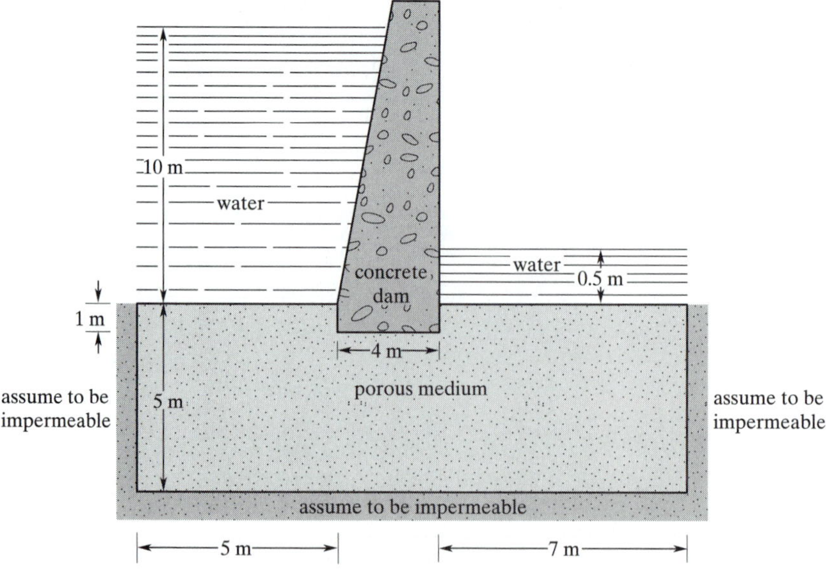


**FIGURE 9.17** The seepage flow of water through a porous medium under a concrete dam.

Enter the **ANSYS** program by using the Launcher. Type **xansys54** on the command line, or consult your system administrator for the appropriate command name to launch ANSYS from your computer system.

Pick **Interactive** from the Launcher menu.

Type **DAM** (or a file name of your choice) in the **Initial Jobname** entry field of the dialog box.

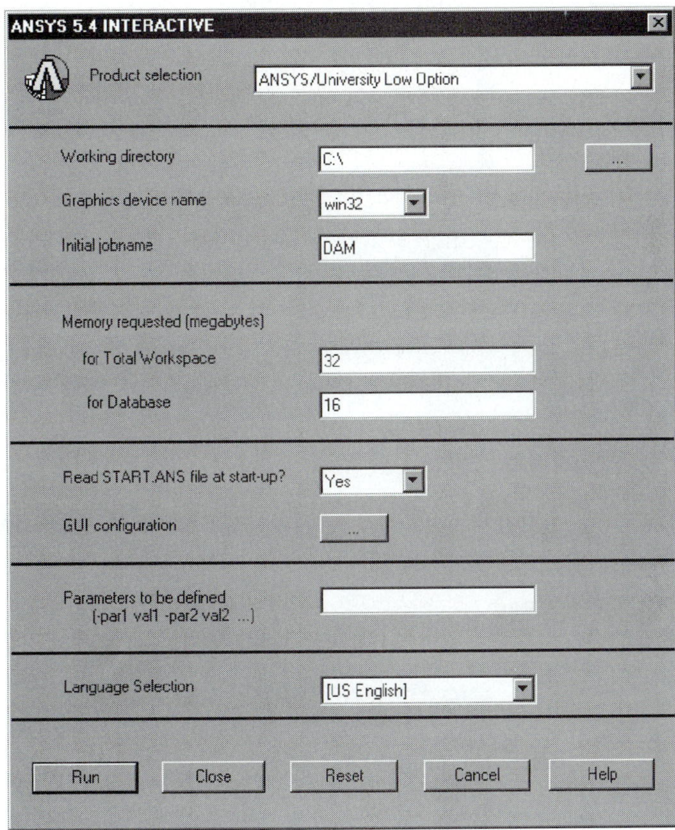


Pick **Run** to start the Graphic User Interface (GUI). A window will open with some disclaimer information. You will eventually be asked to press the **Return** key to start the graphics window and the main menu. Do so in order to proceed.

Create a title for the problem. This title will appear on ANSYS display windows to provide a simple way of identifying the displays. So, issue the following commands:

utility menu: **File** → **Change Title** ...

Change Title

[/TITLE] Enter new title DAM

OK Cancel Help

main menu: **Preprocessor** → **Element Types** → **Add/Edit/Delete** ...

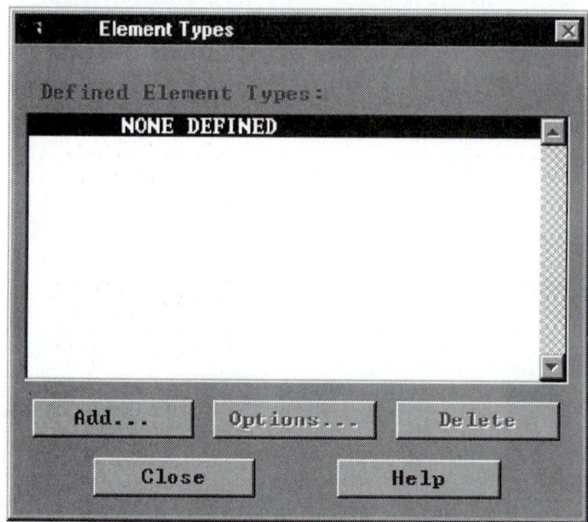


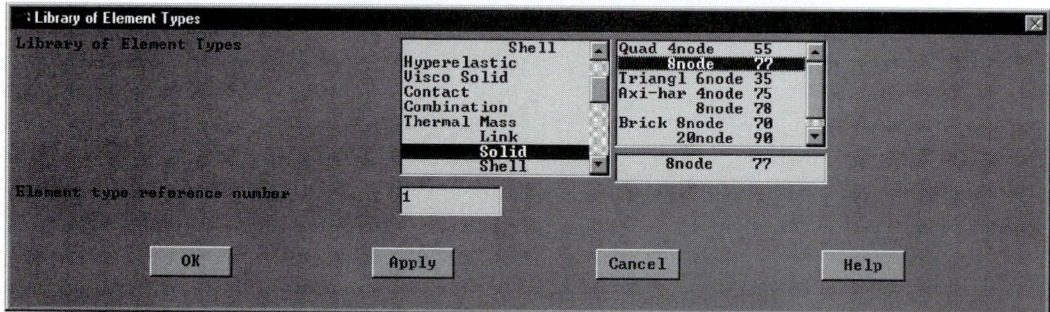


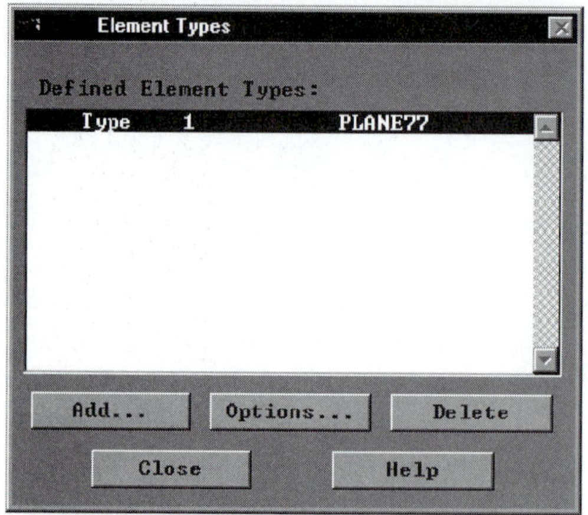


Assign the permeability of the soil with the following commands:

main menu: **Preprocessor** → **Material Props** → **-Constant-Isotropic** ...

Isotropic Material Properties
Isotropic Material Properties
Specify material number 1
OK Cancel Help

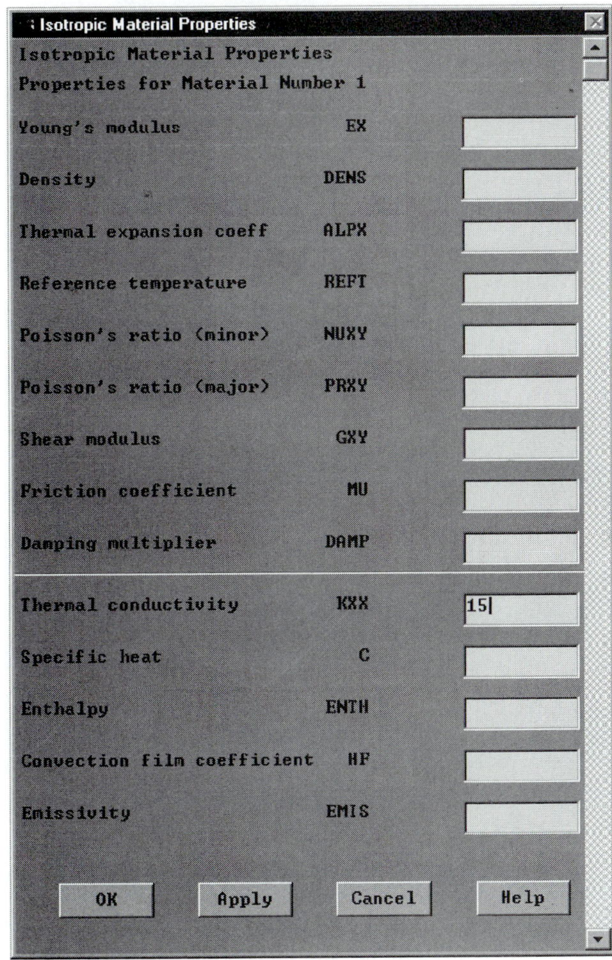


ANSYS Toolbar: **SAVE_DB**

Set up the graphics area (i.e. workplane, zoom, etc.) with the commands:

utility menu: **Workplane** → **Wp Settings** ...

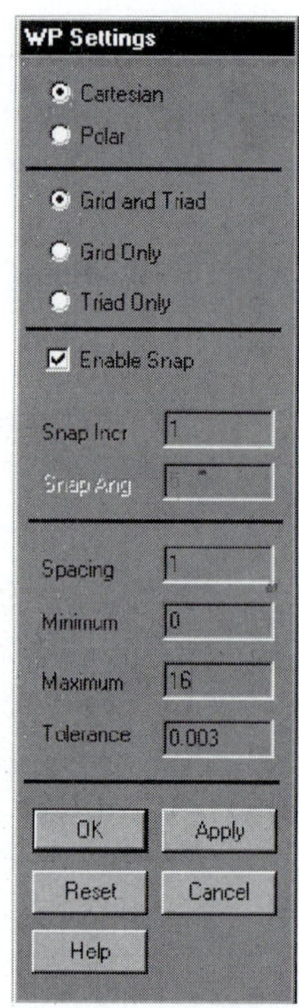


Toggle on the workplane by using the following command:

utility menu: **Workplane → Display Working Plane**

Bring the workplane to view by using the following command:

utility menu: **PlotCtrls → Pan, Zoom, Rotate** ...

Click on the small circle until you bring the workplane to view. Then, create the geometry:

main menu: **Preprocessor → -Modeling-Create → -Areas-Rectangle → By 2 Corners +**

**[WP = 0,0]**

**[Expand the rubber up 5 and right 16]**

**[WP = 5,4]**

**[Expand the rubber up 1 and right 4]**

**OK**

main menu: **Preprocessor** → **-Modeling-Operate** → **-Booleans-Subtract** → **Areas**

Pick Area1 (the large rectangle) and Apply; then, pick Area2 (the small rectangle) and Apply.

**OK**

We now want to mesh the areas to create elements and nodes, but first, we need to specify the element sizes:

main menu: **Preprocessor** → **-Meshing-Size Cntrls** → **-Global- Size** …

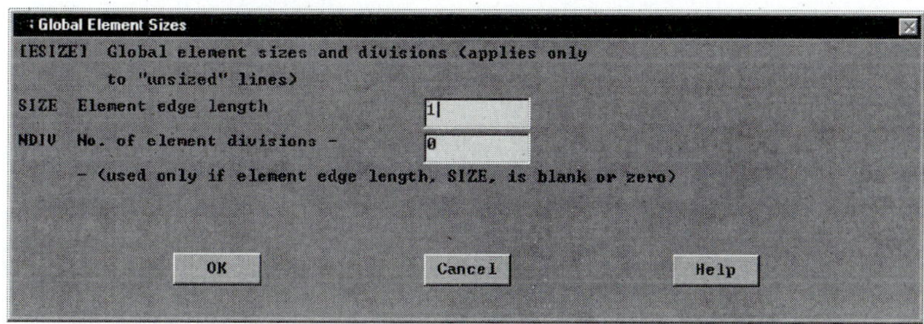


main menu: **Preprocessor** → **-Meshing-Mesh** → **-Areas-Free** +

**Pick All**

Apply boundary conditions with the following commands:

main menu: **Solution** → **-Loads-Apply** → **-Thermal-Temperature** → **On nodes** +

Using the box picking mode, pick all of the nodes attached to the left top edge of the rectangle. Hold down the left button while picking.

**OK**

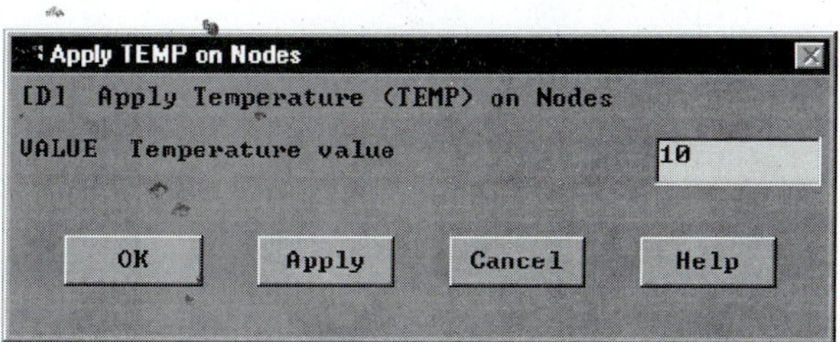


main menu: **Solution** → **-Loads-Apply** → **-Thermal-Temperature** → **On nodes** +

Using the box picking mode, pick all of the nodes attached to the right top edge of the rectangle:

**OK**

Apply TEMP on Nodes

[D] Apply Temperature (TEMP) on Nodes

VALUE Temperature value 0.5

OK Apply Cancel Help

ANSYS Toolbar: **SAVE_DB**

Solve the problem:

main menu: **Solution** → **-Solve-Current LS**

**OK**

**Close** (the solution is done!) window.

**Close** (the /STAT Command) window.

For the postprocessing phase, obtain information such as velocities (see Figure 9.18):

main menu: **General Postproc** → **Plot Results** → **-Vector Plot-Predefined** ...

Vector Plot of Predefined Vectors

[PLVECT] Vector Plot of Predefined Vectors

Item Vector item to be plotted — Flux & gradient — Thermal flux TF / Thermal grad TG — Thermal flux TF

Mode Vector or raster display — Vector Mode / Raster Mode

Loc Vector location for results — Elem Centroid / Elem Nodes

Edge Element edges — Hidden

[/VSCALE] Scaling of Vector Arrows

WN Window Number — Window 1

VRATIO Scale factor multiplier — 1

KEY Vector scaling will be — Magnitude based

OPTION Vector plot based on — Undeformed Mesh

OK Apply Cancel Help

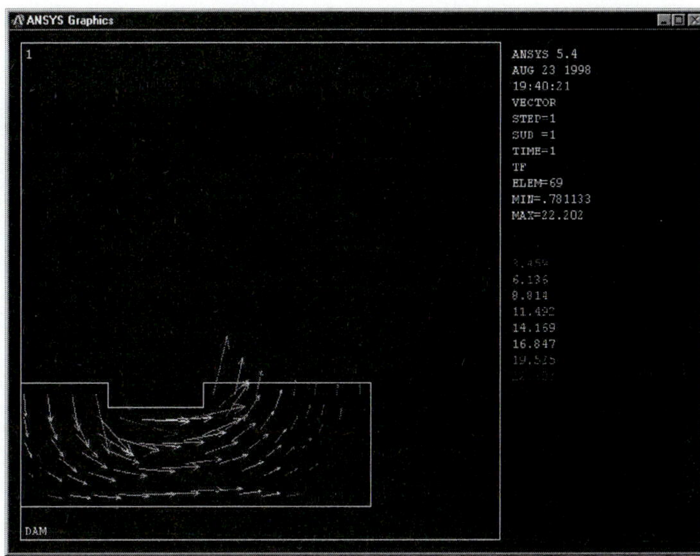


**FIGURE 9.18** The seepage-velocity distribution within the soil.

utility menu: **Plot** → **Areas**

main menu: **General Postproc** → **Path Operations** → **Define Path** → **On Working Plane** +

Pick the two points along the line marked as *A*–*B*, as shown in Figure 9.19.

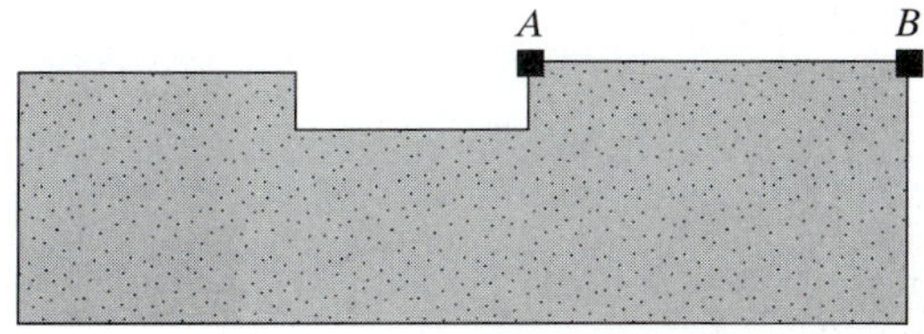


**FIGURE 9.19** Defining the path for path operation.

**OK**

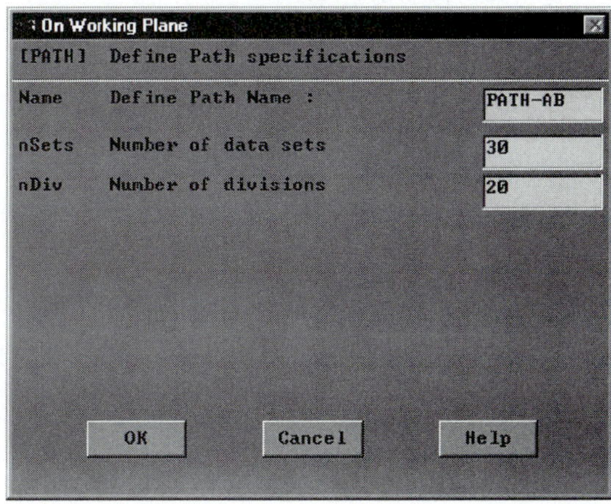
On Working Plane

[PATH] Define Path specifications

| | | |
|---|---|---|
| Name | Define Path Name : | PATH-AB |
| nSets | Number of data sets | 30 |
| nDiv | Number of divisions | 20 |

OK Cancel Help

main menu: **General Postproc** → **Path Operations** → **Map onto Path** ...

Map Result Items onto Path
[PDEF] Map Result Items onto Path
Lab User label for item VELOCITY
Item,Comp Item to be mapped
DOF solution
Flux & gradient
Elem table item
Thermal flux TFX
TFY
TFZ
TFSUM
Thermal grad TGX
TGY
TGZ
TFSUM
Average results across element ☑ Yes
[/PBC] Show boundary condition symbol
Show path on display ☐ No
OK Apply Cancel Help

Now, plot the results (see Figure 9.20):

main menu: **General Postproc** → **Path Operations** → **-Plot Path Items-On Graph** ...

Plot of Path Items on Graph
[PLPATH] Path Plot on Graph
Lab1-6 Path items to be graphed
XG
YG
ZG
S
VELOCITY
OK Apply Cancel Help

Exit and save the results:

ANSYS Toolbar: **QUIT**

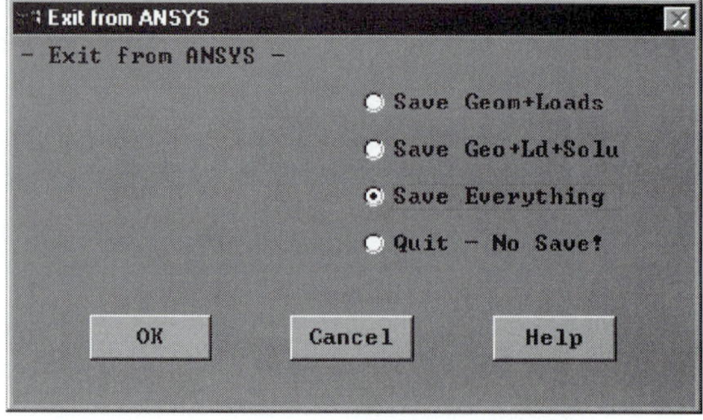

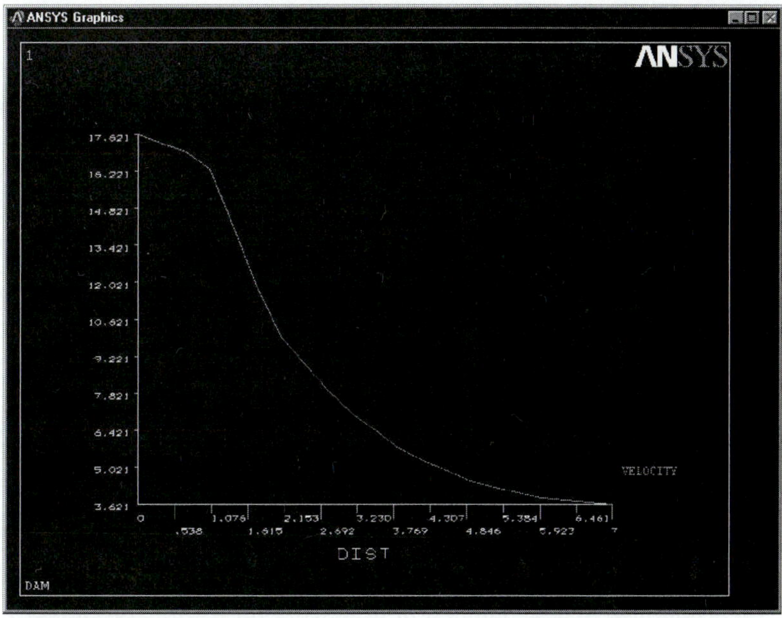


**FIGURE 9.20** The variation of the seepage velocity along path *A–B*.

## 9.5 VERIFICATION OF RESULTS

There are various ways by which you can verify your findings. Consider the flow rate results of Example 9.1, shown in Table 9.1.

Referring to Figure 9.6, elements (2) and (4) are in series; therefore, the flow rate through each element should be equal. Comparing $Q^{(2)}$ to $Q^{(4)}$, we find that this condition is true. Elements (3) and (5) are also in series, and the computed flow rates for these elements are also equal. Moreover, the sum of the flow rates in elements (2) and (3) should equal the flow rate in element (1). This condition is also true.

Let us now turn our attention to Example 9.2. One way of checking for the validity of your FEA findings is to consider the variation of air velocity along path *A–B*, as

TABLE 9.1 Summary of flow rate results for Example 9.1

| Element | Flow Rate |
|---|---|
| 1 | $5.0 \times 10^{-4}$ |
| 2 | $2.79 \times 10^{-4}$ |
| 3 | $2.21 \times 10^{-4}$ |
| 4 | $2.79 \times 10^{-4}$ |
| 5 | $2.21 \times 10^{-4}$ |
| 6 | $5.0 \times 10^{-4}$ |

shown in Figure 9.15. The fluid velocity is at its maximum value at point $A$, and it decreases along path $A$–$B$, approaching the free-stream value. Another check on the validity of our results could come from examining the fluid velocity variation along path $C$–$D$, as shown in Figure 9.16. The air velocity changes from its free-stream value to zero at the forward stagnation point of the cylinder. These results are certainly consistent with the results obtained from applying Euler's equation to an inviscid flow of air around a cylinder.

The results of Example 9.3 can be visually verified in a similar fashion. Consider Figure 9.20, which shows the variation of the seepage velocity along path $A$–$B$. It is clear that the seepage velocities are higher near point $A$ than they are near point $B$. This difference is attributed to the fact that point $A$ lies on the path of the least resistance to the flow, and consequently, more fluid flows near point $A$ than near point $B$. The other check on the validity of the result could come from comparing the seepage flow rates on the dam's upstream side to the seepage flow on the downstream side; of course, they must be equal.

## SUMMARY

At this point you should:

1. know how to solve laminar flow network problems. You should also know that the resistance matrix for laminar pipe flow is given by:

$$[\mathbf{R}]^{(e)} = \begin{bmatrix} \frac{\pi D^4}{128L\mu} & -\frac{\pi D^4}{128L\mu} \\ -\frac{\pi D^4}{128L\mu} & \frac{\pi D^4}{128L\mu} \end{bmatrix}$$

2. know the definitions of streamline and stream function, as well as what they physically represent.
3. know what an irrotational flow is.
4. know that the inviscid flow matrix for a rectangular element is

$$[\mathbf{K}]^{(e)} = \frac{w}{6\ell}\begin{bmatrix} 2 & -2 & -1 & 1 \\ -2 & 2 & 1 & -1 \\ -1 & 1 & 2 & -2 \\ 1 & -1 & -2 & 2 \end{bmatrix} + \frac{\ell}{6w}\begin{bmatrix} 2 & 1 & -1 & -2 \\ 1 & 2 & -2 & -1 \\ -1 & -2 & 2 & 1 \\ -2 & -1 & 1 & 2 \end{bmatrix}$$

and that the inviscid flow matrix for a triangular element is

$$[\mathbf{K}]^{(e)} = \frac{1}{4A}\begin{bmatrix} \beta_i^2 & \beta_i\beta_j & \beta_i\beta_k \\ \beta_i\beta_j & \beta_j^2 & \beta_j\beta_k \\ \beta_i\beta_k & \beta_j\beta_k & \beta_k^2 \end{bmatrix} + \frac{1}{4A}\begin{bmatrix} \delta_i^2 & \delta_i\delta_j & \delta_i\delta_k \\ \delta_i\delta_j & \delta_j^2 & \delta_j\delta_k \\ \delta_i\delta_k & \delta_j\delta_k & \delta_k^2 \end{bmatrix}$$

5. know that the permeability matrix for seepage flow problems is similar to the conductance matrix for two-dimensional conduction problems. The permeability matrix for a rectangular element is

$$[\mathbf{K}]^{(e)} = \frac{k_x w}{6\ell}\begin{bmatrix} 2 & -2 & -1 & 1 \\ -2 & 2 & 1 & -1 \\ -1 & 1 & 2 & -2 \\ 1 & -1 & -2 & 2 \end{bmatrix} + \frac{k_y \ell}{6w}\begin{bmatrix} 2 & 1 & -1 & -2 \\ 1 & 2 & -2 & -1 \\ -1 & -2 & 2 & 1 \\ -2 & -1 & 1 & 2 \end{bmatrix}$$

and that the permeability matrix for a triangular element is

$$[\mathbf{K}]^{(e)} = \frac{k_x}{4A}\begin{bmatrix} \beta_i^2 & \beta_i\beta_j & \beta_i\beta_k \\ \beta_i\beta_j & \beta_j^2 & \beta_j\beta_k \\ \beta_i\beta_k & \beta_j\beta_k & \beta_k^2 \end{bmatrix} + \frac{k_y}{4A}\begin{bmatrix} \delta_i^2 & \delta_i\delta_j & \delta_i\delta_k \\ \delta_i\delta_j & \delta_j^2 & \delta_j\delta_k \\ \delta_i\delta_k & \delta_j\delta_k & \delta_k^2 \end{bmatrix}$$

## REFERENCES


Abbot, I. H., and Von Doenhoff, A. E., *Theory of Wing Sections*, New York, Dover Publications, 1959.

*ANSYS User's Manual: Elements*, Vol. III, Elements, Swanson Analysis Systems, Inc.

Bathe, K., *Finite Element Procedures*, Englewood Cliffs, NJ, Prentice Hall, 1996.

Fox, R. W., and McDonald, A. T., *Introduction to Fluid Mechanics*, 4th ed., New York, John Wiley and Sons, 1992.

Segrlind, L., *Applied Finite Element Analysis*, 2nd ed., New York, John Wiley and Sons, 1984.


## PROBLEMS

1. Oil with dynamic viscosity of $\mu = 0.3\ \text{N}\cdot\text{s/m}^2$ and density of $\rho = 900\ \text{kg/m}^3$ flows through the piping network shown in the accompanying figure. Determine the pressure distribution in the system if the flow rate at node 1 is $20 \times 10^{-4}\ \text{m}^3/\text{s}$. For the given conditions, the flow is laminar throughout the system. How does the flow divide in each branch?

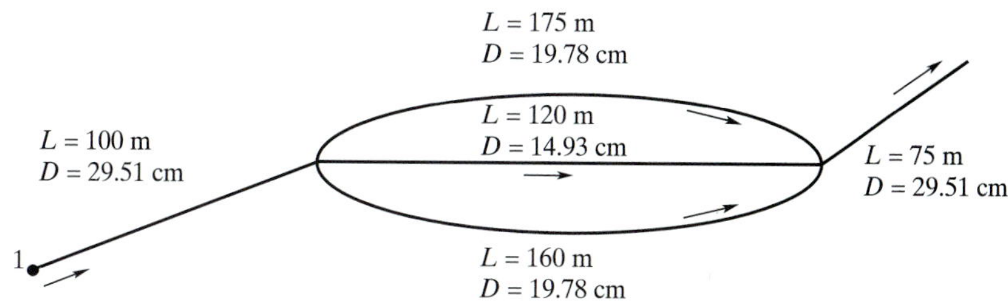


2. Consider the flow of air through the diffuser shown in the accompanying figure. Neglecting the viscosity of air and assuming uniform velocities at the inlet and exit of the diffuser, use ANSYS to compute and plot the velocity distribution within the diffuser.

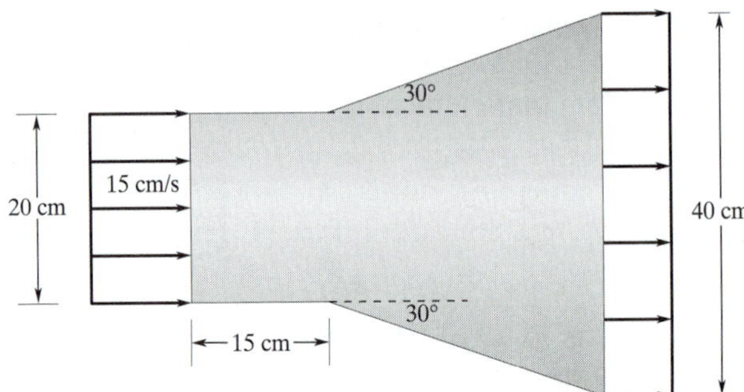


3. Consider the flow of air through a 90° elbow. Assuming ideal flow behavior for air and uniform velocities at the inlet and outlet sections, use ANSYS to compute and plot the velocity distribution within the elbow. The elbow has a uniform depth. Use the continuity equation to obtain the velocity at the outlet.

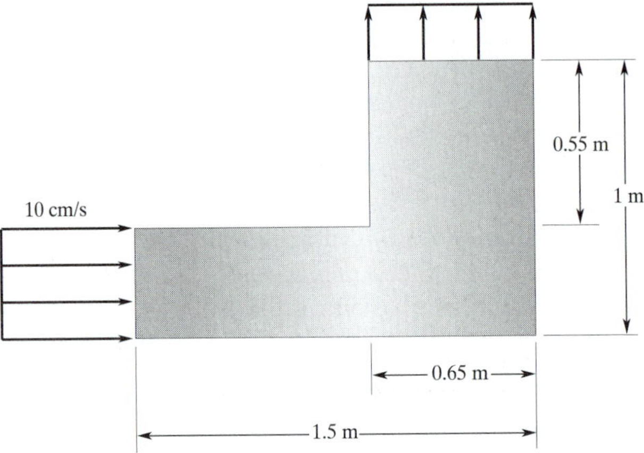


4. Consider the flow of air through the elbow in the accompanying figure. The corners of the elbow are rounded as shown in the figure. Assuming ideal flow behavior for air and uniform velocities at the inlet and outlet sections, use ANSYS to compute and plot the velocity distribution within the elbow. The elbow has a uniform depth. Use the continuity equation to obtain the velocity at the outlet.

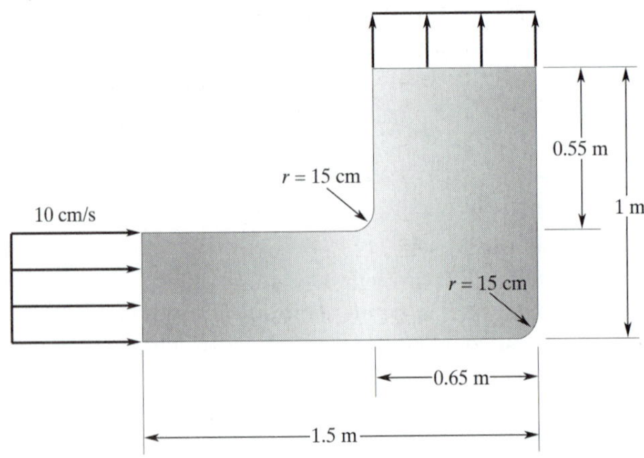

5. Using ANSYS, plot the velocity distributions for the transition-duct fitting shown in the accompanying figure. Plot the results for the combinations of the area ratios and transition angles given in the accompanying table.

| $A_2/A_1$ | | $\theta$ | |
|---|---|---|---|
| 0.1 | 10 | 20 | 45 |
| 0.25 | 10 | 45 | 60 |
| 0.5 | 20 | 45 | 60 |

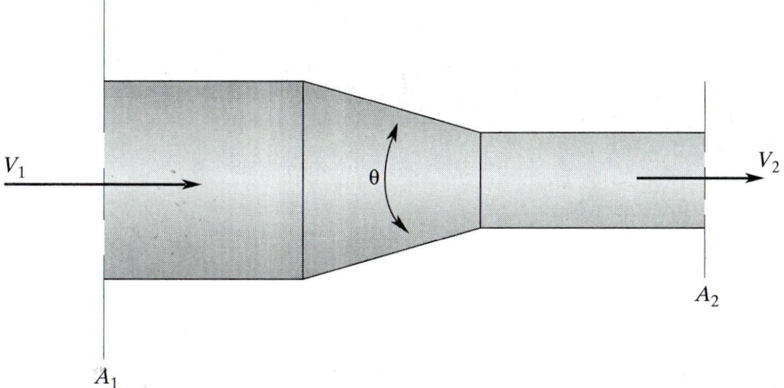


6. Consider the inviscid flow of air past the rounded equilateral triangle shown in the accompanying figure. Perform numerical experiments by changing the velocity of the upstream air and the $r/L$ ratio (for $r/L = 0, r/L = 0.1$, and $r/L = 0.25$) and obtaining the corresponding air velocity distributions over the triangle. Discuss your results.

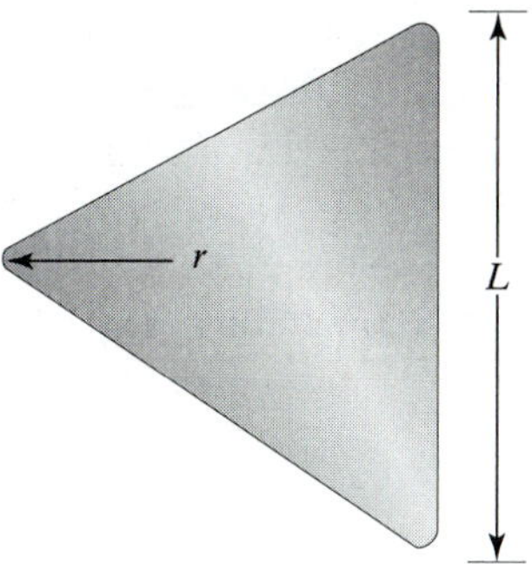


7. Consider the inviscid flow of air past the square rod with rounded corners shown in the accompanying figure. Perform numerical experiments by changing the velocity of the upstream air and the $r/L$ ratio (for $r/L = 0$, $r/L = 0.1$, and $r/L = 0.25$) and obtaining the corresponding air velocity distributions over the square. Discuss your results.

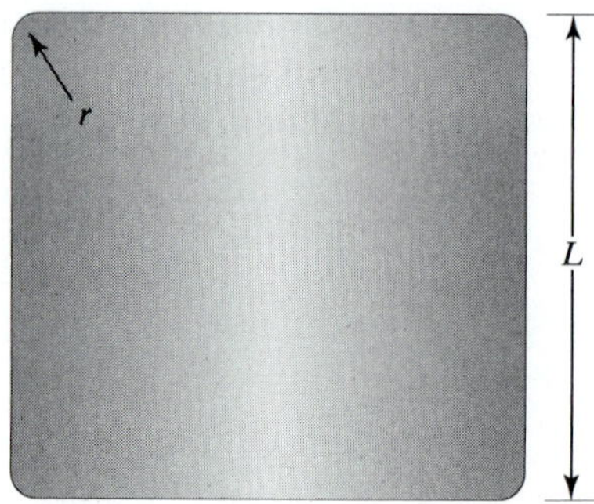


**8.** Consider the inviscid flow of air past a NACA symmetric airfoil. *Abbot and Von Doenhoff (1959)* provide detailed information about NACA symmetric airfoil shapes, including geometric data for NACA symmetric airfoils. Using their geometric data, obtain the velocity distribution over the NACA 0012-airfoil shown in the accompanying figure. Perform numerical experiments by changing the angle of attack and obtaining the corresponding air velocity distributions over the airfoil. Discuss your results.

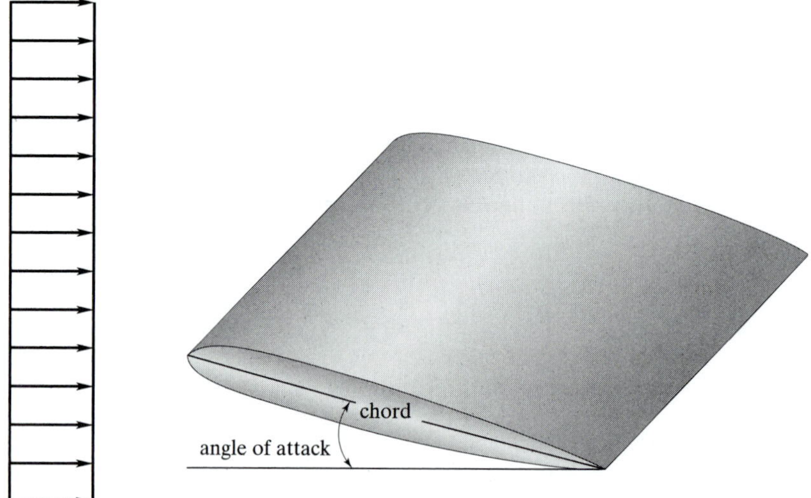

9. Consider the seepage flow of water under the concrete dam shown in the accompanying figure. The permeability of the porous soil under the dam is approximated as k = 45 ft/day. Determine the seepage velocity distribution in the porous soil.

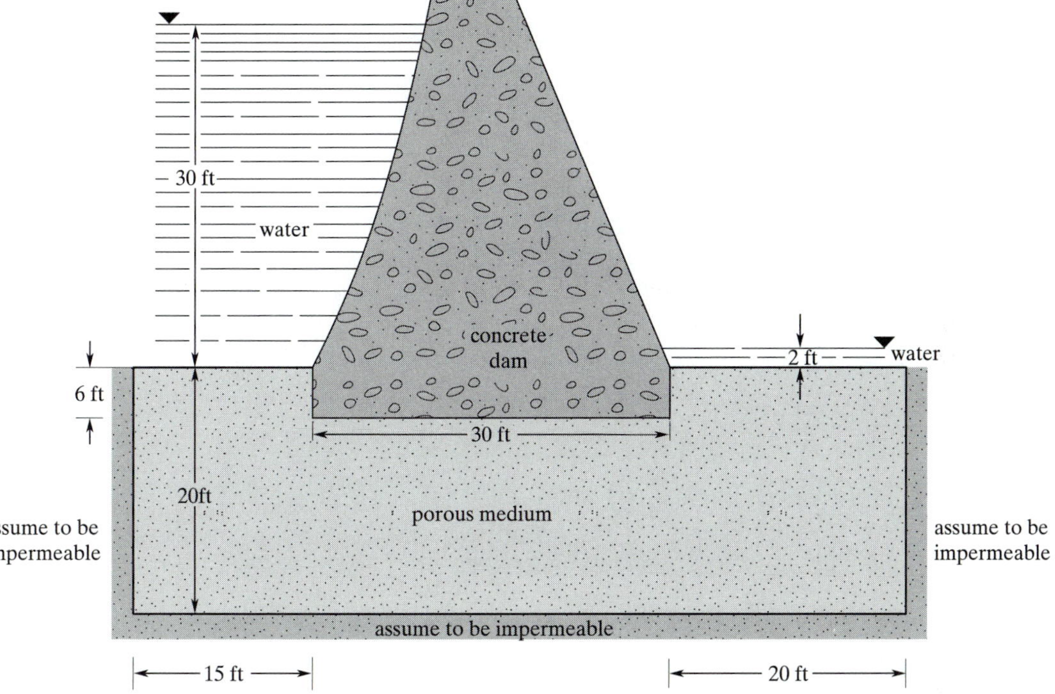

10. Consider the seepage flow of water under the concrete dam shown in the accompanying figure. The permeability of the porous soil under the dam is approximated as $k = 15$ m/day. Determine the seepage velocity distribution in the porous soil.

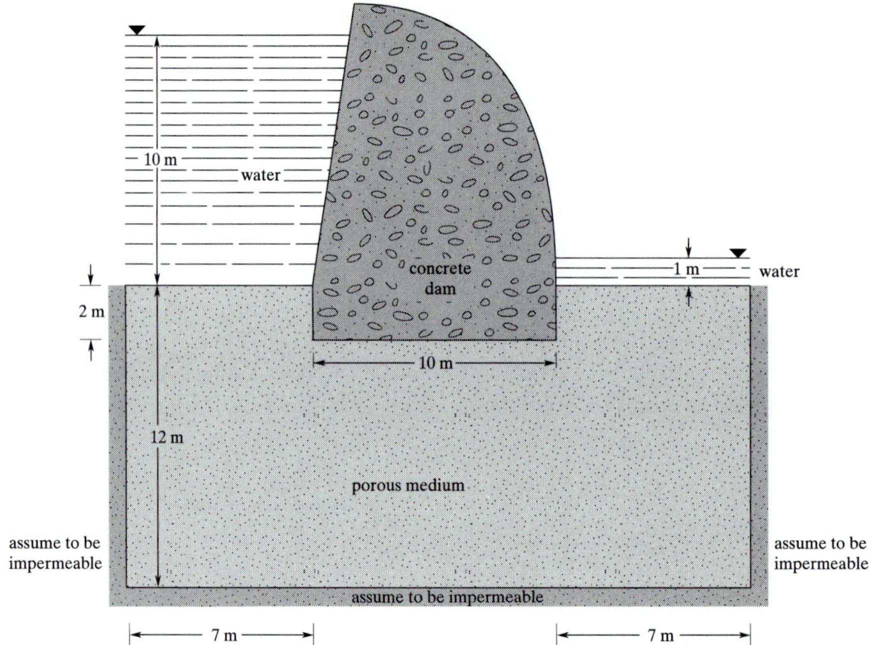

CHAPTER 10

# Three-Dimensional Elements

The main objective of this chapter is to introduce three-dimensional elements. First, we discuss the four-node tetrahedral element and the associated shape functions. Then, we consider the analysis of structural solid problems using the four-node tetrahedral element, including the formulation of an element's stiffness matrix. This section is followed by a discussion of the eight-node brick element and higher order tetrahedral and brick elements. Structural and thermal elements used by ANSYS will be covered next. This chapter also presents basic ideas regarding top-down and bottom-up solid-modeling methods. Finally, hints regarding how to mesh your solid model are given. The main topics of Chapter 10 include the following:

**10.1** The Four-Node Tetrahedral Element
**10.2** Analysis of Three-Dimensional Solid Problems Using Four-Node Tetrahedral Elements
**10.3** The Eight-Node Brick Element
**10.4** The Ten-Node Tetrahedral Element
**10.5** The Twenty-Node Brick Element
**10.6** Example of Three-Dimensional Elements in ANSYS
**10.7** Basic Solid-Modeling Ideas
**10.8** A Thermal Example Using ANSYS
**10.9** A Structural Example Using ANSYS
**10.10** Verification of Results: Error-Estimation Procedures

## 10.1 THE FOUR-NODE TETRAHEDRAL ELEMENT

The four-node tetrahedral element is the simplest three-dimensional element used in the analysis of solid mechanics problems. This element has four nodes, with each node having three translational degrees of freedom in the nodal $X$, $Y$, and $Z$-directions. A typical four-node tetrahedral element is shown in Figure 10.1.

In order to obtain the shape functions for the four-node tetrahedral element, we will follow a procedure similar to the one we followed in Chapter 5 to obtain the triangular

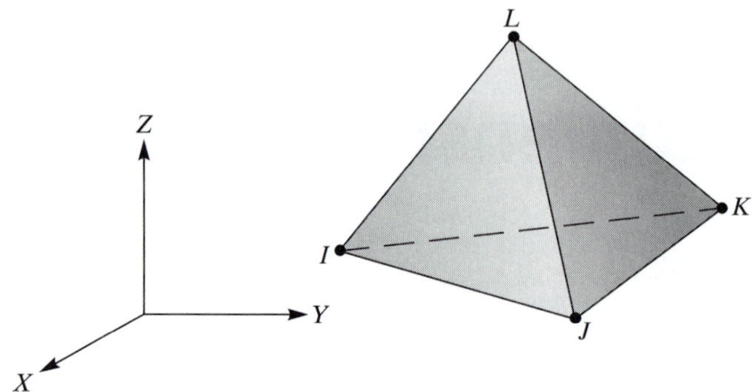


**FIGURE 10.1** A four-node tetrahedral element.

shape functions for two-dimensional problems. We begin by representing the displacement field by the following equations:

$$\begin{aligned} u &= C_{11} + C_{12}X + C_{13}Y + C_{14}Z \\ v &= C_{21} + C_{22}X + C_{23}Y + C_{24}Z \\ w &= C_{31} + C_{32}X + C_{33}Y + C_{34}Z \end{aligned} \tag{10.1}$$

Considering the nodal displacements, we must satisfy the following conditions:

$$\begin{array}{llllll} u = u_I & \text{at} & X = X_I & Y = Y_I & \text{and} & Z = Z_I \\ u = u_J & \text{at} & X = X_J & Y = Y_J & \text{and} & Z = Z_J \\ u = u_K & \text{at} & X = X_K & Y = Y_K & \text{and} & Z = Z_K \\ u = u_L & \text{at} & X = X_L & Y = Y_L & \text{and} & Z = Z_L \end{array}$$

Similarly, we must satisfy the following requirements:

$$\begin{array}{llllll} v = v_I & \text{at} & X = X_I & Y = Y_I & \text{and} & Z = Z_I \\ \vdots & \vdots & \vdots & \vdots & \vdots & \vdots \\ w = w_L & \text{at} & X = X_I & Y = Y_I & \text{and} & Z = Z_I \end{array}$$

Substitution of respective nodal values into Eqs. (10.1) results in 12 equations and 12 unknowns:

$$\begin{aligned} u_I &= C_{11} + C_{12}X_I + C_{13}Y_I + C_{14}Z_I \\ u_J &= C_{11} + C_{12}X_J + C_{13}Y_J + C_{14}Z_J \\ &\vdots \\ w_L &= C_{31} + C_{32}X_L + C_{33}Y_L + C_{34}Z_L \end{aligned} \tag{10.2}$$

Solving for the unknown $C$-coefficients, substituting the results back into Eq. (10.1), and regrouping the parameters, we obtain:

$$\begin{aligned} u &= S_1 u_I + S_2 u_J + S_3 u_K + S_4 u_L \\ v &= S_1 v_I + S_2 v_J + S_3 v_K + S_4 v_L \\ w &= S_1 w_I + S_2 w_J + S_3 w_K + S_4 w_L \end{aligned} \tag{10.3}$$

The shape functions are

$$\begin{aligned} S_1 &= \frac{1}{6V}(a_I + b_I X + c_I Y + d_I Z) \\ S_2 &= \frac{1}{6V}(a_J + b_J X + c_J Y + d_J Z) \\ S_3 &= \frac{1}{6V}(a_K + b_K X + c_K Y + d_K Z) \\ S_4 &= \frac{1}{6V}(a_L + b_L X + c_L Y + d_L Z) \end{aligned} \tag{10.4}$$

where $V$, the volume of the tetrahedral element, is computed from

$$6V = \det\begin{vmatrix} 1 & X_I & Y_I & Z_I \\ 1 & X_J & Y_J & Z_J \\ 1 & X_K & Y_K & Z_K \\ 1 & X_L & Y_L & Z_L \end{vmatrix} \tag{10.5}$$

the $a_I, b_I, c_I, d_I, \ldots,$ and $d_L$-terms are:

$$a_I = \det\begin{vmatrix} X_J & Y_J & Z_J \\ X_K & Y_K & Z_K \\ X_L & Y_L & Z_L \end{vmatrix} \qquad b_I = -\det\begin{vmatrix} 1 & Y_J & Z_J \\ 1 & Y_K & Z_K \\ 1 & Y_L & Z_L \end{vmatrix} \tag{10.6}$$

$$c_I = \det\begin{vmatrix} X_J & 1 & Z_J \\ X_K & 1 & Z_K \\ X_L & 1 & Z_L \end{vmatrix} \qquad d_I = -\det\begin{vmatrix} X_J & Y_J & 1 \\ X_K & Y_K & 1 \\ X_L & Y_L & 1 \end{vmatrix}$$

We can represent the $a_J, b_J, c_J, d_J, \ldots,$ and $d_L$-terms using similar determinants by rotating through the $I, J, K,$ and $L$ subscripts using the right-hand rule. For example,

$$a_J = \det\begin{vmatrix} X_K & Y_K & Z_K \\ X_L & Y_L & Z_L \\ X_I & Y_I & Z_I \end{vmatrix}$$

It is important to note here that for thermal problems, we associate only a single degree of freedom with each node of the four-node tetrahedral element—namely, temperature. The variation of temperature over a four-node tetrahedral element is expressed by

$$T = T_I S_1 + T_J S_2 + T_K S_3 + T_L S_4 \tag{10.7}$$

## 10.2 ANALYSIS OF THREE-DIMENSIONAL SOLID PROBLEMS USING FOUR-NODE TETRAHEDRAL ELEMENTS

You may recall from Chapter 8 that only six independent stress components are needed to characterize the general state of stress at a point. These components are

$$[\boldsymbol{\sigma}]^T = [\sigma_{xx} \quad \sigma_{yy} \quad \sigma_{zz} \quad \tau_{xy} \quad \tau_{yz} \quad \tau_{xz}] \tag{10.8}$$

where $\sigma_{xx}$, $\sigma_{yy}$, and $\sigma_{zz}$ are the normal stresses and $\tau_{xy}$, $\tau_{yz}$, and $\tau_{xz}$ are the shear-stress components. Moreover, we discussed the displacement vector that measures the changes occurring in the position of a point within a body when the body is subjected to a load. You may also recall that the displacement vector $\vec{\delta}$ can be written in terms of its Cartesian components as

$$\vec{\delta} = u(x, y, z)\vec{i} + v(x, y, z)\vec{j} + w(x, y, z)\vec{k} \tag{10.9}$$

The corresponding state of strain at a point was also discussed in Chapter 8. The general state of strain is characterized by six independent components as given by

$$[\boldsymbol{\varepsilon}]^T = [\varepsilon_{xx} \quad \varepsilon_{yy} \quad \varepsilon_{zz} \quad \gamma_{xy} \quad \gamma_{yz} \quad \gamma_{xz}] \tag{10.10}$$

where $\varepsilon_{xx}$, $\varepsilon_{yy}$, and $\varepsilon_{zz}$ are the normal strains and $\gamma_{xy}$, $\gamma_{yz}$, and $\gamma_{xz}$ are the shear-strain components. As previously discussed, the relationship between the strain and the displacement is represented by:

$$\varepsilon_{xx} = \frac{\partial u}{\partial x} \qquad \varepsilon_{yy} = \frac{\partial v}{\partial y} \qquad \varepsilon_{zz} = \frac{\partial w}{\partial z} \tag{10.11}$$

$$\gamma_{xy} = \frac{\partial u}{\partial y} + \frac{\partial v}{\partial x} \qquad \gamma_{yz} = \frac{\partial v}{\partial z} + \frac{\partial w}{\partial y} \qquad \gamma_{xz} = \frac{\partial u}{\partial z} + \frac{\partial w}{\partial x}$$

Equations (10.11) can be represented in matrix form as

$$\{\boldsymbol{\varepsilon}\} = LU \tag{10.12}$$

where

$$\{\boldsymbol{\varepsilon}\} = \begin{Bmatrix} \varepsilon_{xx} \\ \varepsilon_{yy} \\ \varepsilon_{zz} \\ \gamma_{xy} \\ \gamma_{yz} \\ \gamma_{xz} \end{Bmatrix}$$

and

$$LU = \begin{Bmatrix} \dfrac{\partial u}{\partial x} \\ \dfrac{\partial v}{\partial y} \\ \dfrac{\partial w}{\partial z} \\ \dfrac{\partial u}{\partial y} + \dfrac{\partial v}{\partial x} \\ \dfrac{\partial v}{\partial z} + \dfrac{\partial w}{\partial y} \\ \dfrac{\partial w}{\partial x} + \dfrac{\partial u}{\partial z} \end{Bmatrix}$$

$L$ is commonly referred to as the linear-differential operator.

Over the elastic region of a material, there also exists a relationship between the state of stresses and strains, according to the generalized Hooke's Law. This relationship is given by the following equations:

$$\varepsilon_{xx} = \frac{1}{E}\left[\sigma_{xx} - \nu(\sigma_{yy} + \sigma_{zz})\right] \tag{10.13}$$

$$\varepsilon_{yy} = \frac{1}{E}\left[\sigma_{yy} - \nu(\sigma_{xx} + \sigma_{zz})\right]$$

$$\varepsilon_{zz} = \frac{1}{E}\left[\sigma_{zz} - \nu(\sigma_{xx} + \sigma_{yy})\right]$$

$$\gamma_{xy} = \frac{1}{G}\tau_{xy} \quad \gamma_{yz} = \frac{1}{G}\tau_{yz} \quad \gamma_{zx} = \frac{1}{G}\tau_{zx}$$

The relationship between the stress and strain can be expressed in a compact-matrix form as:

$$\{\boldsymbol{\sigma}\} = [\boldsymbol{\nu}]\{\boldsymbol{\varepsilon}\} \tag{10.14}$$

where

$$\{\boldsymbol{\sigma}\} = \begin{Bmatrix} \sigma_{xx} \\ \sigma_{yy} \\ \sigma_{zz} \\ \tau_{xy} \\ \tau_{yz} \\ \tau_{xz} \end{Bmatrix}$$

$$[\boldsymbol{\nu}] = \frac{E}{1+\nu}\begin{bmatrix} \frac{1-\nu}{1-2\nu} & \frac{\nu}{1-2\nu} & \frac{\nu}{1-2\nu} & 0 & 0 & 0 \\ \frac{\nu}{1-2\nu} & \frac{1-\nu}{1-2\nu} & \frac{\nu}{1-2\nu} & 0 & 0 & 0 \\ \frac{\nu}{1-2\nu} & \frac{\nu}{1-2\nu} & \frac{1-\nu}{1-2\nu} & 0 & 0 & 0 \\ 0 & 0 & 0 & \frac{1}{2} & 0 & 0 \\ 0 & 0 & 0 & 0 & \frac{1}{2} & 0 \\ 0 & 0 & 0 & 0 & 0 & \frac{1}{2} \end{bmatrix}$$

$$\{\boldsymbol{\varepsilon}\} = \begin{Bmatrix} \varepsilon_{xx} \\ \varepsilon_{yy} \\ \varepsilon_{zz} \\ \gamma_{xy} \\ \gamma_{yz} \\ \gamma_{xz} \end{Bmatrix}$$

For a solid material under triaxial loading, the strain energy $\Lambda$ is

$$\Lambda^{(e)} = \frac{1}{2}\int_V (\sigma_{xx}\varepsilon_{xx} + \sigma_{yy}\varepsilon_{yy} + \sigma_{zz}\varepsilon_{zz} + \tau_{xy}\gamma_{xy} + \tau_{xz}\tau_{xz} + \tau_{yz}\gamma_{yz})dV \tag{10.15}$$

Or, in a compact-matrix form,

$$\Lambda^{(e)} = \frac{1}{2}\int_V [\boldsymbol{\sigma}]^T\{\boldsymbol{\varepsilon}\}dV \tag{10.16}$$

Substituting for stresses in terms of strains using Hooke's Law, Eq. (10.15) can be written as

$$\Lambda^{(e)} = \frac{1}{2}\int_V \{\boldsymbol{\varepsilon}\}^T[\boldsymbol{\nu}]\{\boldsymbol{\varepsilon}\}dV \tag{10.17}$$

We will now use the four-node tetrahedral element to formulate the stiffness matrix. Recall that this element has four nodes, with each node having three translational degrees of freedom in the nodal $x$, $y$, and $z$-directions. The displacements $u$, $v$, and $w$ in terms of the nodal values and the shape functions are represented by

$$\{\mathbf{u}\} = [\mathbf{S}]\{\mathbf{U}\} \tag{10.18}$$

where

$$\{\mathbf{u}\} = \begin{Bmatrix} u \\ v \\ w \end{Bmatrix}$$

$$[\mathbf{S}] = \begin{bmatrix} S_1 & 0 & 0 & S_2 & 0 & 0 & S_3 & 0 & 0 & S_4 & 0 & 0 \\ 0 & S_1 & 0 & 0 & S_2 & 0 & 0 & S_3 & 0 & 0 & S_4 & 0 \\ 0 & 0 & S_1 & 0 & 0 & S_2 & 0 & 0 & S_3 & 0 & 0 & S_4 \end{bmatrix}$$

$$\{\mathbf{U}\} = \begin{Bmatrix} u_I \\ v_I \\ w_I \\ u_J \\ v_J \\ w_J \\ u_K \\ v_K \\ w_K \\ u_L \\ v_L \\ w_L \end{Bmatrix}$$

The next few steps are similar to the steps we took to derive the stiffness matrix for plane-stress situations in Chapter 8, except more terms are involved in this case. We begin by relating the strains to the displacement field and, in turn, to the nodal displacements through the shape functions. We need to take the derivatives of the components of the displacement field with respect to the $x$, $y$, and $z$-coordinates according to the strain-displacement relations given by Eq. (10.12). The operation results in:

$$\begin{Bmatrix} \varepsilon_{xx} \\ \varepsilon_{yy} \\ \varepsilon_{zz} \\ \gamma_{xy} \\ \gamma_{yz} \\ \gamma_{xz} \end{Bmatrix} = \begin{bmatrix} \frac{\partial S_1}{\partial x} & 0 & 0 & \frac{\partial S_2}{\partial x} & 0 & 0 & \frac{\partial S_3}{\partial x} & 0 & 0 & \frac{\partial S_4}{\partial x} & 0 & 0 \\ 0 & \frac{\partial S_1}{\partial y} & 0 & 0 & \frac{\partial S_2}{\partial y} & 0 & 0 & \frac{\partial S_3}{\partial y} & 0 & 0 & \frac{\partial S_4}{\partial y} & 0 \\ 0 & 0 & \frac{\partial S_1}{\partial z} & 0 & 0 & \frac{\partial S_2}{\partial z} & 0 & 0 & \frac{\partial S_3}{\partial z} & 0 & 0 & \frac{\partial S_3}{\partial z} \\ \frac{\partial S_1}{\partial y} & \frac{\partial S_1}{\partial x} & 0 & \frac{\partial S_2}{\partial y} & \frac{\partial S_2}{\partial x} & 0 & \frac{\partial S_3}{\partial y} & \frac{\partial S_3}{\partial x} & 0 & \frac{\partial S_4}{\partial y} & \frac{\partial S_4}{\partial x} & 0 \\ 0 & \frac{\partial S_1}{\partial z} & \frac{\partial S_1}{\partial y} & 0 & \frac{\partial S_2}{\partial z} & \frac{\partial S_1}{\partial y} & 0 & \frac{\partial S_3}{\partial z} & \frac{\partial S_3}{\partial y} & 0 & \frac{\partial S_4}{\partial z} & \frac{\partial S_4}{\partial y} \\ \frac{\partial S_1}{\partial z} & 0 & \frac{\partial S_1}{\partial x} & \frac{\partial S_2}{\partial z} & 0 & \frac{\partial S_2}{\partial x} & \frac{\partial S_3}{\partial z} & 0 & \frac{\partial S_3}{\partial x} & \frac{\partial S_4}{\partial z} & 0 & \frac{\partial S_4}{\partial x} \end{bmatrix} \begin{Bmatrix} u_I \\ v_I \\ w_I \\ u_J \\ v_J \\ w_J \\ u_K \\ v_K \\ w_K \\ u_L \\ v_L \\ w_L \end{Bmatrix} \tag{10.19}$$

Substituting for the shape functions using the relations of Eq. (10.4) and differentiating, we have

$$\{\boldsymbol{\varepsilon}\} = [\mathbf{B}]\{\mathbf{U}\} \tag{10.20}$$

where

$$[\mathbf{B}] = \frac{1}{6V}\begin{bmatrix} b_I & 0 & 0 & b_J & 0 & 0 & b_K & 0 & 0 & b_L & 0 & 0 \\ 0 & c_I & 0 & 0 & c_J & 0 & 0 & c_K & 0 & 0 & c_L & 0 \\ 0 & 0 & d_I & 0 & 0 & d_J & 0 & 0 & d_K & 0 & 0 & d_L \\ c_I & b_I & 0 & c_J & b_J & 0 & c_K & b_K & 0 & c_L & b_L & 0 \\ 0 & d_I & c_I & 0 & d_J & c_J & 0 & d_K & c_K & 0 & d_L & c_L \\ d_I & 0 & b_I & d_J & 0 & b_J & d_K & 0 & b_K & d_L & 0 & b_L \end{bmatrix}$$

and the volume $V$ and the $b$, $c$, and $d$-terms are given by Eqs. (10.5) and (10.6). Substituting into the strain energy equation for the strain components in terms of the displacements, we obtain

$$\Lambda^{(e)} = \frac{1}{2}\int_V \{\boldsymbol{\varepsilon}\}^T[\boldsymbol{\nu}]\{\boldsymbol{\varepsilon}\}\,dV = \frac{1}{2}\int_V [\mathbf{U}]^T[\mathbf{B}]^T[\boldsymbol{\nu}][\mathbf{B}][\mathbf{U}]\,dV \tag{10.21}$$

Differentiating with respect to the nodal displacements yields

$$\frac{\partial \Lambda^{(e)}}{\partial U_k} = \frac{\partial}{\partial U_k}\left(\frac{1}{2}\int_V [\mathbf{U}]^T[\mathbf{B}]^T[\boldsymbol{\nu}][\mathbf{B}][\mathbf{U}]\,dV\right) \quad \text{for } k = 1, 2, \ldots, 12 \tag{10.22}$$

Evaluation of Eq. (10.22) results in the expression $[\mathbf{K}]^{(e)}\{\mathbf{U}\}$ and, subsequently, the expression for the stiffness matrix, which is

$$[\mathbf{K}]^{(e)} = \int_V [\mathbf{B}]^T[\boldsymbol{\nu}][\mathbf{B}]\,dV = V[\mathbf{B}]^T[\boldsymbol{\nu}][\mathbf{B}] \tag{10.23}$$

where $V$ is the volume of the element. Note that the resulting stiffness matrix will have the dimensions of $12 \times 12$.

### Load Matrix

The load matrix for three-dimensional problems is obtained by using a procedure similar to the one described in Section 8.3. The load matrix for a tetrahedral element is a $12 \times 1$ matrix. For a concentrated-loading situation, the load matrix is formed by placing the components of the load at appropriate nodes in appropriate directions. For a distributed load, the load matrix is computed from the equation

$$\{\mathbf{F}\}^{(e)} = \int_A [\mathbf{S}]^T\{\mathbf{p}\}\,dA \tag{10.24}$$

where

$$\{\mathbf{p}\} = \begin{Bmatrix} p_x \\ p_y \\ p_z \end{Bmatrix}$$

and $A$ represents the surface over which the distributed-load components are acting. The surfaces of the tetrahedral element are triangular in shape. Assuming that the distributed load acts on the $I-J-K$ surface, the load matrix becomes:

$$\{\mathbf{F}\}^{(e)} = \frac{A_{I\text{-}J\text{-}K}}{3}\begin{Bmatrix} p_x \\ p_y \\ p_z \\ p_x \\ p_y \\ p_z \\ p_x \\ p_y \\ p_z \\ 0 \\ 0 \\ 0 \end{Bmatrix} \tag{10.25}$$

The load matrix for a distributed load acting on the other surfaces of the tetrahedral element is obtained in a similar fashion.

## 10.3 THE EIGHT-NODE BRICK ELEMENT

The eight-node brick element is the next simple three-dimensional element used in the analysis of solid mechanics problems. Each of the eight nodes of this element has three translational degrees of freedom in the nodal $x$-, $y$-, and $z$-directions. A typical eight-node brick element is shown in Figure 10.2.

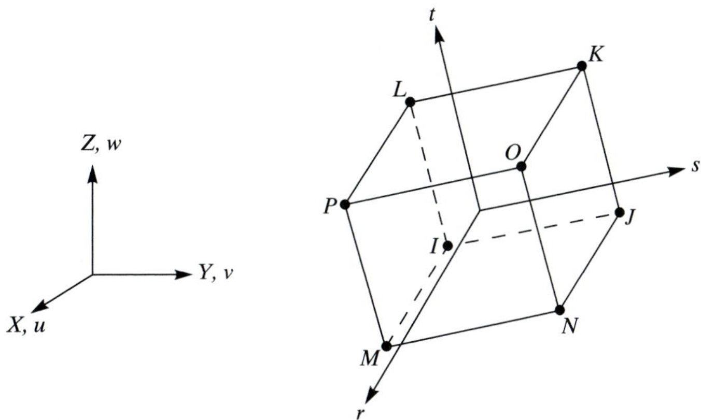


**FIGURE 10.2** An eight-node brick element.

The element's displacement field in terms of the nodal displacements and the shape functions can be written as:

$$u = \frac{1}{8}\left(u_I(1-s)(1-t)(1-r) + u_J(1+s)(1-t)(1-r)\right) \tag{10.26}$$

$$+ \frac{1}{8}\left(u_K(1+s)(1+t)(1-r) + u_L(1-s)(1+t)(1-r)\right)$$

$$+ \frac{1}{8}\left(u_M(1-s)(1-t)(1+r) + u_N(1+s)(1-t)(1+r)\right)$$

$$+ \frac{1}{8}\left(u_O(1+s)(1+t)(1+r) + u_P(1-s)(1+t)(1+r)\right)$$

$$v = \frac{1}{8}\left(v_I(1-s)(1-t)(1-r) + v_J(1+s)(1-t)(1-r)\right) \tag{10.27}$$

$$+ \frac{1}{8}\left(v_K(1+s)(1+t)(1-r) + v_L(1-s)(1+t)(1-r)\right)$$

$$+ \frac{1}{8}\left(v_M(1-s)(1-t)(1+r) + v_N(1+s)(1-t)(1+r)\right)$$

$$+ \frac{1}{8}\left(v_O(1+s)(1+t)(1+r) + v_P(1-s)(1+t)(1+r)\right)$$

$$w = \frac{1}{8}\left(w_I(1-s)(1-t)(1-r) + w_J(1+s)(1-t)(1-r)\right) \tag{10.28}$$

$$+ \frac{1}{8}\left(w_K(1+s)(1+t)(1-r) + w_L(1-s)(1+t)(1-r)\right)$$

$$+ \frac{1}{8}\left(w_M(1-s)(1-t)(1+r) + w_N(1+s)(1-t)(1+r)\right)$$

$$+ \frac{1}{8}\left(w_O(1+s)(1+t)(1+r) + w_P(1-s)(1+t)(1+r)\right)$$

In a similar fashion, for thermal problems, the spatial variation of temperature over an element is represented by:

$$T = \frac{1}{8}\left(T_I(1-s)(1-t)(1-r) + T_J(1+s)(1-t)(1-r)\right) \tag{10.29}$$

$$+ \frac{1}{8}\left(T_K(1+s)(1+t)(1-r) + T_L(1-s)(1+t)(1-r)\right)$$

$$+\frac{1}{8}\left(T_M(1-s)(1-t)(1+r)+T_N(1+s)(1-t)(1+r)\right)$$

$$+\frac{1}{8}\left(T_O(1+s)(1+t)(1+r)+T_P(1-s)(1+t)(1+r)\right)$$

## 10.4 THE TEN-NODE TETRAHEDRAL ELEMENT

The ten-node tetrahedral element, shown in Figure 10.3, is a higher order version of the three-dimensional linear tetrahedral element. When compared to the four-node tetrahedral element, the ten-node tetrahedral element is better suited for and more accurate in modeling problems with curved boundaries.

For solid problems, the displacement field is represented by:

$$u = u_I(2S_1-1)S_1 + u_J(2S_2-1)S_2 + u_K(2S_3-1)S_3 + u_L(2S_4-1)S_4 + 4\left(u_M S_1 S_2 + u_N S_2 S_3 + u_O S_1 S_3 + u_P S_1 S_4 + u_Q S_2 S_4 + u_R S_3 S_4\right) \quad (10.30)$$

$$v = v_I(2S_1-1)S_1 + v_J(2S_2-1)S_2 + v_K(2S_3-1)S_3 + v_L(2S_4-1)S_4 + 4\left(v_M S_1 S_2 + v_N S_2 S_3 + v_O S_1 S_3 + v_P S_1 S_4 + v_Q S_2 S_4 + v_R S_3 S_4\right) \quad (10.31)$$

$$w = w_I(2S_1-1)S_1 + w_J(2S_2-1)S_2 + w_K(2S_3-1)S_3 + w_L(2S_4-1)S_4 + 4\left(w_M S_1 S_2 + w_N S_2 S_3 + w_O S_1 S_3 + w_P S_1 S_4 + w_Q S_2 S_4 + w_R S_3 S_4\right) \quad (10.32)$$

In similar fashion, the spatial distribution of temperature over an element is given by:

$$T = T_I(2S_1-1)S_1 + T_J(2S_2-1)S_2 + T_K(2S_3-1)S_3 + T_L(2S_4-1)S_4 + 4\left(T_M S_1 S_2 + T_N S_2 S_3 + T_O S_1 S_3 + T_P S_1 S_4 + T_Q S_2 S_4 + T_R S_3 S_4\right) \quad (10.33)$$

## 10.5 THE TWENTY-NODE BRICK ELEMENT

The twenty-node brick element, shown in Figure 10.4, is a higher order version of the three-dimensional eight-node brick element. This element is more capable and more accurate for modeling problems with curved boundaries than the eight-node brick element.

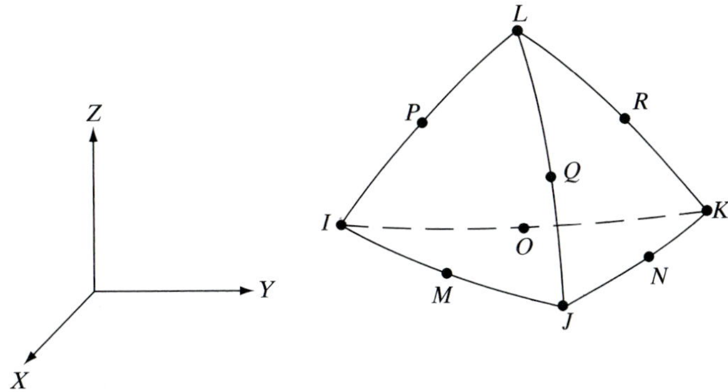


**FIGURE 10.3** A ten-node tetrahedral element.

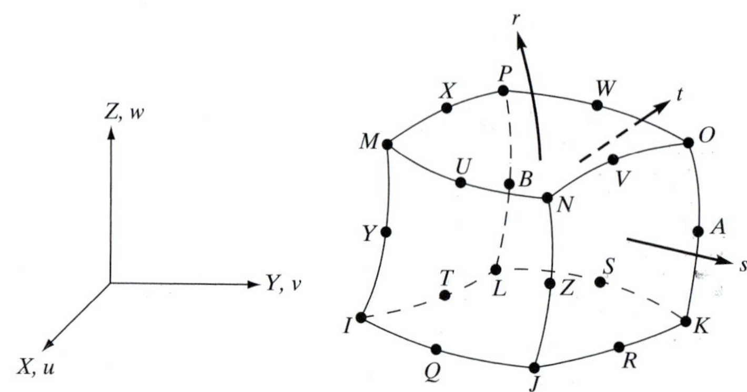


**FIGURE 10.4** A twenty-node brick element.

For solid mechanics problems, the displacement field is given by:

$$
\begin{aligned}
u = {} & \frac{1}{8}\left(u_I(1-s)(1-t)(1-r)(-s-t-r-2) + u_J(1+s)(1-t)(1-r)(s-t-r-2)\right) \\
& + \frac{1}{8}\left(u_K(1+s)(1+t)(1-r)(s+t-r-2) + u_L(1-s)(1+t)(1-r)(-s+t-r-2)\right) \\
& + \frac{1}{8}\left(u_M(1-s)(1-t)(1+r)(-s-t+r-2) + u_N(1+s)(1-t)(1+r)(s-t+r-2)\right) \\
& + \frac{1}{8}\left(u_O(1+s)(1+t)(1+r)(s+t+r-2) + u_P(1-s)(1+t)(1+r)(-s+t+r-2)\right) \\
& + \frac{1}{4}\left(u_Q(1-s^2)(1-t)(1-r) + u_R(1+s)(1-t^2)(1-r)\right) \\
& + \frac{1}{4}\left(u_S(1-s^2)(1+t)(1-r) + u_T(1-s)(1-t^2)(1-r)\right) \\
& + \frac{1}{4}\left(u_U(1-s^2)(1-t)(1+r) + u_V(1+s)(1-t^2)(1+r)\right) \\
& + \frac{1}{4}\left(u_W(1-s^2)(1+t)(1+r) + u_X(1-s)(1-t^2)(1+r)\right) \\
& + \frac{1}{4}\left(u_Y(1-s)(1-t)(1-r^2) + u_Z(1+s)(1-t)(1-r^2)\right) \\
& + \frac{1}{4}\left(u_A(1+s)(1+t)(1-r^2) + u_B(1-s)(1+t)(1-r^2)\right) \qquad (10.34)
\end{aligned}
$$

The $v$- and $w$-components of the displacement are similar to the $u$-component:

$$v = \frac{1}{8}(v_I(1-s)(1-t)(1-r)(-s-t-r-2) + v_J(1+s)(1-t)(1-r)(s-t-r-2))$$

$$+\frac{1}{8}(v_K(1+s)(1+t)(1-r)(s+t-r-2) + \ldots)$$

$$\ldots$$

$$w = \frac{1}{8}(w_I(1-s)(1-t)(1-r)(-s-t-r-2) + w_J(1+s)(1-t)(1-r)(s-t-r-2))$$

$$+\frac{1}{8}(w_K(1+s)(1+t)(1-r)(s+t-r-2) + \ldots) \tag{10.35}$$

$$\ldots$$

For heat transfer problems, the spatial variation of temperature over an element is given by:

$$T = \frac{1}{8}(T_I(1-s)(1-t)(1-r)(-s-t-r-2) + T_J(1+s)(1-t)(1-r)(s-t-r-2))$$

$$+\frac{1}{8}(T_K(1+s)(1+t)(1-r)(s+t-r-2) + \ldots) \tag{10.36}$$

$$\ldots$$

## 10.6 EXAMPLES OF THREE-DIMENSIONAL ELEMENTS IN ANSYS*

ANSYS offers a broad variety of elements for the analysis of three-dimensional problems. Some examples of three-dimensional elements in ANSYS are presented next.

### Thermal-Solid Elements

**SOLID70** is a three-dimensional element used to model conduction heat transfer problems. It has eight nodes, with each node having a single degree of freedom—temperature—as shown in Figure 10.5. The element's faces are shown by the circled numbers.

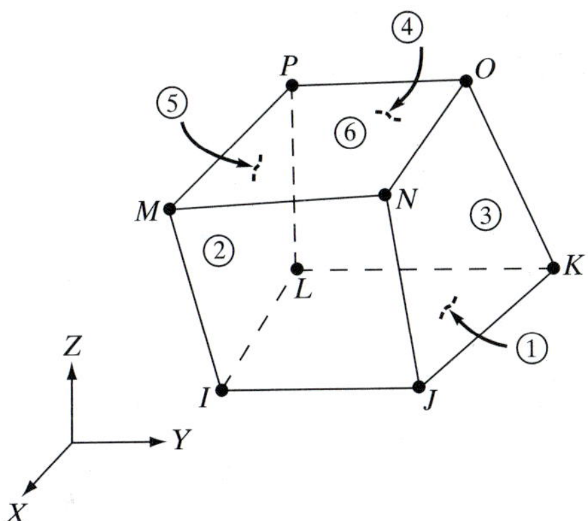


**FIGURE 10.5** The SOLID70 element used by ANSYS.

*Materials were adapted with permission from ANSYS documents.

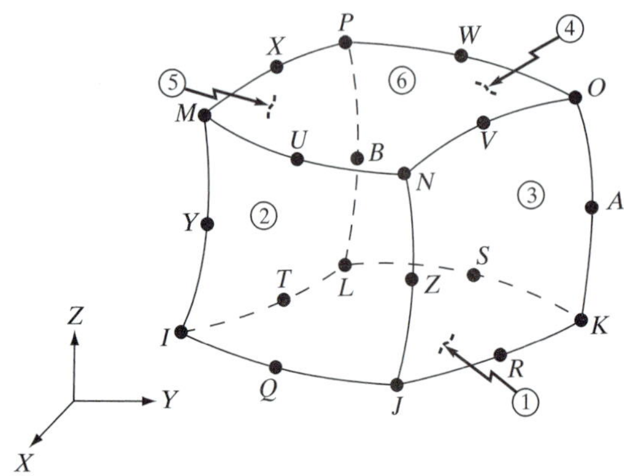


**FIGURE 10.6** The SOLID90 element used by ANSYS.

Convection or heat fluxes may be applied to the element's surfaces. In addition, heat-generation rates may be applied at the nodes. This element may be used to analyze steady-state or transient problems.

The solution output consists of nodal temperatures and other information, such as average face temperature, temperature-gradient components, the vector sum at the centroid of the element, and the heat-flux components.

**SOLID90** is a twenty-node brick element used to model steady-state or transient conduction heat transfer problems. This element is more accurate than the SOLID70 element, but it requires more solution time. Each node of the element has a single degree of freedom—temperature—as shown in Figure 10.6. This element is well suited to model problems with curved boundaries. The required input data and the solution output are similar to the data format of the SOLID70 elements.

### Structural-Solid Elements

**SOLID45** is a three-dimensional brick element used to model isotropic solid problems. It has eight nodes, with each node having three translational degrees of freedom in the nodal *x*-, *y*-, and *z*-directions, as shown in Figure 10.7 (The element's faces are shown by

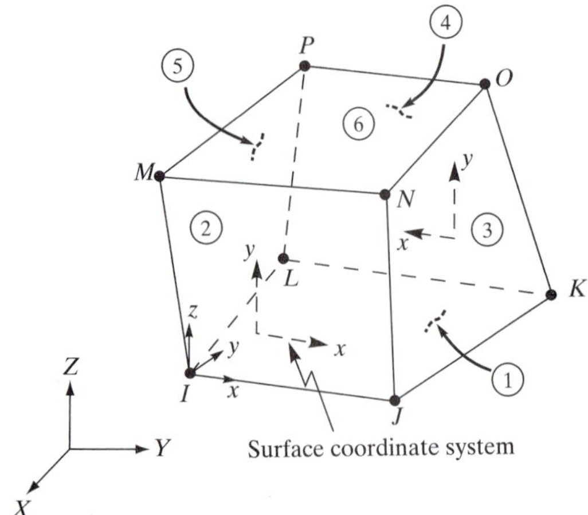


**FIGURE 10.7** The SOLID45 element used by ANSYS.

the circled numbers.) Distributed surface loads (pressures) may be applied to the element's surfaces. This element may be used to analyze large-deflection, large-strain, plasticity, and creep problems.

The solution output consists of nodal displacements. Examples of additional elemental output include normal components of the stresses in $x$, $y$, and $z$-directions; shear stresses; and principal stresses. The element's stress directions are parallel to the element's coordinate systems.

**SOLID65** is used to model reinforced-concrete problems or reinforced composite materials, such as fiberglass. This element is similar to the SOLID45 elements, and it has eight nodes, with each node having three translational degrees of freedom in the nodal $x$, $y$, and $z$-directions, as shown in Figure 10.8. The element may be used to analyze cracking in tension or crushing in compression. The element can also be used to analyze problems with or without reinforced bars. Up to three rebar specifications may be defined. The rebars are capable of plastic deformation and creep. The element has one solid material and up to three rebar materials. Rebar specifications include the material number; the volume ratio, which is defined as the ratio of the rebar volume to the total element volume; and the orientation angles. The rebar orientation is defined by two angles measured with respect to the element's coordinate system. The rebar capability is removed by assigning a zero value to the rebar material number.

The solution output consists of nodal displacements. Examples of additional elemental output include the normal components of the stresses in $x$, $y$, and $z$-directions, shear stresses, and principal stresses. The element's stress directions are parallel to the element's coordinate system.

**SOLID72** is a four-node tetrahedral element, with each node having three translational degrees of freedom in the nodal $x$, $y$, and $z$-directions, as well as rotations about the nodal $x$, $y$, and $z$-directions, as shown in Figure 10.9. As in previous examples, the element's faces are shown by the circled numbers. Distributed surface loads (pressures) may be applied to the element's surfaces.

The solution output is similar to that of other structural-solid elements.

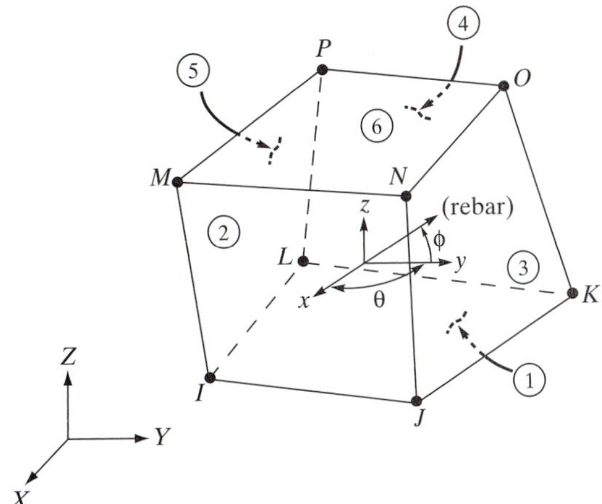


**FIGURE 10.8** The SOLID65 element used by ANSYS.

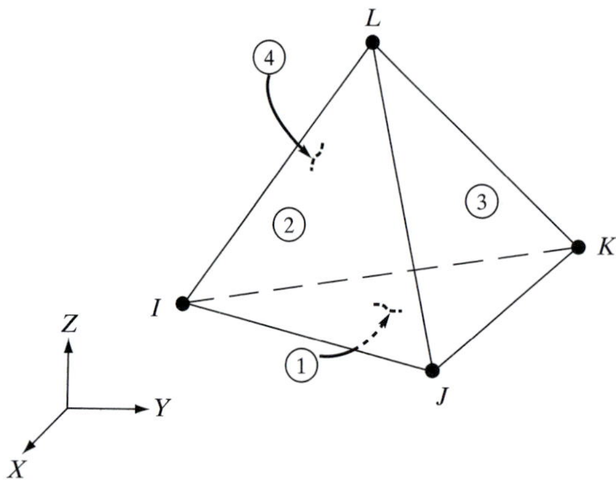


**FIGURE 10.9** The SOLID72 element used by ANSYS.

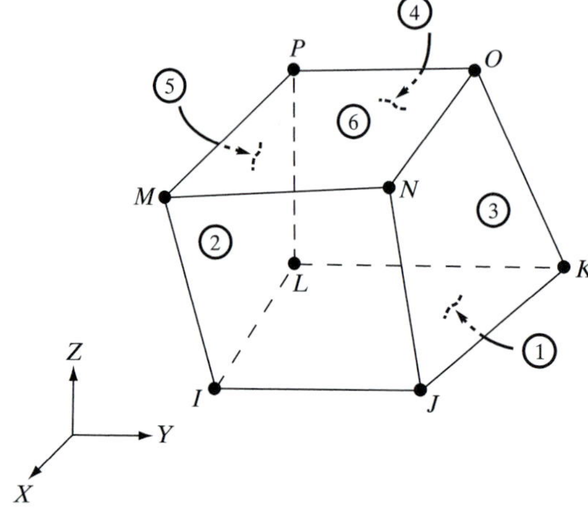


**FIGURE 10.10** The SOLID73 element used by ANSYS.

**SOLID73** is an eight-node brick element that has three translational degrees of freedom in the nodal $x$, $y$, and $z$-directions, as well as rotations about the nodal $x$, $y$, and $z$-directions, as shown in Figure 10.10. The input data and the solution output are similar to those of elements discussed previously.

**SOLID92** is a ten-node tetrahedral element that is more accurate than the SOLID72 element, but it requires more solution time. Each node has three translational degrees of freedom in the nodal $x$-, $y$-, and $z$-directions, as shown in Figure 10.11. This element may be used to analyze large-deflection, large-strain, plasticity, and creep problems.

## 10.7 BASIC SOLID-MODELING IDEAS*

There are two ways to create a solid model of an object under investigation: *bottom-up modeling* and *top-down modeling*. With *bottom-up modeling*, you start by defining keypoints first, then lines, areas, and volumes in terms of the defined keypoints. You can

*Materials were adapted with permission from ANSYS documents.

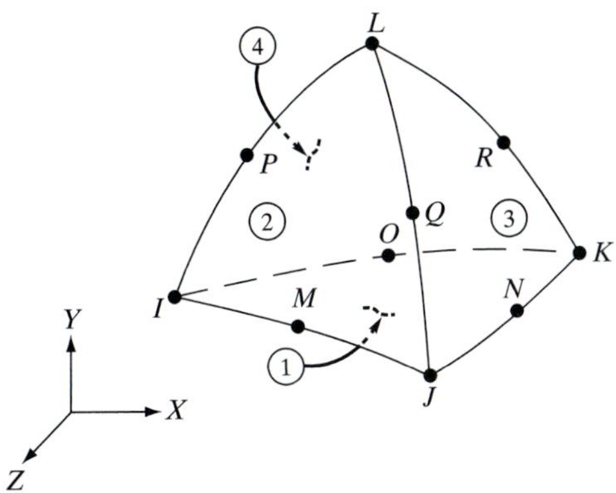


**FIGURE 10.11** The SOLID92 element used by ANSYS.

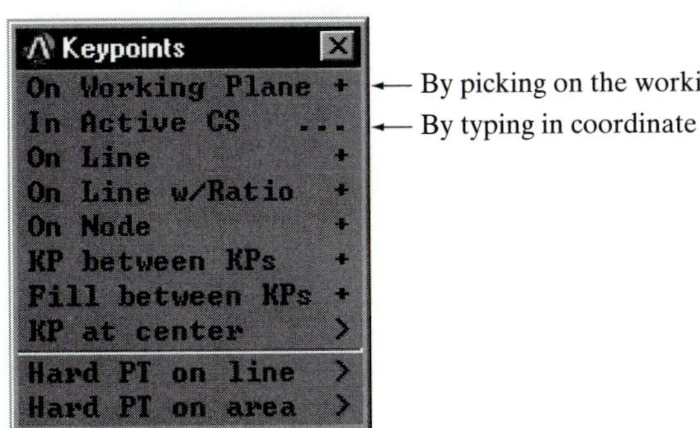


**FIGURE 10.12** The Keypoints menu.

define keypoints on the working plane by the picking method or you can enter, in appropriate fields, the coordinates of the keypoints in terms of the active coordinate system. The keypoints menu is shown in Figure 10.12, and the command for creating keypoints is:

main menu: **Preprocessor** → **-Modeling-Create** → **Keypoints**

Lines, next in the hierarchy of bottom-up modeling, are used to represent the edges of an object. ANSYS provides four options for creating lines, as shown in Figure 10.13.

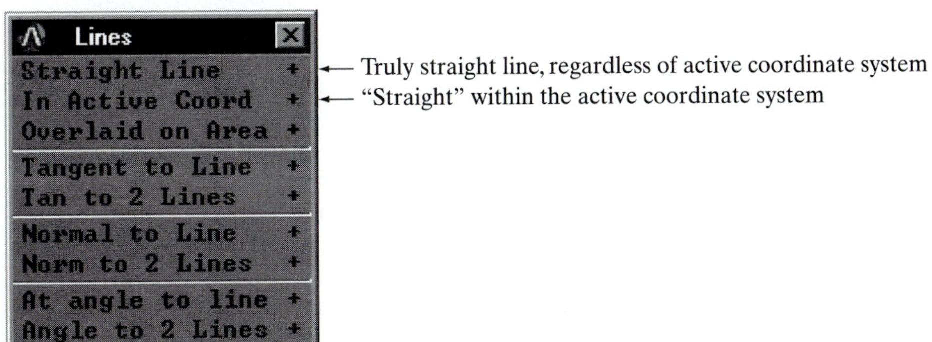


**FIGURE 10.13** The lines menu.

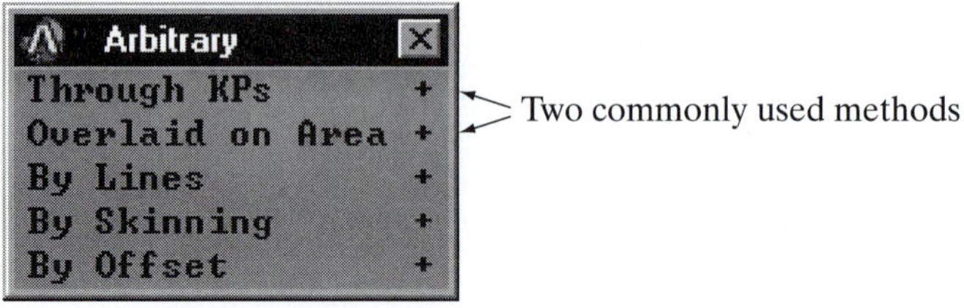


**FIGURE 10.14** The Area-Arbitrary sub-menu.

With the splines options, you can create a line of arbitrary shape from a spline fit to a series of keypoints. You can then use the created line(s) to generate a surface with an arbitrary shape. The command for creating lines is:

main menu: **Preprocessor** → **-Modeling-Create** → **Lines**

Using bottom-up modeling, you can define areas using the Area-Arbitrary submenu, as shown in Figure 10.14. The command for defining areas is:

main menu: **Preprocessor** → **-Modeling-Create** → **-Areas-Arbitrary**

There are five other ways by which you can create areas: (1) dragging a line along a path, (2) rotating a line about an axis; (3) creating an area fillet, (4) skinning a set of lines, and (5) offsetting areas. With the drag and rotate options, you can generate an area by dragging (sweeping) a line along another line (path) or by rotating a line about another line (axis of rotation). With the area-fillet command, you can create a constant-radius fillet tangent to two other areas. You can generate a smooth surface over a set of lines by using the skinning command. Using the area-offset command, you can generate an area by offsetting an existing area. These operations are all shown in Figure 10.15.

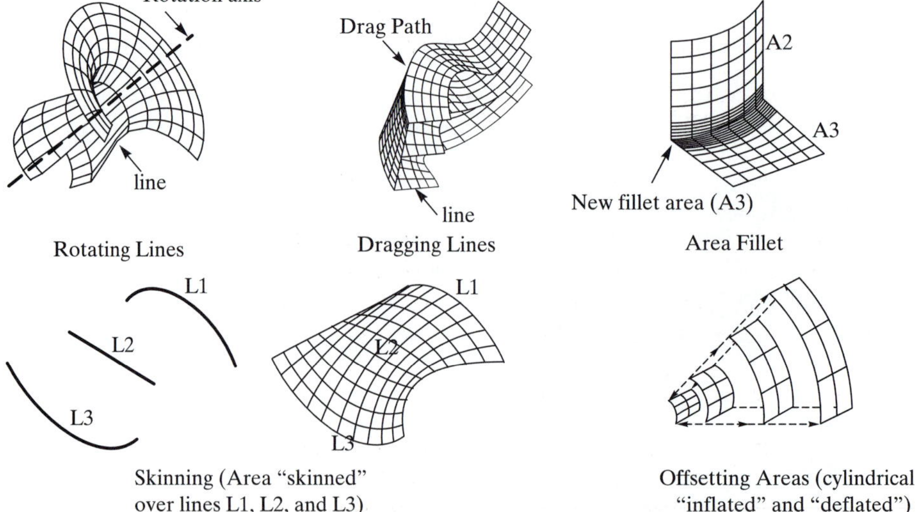


**FIGURE 10.15** Additional area-generation methods.

**FIGURE 10.16** The Volume submenu.

You can define volumes using the bottom-up method by selecting the Volume submenu, as shown in Figure 10.16. The command for defining volume is:

main menu: **Preprocessor** → **-Modeling-Create** → **-Volumes-Arbitrary**

As with areas, you can also generate volumes by dragging (sweeping) an area along a line (path) or by rotating an area about a line (axis of rotation).

With top-down modeling, you can create three-dimensional solid objects using volume primitives. ANSYS provides the following three-dimensional primitives: block, prism, cylinder, cone, sphere, and torus, as shown in Figure 10.17.

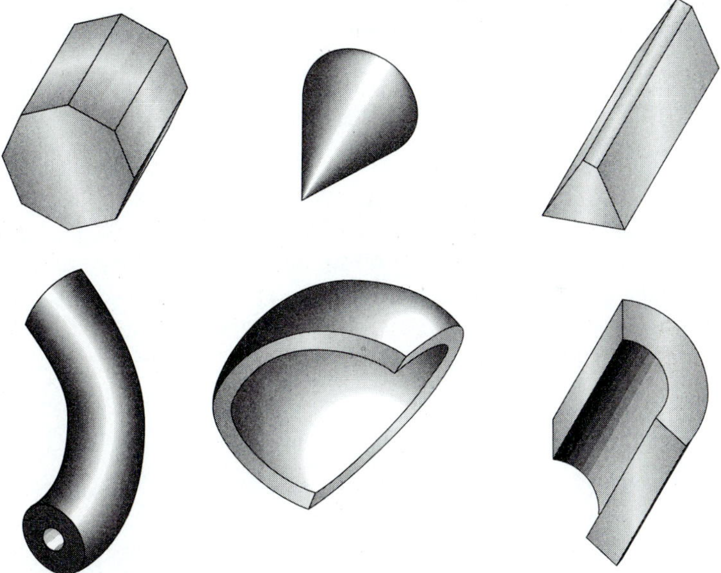

**FIGURE 10.17** Examples of three-dimensional primitives.

Keep in mind that when you create a volume using primitives, ANSYS automatically generates and assigns numbers to areas, lines, and keypoints that bound the volume.

Regardless of how you generate areas or volumes, you can use Boolean operations to add or subtract entities to create a solid model.

### Meshing Control

So far, you have been using global element size to control the size of elements in your model. The global-element-size dialog box allows you to specify the size of an element's edge length in the units of your model's dimensions. Let us consider other ways of controlling not only the size of elements, but also their shapes. Setting the element shape

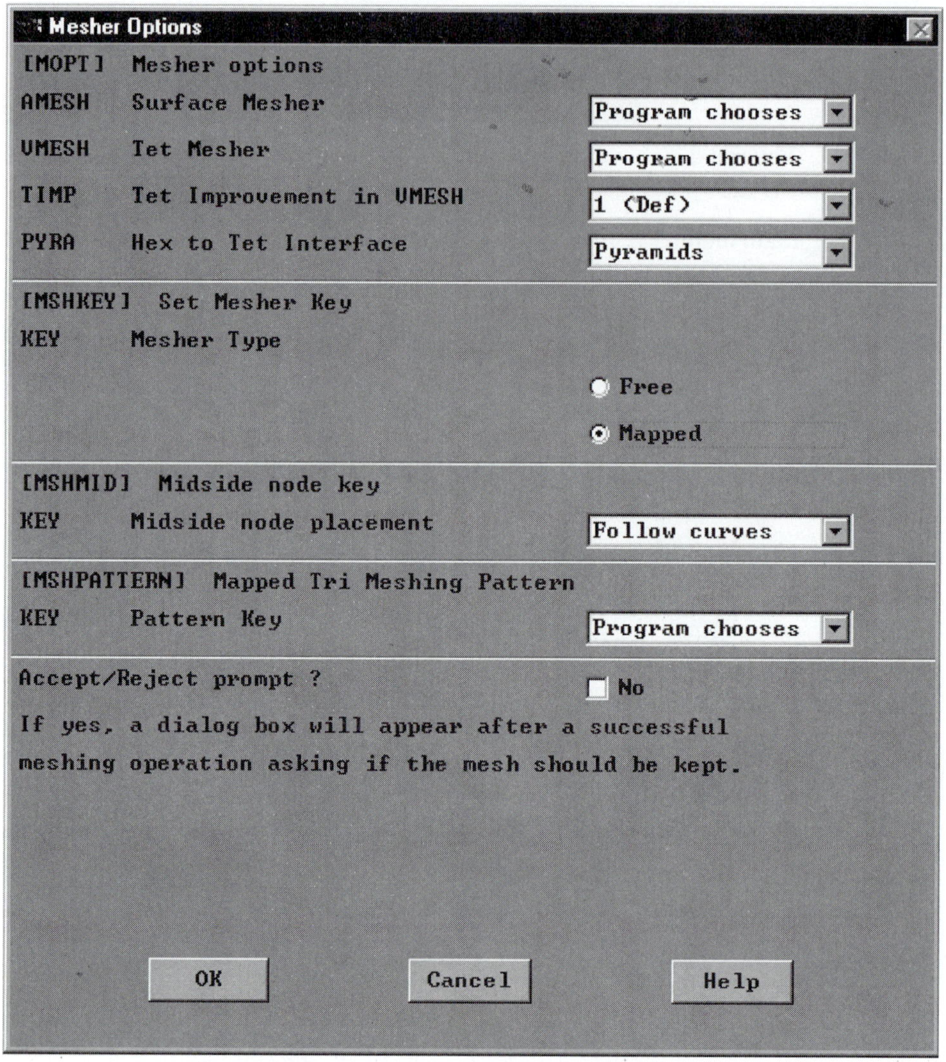

**FIGURE 10.18** The dialog box for element shape.

prior to meshing is important when using elements that can take on two shapes. For example, PLANE82 can take on triangular or quadrilateral shapes. Use the following command to see the dialog box for the meshing options (see Figure 10.18):

main menu: **Preprocessor** → **-Meshing-Mesher Opts** …

### Free Meshing Versus Mapped Meshing

*Free meshing* uses either mixed-area element shapes, all triangular-area elements, or all tetrahedral-volume elements. You may want to avoid using lower order triangular and

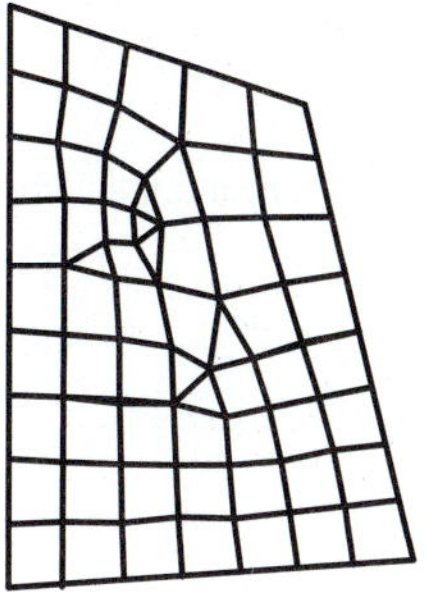

Free meshing

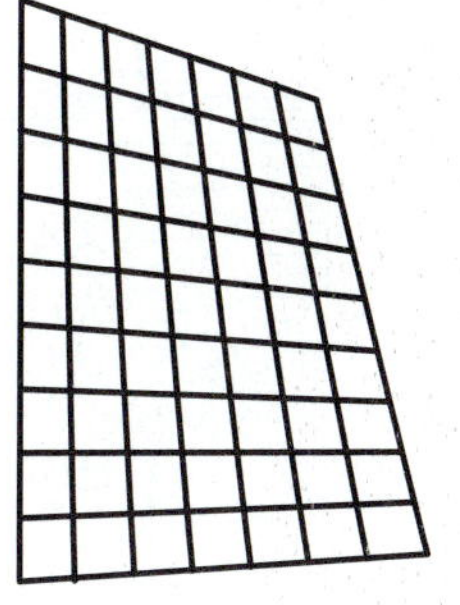

Mapped meshing

**FIGURE 10.19** An illustration of the difference between free and mapped meshing.

tetrahedral elements (those without midside nodes) in analysis of structures, when possible. On the other hand, *mapped meshing* uses all quadrilateral-area elements and all hexahedral-volume elements. Figure 10.19 illustrates the difference between free and mapped meshing.

There are, however, some requirements that need to be met for mapped meshing. The mapped-area mesh requires that the area has three or four sides, equal numbers of elements on opposite sides, and an even number of elements for three-sided areas. If an area is bounded by more than four lines, then you need to use the *combine* command or the *concatenate* command to combine (reduce) the number of lines to four. The mapped-volume requirements are that the volume must be bound by four, five, or six sides, have an equal number of elements on the opposite side, and have an even numbers of elements if a five-sided prism or tetrahedron volume is involved. For volumes, you can *add* or *concatenate* areas to reduce the number of areas bounding a volume. Concatenation should be the last step before meshing. You cannot perform any other solid-modeling operations on concatenated entities. To concatenate, issue the following command (see Figure 10.20):

main menu: **Preprocessing** → **-Meshing-Concatenate**

**FIGURE 10.20** Concatenate dialog box.

Figure 10.21 shows an example of free and mapped meshing for an area. As a general rule, you want to avoid meshing poorly shaped regions and avoid creating extreme element-size transitions in your model. Examples of these situations are given in Figure 10.22.

If you are unhappy with the results of a meshed region, you can use the *clear* command to delete the nodes and elements associated with a corresponding solid-model entity. To issue the clear command, use the following sequence:

main menu: **Preprocessor** → **-Meshing-Clear**

With the aid of an example, we will now demonstrate how to create a solid model of a heat sink, using area and extrusion commands.

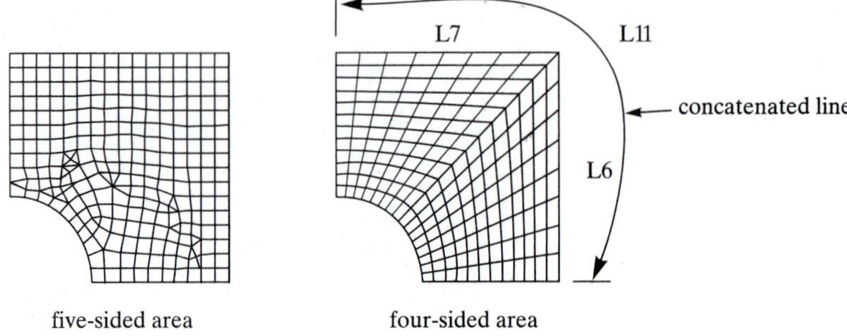


**FIGURE 10.21** An example of free meshing and mapped meshing for an area.

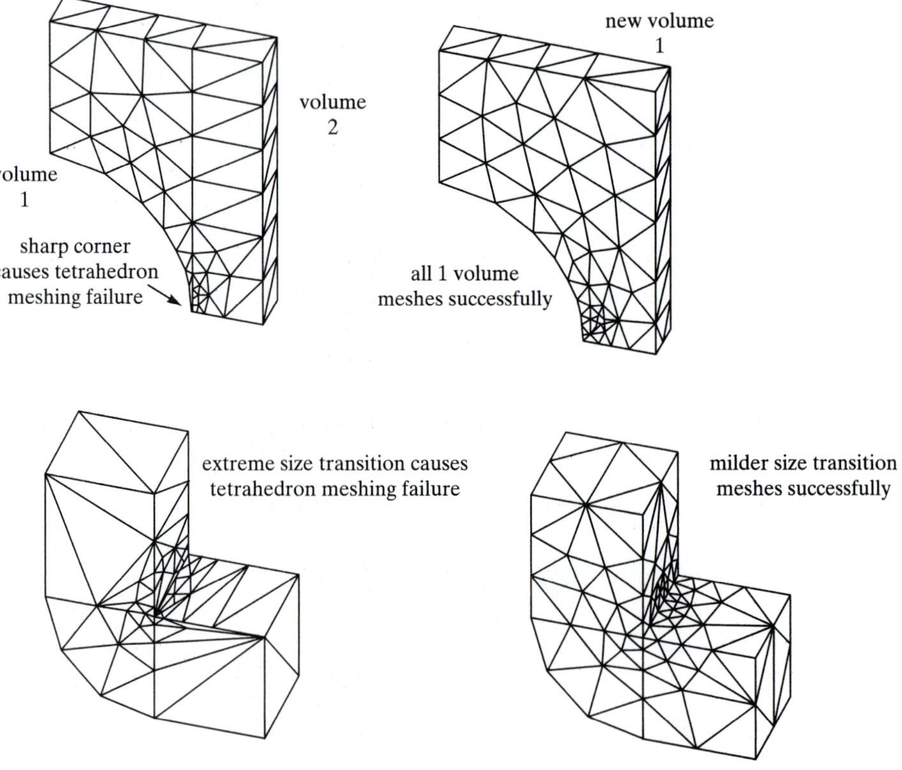


**FIGURE 10.22** Examples of undesirable meshing situations.

## EXAMPLE 10.1

Aluminum heat sinks are commonly used to dissipate heat from electronic devices. Consider an example of a heat sink used to cool a personal-computer microprocessor chip. The front view of the heat sink is shown in Figure 10.23. Using ANSYS, generate the solid model of the heat sink. Because of the symmetry, model only a quarter of the heat sink by extruding the shown frontal area by 20.5 mm.

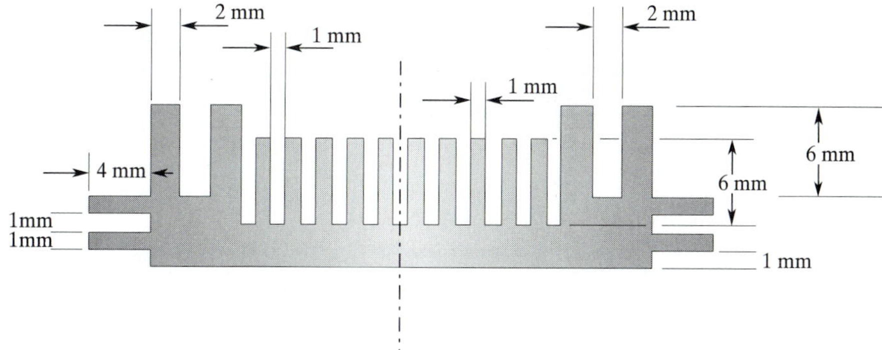


**FIGURE 10.23** The front view of the heat sink in Example 10.1.

Enter the **ANSYS** program by using the Launcher. Type **xansys54** on the command line, or consult your system administrator for the appropriate command name to launch ANSYS on your computer system.

Pick **Interactive** from the Launcher menu.

Type **Fin** (or a file name of your choice) in the **Initial Jobname** entry field of the dialog box.

ANSYS 5.4 INTERACTIVE

Product selection: ANSYS/University Low Option

Working directory: C:\

Graphics device name: win32

Initial jobname: Fin

Memory requested (megabytes)
for Total Workspace: 32
for Database: 16

Read START.ANS file at start-up? Yes

GUI configuration ...

Parameters to be defined
(-par1 val1 -par2 val2 ...)

Language Selection: [US English]

Run | Close | Reset | Cancel | Help

Pick **Run** to start the Graphic User Interface (GUI). A window will open with some disclaimer information. You will eventually be asked to press the **Return** key to start the graphics window and the main menu. Do so in order to proceed.

Create a title for the problem. This title will appear on ANSYS display windows to provide a simple way of identifying the displays. To create a title, issue the following command:

utility menu: **File** → **Change Title** ...

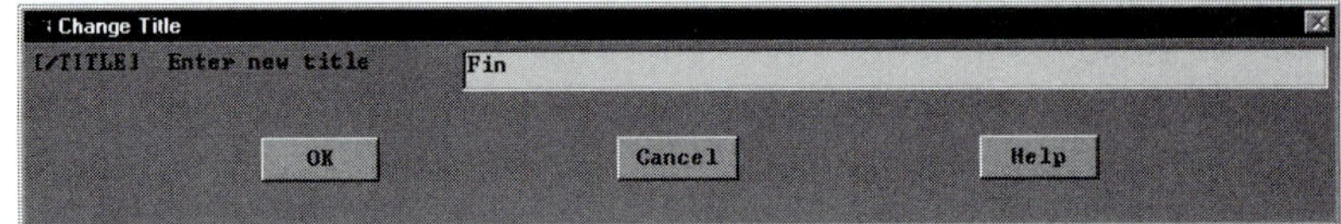


Set up the graphics area (i.e., workplane, zoom, etc.) with the following commands:

utility menu: **Workplane** → **Wp Settings** ...

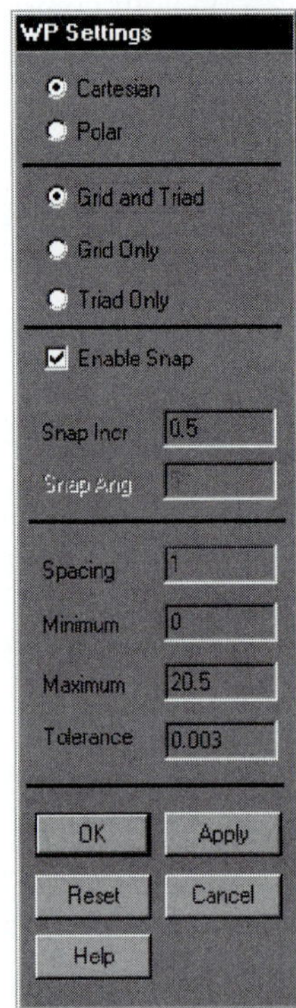


Toggle on the workplane by using the following command:

utility menu: **Workplane** → **Display Working Plane**

Bring the workplane to view by using the following command:

utility menu: **PlotCtrls** → **Pan, Zoom, Rotate** ...

Click on the small circle until you bring the workplane to view. Then, create the geometry with the following command:

main menu: **Preprocessor** → **-Modeling-Create** → **-Areas-Rectangle** → **By 2 Corners +**

**[WP = 4,0]**

**[Expand the rubber band up 2.5 and right 16.5]**

**[WP = 0,1]**

**[Expand the rubber band up 1.0 and right 4.0]**

**[WP = 0,3]**

**[Expand the rubber band up 1 and right 4.0]**

**[WP = 4,2.5]**

**[Expand the rubber band up 1.5 and right 6.0]**

**[WP = 4,4]**

**[Expand the rubber band up 6 and right 2]**

**[WP = 8,4]**

**[Expand the rubber band up 6.0 and right 2.0]**

**[WP = 11,2.5]**

**[Expand the rubber band up 6.0 and right 1.0]**

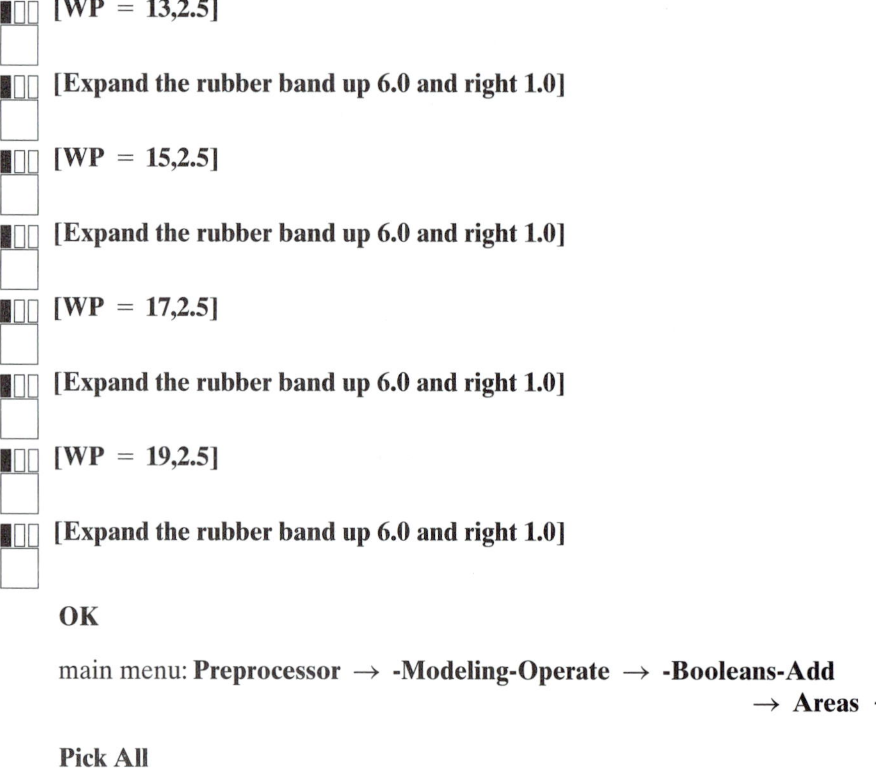

**[WP = 13,2.5]**

**[Expand the rubber band up 6.0 and right 1.0]**

**[WP = 15,2.5]**

**[Expand the rubber band up 6.0 and right 1.0]**

**[WP = 17,2.5]**

**[Expand the rubber band up 6.0 and right 1.0]**

**[WP = 19,2.5]**

**[Expand the rubber band up 6.0 and right 1.0]**

**OK**

main menu: **Preprocessor → -Modeling-Operate → -Booleans-Add → Areas +**

**Pick All**

main menu: **Preprocessor → -Modeling-Operate → Extrude/Sweep → -Areas-Along Normal +**

Pick or enter the area to be extruded, and then press the **Apply** button:

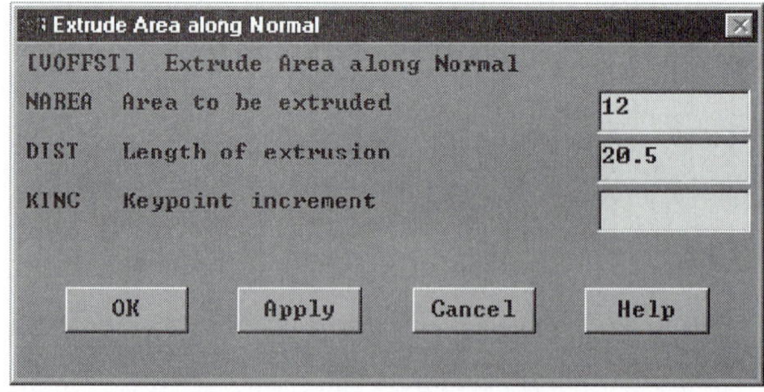


utility menu: **PlotCtrls → Pan, Zoom, Rotate** …

Press the **Iso** (Isometric view) button. You should then see the image in Figure 10.24.

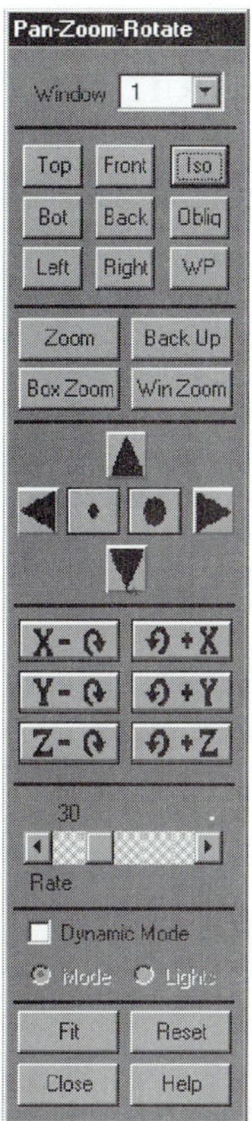


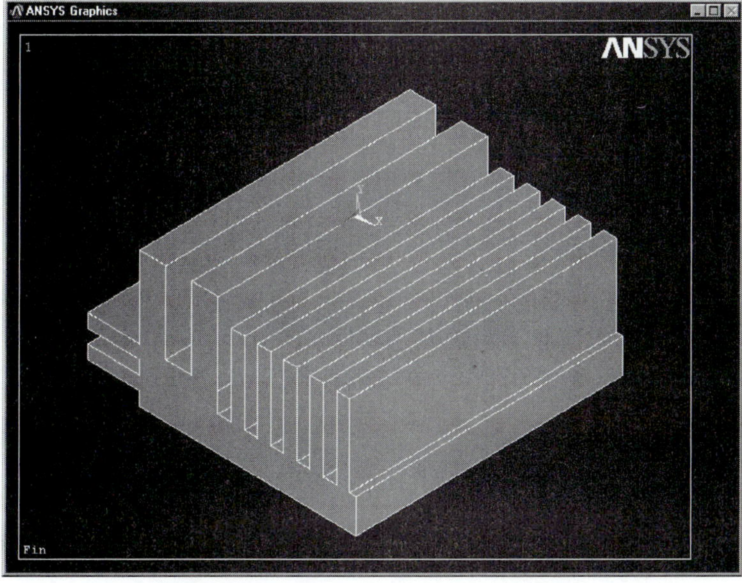


**FIGURE 10.24** Isometric view of the heat sink.

Exit and save your results:

ANSYS Toolbar: **QUIT**

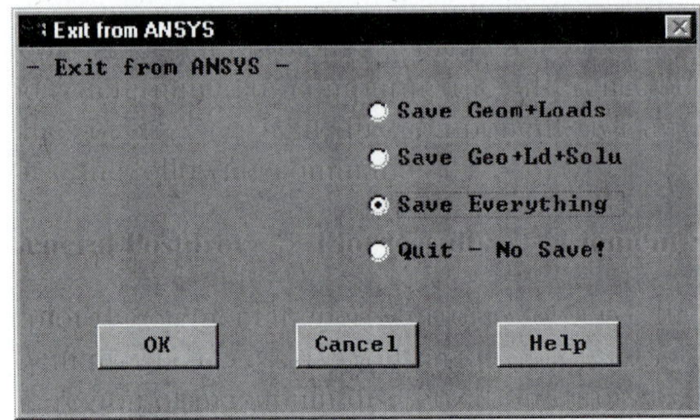


This example has demonstrated how to extrude an area along a normal direction to create a volume.

## 10.8 A THERMAL EXAMPLE USING ANSYS

### EXAMPLE 10.2

A section of an aquarium wall with a viewing window has the dimensions shown in Figure 10.25. The wall is constructed from concrete and other insulating materials, with an average thermal conductivity of $k = 0.81$ Btu/hr · ft · °F. The section of the wall has a viewing window that is made of a six-inch-thick clear plastic with a thermal conductivity of $k = 0.195$ Btu/hr · ft · °F. The inside air temperature is kept at 70°F, with a corresponding heat transfer coefficient of $h = 1.46$ Btu/hr · ft$^2$ · °F. Assuming a water-tank temperature of 50°F and a corresponding heat transfer coefficient of $h = 10.5$ Btu/hr · ft$^2$ · °F, use ANSYS to plot the temperature distribution within the wall section. Note that the main purpose of this example is to show the selection capabilities of ANSYS and to show how to move the working plane when constructing three-dimensional models. Recall that the heat loss through such a wall may be obtained with reasonable accuracy from the equation $q = U_{\text{overall}}(T_{\text{inside}} - T_{\text{water}})$ and by calculating the overall $U$-factor for the wall.

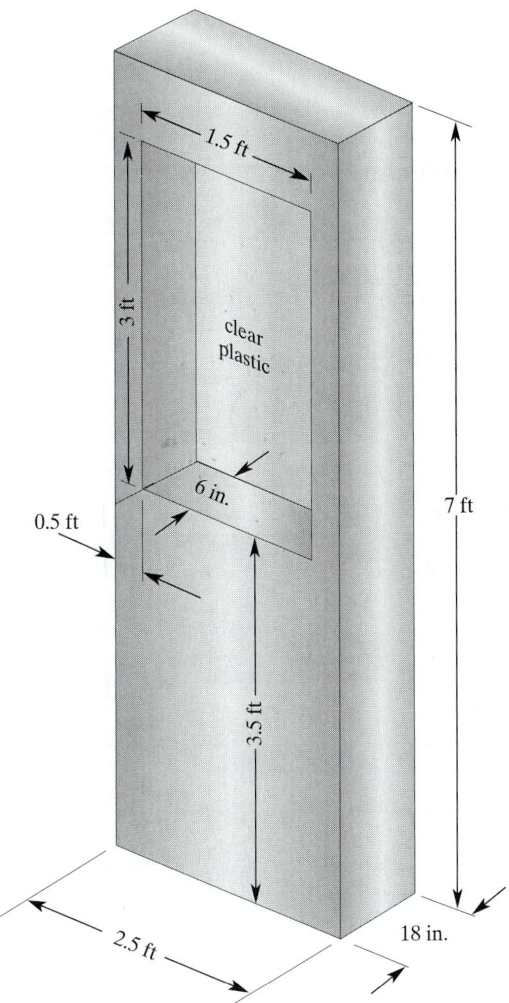


**FIGURE 10.25** Dimensions of the wall and the clear-plastic viewing window of Example 10.2.

Enter the **ANSYS** program by using the Launcher. Type **xansys54** on the command line, or consult your system administrator for the appropriate command name to launch ANSYS on your computer system.

Pick **Interactive** from the Launcher menu.

Type **Wall** (or a file name of your choice) in the **Initial Jobname** entry field of the dialog box.

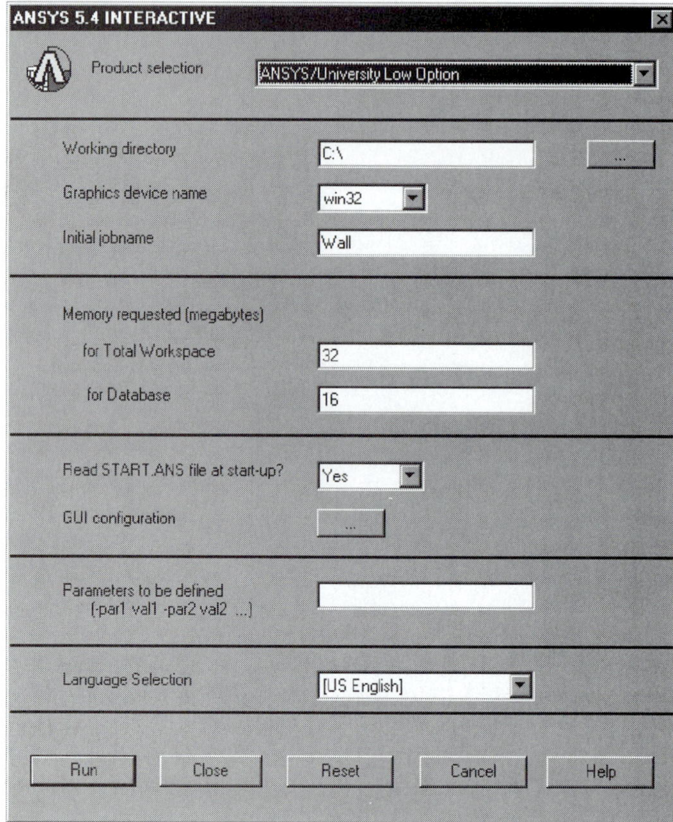


Pick **Run** to start the Graphic User Interface (GUI). A window will open with some disclaimer information. You will eventually be asked to press the **Return** key to start the graphics window and the main menu. Do so in order to proceed.

Create a title for the problem. This title will appear on ANSYS display windows to provide a simple way of identifying the displays. To create a title, issue the following command:

utility menu: **File** → **Change Title ...**

Change Title
[/TITLE] Enter new title Wall
OK Cancel Help

main menu: **Preprocessor** → **Element Types** → **Add/Edit/Delete ...**

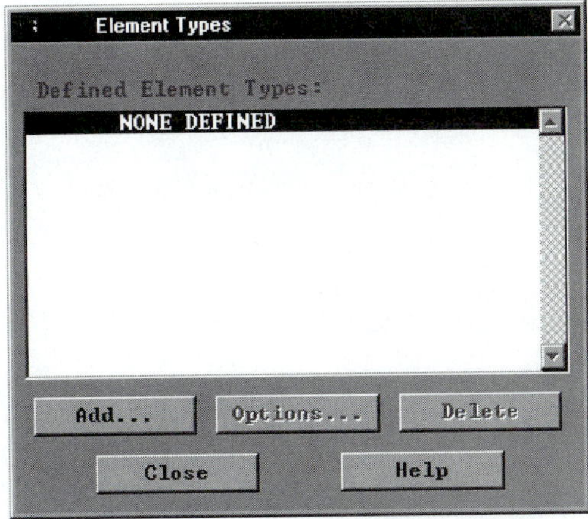




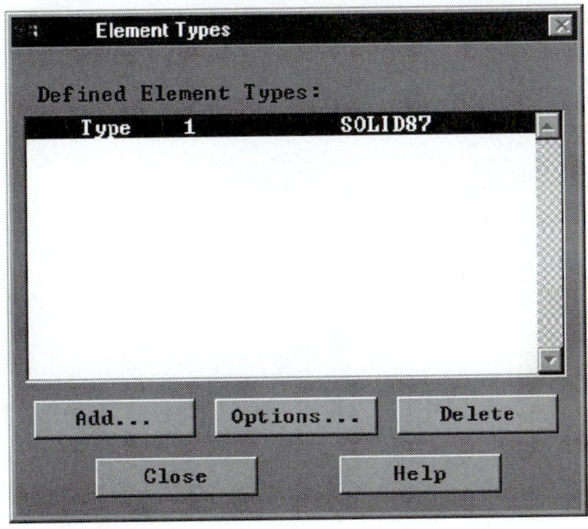

Assign thermal conductivity values for concrete and plastic with the following commands:

main menu: **Preprocessor** → **Material Props** → **-Constant-Isotropic ...**

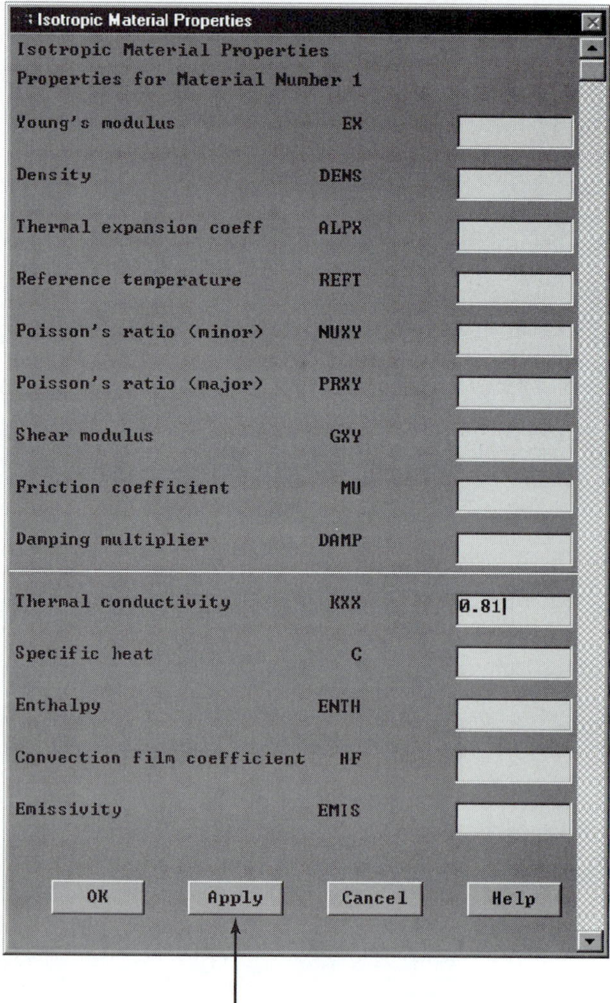

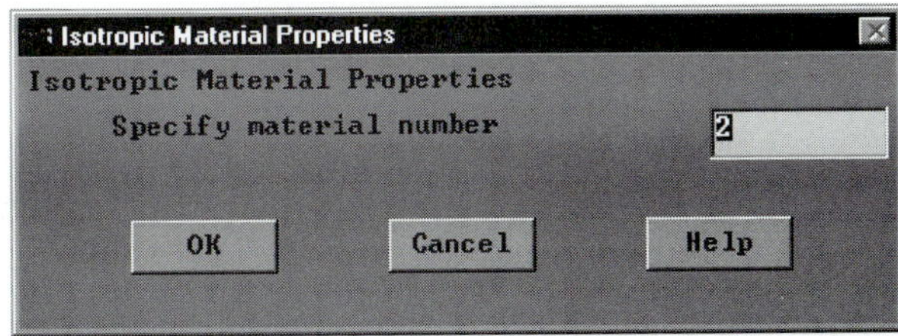


Isotropic Material Properties

Isotropic Material Properties
Properties for Material Number 2

| | | |
|---|---|---|
| Young's modulus | EX | |
| Density | DENS | |
| Thermal expansion coeff | ALPX | |
| Reference temperature | REFT | |
| Poisson's ratio (minor) | NUXY | |
| Poisson's ratio (major) | PRXY | |
| Shear modulus | GXY | |
| Friction coefficient | MU | |
| Damping multiplier | DAMP | |
| Thermal conductivity | KXX | 0.195 |
| Specific heat | C | |
| Enthalpy | ENTH | |
| Convection film coefficient | HF | |
| Emissivity | EMIS | |

OK Apply Cancel Help

ANSYS Toolbar: **SAVE_DB**

Set up the graphics area (i.e., workplane, zoom, etc.) with the following commands:

utility menu: **Workplane** → **Wp Settings ...**

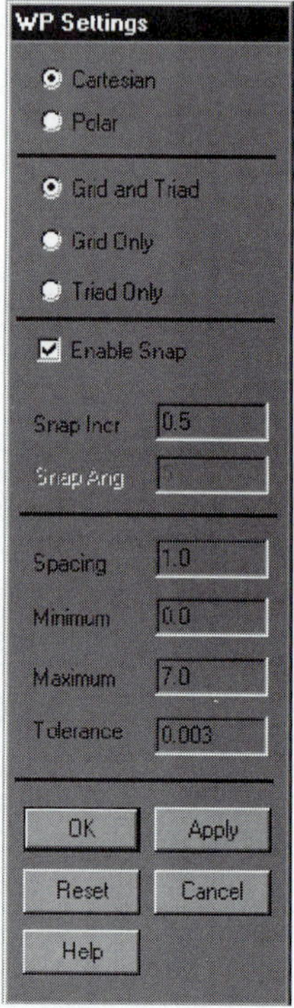


Toggle on the workplane by using the following command:

utility menu: **Workplane** → **Display Working Plane**

Bring the workplane to view by using the following command:

utility menu: **PlotCtrls** → **Pan, Zoom, Rotate ...**

Click on the small circle until you bring the workplane to view. Then, press the **Iso** (Isometric view) button. Next, create the geometry with the following commands:

main menu: **Preprocessor** → **-Modeling-Create** → **-Volumes-Block** → **By 2 Corners & Z +**

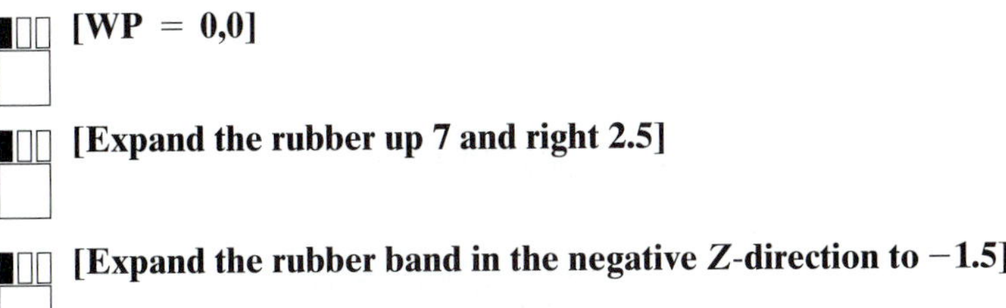

**[WP = 0,0]**

**[Expand the rubber up 7 and right 2.5]**

**[Expand the rubber band in the negative $Z$-direction to −1.5]**

Create a volume, to be removed later, for the plastic volume:

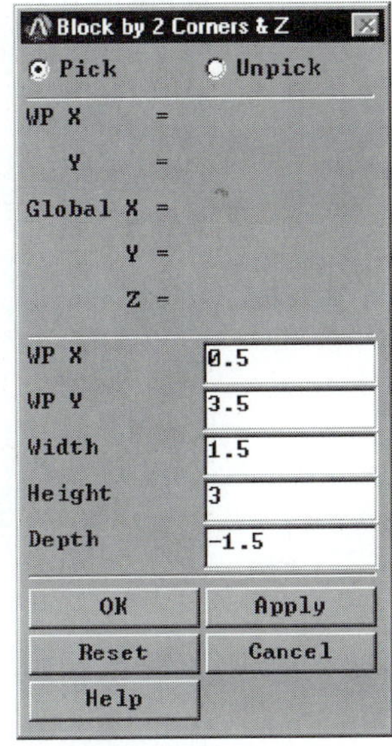


**OK**

main menu: **Preprocessor** → **-Modeling-Operate** → **-Booleans-Subtract** → **Volumes +**

Pick Volume1 and **Apply**; then pick Volume2 and **Apply.**

**OK**

utility menu: **Plot** → **Volumes**

Create the plastic volume with the following command:

utility menu: **WorkPlane → Offset WP by Increments ...**

In the X, Y, Z Offsets box, type in **[0, 0, −0.5]**.

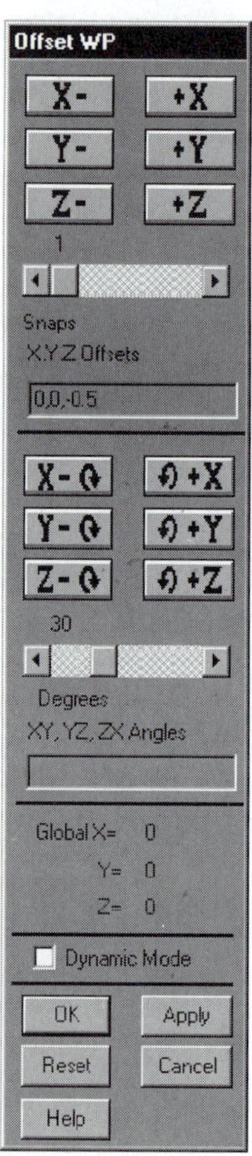


**OK**

Now, issue the following commands:

main menu: **Preprocessor → -Modeling-Create → -Volumes-Block → By 2 Corners & Z +**

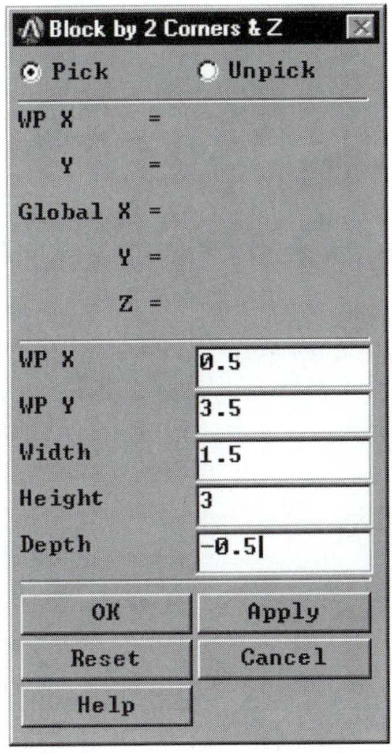


**OK**

main menu: **Preprocessor** → **-Modeling-Operate** → **-Booleans-Glue** → **Volumes** +

**Pick All**

We now want to mesh the volumes to create elements and nodes, but first, we need to specify the element sizes. So, issue the following commands:

main menu: **Preprocessor** → **-Meshing-Size-Cntrls** → **Global-Size** ...

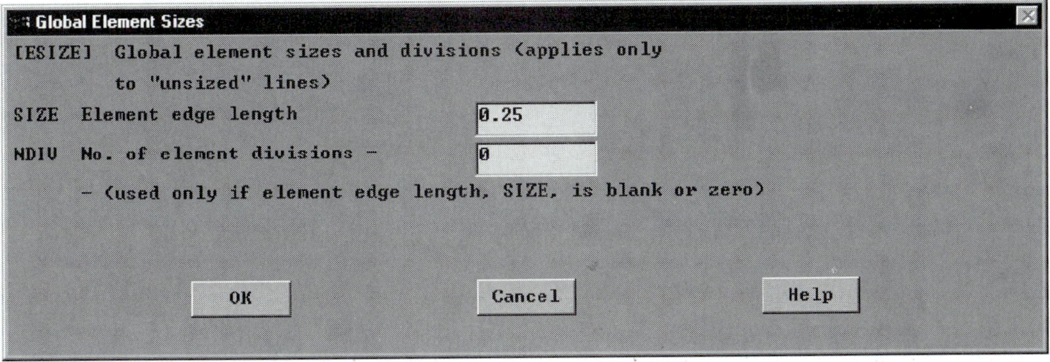

We also need to specify material attributes for the concrete and the plastic volumes before we proceed with meshing. To do so, we issue the following commands:

main menu: **Preprocessor** → **-Attributes-Define** → **Picked Volumes** +

**[Pick the concrete part of the wall volume]**

**[Apply anywhere in the ANSYS graphics window]**

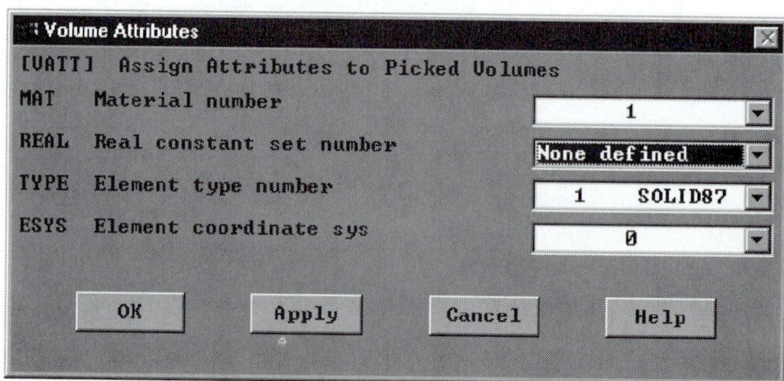


main menu: **Preprocessor** → **-Attributes-Define** → **Picked Volumes** +

**[Pick the plastic volume]**

**[Apply anywhere in the ANSYS graphics window]**

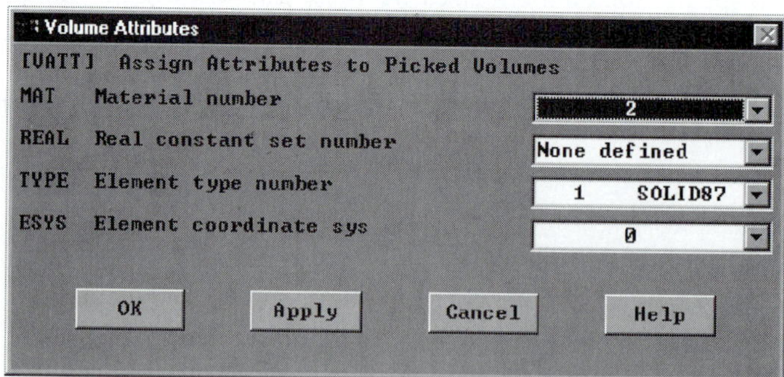


ANSYS Toolbar: **SAVE_DB**

We can proceed to mesh by issuing the following commands:

main menu: **Preprocessor** → **-Meshing-Mesh** → **-Volumes-Free** +

**Pick All**

If you exceed the maximum number of elements allowed in the educational version of ANSYS try the following:

main menu: **Preprocessor** → **-Meshing-Size Cntrl** → **-Smart Size-basic** …

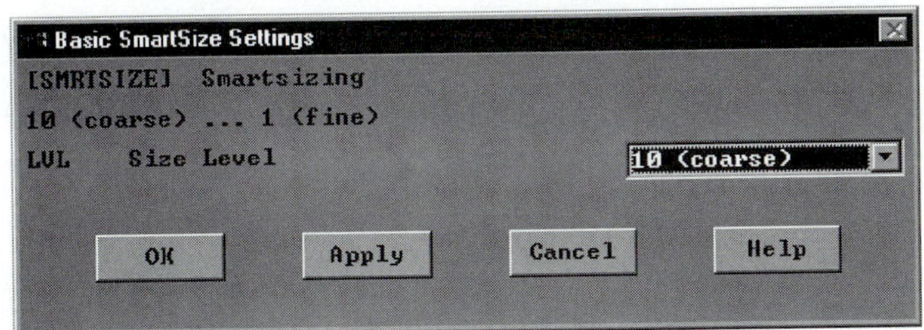


Now try to mesh the volumes again. To apply the boundary conditions, we first select the interior surfaces of the wall, including the clear plastic:

utility menu: **Select** → **Entities** …

In the **Min, Max** field, type: **[0, − 0.5]**

**OK**

utility menu: **Plot** → **Areas**

main menu: **Solution** → **-Loads-Apply** → **-Thermal-Convection** → **On Areas +**

**Pick All**

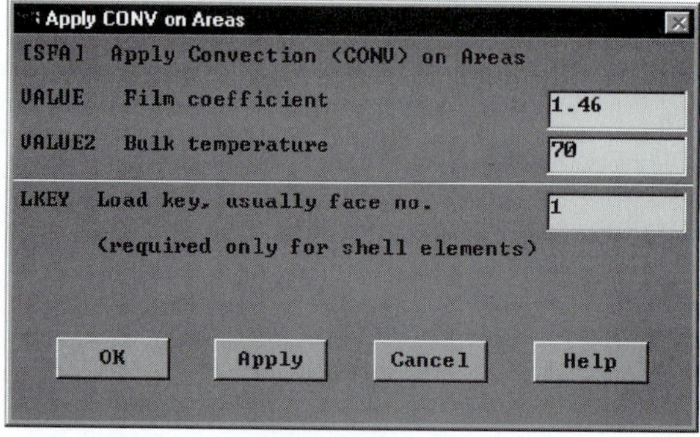


utility menu: **Select** → **Everything**

utility menu: **Select** → **Entities** ...

In the **Min, Max** field, type: **[−1.0, −1.5]**

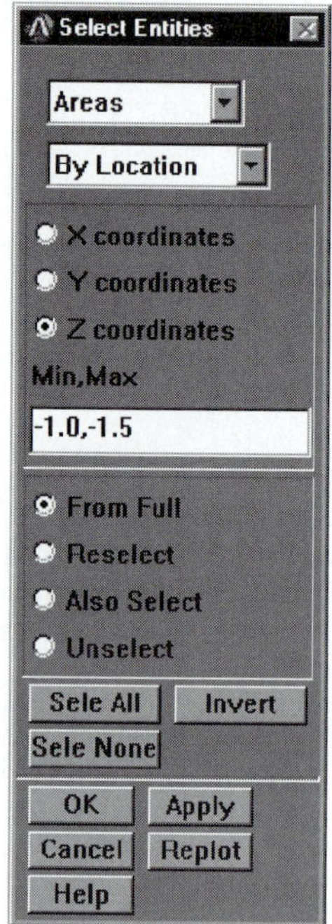


**OK**

utility menu: **Plot** → **Areas**

main menu: **Solution** → **-Loads-Apply** → **-Thermal-Convection** → **On Areas +**

**Pick All** to specify the convection coefficient and temperature:

Apply CONV on Areas

[SFA] Apply Convection (CONV) on Areas

VALUE Film coefficient 10.5

VALUE2 Bulk temperature 50

LKEY Load key, usually face no. 1

(required only for shell elements)

OK Apply Cancel Help

To see the applied boundary conditions, use the following commands:

utility menu: **PlotCntrls** → **Symbols** ...

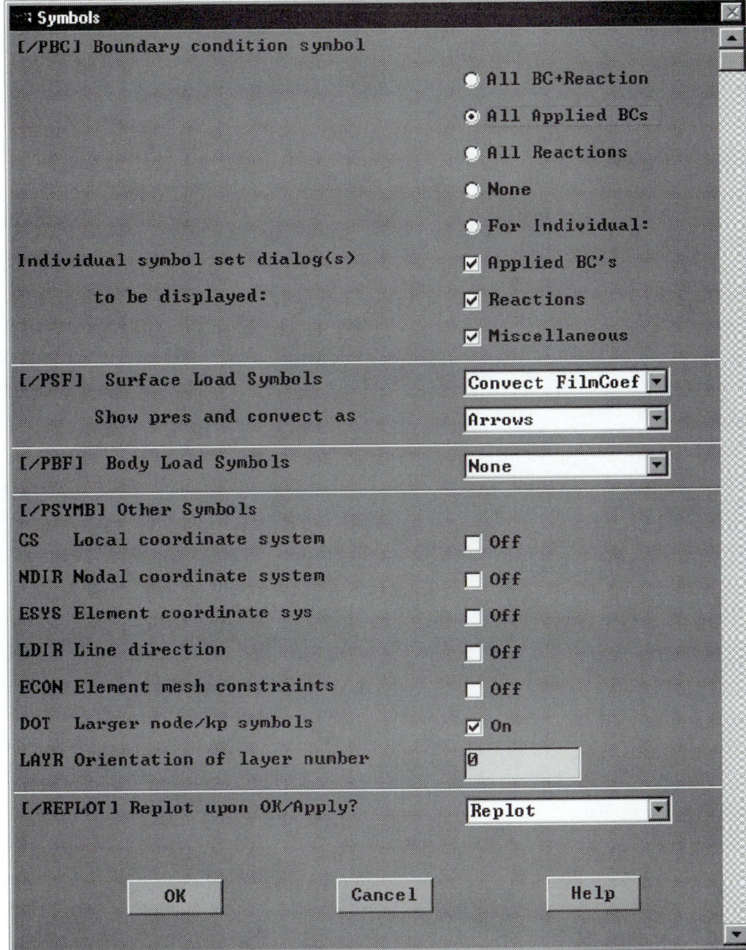

utility menu: **Select** → **Everything**

utility menu: **Plot** → **Areas**

ANSYS Toolbar: **SAVE_DB**

Solve the problem:

main menu: **Solution** → **-Solve-Current LS**

**OK**

**Close** (the solution is done!) window.

**Close** (the /STAT Command) window.

For the postprocessing phase, obtain information such as nodal temperatures and heat fluxes with the following commands (see Figure 10.26 and Figure 10.27):

main menu: **General Postproc** → **Plot Results** → **-Contour Plot-Nodal Solu ...**

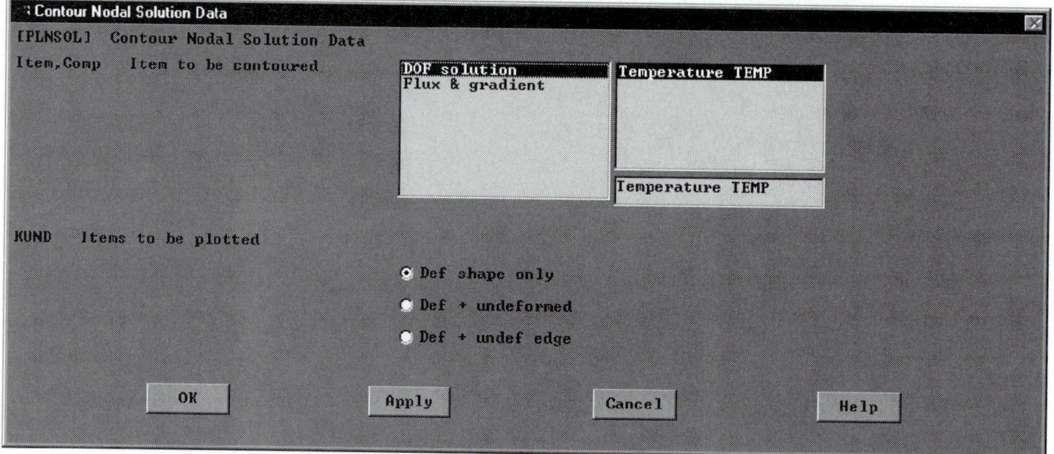

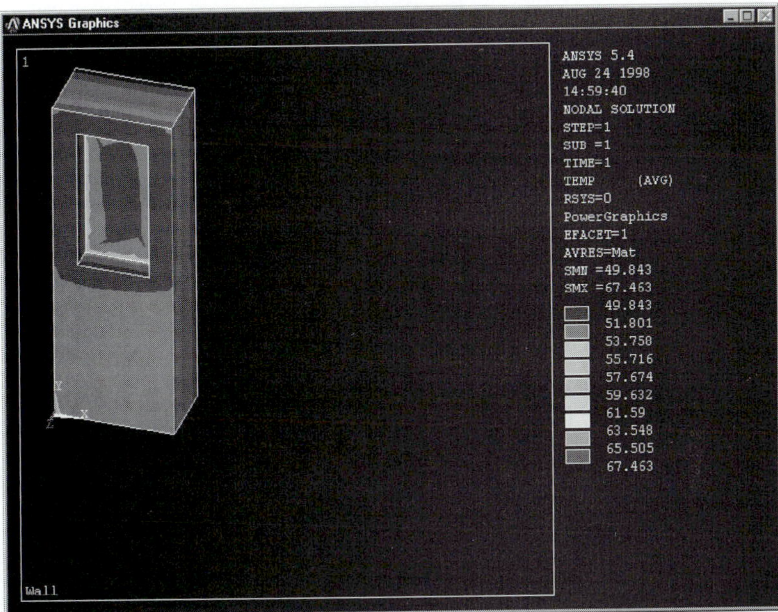


**FIGURE 10.26** Temperature contour plot.

main menu: **General Postproc** → **Plot Results** → **-Vector Plot-Predefined ...**

Vector Plot of Predefined Vectors

[PLVECT] Vector Plot of Predefined Vectors

Item Vector item to be plotted — Flux & gradient — Thermal flux TF / Thermal grad TG — Thermal flux TF

Mode Vector or raster display — Vector Mode / Raster Mode

Loc Vector location for results — Elem Centroid / Elem Nodes

Edge Element edges — Hidden

[/VSCALE] Scaling of Vector Arrows

WN Window Number — Window 1

VRATIO Scale factor multiplier — 1

KEY Vector scaling will be — Magnitude based

OPTION Vector plot based on — Undeformed Mesh

OK Apply Cancel Help

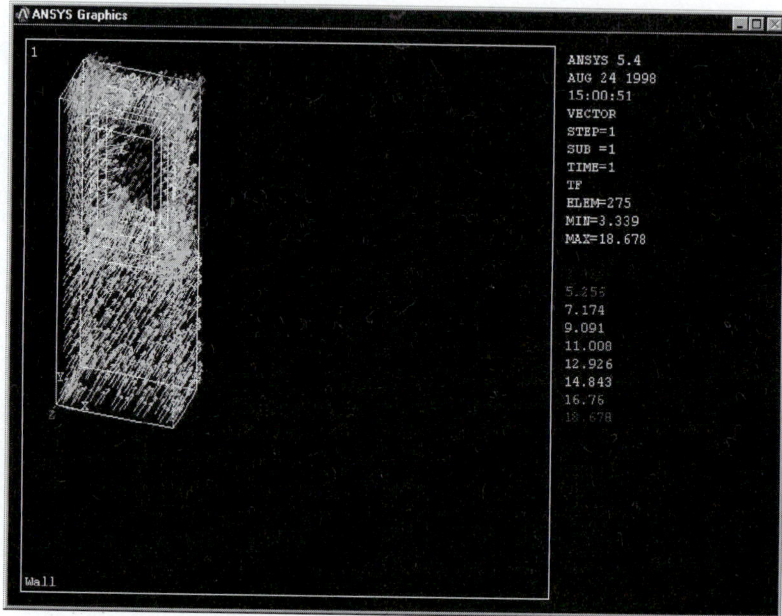


**FIGURE 10.27** The heat flow vectors.

Exit and save your results:

ANSYS Toolbar: **QUIT**

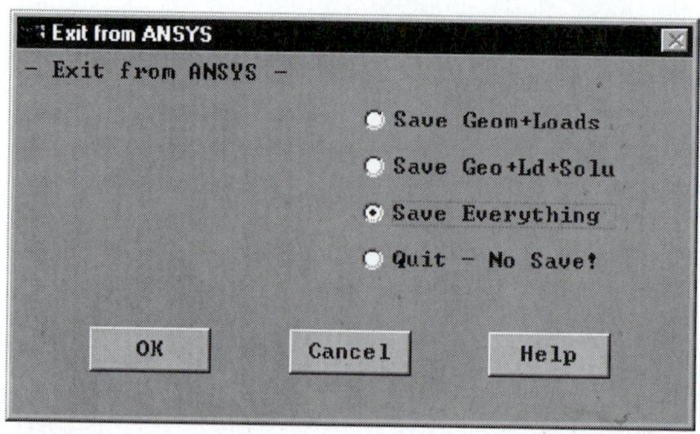

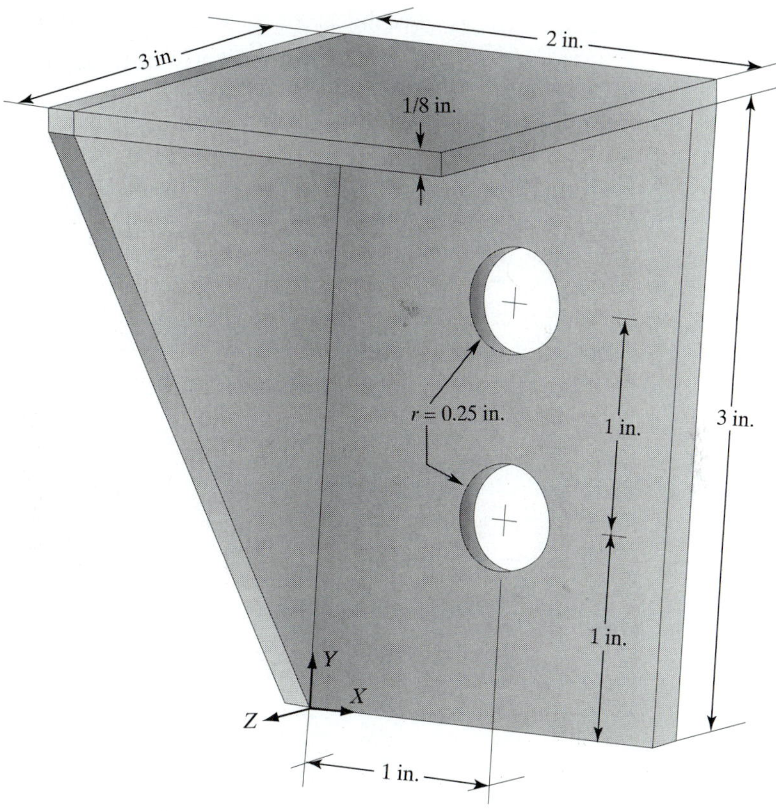


**FIGURE 10.28** Dimensions of the bracket in Example 10.3.

## 10.9 A STRUCTURAL EXAMPLE USING ANSYS

### EXAMPLE 10.3

The bracket shown in Figure 10.28 is subjected to a distributed load of 50 $\text{lb/in}^2$ on the top surface. It is fixed around the hole surfaces. The bracket is made of steel, with a modulus elasticity of $29 \times 10^6\ \text{lb/in}^2$ and $\nu = 0.3$. Plot the deformed shape. Also, plot the von Mises stress distribution in the bracket.

The following steps demonstrate how to create the geometry of the problem, choose the appropriate element type, apply boundary conditions, and obtain nodal results:

Enter the **ANSYS** program by using the Launcher. Type **xansys54** on the command line, or consult your system administrator for the appropriate command name to launch ANSYS on your computer system.

Pick **Interactive** from the Launcher menu.

Type **Brack3D** (or a file name of your choice) in the **Initial Jobname** entry field of the dialog box.

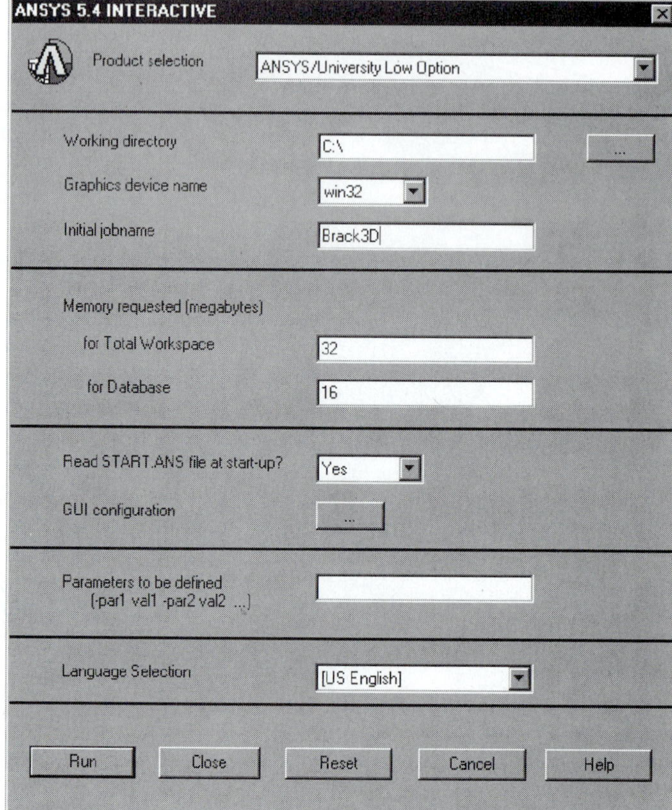


Pick **Run** to start the Graphic User Interface (GUI). A window will open with some disclaimer information. You will eventually be asked to press the **Return** key to start the graphics window and the main menu. Do so in order to proceed.

Create a title for the problem. This title will appear on ANSYS display windows to provide a simple way of identifying the displays. So, issue the following commands:

utility menu: **File** → **Change Title ...**

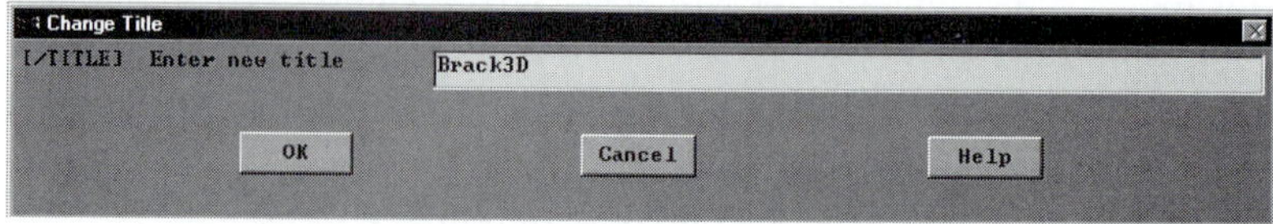


main menu: **Preprocessor** → **Element Type** → **Add/Edit/Delete ...**

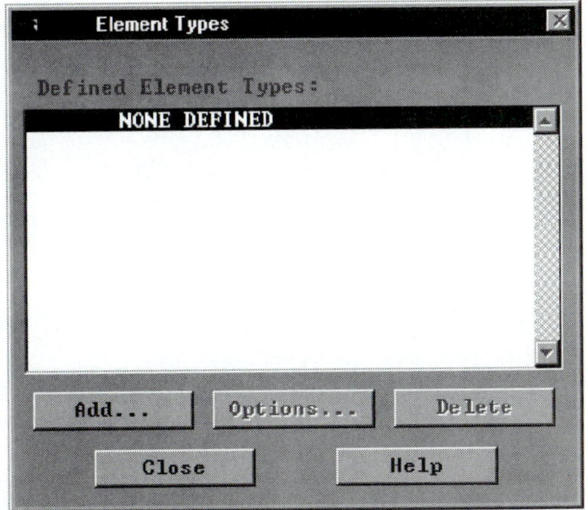


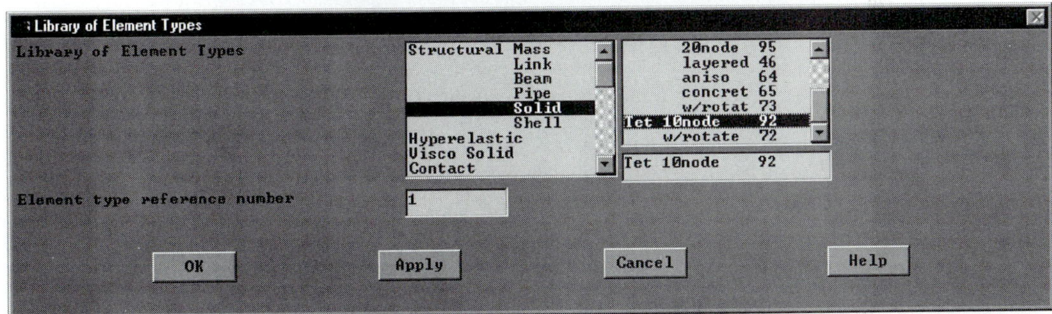


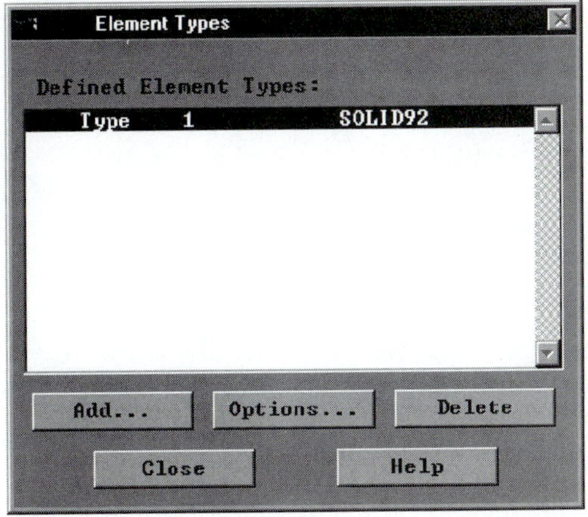


Assign the modulus of elasticity and Poisson's ratio with the following commands:

main menu: **Preprocessor** → **Material Props** → **-Constant-Isotropic ...**

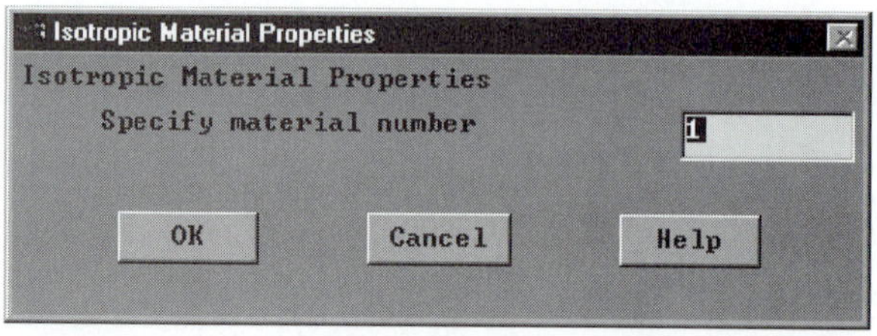

Isotropic Material Properties

Isotropic Material Properties

Specify material number 1

OK Cancel Help

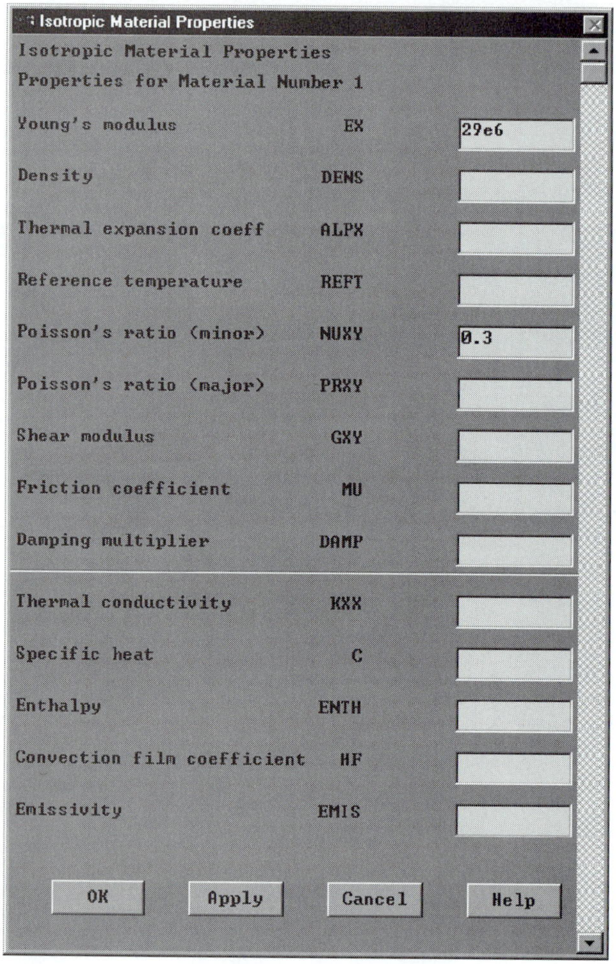

Isotropic Material Properties

Isotropic Material Properties

Properties for Material Number 1

| | | |
|---|---|---|
| Young's modulus | EX | 29e6 |
| Density | DENS | |
| Thermal expansion coeff | ALPX | |
| Reference temperature | REFT | |
| Poisson's ratio (minor) | NUXY | 0.3 |
| Poisson's ratio (major) | PRXY | |
| Shear modulus | GXY | |
| Friction coefficient | MU | |
| Damping multiplier | DAMP | |
| Thermal conductivity | KXX | |
| Specific heat | C | |
| Enthalpy | ENTH | |
| Convection film coefficient | HF | |
| Emissivity | EMIS | |

OK Apply Cancel Help

ANSYS Toolbar: **SAVE_DB**

Set up the graphics area (i.e. workplane, zoom, etc.) with the following commands:

utility menu: **Workplane** → **WP Settings** …

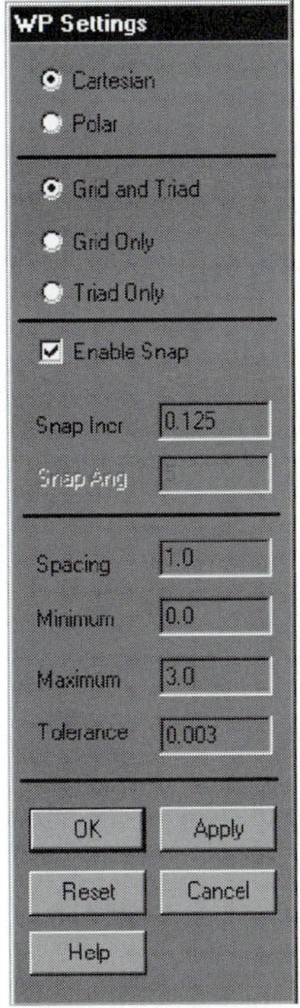


Toggle on the workplane by using the following command:

utility menu: **Workplane** → **Display Working Plane**

Bring the workplane to view by using the following command:

utility menu: **PlotCtrls** → **Pan, Zoom, Rotate** ...

Click on the small circle until you bring the workplane to view. Then, press the **Iso** (Isometric view) button. Next, create the vertical plate by issuing the following commands:

main menu: **Preprocessor** → **-Modeling-Create** → **-Volumes-Block** → **By 2 Corners & Z** +

**[WP = 0,0]**

**[Expand the rubber band up 3.0 and to the right 2.0]**

**[Expand the rubber band in the negative *Z*-direction by 0.125]**

**OK**

To create the holes, first we must create two cylinders, with the following commands:

main menu: **Preprocessor** → **-Modeling-Create** → **-Volumes-Cylinder** → **Solid Cylinder +**

On the workplane, pick the following locations:

**[WP = 1,1]**

**[Expand the circle to rad = 0.25]**

**[Expand the cylinder to a length of 0.125 in the negative *Z*-direction]**

**[WP = 1,2]**

**[Expand the circle to rad = 0.25]**

**[Expand the cylinder to a length of 0.125 in the negative *Z*-direction]**

**OK**

Now, create the holes by subtracting from the vertical plate the volume of cylinders, with the following commands:

main menu: **Preprocessor** → **-Modeling-Operate** → **-Booleans-Subtract** → **Volumes +**

Pick Volume-1 (the vertical plate) and **Apply**; then, pick Volume-2 and Volume-3 (the cylinders) and **Apply.**

**OK**

utility menu: **Plot** → **Volumes**

ANSYS Toolbar: **SAVE_DB**

Move and rotate the workplane and create the top plate with the following command:

utility menu: **Workplane** → **Offset WP by Increments** ...

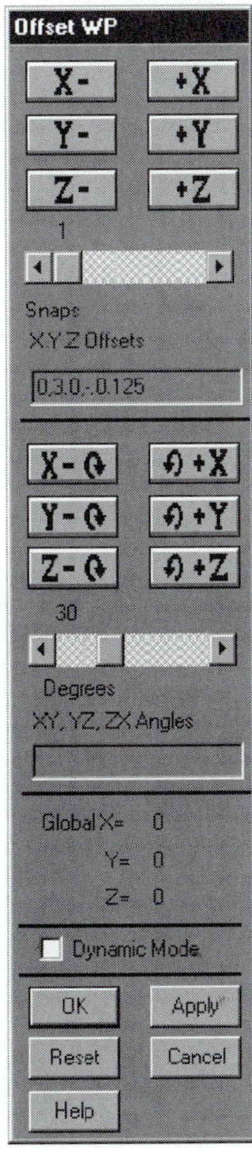


In the X, Y, Z Offsets box, type in **[0, 3.0, −0.125]**, and then **Apply.** To rotate the WP, move the Degrees Slider bar to **90** and then press the **+X rotation** button.

**OK**

utility menu: **PlotCtrls** → **Pan, Zoom, Rotate** ...

Press the **Bot** (bottom view) button and issue the following commands:

main menu: **Preprocessor** → **-Modeling-Create** → **-Volumes-Block** → **By 2 Corners & Z+**

**[WP = 0,0]**

**[In the active workplane, expand the rubber band to 3.0 and 2.0]**

**[Expand the rubber band in the negative Z-direction by 0.125]**

**OK**

utility menu: **WorkPlane** → **Align WP With** → **Global Cartesian**

utility menu: **Plot** → **Volumes**

utility menu: **WorkPlane** → **Offset WP by Increments** ...

In the X, Y, Z Offsets box, type in **[0, 0, −0.125]**, then **Apply.** Rotate the workplane about the Y-axis. Move the Degrees Slider bar to **90** and then press the **−Y rotation** button.

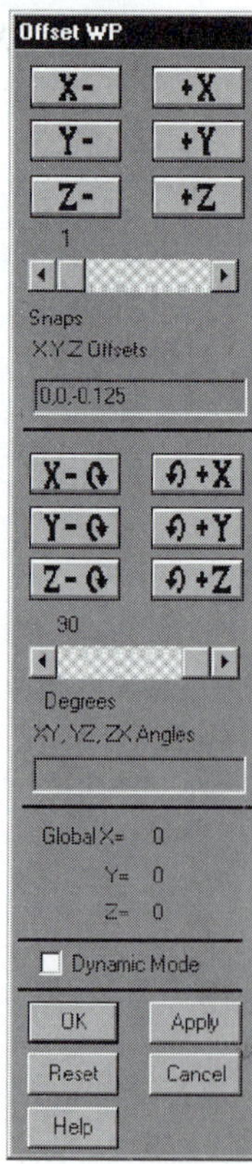


**OK**

utility menu: **PlotCtrls → Pan, Zoom, Rotate** ...

Change the view to **Left** and issue the following commands:

main menu: **Preprocessor → -Modeling-Create → -Volumes-Prism → By Vertices +**

**[WP = 0,0]**

**[WP = 0, 3.125]**

**[WP = 3, 3.125]**

**[WP = 3.0, 3.0]**

**[WP = 0.125, 0]**

**[WP = 0,0]**

Change the view to the isometric view by pressing the **Iso** button:

**[Stretch the rubber band 0.125 in the *Z*-direction.]**

**OK**

utility menu: **Plot → Volumes**

utility menu: **PlotCtrls → Pan, Zoom, Rotate** ...

Toggle on the dynamic mode; hold down the right button on the mouse and rotate the object as desired. Then, issue the following commands:

main menu: **Preprocessor → -Modeling-Operate → -Booleans-Add → Volumes +**

**Pick All**

We now want to mesh the volumes to create elements and nodes, but first, we need to specify the element sizes. So, issue the following commands:

main menu: **Preprocessor → -Meshing-Size Cntrls → -Smart Size-Basic** ...

Basic SmartSize Settings
[SMRTSIZE] Smartsizing
10 (coarse) ... 1 (fine)
LVL Size Level 6 (default)
OK Apply Cancel Help

ANSYS Toolbar: **SAVE_DB**

main menu: **Preprocessor** → **-Meshing-Mesh** → **-Volumes-Free** +

**Pick All**

**Close**

ANSYS Toolbar: **SAVE_DB**

Now, we need to apply boundary conditions. First, we will fix the periphery of the holes by using the following command:

utility menu: **PlotCtrls** → **Pan, Zoom, Rotate** ...

Choose the **Front** view and issue the following commands:

main menu: **Solution** → **-Loads-Apply** → **-Structural-Displacement** → **On Keypoints** +

Change the picking mode to "circle" by toggling on the • **Circle** feature. Now, starting at the center of the holes, stretch the rubber band until you are just outside the holes and apply:

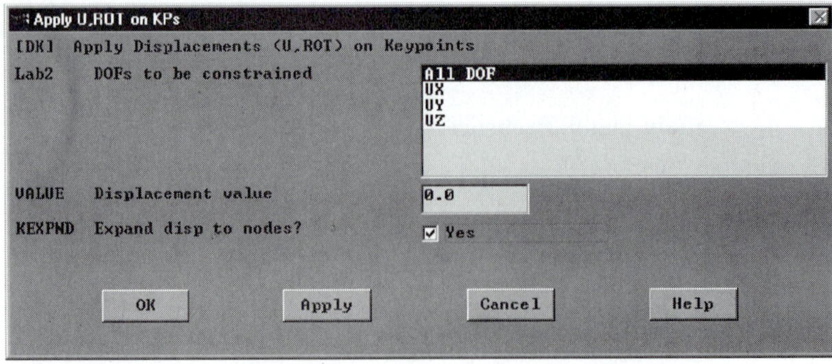

Choose the isometric view and issue the following commands:

utility menu: **Select** → **Entities** ...

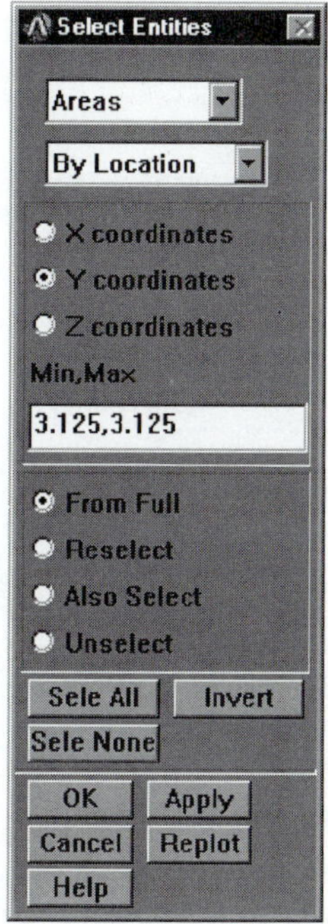


In the **Min, Max** field, type **[3.125, 3.125]**.

**OK**

utility menu: **Plot** → **Areas**

main menu: **Solution** → **-Loads-Apply** → **-Structural-Pressure** → **On Areas** +

**Pick All** to specify the distributed load (pressure) value:

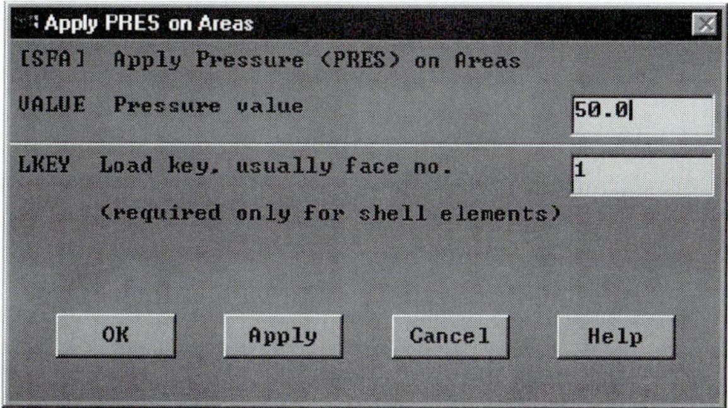

To see the applied boundary conditions, use the following commands:

utility menu: **PlotCtrls** → **Symbols** ...

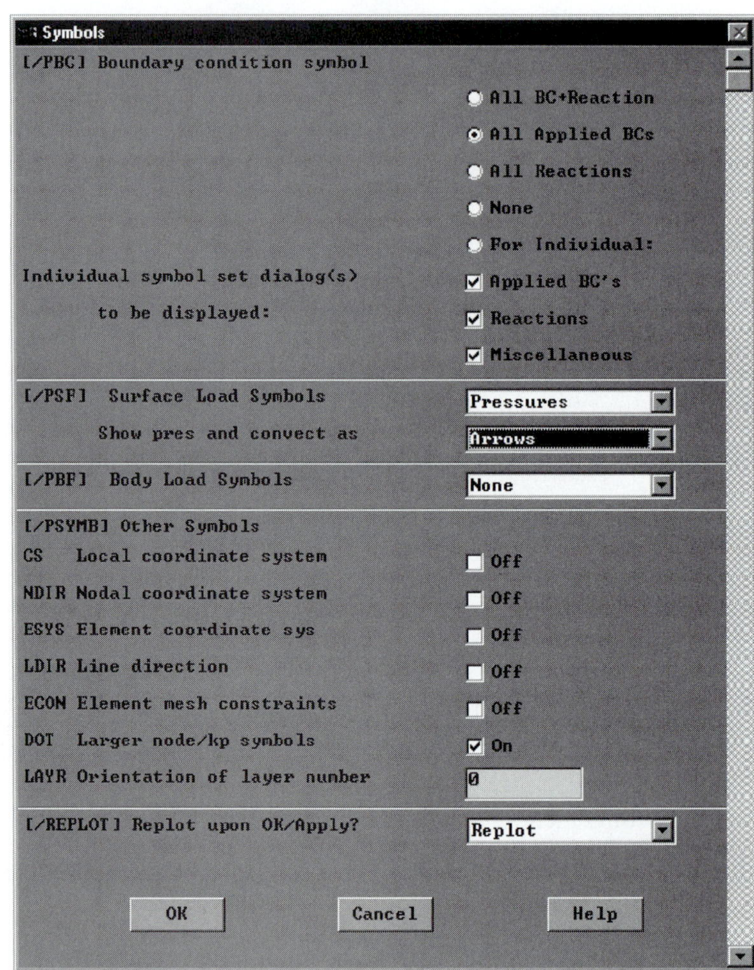


utility menu: **Select** → **Everything**

utility menu: **Plot** → **Areas**

ANSYS Toolbar: **SAVE_DB**

Solve the problem:

main menu: **Solution** → **-Solve-Current LS**

**OK**

**Close** (the solution is done!) window.

**Close** (the /STAT Command) window.

In the postprocessing phase, first plot the deformed shape by using the following commands (see Figure 10.29):

main menu: **General Postproc** → **Plot Results** → **Deformed Shape ...**

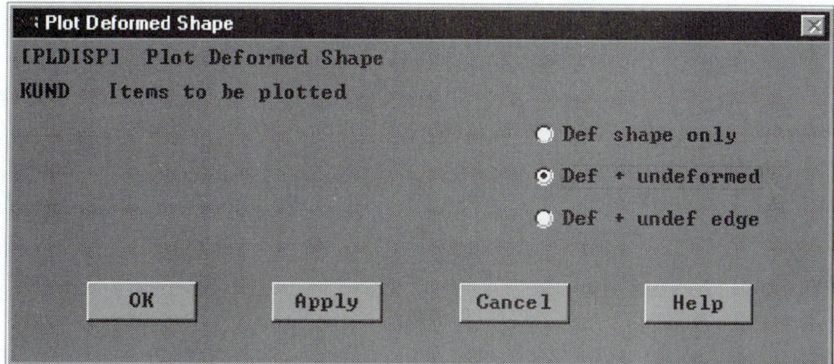


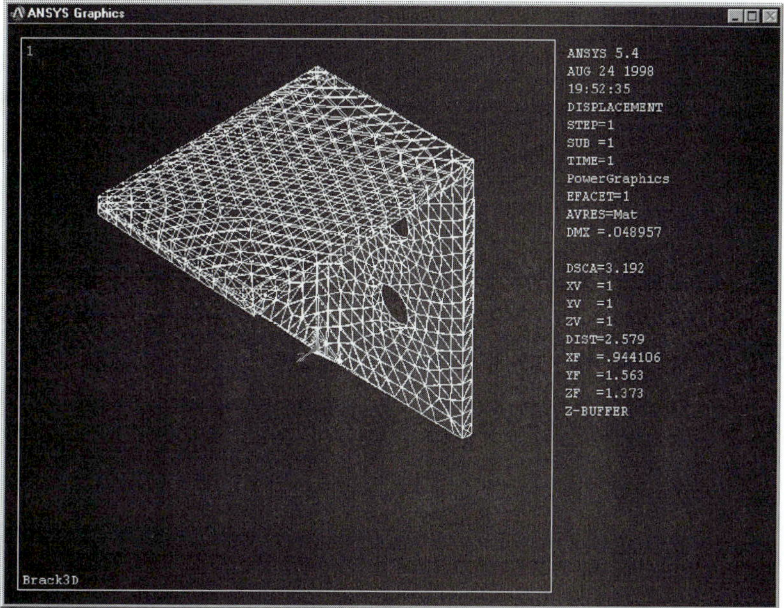


**FIGURE 10.29** The deformed shape of the bracket.

Plot the von Mises stresses by using the following commands (see Figure 10.30):

main menu: **General Postproc** → **Plot Results** → **-Contour Plot-Nodal Solu** ...

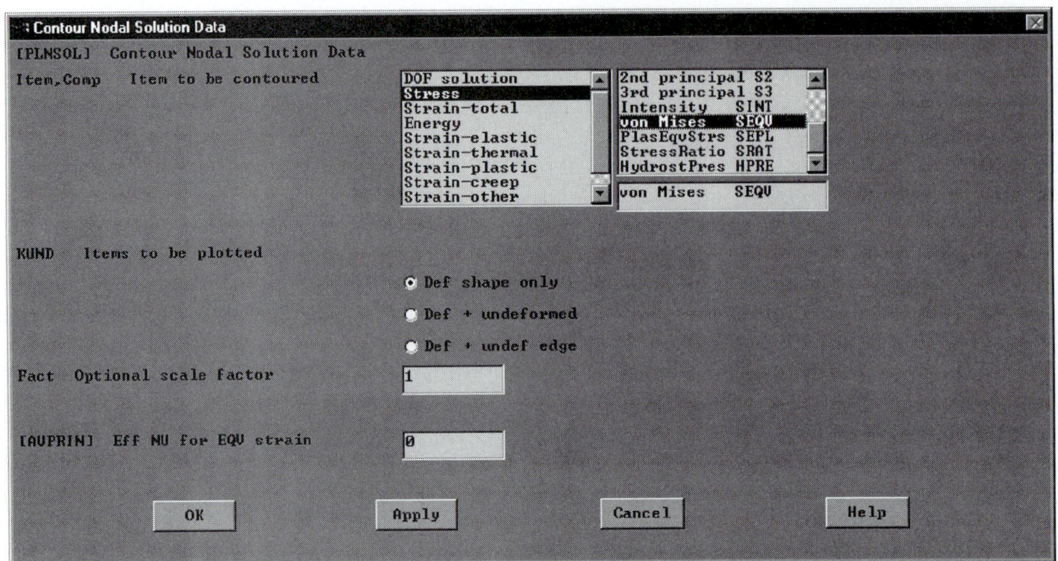


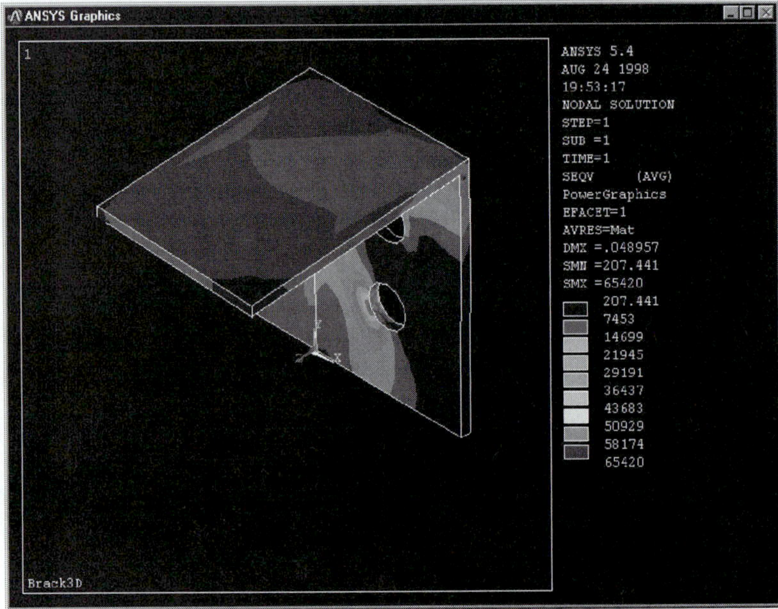


**FIGURE 10.30** The von Mises stress distribution within the bracket.

Exit and save your results:

ANSYS Toolbar: **QUIT**

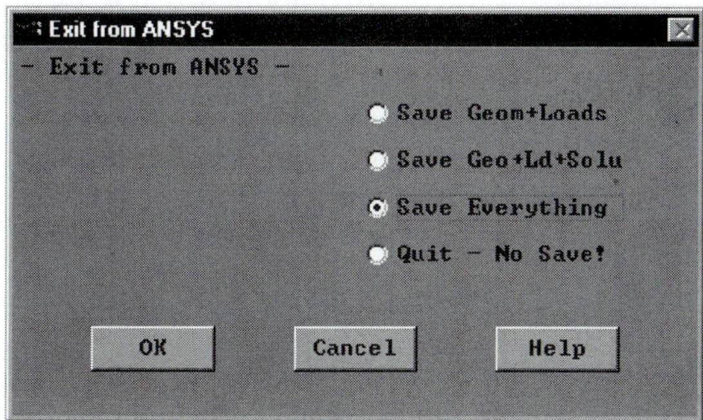


## 10.10 VERIFICATION OF RESULTS: ERROR-ESTIMATION PROCEDURES

Up to this point, we have discussed how to use fundamental principles, such as statics equilibrium conditions or the conservation of energy, to check for the validity of results. We have also noted that when economically feasible or practical, the experimental verification of a finite element model is the best way to check for the validity of results. Moreover, it has been pointed out that the element size affects the accuracy of your results. Now, consider how you know whether the element sizes associated with a meshed model are fine enough to produce good results. A simple way to find out is to first model a problem with a certain number of elements and then compare its results to the results of a model that you create with twice as many elements. In other words, double the number of original elements and compare the results of the analysis. If no significant difference between the results of the two meshes is detected, then the meshing is adequate. If substantially different results are obtained, then further mesh refinement might be necessary.

The ANSYS program offers error-estimation procedures that calculate the level of solution error due to mesh discretization. Error calculations used by ANSYS are based on discontinuity of stresses (or heat fluxes) along interelemental boundaries. Because neighboring elements share common nodes, the difference between the nodal stresses calculated for each element results in a discontinuous stress solution from element to element. The degree of discontinuity is based on both the mesh discretization and the severity of the stress gradient. Therefore, it is the difference in stresses from element to element that forms the basis for error calculations in ANSYS.

Error calculations in ANSYS are presented in three different forms: (1) the elemental-energy error (SERR for structural problems and TERR for thermal problems), which measures the error in each element based on the differences between averaged and unaveraged nodal stress or thermal flux values; (2) the percent error in energy norm (SEPC for structural problems and TEPC for thermal problems), which is a global measure of error energy in the model that is based on the sum of the elemental-error energies; and (3) the nodal-component value deviation (SDSG for structural problems and TDSG for thermal problems), which measures the local error quantity for each element and is determined by computing the difference between the averaged and unaveraged values of stress or heat flux components for an element. To display error distributions, use the following commands:

main menu: **General Postproc → Plot Results → Element Solu**

You can select and plot the elemental-energy error to observe the high-error regions where mesh refinement may be necessary. You can also plot SDSG (or TDSG) to identify and quantify the region of maximum discretization errors by using the following command:

main menu: **General Postproc → Element Table → Define Table**

The elemental-energy error or the nodal-component deviations can be listed as well by using the following command:

main menu: **General Postproc → Element Table → List Element Table**

Examples of ANSYS elements that include error estimations are given in Table 10.1. Note that ANSYS stress-contour plots and listings give the upper and the lower error bounds based on SDSG or TDSG calculations. The estimated-error bound of plotted stresses is denoted by SMXB or SMNB labels in the graphics-status area.

To make the task of mesh evaluation and refinement simpler, ANSYS offers adaptive meshing, which is a process that automatically evaluates mesh-discretization error and performs mesh refinement to reduce the error. The adaptive meshing performed

TABLE 10.1 Examples of ANSYS elements that include error estimations

| Structural Solids | Thermal Solids |
|---|---|
| PLANE2 | PLANE35 |
| PLANE42 | PLANE55 |
| PLANE82 | PLANE77 |
| SOLID45 | SOLID70 |
| SOLID92 | SOLID87 |
| SOLID95 | SOLID90 |

by the ADAPT program of ANSYS will perform the following tasks: (1) it will generate an initial mesh and solve the model; (2) based on error calculations, it will determine if mesh refinement is needed; (3) if mesh refinement is necessary, it will automatically refine the mesh and solve the new model; and (4) it will refine the mesh until a loop limit or an acceptable error limit has been reached. To start the adaptive-meshing program, issue the following command:

main menu: **Solution** → **-Solve-Adaptive Mesh** ...

Note that to begin the first run of the adaptive-meshing program, you need to create the initial model by defining the element type, material property, and so on.

## SUMMARY

At this point you should:

1. know how the shape functions for a tetrahedral element are derived.
2. know how the stiffness matrix and load matrix for a tetrahedral element are derived.
3. be familiar with the eight-node brick element and its higher order counterpart, the twenty-node brick element.
4. be familiar with some of the structural-solid and thermal elements available through ANSYS.
5. understand the difference between the top-down and bottom-up solid-modeling methods.
6. be able to find ways to verify your FEA results.

## REFERENCES

*ANSYS User's Manual: Procedures*, Vol. I, Swanson Analysis Systems, Inc.

*ANSYS User's Manual: Commands*, Vol. II, Swanson Analysis Systems, Inc.

*ANSYS User's Manual: Elements*, Vol. III, Swanson Analysis Systems, Inc.

Chandrupatla, T., and Belegundu, A., *Introduction to Finite Elements in Engineering*, Englewood Cliffs, N. J., Prentice Hall, 1991.

Zienkiewicz, O. C., *The Finite Element Method*, 3d. ed., New York, McGraw-Hill, 1977.

## PROBLEMS

1. For a tetrahedral element, derive an expression for the stress components in terms of the nodal displacement solutions. How are the three principal stresses computed from the calculated stress component values?

2. Use ANSYS to create the solid model of the object shown in the accompanying figure. Use the dynamic-mode option to view the object from various directions. Plot the solid object in its isometric view.

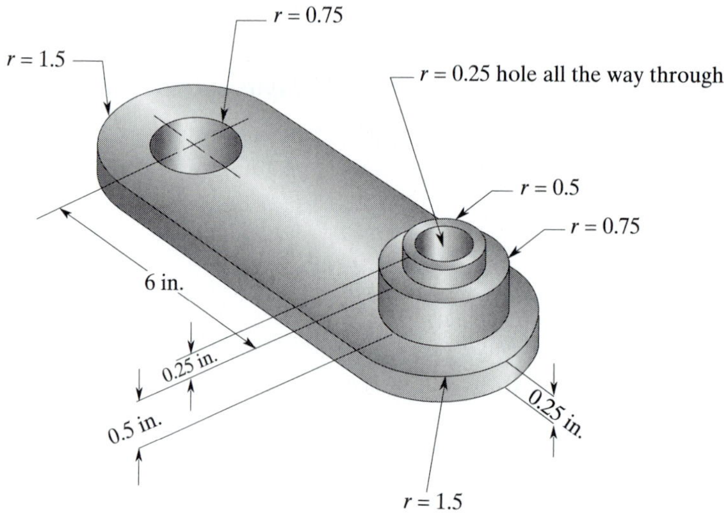


3. Use ANSYS to create a solid model of a foot-long section of a pipe with the internal longitudinal fins shown in the accompanying figure. Use the dynamic-mode option to view the object from various directions. Plot the object in its isometric view.

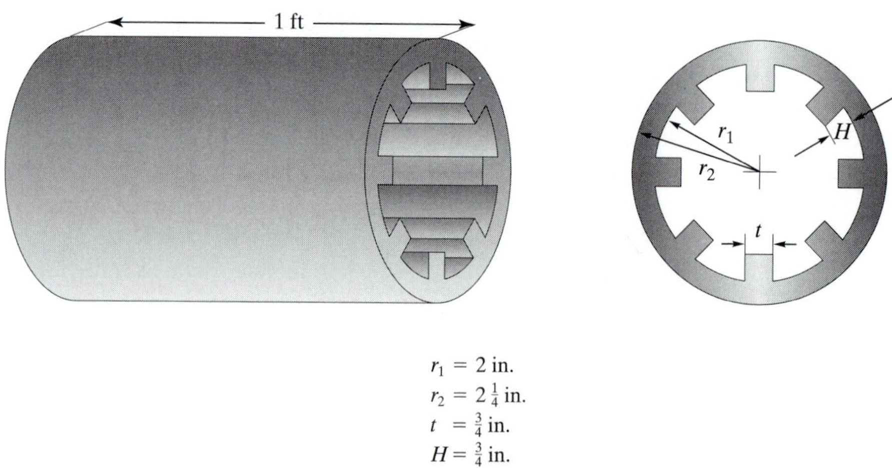


$r_1 = 2$ in.
$r_2 = 2\frac{1}{4}$ in.
$t = \frac{3}{4}$ in.
$H = \frac{3}{4}$ in.

4. Use ANSYS to create the solid model of the wall-mount piping support bracket shown in the accompanying figure. Use the dynamic-mode option to view the object from various directions. Plot the solid object in its isometric view.

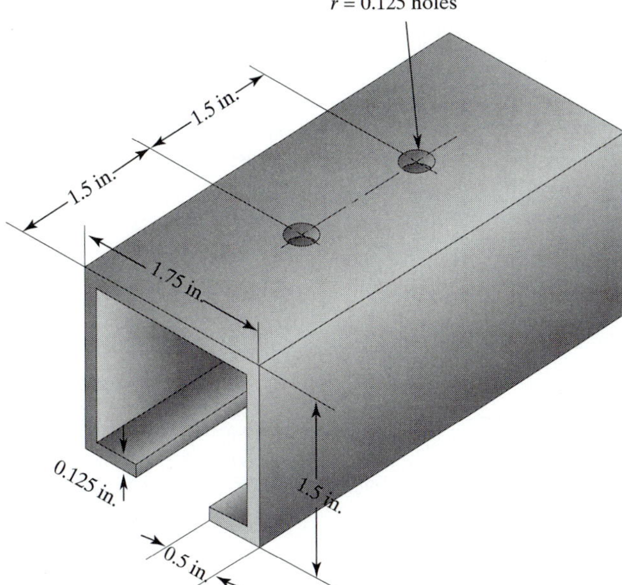


5. Use ANSYS to create the solid model of the heat exchanger shown in the accompanying figure. Use the dynamic-mode option to view the object from various directions. Plot the model of the heat exchanger in its isometric view.

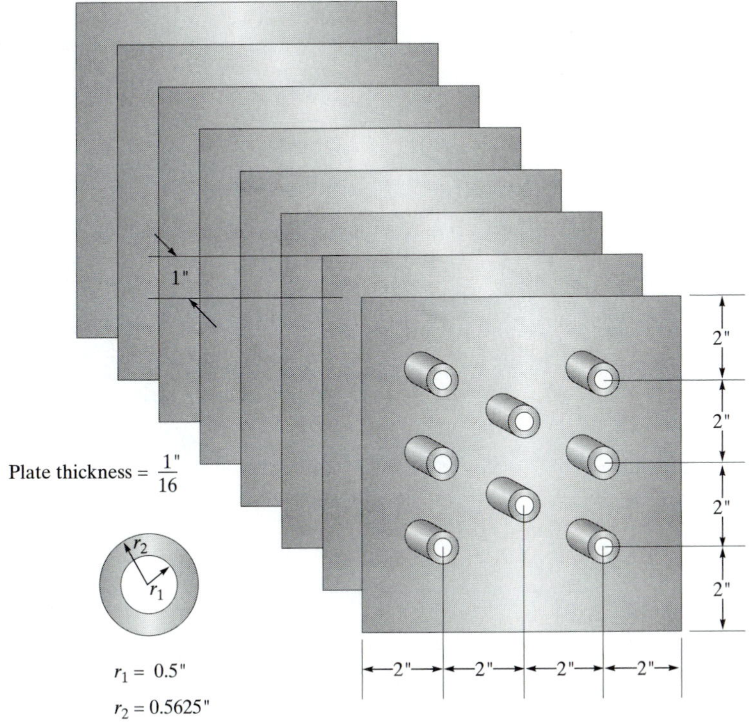

6. Use ANSYS to create the solid model of the wheel shown in the accompanying figure. Use the dynamic-mode option to view the object from various directions. Plot the object in its isometric view.

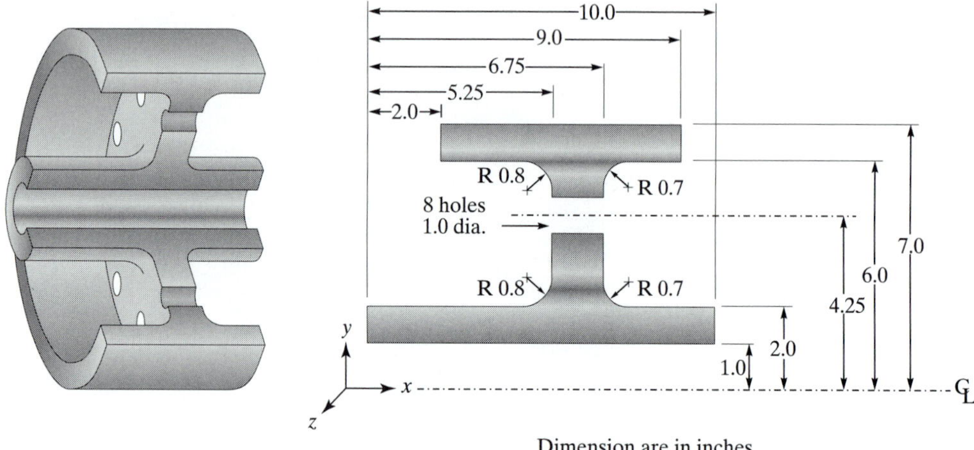


Dimension are in inches.

7. Use ANSYS to create a solid model of a 100-mm-long section of a pipe with the internal longitudinal fins shown in the accompanying figure. Use the dynamic-mode option to view the object from various directions. Plot the object in its isometric view.

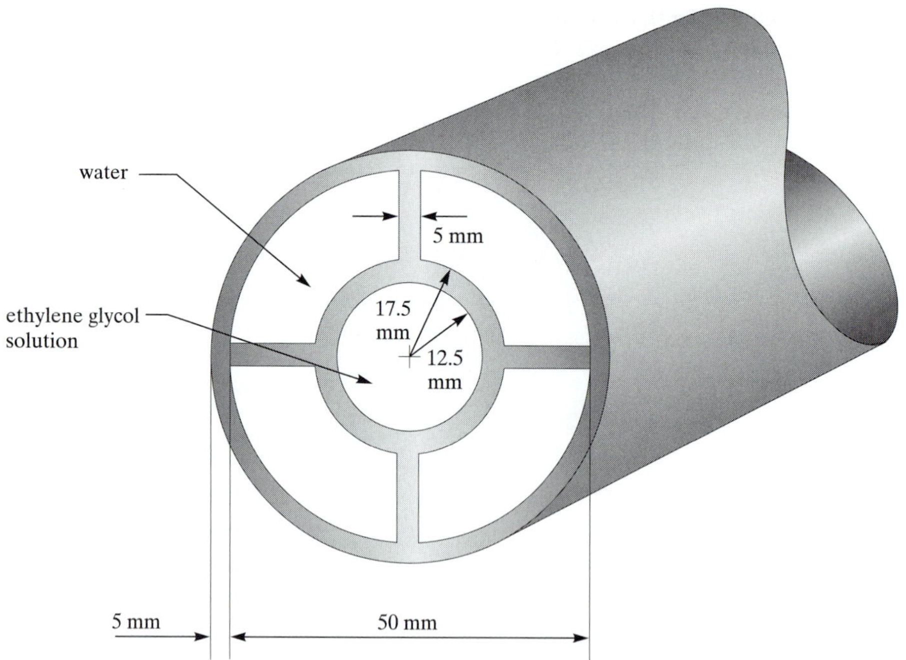


8. Using ANSYS, calculate and plot the principal stress distributions in the support component shown in the accompanying figure. The bracket is made of steel. It is fixed around the hole surfaces.

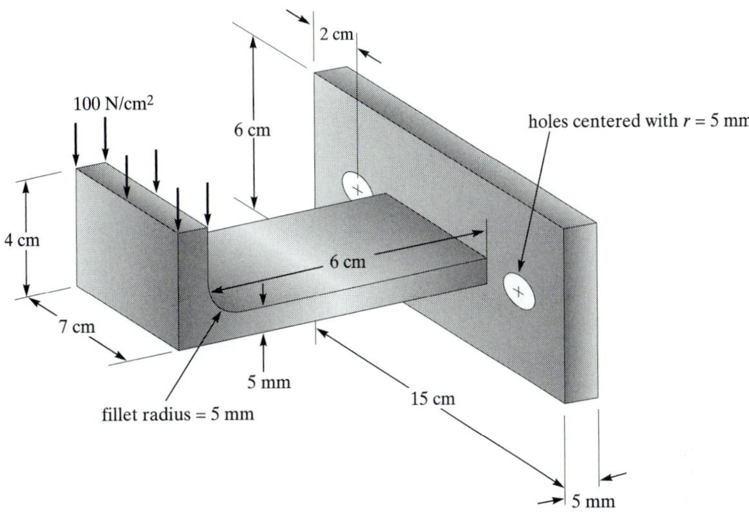


**9.** Using ANSYS, calculate and plot the von Mises stress distribution in the traffic signpost shown in the accompanying figure. The post is made of steel, and the sign is subjected to a wind gust of 60 miles/hr. Use the drag force relation $F_D = C_D A \frac{1}{2} \rho U^2$ to calculate the load caused by the wind, where $F_D$ is the load, $C_D = 1.18$, $\rho$ represents the density of air, $U$ is the wind speed, and $A$ gives the frontal area of the sign. Distribute the load on the section of the post covered by the sign. Could you model this problem as a simple cantilever beam and thus avoid creating an elaborate finite element model? Explain.

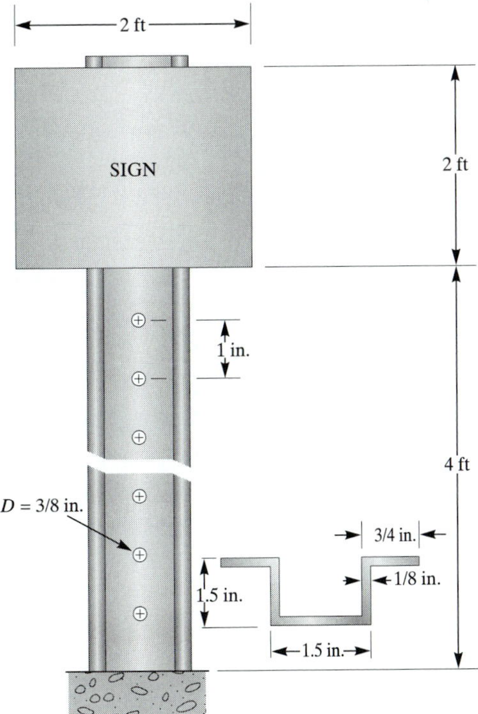

**10.** Determine the temperature distribution inside the aluminum heat sink in Example 10.1 if the surrounding air is at 25°C, with a corresponding heat transfer coefficient $h = 20 \text{ W/m}^2 \cdot \text{K}$. The heat sink sits on a chip that dissipates approximately 200 W. Extrude the frontal area shown in the accompanying figure 20.5 mm to create a quarter model of the heat sink.

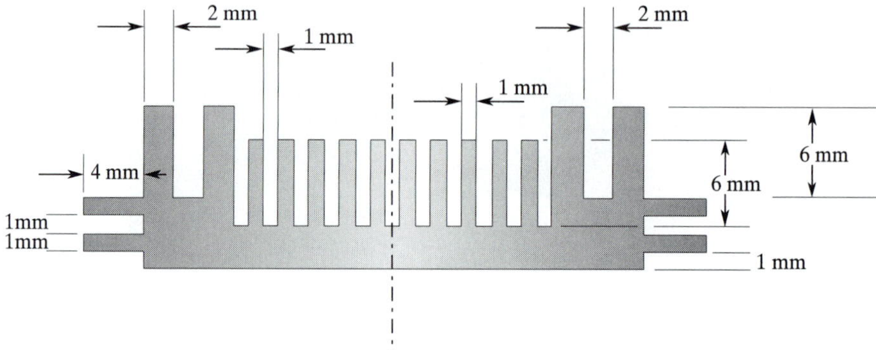


The front view of the heat sink in Problem 10.

**11.** Imagine that by mistake, an empty coffee pot has been left on a heating plate. Assuming that the heater puts approximately 200 Watts into the bottom of the pot, determine the temperature distribution within the glass if the surrounding air is at 25°C, with a corresponding heat transfer coefficient $h = 15 \text{ W/m}^2 \cdot \text{K}$. The pot is cylindrical in shape, with a diameter of 14 cm and height of 14 cm, and the glass is 3 mm thick.

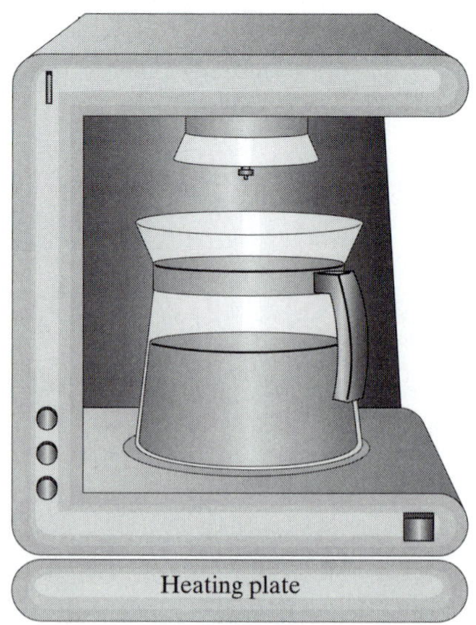

**12.** Using ANSYS, generate a three-dimensional model of a socket wrench. Take measurements from an actual socket. Use solid-cylinder, hexagonal-prism, and block primitives to construct the model. Make reasonable assumptions regarding loading and boundary conditions, and perform stress analysis. Plot the von Mises stresses. Discuss the type and magnitude of loading that could cause failure.

**13.** During the winter months, the inside air temperature of a room is to be kept at 70°F. However, because of the location of a heat register under the window, the temperature distribution of the warm air along the window base is nonuniform. Assume a linear temperature variation from 80°F to 90°F (over a foot long section) with a corresponding heat transfer coefficient $h = 1.46$ Btu/hr • ft$^2$ • °F. Also, assume an outside air temperature of 10°F and a corresponding $h = 6$ Btu/hr • ft$^2$ • °F. Using ANSYS, determine the temperature distribution in the window assembly, as shown in the accompanying figure. What is the overall heat loss through the window assembly?

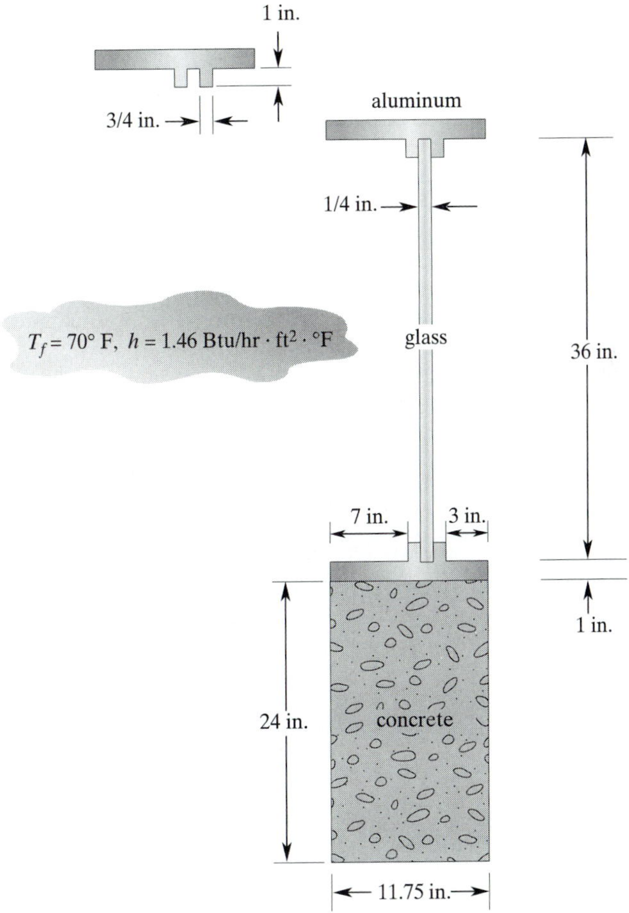


**14.** Using ANSYS, calculate and plot the principal stress distributions in the link component shown in the accompanying figure. The link is made of steel.

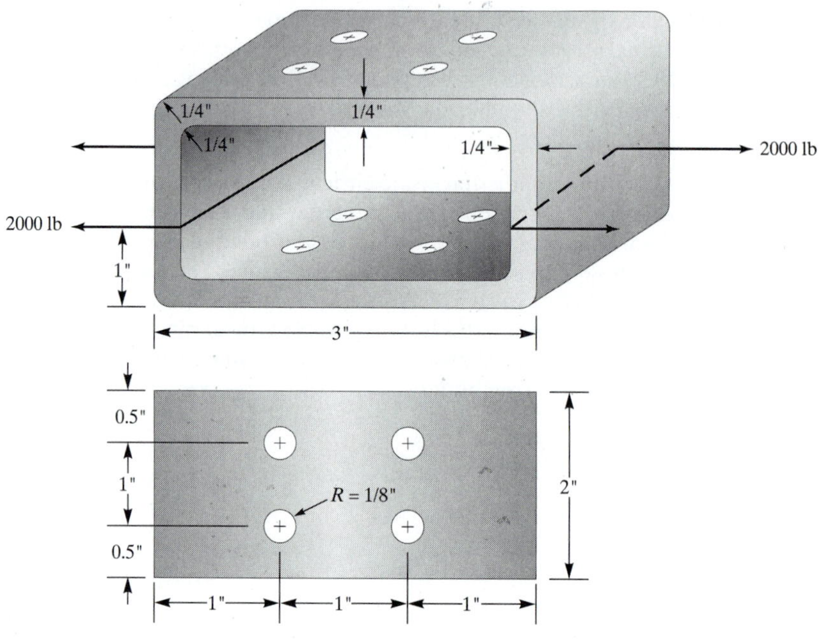


15. **Design Problem** Referring to one of the design problems in Chapter 8 (Problem 22), each student is to design and construct a structural model from a $\frac{3}{8}'' \times 6'' \times 6''$ sheet of plexiglas material that adheres to the specifications and rules given in Problem 22. Additionally, for this project, the model may have any cross-sectional shape. Examples of some common sections are shown in the accompanying figure.

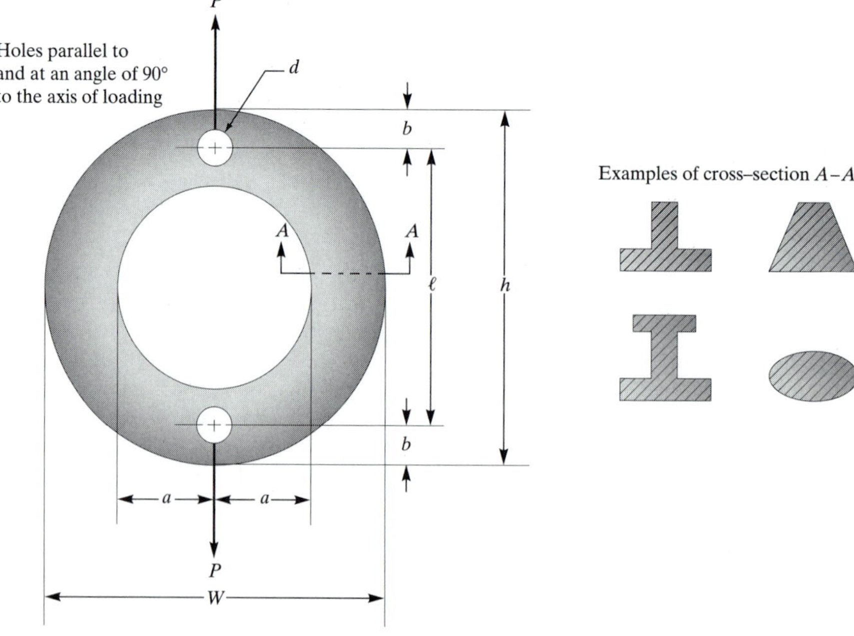

16. **Design Problem** Using a three-dimensional beam element in ANSYS, you are to size the cross sections of members of the frame shown in the accompanying figure. Use hollow tubes. The frame is to support the weight of a traffic light and withstand a wind gust of 80 miles/hr. Write a brief report discussing your final design.

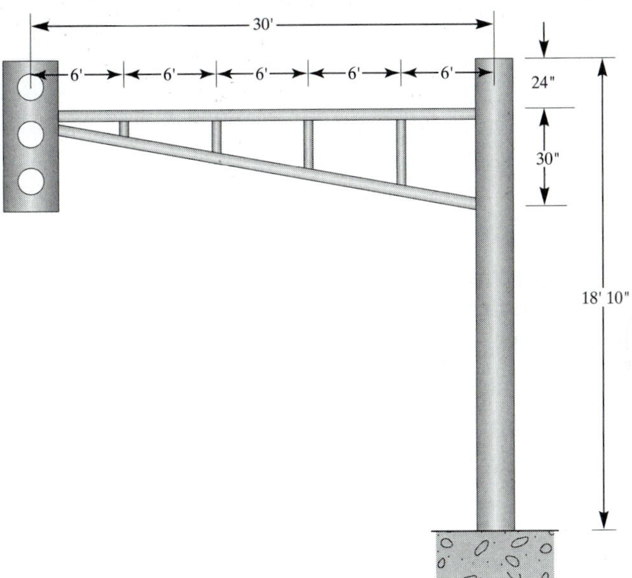

C H A P T E R 1 1

# Design Optimization

The objectives of this chapter are to introduce the basic design optimization ideas and the parametric design language of ANSYS. The main topics discussed in Chapter 11 include the following:

**11.1** Introduction to Design Optimization

**11.2** The Parametric Design Language of ANSYS

**11.3** An Example Using ANSYS

## 11.1 INTRODUCTION TO DESIGN OPTIMIZATION

Optimization means minimization or maximization. There are two broad types of design: a functional design and an optimized design. A functional design is one that meets all of the preestablished design requirements, but allows for improvements to be made in certain areas of the design. To better understand the concept of a functional design, we will consider an example. Let us assume that we are to design a ten-foot-tall ladder to support a person who weighs 300 pounds with a certain factor of safety. We will come up with a design that consists of a steel ladder that is ten feet tall and can safely support the load of 300 lb at each step. The ladder would cost a certain amount of money. This design would satisfy all of the requirements, including those of the strength and the size and, thus, constitutes a functional design. Before we can consider improving our design, we need to ask ourselves what criterion should we use to optimize the design? Design optimization is always based on some criterion such as cost, strength, size, weight, reliability, noise, or performance. If we use the weight as an optimization criterion, then the problem becomes one of minimizing the weight of the ladder. For example, we may consider making the ladder from aluminum. We could also perform stress analysis on the new ladder to see if we could remove material from certain sections of the ladder without compromising the loading and factor of safety requirements.

Another important fact to keep in mind is that while an engineering system consists of various components, optimizing individual components that make up a system does not necessarily lead to an optimized system. For example, consider a thermal-fluid system such as a refrigerator. Optimizing the individual components independently—such as the compressor, the evaporator, or the condenser—with respect to some criterion does not lead to an optimized overall system.

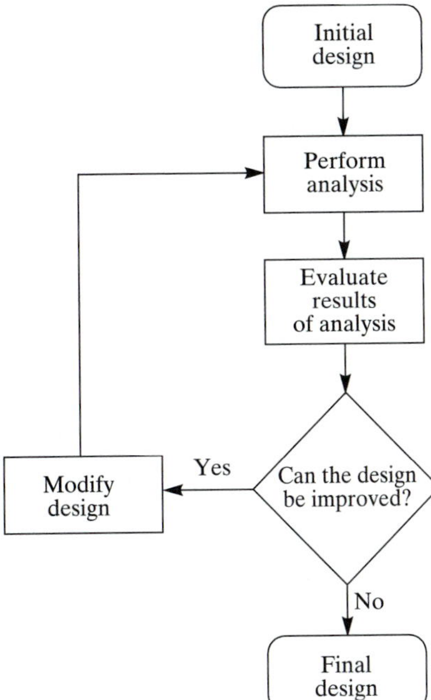


**FIGURE 11.1** An optimization procedure.

This chapter presents some basic ideas in design optimization of a component. We will focus only on weight as an optimization criterion. Traditionally, improvements in a design come from the process of starting with an initial design, performing an analysis, looking at results, and deciding whether or not we can improve the initial design. This procedure is shown in Figure 11.1.

In the past few decades, the optimization process has grown into a discipline that ranges from linear to nonlinear programming techniques. As is the case with any discipline, the optimization field has its own terminology. We will use the next two examples to introduce the fundamental concepts of optimization and its terminology.

---

## EXAMPLE 11.1

Assume that you have been asked to look into purchasing some storage tanks for your company, and for the purchase of these tanks, you are given a budget of $1680. After some research, you find two tank manufacturers that meet your requirements. From Manufacturer A, you can purchase 16-ft$^3$-capacity tanks that cost $120 each. Moreover, this type of tank requires a floor space of 7.5 ft$^2$. Manufacturer B makes 24-ft$^3$-capacity tanks that cost $240 each and that require a floor space of 10 ft$^2$. The tanks will be placed in a section of a lab that has 90 ft$^2$ of floor space available for storage. You are looking for the greatest storage capability within the budgetary and floor-space limitations. How many of each tank must you purchase?

First, we need to define the *objective function*, which is the function that we will attempt to minimize or maximize. In this example, we want to maximize storage capacity. We can represent this requirement mathematically as

$$\text{Maximize } Z = 16x_1 + 24x_2 \tag{11.1}$$

subject to the following constraints:

$$120x_1 + 240x_2 \leq 1680 \tag{11.2}$$

$$7.5x_1 + 10x_2 \leq 90 \tag{11.3}$$

$$x_1 \geq 0 \tag{11.4}$$

$$x_2 \geq 0 \tag{11.5}$$

In Eq. (11.1), $Z$ is the objective function, while the variables $x_1$ and $x_2$ are called *design variables*. The limitations imposed by the inequalities in (11.2)–(11.5) are referred to as a set of constraints. Although there are specific techniques that deal with solving linear programming problems (the objective function and constraints are linear), we will solve this problem graphically to illustrate some additional concepts. The inequalities in (11.2)–(11.5) are plotted in Figure 11.2.

The shaded region shown in Figure 11.2 is called a *feasible solution region*. Every point within this region satisfies the constraints. However, our goal is to maximize the objective function given by Eq. (11.1). Therefore, we need to move the objective function over the feasible region and determine where its value is maximized. It can be

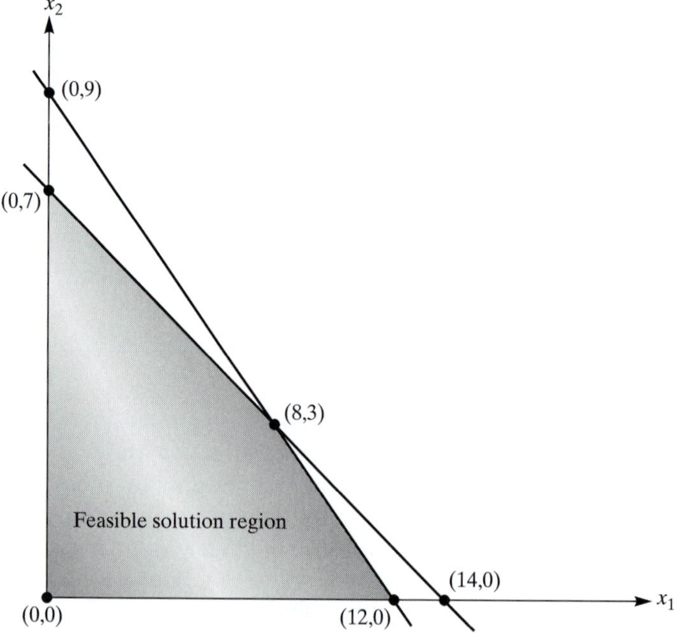


**FIGURE 11.2** The feasible solution region for Example 11.1.

TABLE 11.1 Value of the objective function at the cornerpoints of the feasible region

| Cornerpoints $(x_1, x_2)$ | Value of $Z = 16x_1 + 24x_2$ |
|---|---|
| 0,0 | 0 |
| 0,7 | 168 |
| 12,0 | 192 |
| 8,3 | 200 (max.) |

shown that the maximum value of the objective function will occur at one of the cornerpoints of the feasible region. By evaluating the objective function at the cornerpoints of the feasible region, we see that the maximum value occurs at $x_1 = 8$ and $x_2 = 3$. This evaluation is shown in Table 11.1.

Thus, we should purchase eight of the 16-$\text{ft}^3$ tanks from Manufacturer A and three of the 24-$\text{ft}^3$ tanks from Manufacturer B to maximize the storage capacity within the given constraints.

---

Let us now consider a nonlinear example to demonstrate some additional terms.

## EXAMPLE 11.2

Consider a wooden cantilever beam with rectangular cross section subject to the point loads shown in Figure 11.3. To satisfy safety requirements, the average stress in the beam is not to exceed a value of 30 MPa. Furthermore, the maximum deflection of the beam must be kept under 1 cm. Additional spatial restrictions limit the size of the cross section according to the limits 5 cm $\leq x_1 \leq$ 15 cm and 20 cm $\leq x_2 \leq$ 40 cm. We are interested in sizing the cross section so that it results in a minimum weight of the beam.

This problem is modeled by the objective function:

$$\text{Minimize } W = \rho g x_1 x_2 L \tag{11.6}$$

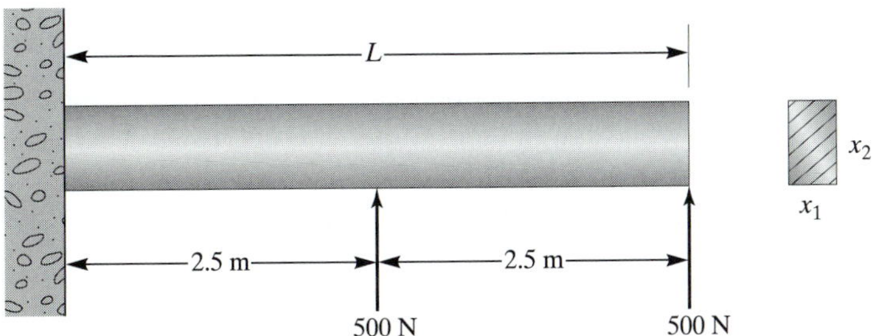


**FIGURE 11.3** A schematic of the beam in Example 11.2.

Assuming constant material density, the problem then becomes one of minimizing the volume:

$$\text{Minimize } V = x_1 x_2 L \tag{11.7}$$

The constraints for this problem are:

$$\sigma_{max} \leq 30 \text{ MPa} \tag{11.8}$$

$$\delta_{max} \leq 1 \text{ cm} \tag{11.9}$$

$$5 \text{ cm} \leq x_1 \leq 15 \text{ cm} \tag{11.10}$$

$$20 \text{ cm} \leq x_2 \leq 40 \text{ cm} \tag{11.11}$$

The variables $x_1$ and $x_2$ are called *design variables*; the $\sigma$-variable for stress and the $\delta$-variable for deflection are called *state variables.* We will solve this problem using ANSYS, but first, let us look at the parametric design language and optimization routines of ANSYS.

## 11.2 THE PARAMETRIC DESIGN LANGUAGE OF ANSYS*

You can define your own variables or choose one of the ANSYS-supplied parameters. User-named parameters, however, must adhere to the following rules: (1) User-named parameters must consist of one to eight characters and must begin with a letter; (2) a parameter may be assigned a numeric value, a character value, or another parameter, as long as the value of the other parameter is currently known to ANSYS; and (3) parameters can be of a scalar type or represent an array of values. Scalar parameters may be defined by using the following command:

utility menu: **Parameters → Scalar Parameters**

To use a parameter, input the parameter's name in the field where ANSYS expects a value. For example, to assign a modulus of elasticity value of $29 \times 10^6$ lb/in$^2$ to a machine part made of steel, you can define a parameter with the name STEEL and assign a value of 29e6 to it.

ANSYS allows the user to define up to 400 parameters. You can define character parameters by placing the characters in single quotes. For example, if you want to define a parameter by the name of Element and assign the characters PLANE42 to it, you can do so by typing: Element = 'PLANE42'. You can obtain predefined parameters by using the following commands:

utility menu: **Parameters → Get Scalar Data → Parameters**

You can also use thousands of ANSYS-supplied values as parameters. For example, you can retrieve nodal coordinates, node numbers, nodal displacements, nodal stresses, an element volume, etc. and assign them to parameters. You can access the ANSYS-supplied parameters by using the command

utility menu: **Parameters → Get Scalar Data**

or by using the command

utility menu: **Parameters → Get Vector Data**

*Materials were adapted with permission from ANSYS documents.

You can list the parameters that have been defined by using the following command:

utility menu: **List** → **Status** → **Parameters** → **Named Parameters**

You can use already-defined parameters to form an expression—for example: Area = Length* Width. When using parametric expressions in a command field, use parentheses to force operations to occur in the desired order. ANSYS also offers built-in functions that are a set of mathematical operations that return a single value. Examples include: SIN, COS, LOG, EXP, SQRT, ABS. To make use of these functions, use the following command:

utility menu: **Parameters** → **Array Operations** → **Vector Functions**

Once you define the model in terms of design parameters, then you can run ANSYS's design-optimization routines interactively with the Graphical User Interface or by using a batch file. The batch mode is generally preferable because it offers a much quicker way to perform analyses. Up to this point, we have been running ANSYS interactively using the GUI. Using a text editor, you can also create an ANSYS batch file with all of the necessary commands to generate a model. The batch file is then submitted to ANSYS as a batch job. The usual procedure for design optimization consists of the following eight main steps:

1. *Create an analysis file to be used during looping.* You begin by initializing the design variables, building the model parametrically, and obtaining a solution. You then need to retrieve and assign to parameters the values that will be used as state variables and objective functions.
2. *Enter OPT and specify the analysis file.* At this point, you are ready to enter the OPT processor to begin optimization.
3. *Declare optimization variables.* Here, you define the objective function and specify which variables are design variables and which are state variables. ANSYS allows you to define only one objective function. You can use up to 60 design variables and up to 100 state variables in your model.
4. *Choose an optimization procedure.* The ANSYS program offers several different optimization procedures. The procedures are divided into *methods* and *tools*. The optimization methods of ANSYS deal with minimizing a single objective function. On the other hand, the optimization tools are techniques to measure and understand the design space of a problem. For a complete list of procedures available with ANSYS, along with the relevant theory behind each procedure, see ANSYS on-line documents. You can also supply your own external procedure to ANSYS to be used during the optimization phase.
6. *Specify optimization looping controls.* Here, you specify the maximum number of iterations to be used with an optimization procedure.
7. *Initiate optimization analysis.*
8. *Review the resulting design sets and postprocess results.*

Throughout the book, up to this point, we have explained how to use ANSYS Interactively. We now introduce the required steps to create a batch file. We will then create a batch file for Example 11.2, and using this problem, we will demonstrate the optimization steps.

### Batch Files

You may recall from studying Chapter 6 that when you first enter ANSYS, you are at the Begin Level. From the Begin level, you can enter one of the ANSYS processors. Commands that give you entry to a processor always start with a slash (/). For example, the **/PREP7** command gives general access to the ANSYS preprocessor. You gain access to the general postprocessor by issuing the command/**POST1**. To move from one processor to another, you must first return to the Begin Level by exiting the processor you are currently in. Only then can you access another processor. To leave a processor and return to the Begin Level, you must issue the **FINISH** command.

The fundamental tool used to enter data and control the ANSYS program is the *command*. Some commands can be used only at certain places in your batch file, while others may be used in other processors. For example, you cannot use the **/PREP7** model-generating commands in other processors. The command format consists of one or more fields separated by commas. The first field always contains the command name. A command argument may be skipped by not specifying any data between the commas. In such cases, ANSYS substitutes the default data for the argument.

For long programs and to keep track of the flow of the batch file, you can document the batch file by placing comments within the file. A comment is indicated by exclamation mark (!), and thus, information beyond the exclamation point is interpreted as comments by ANSYS.

## 11.3 AN EXAMPLE USING ANSYS

We will now solve Example 11.2 using ANSYS. The batch file is as follows:

```
/PREP7              ! Gain access to the preprocessor
!
!      Initialize design variable parameters:
!
X1=2.0              ! Initialize variable X1, the width of the cross section
X2=3.0              ! Initialize variable X2, the height of the cross section
!
!      Define element type, area, area moment of inertia:
!
ET,1,BEAM3          ! Define element type; two-dimensional beam element selected
                    ! (ET=Element Type, element reference number=1)
AREA=X1*X2          ! Define the beam's cross-sectional area
IZZ=(X1*(X2**3))/12       ! Second moment of the area about the Z-axis
!
!      Assign real constants, modulus of elasticity:
!
R,1,AREA,IZZ,X2           ! Assign area, area moment of inertia, cross-
                            sectional height
                          ! (R=Real Constant-designating geometry properties
                            such as area, area moment of inertia, real constant
                            reference number=1)
MP,EX,1,30E6              ! Assign value of the modulus of elasticity
                          ! (MP=Material Property, EX=modulus of elasticity,
                            material reference number=1, value of modulus of
                            elasticity)
```

```
!
!     Create the geometry of the problem:
!
N,1,0,0                  ! Define node 1 at location X=0, Y=0
N,2,2.5,0                ! Define node 2 at location X=2.5, Y=0
N,3,5.0,0                ! Define node 3 at location X=5.0, Y=0
E,1,2               ! Define element 1 as having node 1 and node 2
E,2,3               ! Define element 2 as having node 2 and node 3
FINISH              ! Return to the Begin Level to access other processors
!
!     Apply boundary conditions, apply loading, and obtain solution:
!
/SOLU               ! Enter the Solver processor


ANTYPE,STATIC       ! Analysis type is static
D,1,ALL,0           ! Apply boundary conditions; all displacements at node
                      are zero
F,2,FY,500          ! Apply 500 lb at node 2 in the positive Y-direction
F,3,FY,500          ! Apply 500 lb at node 3 in the positive Y-direction
SOLVE               ! Solve the problem
FINISH              ! Return to the Begin Level to access other processors
!
!     Retrieve results parametrically:
!
/POST1                          ! Enter postprocessing
NSORT,U,Y                       ! Sort nodes based on UY deflection
*GET,DELTAMAX,SORT,,MAX         ! Assign DELTAMAX = maximum deflection
ETABLE,VOLU,VOLU                ! VOLU = volume of each element
ETABLE,SMAX_I,NMISC,1       ! SMAX_I = maximum stress at end I of each
                                      element

ETABLE,SMAX_J,NMISC,3       ! SMAX_J = maximum stress at end J of each
                                      element

SSUM
*GET,VOLUME,SSUM,ITEM,VOLU  ! Parameter VOLUME = total volume
ESORT,ETAB,SMAX_I,,1        ! Sorts elements based on absolute value of
                              SMAX_I

*GET,SMAXI,SORT,,MAX        ! Parameter SMAX_I = maximum value of SMAX_I
ESORT,ETAB,SMAX_J,,1        ! Sorts elements based on absolute value of
                              SMAX_J

*GET,SMAXJ,SORT,,MAX        ! Parameter SMAX_J = maximum value of SMAX_J
SMAX=SMAXI>SMAXJ            ! Parameter SMAX = the greater of SMAXI and
                              SMAXJ

FINISH
!
!     Establish parameters for optimization:
!
/OPT
OPVAR,X1,DV,0.05,0.15             ! Parameter X1 is a design variable
OPVAR,X2,DV,0.2,0.4               ! Parameter X2 is a design variable
OPVAR,DELTAMAX,SV,0,0.01          ! Parameter DELTAMAX is a state variable
OPVAR,SMAX,SV,0,30000000          ! Parameter SMAX is a state variable
```

```
OPVAR,VOLUME,OBJ              ! VOLUME is the objective function
OPTYPE,SUBP                   ! Use subproblem approximation method
OPSUBP,100                    ! Maximum number of iterations
OPEXE                         ! Initiate optimization
OPTYPE,SWEEP                  ! Sweep evaluation tool
OPSWEEP,BEST,5                ! 5 evaluations per DV at best design set
OPEXE                         ! Initiate optimization looping
FINISH
```

The optimization procedure used in the problem we just solved included the subproblem approximation method, which is an advanced zero-order method, and the sweep-generation technique, which varies one design variable at a time over its full range using uniform design variable increments.

An edited version of the output is shown next.

```
***** ANSYS ANALYSIS DEFINITION (PREP7) *****

    PARAMETER X1 = 0.1000000

    PARAMETER X2 = 0.3000000

    ELEMENT TYPE 1 IS BEAM3 2-D ELASTIC BEAM
        KEYOPT(1-12)= 0 0 0 0 0 0 0 0 0 0 0 0

    CURRENT NODAL DOF SET IS UX UY ROTZ
    TWO-DIMENSIONAL MODEL

    PARAMETER AREA = 0.3000000E-01

    PARAMETER IZZ = 0.2250000E-03

    REAL CONSTANT SET 1 ITEMS 1 TO 6
0.30000E-01 0.22500E-03 0.30000 0. 0. 0.

    MATERIAL 1 EX = 0.1300000E+11

    NODE 1 KCS= 0 X,Y,Z= 0. 0. 0.

    NODE 2 KCS= 0 X,Y,Z= 2.5000 0. 0.

    NODE 3 KCS= 0 X,Y,Z= 5.0000 0. 0.

    ELEMENT 1 1 2

    ELEMENT 2 2 3

***** ROUTINE COMPLETED ***** CP = 0.551
```

```
***** ANSYS SOLUTION ROUTINE *****

 PERFORM A STATIC ANALYSIS
 THIS WILL BE A NEW ANALYSIS

 SPECIFIED CONSTRAINT UX FOR SELECTED NODES 1 TO 1
 BY 1
 REAL= 0. IMAG= 0.
 ADDITIONAL DOFS= UY ROTZ

 SPECIFIED NODAL LOAD FY FOR SELECTED NODES 2 TO 2
 BY 1

 REAL= 500.000000 IMAG= 0.

 SPECIFIED NODAL LOAD FY FOR SELECTED NODES 3 TO 3
 BY 1

 REAL= 500.000000 IMAG= 0.

****** ANSYS SOLVE COMMAND *****

    S O L U T I O N   O P T I O N S

 PROBLEM DIMENSIONALITY.........2-D
 DEGREES OF FREEDOM.... UX UY ROTZ
 ANALYSIS TYPE.............STATIC (STEADY-STATE)

    L O A D   S T E P   O P T I O N S

 LOAD STEP NUMBER............. 1
 TIME AT END OF THE LOAD STEP..... 1.0000
 NUMBER OF SUBSTEPS........... 1
 STEP CHANGE BOUNDARY CONDITIONS....... NO
  PRINT OUTPUT CONTROLS..........NO PRINTOUT
 DATABASE OUTPUT CONTROLS..........ALL DATA WRITTEN
                  FOR THE LAST SUBSTEP
```

```
***** ANSYS RESULTS INTERPRETATION (POST1) *****

 SORT ON ITEM=U  COMPONENT=Y  ORDER= 0 KABS= 0 NMAX= 3

 SORT COMPLETED FOR 3 VALUES.

 *GET DELTAMAX FROM SORT ITEM=MAX    VALUE= 0.934829060E-02

 STORE VOLU FROM ITEM=VOLU FOR ALL SELECTED ELEMENTS

 STORE SMAX_I FROM ITEM=NMIS COMP= 1 FOR ALL SELECTED ELEMENTS

 STORE SMAX_J FROM ITEM=NMIS COMP= 3 FOR ALL SELECTED ELEMENTS

 SUM ALL THE ACTIVE ENTRIES IN THE ELEMENT TABLE

 TABLE LABEL TOTAL
 VOLU    0.150000
 SMAX_I   0.333333E+07
 SMAX_J     833333.

 *GET VOLUME FROM SSUM ITEM=ITEM VOLU          VALUE= 0.150000000

 SORT ON ITEM=ETAB COMPONENT=SMAX ORDER= 0 KABS= 1 NMAX= 2

 SORT COMPLETED FOR   2 VALUES.

 *GET SMAXI   FROM SORT ITEM=MAX    VALUE= 2500000.00

 SORT ON ITEM=ETAB COMPONENT=SMAX ORDER= 0 KABS= 1 NMAX= 2

 SORT COMPLETED FOR   2 VALUES.

 *GET SMAXJ  FROM SORT ITEM=MAX     VALUE= 833333.333

 PARAMETER SMAX  =  2500000.

 EXIT THE ANSYS POST1 DATABASE PROCESSOR
```

```
***** ANSYS OPTIMIZATION ANALYSIS (OPT) *****

9 Parameters exist and design set No. 1 is established.

DV NAME= X1       MIN= 0.50000E-01  MAX= 0.15000     TOLER= 0.10000E-02

DV NAME= X2       MIN= 0.20000    MAX= 0.40000      TOLER= 0.20000E-02

SV NAME= DELTAMAX MIN=       0.    MAX= 0.10000E-01   TOLER= 0.10000E-03

SV NAME= SMAX  MIN=  0.       MAX= 0.30000E+08    TOLER= 0.30000E+06

Default OBJ tolerance set to 0.01*(CURRENT PARAMETER VALUE) = 1.5E-03.

OBJ NAME= VOLUME    TOLER= 0.15000E-02

ACTIVE OPTIMIZATION IS THE SUBPROBLEM APPROXIMATION METHOD

SUBPROBLEM APPROXIMATION OPTIMIZATION WILL PERFORM A MAXIMUM OF
100 ITERATIONS
UPON EXECUTION WITH A MAXIMUM OF 7 SEQUENTIAL INFEASIBLE SOLUTIONS

RUN OPTIMIZATION (SUBPROBLEM APPROXIMATION) WITH A MAXIMUM OF 100
ITERATIONS
AND 7 ALLOWED SEQUENTIAL INFEASIBLE SOLUTIONS.

>>> BEGIN SUBPROBLEM APPROXIMATION ITERATION 1 OF 100 (MAX) <<<

>>>>>> SOLUTION HAS CONVERGED TO POSSIBLE OPTIMUM <<<<<<
 (BASED ON OBJ TOLERANCE BETWEEN FINAL TWO DESIGNS)

FINAL VARIABLES ARE

       SET 6
      (FEASIBLE)
DELTAMAX(SV)   0.93351E-02
SMAX (SV) 0.29249E+07
X1  (DV)  0.62267E-01
X2  (DV)  0.35148
VOLUME (OBJ) 0.10943
```

```
****** DESIGN SENSITIVITY SUMMARY TABLE ******

      VOLUME DELTAMAX SMAX
 X1  1.690  -0.1304  -0.4032E+08
 X2 0.4513 -0.7969E-01  -0.1533E+08

ACTIVE OPTIMIZATION TOOL IS SWEEP EVALUATION

SWEEP OPTIMIZATION TOOL WILL PERFORM 5 ITERATIONS PER DESIGN
VARIABLE UPON
EXECUTION ABOUT BEST DESIGN SET BASED ON 2 CURRENT DESIGN VARIABLES

RUN OPTIMIZATION (SWEEP DESIGNS) WITH A MAXIMUM OF 10 ITERATIONS
ABOUT DESIGN SET 6.

>>>BEGIN SWEEP ITERATION 1 OF 10 <<<

>>>>>>SWEEP SELECTION OPTIMIZATION COMPLETED AFTER 10 ITERATIONS<<<<<<
```

```
BEST VARIABLES ARE:

       SET 15
        (FEASIBLE)

DELTAMAX(SV)       0.94544E-02
SMAX   (SV)        0.29498E+07
X1     (DV)        0.62267E-01
X2     (DV)        0.35000
VOLUME (OBJ)       0.10897
```

```
*** EXIT FROM ANSYS DESIGN OPTIMIZATION (/OPT) ***

***** ROUTINE COMPLETED ***** CP = 4.035

***** END OF INPUT ENCOUNTERED *****

NUMBER OF WARNING MESSAGES ENCOUNTERED= 0
NUMBER OF ERROR MESSAGES ENCOUNTERED= 0
```

The above ANSYS output should give you a good idea of the steps that the program follows to move toward an optimized solution.

## SUMMARY

At this point you should:

1. have a good understanding of the fundamental concepts in design optimization, including the definitions of objective function, constraints, state variables, and design variables. You should also know what is meant by a feasible solution region.
2. know how to define and retrieve user-defined and ANSYS-supplied parameters.
3. know the basic steps involved in the optimization process of ANSYS.
4. be familiar with the creation of batch files.

## REFERENCES

*ANSYS User's Manual: Introduction to ANSYS*, Vol. I, Swanson Analysis Systems, Inc.

*ANSYS User's Manual: Procedures*, Vol. I, Swanson Analysis Systems, Inc.

*ANSYS User's Manual: Commands*, Vol. II, Swanson Analysis Systems, Inc.

*ANSYS User's Manual: Elements*, Vol. III, Swanson Analysis Systems, Inc.

Hillier, F. S., and Lieberman, G. J., *Introduction to Operations Research*, 6th ed., New York, McGraw-Hill, 1995.

Rekaitis, G. V., Ravindran A., and Ragsdell, K. M, *Engineering Optimization—Methods and Applications*, New York, John Wiley and Sons, 1983.

APPENDIX A

# Mechanical Properties of Some Materials

Mechanical Properties of Some Materials (SI Units)

| Material | Density ($kg/m^3$) | Modulus of Elasticity (GPa) | Modulus of Rigidity (GPa) | Poisson's Ratio | Yield Strength (Mpa) | Ultimate Strength (MPa) |
|---|---|---|---|---|---|---|
| Aluminum (2014–T6) | 2790 | 73.1 | 27 | 0.35 | 414 | 469 |
| ALUMINUM (6061–T6) | 2710 | 68.9 | 26 | 0.35 | 255 | 290 |
| Cast Iron (gray ASTM 20) | 7190 | 67.0 | 27 | 0.28 | – | 179 (669 Comp.) |
| Concrete (low strength) | 2380 | 22.1 | – | 0.15 | – | – |
| Concrete (high strength) | 2380 | 29 | – | 0.15 | – | – |
| Steel (structural A36) | 7850 | 200 | 75 | 0.32 | 250 | 400 |
| Steel (stainless 304) | 7860 | 193 | 75 | 0.27 | 207 | 517 |
| Steel (tool L2) | 8160 | 200 | 78 | 0.32 | 703 | 800 |
| Wood (Douglas Fir) | 470 | 13.1 | – | 0.29 | – | 2.1 (26 Comp.) |

*Reference:* Hibbeler, R. C., *Mechanics of Materials*, 2d. ed, New York, Macmillan, 1994.

APPENDIX B

# Thermophysical Properties of Some Materials

Thermophysical Properties of Some Materials (at room temperature or at the specified temperature) (SI units)

| Material | Density ($kg/m^3$) | Specific Heat (J/kg · K) | Thermal Conductivity (W/m · K) |
|---|---|---|---|
| Aluminum (alloy 1100) | 2740 | 896 | 221 |
| Asphalt | 2110 | 920 | 0.74 |
| Cement | 1920 | 670 | 0.029 |
| Clay | 1000 | 920 | |
| Fireclay Brick | 1790 @ 373 K | 829 | 1.0 @ 473 K |
| Glass (soda lime) | 2470 | 750 | 1.0 @ 366 K |
| Glass (lead) | 4280 | 490 | 1.4 |
| Glass (pyrex) | 2230 | 840 | 1.0 @ 366 K |
| Iron (cast) | 7210 | 500 | 47.7 @ 327 K |
| Iron (wrought) | 7700 @ 373 K | | 60.4 |
| Paper | 930 | 1300 | 0.13 |
| Steel (mild) | 7830 | 500 | 45.3 |
| Wood (ash) | 690 | | 0.172 @ 323 K |
| Wood (mahogany) | 550 | | 0.13 |
| Wood (oak) | 750 | 2390 | 0.176 |
| Wood (pine) | 430 | | 0.11 |

*Reference:* *ASHRAE Handbook: Fundamental Volume*, American Society of Heating, Refrigerating, and Air-Conditioning Engineers, Atlanta, 1993.

APPENDIX C

# Conversion Factors

| Conversion Factors | | |
|---|---|---|
| Quantity | SI → US Customary | US Customary → SI |
| **Length** | 1 mm = 0.03937 in | 1 in = 25.4 mm |
| | 1 mm = 0.00328 ft | 1 ft = 304.8 mm |
| | 1 cm = 0.39370 in | 1 in = 2.54 cm |
| | 1 cm = 0.0328 ft | 1 ft = 30.48 cm |
| | 1 m = 39.3700 in | 1 in = 0.0254 m |
| | 1 m = 3.28 ft | 1 ft = 0.3048 |
| **Area** | 1 $mm^2$ = 1.55E−3 $in^2$ | 1 $in^2$ = 645.16 $mm^2$ |
| | 1 $mm^2$ = 1.0764E−5 $ft^2$ | 1 $ft^2$ = 92903 $mm^2$ |
| | 1 $cm^2$ = 0.155 $in^2$ | 1 $in^2$ = 6.4516 $cm^2$ |
| | 1 $cm^2$ = 1.07E−3 $ft^2$ | 1 $ft^2$ = 929.03 $cm^2$ |
| | 1 $m^2$ = 1550 $in^2$ | 1 $in^2$ = 6.4516E−4 $m^2$ |
| | 1 $m^2$ = 10.76 $ft^2$ | 1 $ft^2$ = 0.0929 $m^2$ |
| **Volume** | 1 $mm^3$ = 6.1024E−5 $in^3$ | 1 $in^3$ = 16387 $mm^3$ |
| | 1 $mm^3$ = 3.5315E−8 $ft^3$ | 1 $ft^3$ = 28.317E6 $mm^3$ |
| | 1 $cm^3$ = 0.061024 $in^3$ | 1 $in^3$ = 16.387 $cm^3$ |
| | 1 $cm^3$ = 3.5315E−5 $ft^3$ | 1 $ft^3$ = 28317 $cm^3$ |
| | 1 $m^3$ = 61024 $in^3$ | 1 $in^3$ = 1.6387E−5 $m^3$ |
| | 1 $m^3$ = 35.315 $ft^3$ | 1 $ft^3$ = 0.028317 $m^3$ |
| **Second Moment of Area $(length)^4$** | 1 $mm^4$ = 2.402E−6 $in^4$ | 1 $in^4$ = 416.231E3 $mm^4$ |
| | 1 $mm^4$ = 115.861E−12 $ft^4$ | 1 $ft^4$ = 8.63097E9 $mm^4$ |
| | 1 $cm^4$ = 24.025E−3 $in^4$ | 1 $in^4$ = 41.623 $cm^4$ |
| | 1 $cm^4$ = 1.1586E−6 $ft^4$ | 1 $ft^4$ = 863110 $cm^4$ |
| | 1 $m^4$ = 2.40251E6 $in^4$ | 1 $in^4$ = 416.231E−9 $m^4$ |
| | 1 $m^4$ = 115.86 $ft^4$ | 1 $ft^4$ = 8.631E−3 $m^4$ |

Conversion Factors (*continued*)

| Quantity | SI → US Customary | US Customary → SI |
|---|---|---|
| **Mass** | 1 kg = 68.521E−3 slug<br>1 kg = 2.2046 lbm | 1 slug = 14.593 kg<br>1 lbm = 0.4536 kg |
| **Density** | 1 kg/m$^3$ = 0.001938 slug/ft$^3$<br>1 kg/m$^3$ = 0.06248 lbm/ft$^3$ | 1 slug/ft$^3$ = 515.7 kg/m$^3$<br>1 lbm/ft$^3$=16.018 kg/m$^3$ |
| **Force** | 1 N = 224.809E−3 lbf | 1 lbf = 4.448 N |
| **Moment** | 1 N • m = 8.851 in • lb<br>1 N • m = 0.7376 ft • lb | 1 in • lb = 0.113 N • m<br>1 ft • lb = 1.356 N • m |
| **Pressure, Stress, Modulus of Elasticity, Modulus of Rigidity** | 1 Pa = 145.0377E−6 lb/in$^2$<br>1 Pa = 20.885E−3 lb/ft$^2$<br>1 KPa = 145.0377E−6 Ksi | 1 lb/in$^2$ = 6.8947E3 Pa<br>1 lb/ft$^2$ = 47.880 Pa<br>1 Ksi = 6.8947E3 KPa |
| **Work, Energy** | 1 J = 0.7375 ft • lb<br>1 KW•hr = 3.41214E3 Btu | 1 ft • lb = 1.3558 J<br>1 Btu = 293.071E−6 |
| **Power** | 1 W = 0.7375 ft • lb/sec<br>1 KW = 3.41214E3 Btu/hr<br>1 KW = 1.341 hp | 1 ft • lb/sec = 1.3558 W<br>1 Btu/hr = 293.07E−6 KW<br>1 hp = 0.7457 KW |
| **Temperature** | $°C = \frac{5}{9}(°F - 32)$ | $°F = \frac{9}{5}°C + 32$ |

# INDEX

## A